INTRODUCTION TO
THE THEORY OF
TOPOLOGICAL RINGS
AND MODULES

PURE AND APPLIED MATHEMATICS

A Program of Monographs, Textbooks, and Lecture Notes

MONOGRAPHS AND TEXTBOOKS IN
PURE AND APPLIED MATHEMATICS

Additional Volumes in Preparation

INTRODUCTION TO THE THEORY OF TOPOLOGICAL RINGS AND MODULES

V. I. Arnautov
Moldovan Academy of Sciences
Kishinev, Moldova

S. T. Glavatsky
Moscow M. V. Lomonosov State University
Moscow, Russia

A. V. Mikhalev
Moscow M. V. Lomonosov State University
Moscow, Russia

Marcel Dekker, Inc.   New York • Basel • Hong Kong

Library of Congress Cataloging-in-Publication Data

Arnautov, V. I.
  Introduction to the theory of topological rings and modules /
V. I. Arnautov, Sergei T. Glavatsky, Alexander V. Mikhalev.
    p. cm. — (Monographs and textbooks in pure and applied
mathematics ; 197)
  Includes index.
  ISBN 0-8247-9323-4 (pbk. : alk. paper)
  1. Topological rings. 2. Modules (Algebra) I. Glavatsky, Sergei
T. II. Mikhalev, A. V. (Aleksandr Vasil´evich)
III. Title. IV. Series.
QA251.5.A75 1995
512'.4—dc20   95-41065
                                                                      CIP

The publisher offers discounts on this book when ordered in bulk quantities. For more information, write to Special Sales/Professional Marketing at the address below.

This book is printed on acid-free paper.

MARCEL DEKKER, INC.
270 Madison Avenue, New York, New York 10016

Current printing (last digit):
10 9 8 7 6 5 4 3 2 1

PRINTED IN THE UNITED STATES OF AMERICA

Preface

Since the 1940s, systematic investigation of topological rings has been activ carried out using the frame of topological algebra. Several parts of the theory topological rings have been exposed in mathematical texts. For examp topological fields (of real, complex, p-adic numbers, etc.) are under analy from different points of view while taking into account complexly their algebraic, topological, metrical, ordered, and other structures.

One of the first fundamental results in the theory of topological rings was obtained by L. S. Pontryagin in the classification of locally compact skew fields and was included in his famous book [326] on topological groups. (Note that the problem of classification of locally compact, connected skew fields was posed by A. N. Kolmogorov in connection with his classification of continuous geometries.) Some properties of topological rings and modules were also noted in books [82, 243]. Intensive research during the last fifty years has been carried out in the field of normed and Banach algebras as well; those algebras form one of the most important classes of topological rings (see, for example [146, 176–178, 301]). The theory of topological linear spaces [83], one of many rich chapters on functional analysis, is also a good introduction to the theory of topological modules. Another source of topological modules is the theory of topological Abelian groups, in particular, duality theory [326].

Some parts of the theory of topological rings were systematically investigated in a number of review papers [155, 210, 248, 392, 393, 486] as well as monographs [61, 62, 423, 468, 469, 497, 498]. Note that many useful results can be found in a series of books and monographs [82, 83, 159, 181, 241, 252, 253, 285, 286, 350, 395, 477] as well as review papers [22, 89, 292].

The purpose of this book is to introduce the reader to the general theory of topological rings and modules. The first four chapters are based on a similar presentation of our earlier books [61, 62].

Investigation of topological algebraic systems (in particular, topological rings and modules) requires a knowledge of the basics of the set theory, algebra, and topology. For the convenience of the reader, the basic concepts and results from the mentioned fields of mathematics were included in the "Preliminaries." The proofs of those results can be found, for instance, in the following monographs on the set theory [7, 80]; on algebra [64, 81, 133, 179, 195, 196, 236, 243-247, 373, 441, 442]; and on topology [7, 23, 82, 131, 213].

Chapter 1 provides fundamental concepts and basic results on topological rings and modules. Under detailed consideration are bounded subsets, topological divisors of zero, topologically nilpotent elements, and minimal topologies. There are also many examples of topologies on rings and modules that illustrate the main topics.

Chapter 2 discusses the problem of the existence of real-valued pseudo-norms on rings and modules which define topologies on them.

In Chapters 3 and 4, the main focus is the following constructions of topological rings and modules: completion; direct, semidirect, and subdirect products; and inverse limits. Topological and algebraic properties of these constructions (such as completeness, compactness, and invertibility of elements) are considered. Again, there are many examples that illustrate and further define the main topics.

Chapter 5 is devoted to the problems of nondiscrete topologizations of rings and modules. In Chapter 6 we explore the problems of extensions of topologies on a ring onto certain over-rings.

In the Appendix is a List of Open Problems which is based mainly on a similar list [413]. We have also compiled an extensive list of References, including research papers and monographs on topological algebra that are pertinent to the questions discussed in the book. It represents an up-to-date state of this field of mathematics.

This book is addressed primarily to students and postgraduate students, but it can also serve as a reference book for more advanced mathematicians interested in topological algebra. It is a suitable text for first year postgraduate courses.

V. I. Arnautov
S. T. Glavatsky
A. V. Mikhalev

Contents

Preliminaries

The actual prerequisites for a reader are quite few. We suppose that he is familiar with some basic facts from set theory, algebra and topology which, nevertheless, for the sake of convenience are included in this section. With full details they could be found, for example:

on set theory in [7, 80];

on algebra in [64, 81, 133, 179, 195, 196, 236, 243–247, 373, 441, 442];

on topology in [7, 23, 82, 131, 213].

A. Set Theory

A.1. Denote over $\{x \mid \mathcal{P}\}$ the set of all x satisfying the condition $\mathcal{P}$, and over $\{a\}$ the set consisting of only one element a, over $\emptyset$ - the empty set.

A.2. $A \subseteq B$ $(A \subset B)$ means that A is a subset of the set B (and different from B), $x \in A$ means that the element x belongs to the set A.

A.3. Over $\bigcup$, $\bigcap$ denote the operations of union and intersection of subsets, over $A \backslash B$ the complement of B relative to A, i.e. $\{x \in A \mid x \notin B\}$, over $A \times B$ the product of A and B, i.e. $\{(a, b) \mid a \in A, b \in B\}$.

A.4. Let $\varphi : A \longrightarrow B$ be a mapping from A to B, $A_1 \subseteq A$ and $B_1 \subseteq B$, then put:

$\varphi(A_1) = \{\varphi(x) \mid x \in A_1\}$ (the image of the subset A_1 by the mapping φ);

$\varphi|_{A_1} : A_1 \to B$ (the restriction of φ on A_1);

$\varphi^{-1}(B_1) = \{x \in A \mid \varphi(x) \in B_1\}$ (the full inverse image of the subset B_1 by the mapping φ).

A.5. A mapping $\varphi : A \to B$ is called:

an injection, if $\varphi(a) \neq \varphi(c)$ for any $a \neq c$;

a surjection, if $\varphi(A) = B$;

a bijection, if φ is both an injection and surjection.

Let $\varphi : A \to B$, $\psi : B \to C$ be mappings of sets, then define the mapping $\psi \circ \varphi : A \to C$ as follows:

$$(\psi \circ \varphi)(a) = \psi(\varphi(a)), \quad \text{for any} \quad a \in A.$$

The mapping $\psi \circ \varphi$ is called a composition of φ and ψ. If φ and ψ both are surjections (injections, bijections), then $\psi \circ \varphi$ is a surjection (injection, bijection, respectively), too.

A.6. Let $\varphi : A \to B$ and $A_\gamma \subseteq A$ for $\gamma \in \Gamma$. If $\varphi^{-1}(\varphi(A_\gamma)) = A_\gamma$ for all $\gamma \in \Gamma$ except, possibly, one $\gamma_0 \in \Gamma$, then

$$\varphi\left(\bigcap_{\gamma \in \Gamma} A_\gamma\right) = \bigcap_{\gamma \in \Gamma} \varphi(A_\gamma).$$

PROOF. Obviously, $\varphi(\bigcap_{\gamma \in \Gamma} A_\gamma) \subseteq \bigcap_{\gamma \in \Gamma} \varphi(A_\gamma)$. Let now $x \in \varphi(A_\gamma)$ for all $\gamma \in \Gamma$. Then $x = \varphi(a)$, where $a \in A_{\gamma_0}$. Since $a \in \varphi^{-1}(x) \subseteq \varphi^{-1}(\varphi(A_\gamma)) = A_\gamma$ for all $\gamma_0 \neq \gamma \in \Gamma$, then $a \in \bigcap_{\gamma \in \Gamma} A_\gamma$, i.e. $x = \varphi(a) \in \varphi(\bigcap_{\gamma \in \Gamma} A_\gamma)$.

A.7. $|S|$ denotes the cardinality of the a S, $\aleph_0$ the cardinality of a countable set, $\mathfrak{c}$ the continuum cardinality. For every infinite set X the cardinality of the set of all mappings of X into itself is equal to $2^{|X|}$.

Let $\mathfrak{m}$ be a cardinal number, then over $\omega(\mathfrak{m})$ we denote the minimal transfinite number of the cardinality $\mathfrak{m}$, over ω_0 - the minimal countable transfinite number, i.e. $\omega_0 = \omega(\aleph_0)$.

A.8. Over $\mathbb{N}$, $\mathbb{Z}$, $\mathbb{Q}$, $\mathbb{R}$, $\mathbb{C}$ and $\mathbb{H}$ denote the sets of all natural, integer, rational, real, complex numbers and the set of quaternions (see B.19), respectively.

A.9. A partially ordered set $(E, \leq)$ is a non-empty set E with the relation $\leq$ between some of its elements such that:

$a \leq a$;

from $a \leq b$ and $b \leq c$ follows that $a \leq c$;

from $a \leq b$ and $b \leq a$ follows that $a = b$, for any $a, b, c \in E$.

If the relation of the partial order $\leq$ is defined for any $x, y \in E$, then $(E, \leq)$ is called linearly ordered.

Let $(E, \leq)$ be a linearly ordered set, and $a, b \in E$, then denote: $[a; b] = \{x \in E \mid a \leq x \leq b\}$ and $(a; b) = \{x \in E \mid a < x < b\}$ (the last subset is called an interval).

A.10. An element a of a partially ordered set $(E, \leq)$ is called maximal (minimal), if from $a \leq b$ ($b \leq a$) for some element $b \in E$ follows that $a = b$.

An element a of a partially ordered set is called the greatest (smallest), if $x \leq a$ ($a \leq x$) for any $x \in E$. It is clear, that the greatest (smallest) element, if it exists, is also a maximal (minimal) element.

A.11. A partially ordered set $(E, \leq)$ is said to be inductive if for any linearly ordered subset A there exists an element $a \in E$ such that $b \leq a$ for any $b \in A$.

Zorn's Lemma. Let $(E, \leq)$ be an inductive partially ordered set, then for any $c \in E$ there exists a maximal element $a \in E$ such that $c \leq a$.

A.12. A family $\mathcal{F}$ of subsets of a set X is called a filter of X, if:

$\emptyset \notin \mathcal{F}$;

from $A \subseteq B \subseteq X$ and $A \in \mathcal{F}$ follows that $B \in \mathcal{F}$;

from $A, B \in \mathcal{F}$ follows that $A \bigcap B \in \mathcal{F}$.

A family $\mathcal{B}$ of subsets of a set X is called a basis of a filter $\mathcal{F}$ if:

$\mathcal{B} \subseteq \mathcal{F}$ and for any $A \in \mathcal{F}$ there exists $B \in \mathcal{B}$ such that $B \subseteq A$.

A family $\mathcal{B}$ of subsets of a set X is a basis of some filter of X if and only if:

$\emptyset \notin \mathcal{B}$ and for any $A, B \in \mathcal{B}$ there exists $C \in \mathcal{B}$ such that $C \subseteq A \bigcap B$.

A.13. Let X be a set and Φ be the set of all filters of X. For $\mathcal{A}, \mathcal{F} \in \Phi$ put $\mathcal{A} \leq \mathcal{F}$ if and only if $\mathcal{A} \subseteq \mathcal{F}$. It can be easily verified that $(\Phi, \leq)$ is an inductive partially ordered set with respect to the above defined relation $\leq$. Then from Zorn's Lemma (see A.11) follows that there are maximal elements in Φ. They are called ultrafilters of X.

A filter $\mathcal{F}$ is called free if $\bigcap_{A \in \mathcal{F}} A = \emptyset$.

Let $\mathcal{F}$ be an ultrafilter of a set X, and $A_1, \ldots, A_n \subseteq X$. If $\bigcup_{i=1}^{n} A_i \in \mathcal{F}$, then $A_k \in \mathcal{F}$ for some $1 \leq k \leq n$.

A filter $\mathcal{F}$ of a set X is an ultrafilter if and only if $A \in \mathcal{F}$ or $X \backslash A \in \mathcal{F}$ for any subset A of X.

A.14. By induction by $\max\{i, j\}$ it is easily proved that the mapping $\rho : \mathbb{N} \times \mathbb{N} \longrightarrow \mathbb{N}$ defined as:

$$\rho(i, j) = \begin{cases} (i-1)^2 + 2j - 1, & \text{if} \quad j \leq i; \\ (j-1)^2 + 2i, & \text{if} \quad i < j, \end{cases}$$

is a bijection with the following properties:

a) $\rho(i_1, j_1) \leq \rho(i_2, j_2)$, if $i_1 \leq i_2$, $j_1 \leq j_2$;

b) $\max\{i, j\} \leq \rho(i, j) \leq \max\{i^2, j^2\}$, for $i, j \in \mathbb{N}$.

A.15. Let X be a set and σ some binary relation on X (i.e. for certain pairs of elements $x, y \in X$ is defined that x is related to y with respect to σ what is denoted over, $x\sigma y$). We say that a binary relation σ is an equivalence relation if the following conditions are satisfied:

$x\sigma x$ for each $x \in X$;

if $x\sigma y$ and $y\sigma z$, then $x\sigma z$ and $y\sigma x$ for any $x, y, z \in X$.

It is clear that if in X is defined an equivalence relation, then it defines a partition of X into pairwise disjoint subsets each of which consists of pairwise equivalent elements (i.e. related to each other with respect to the equivalence relation).

Vice versa, if on X is defined a partition into pairwise disjoint subsets, then we can define a relation on X such that two elements are related to each other if and only if they are contained in same subset of the partition. It's easy to see that it is an equivalence relation on X.

Let σ be an equivalence relation on a set X and $\{X_j \mid j \in \Gamma\}$ be its partition defined by σ, then the set $\{X_j \mid j \in \Gamma\}$ is called a factor–set of X by σ.

B. Algebra

B.1. A monoid is a semigroup with the unitary element, i.e. a set M provided with the associative binary operation of multiplication and containing such element e that $e \cdot a = a \cdot e = a$ for every $a \in M$. If any element a of a monoid M is invertible, i.e. $a \cdot b = b \cdot a = e$ for some $b \in M$, then M is called a group.

B.2. An Abelian group, as usual, is a commutative group with additive form of recording of an operation (unless otherwise stated), i.e. a set G provided with associative and commutative binary operation " $+$ " and containing element 0 such that $g + 0 = g$ for any $g \in G$, and that for any $b \in G$ there exists an element $c \in G$ such that $b + c = 0$ (this element c is denoted over $-b$). An Abelian group G is said to be divisible if for any $g \in G$ and any natural number n there exists an element $b \in G$ such that $g = nb$, where $nb = \underbrace{b + \ldots + b}_{n \text{ times}}$.

B.3. Let G be an Abelian group, $A, B \subseteq G$ and $n \in \mathbb{N}$, then put:

$$-A = \{-a \mid a \in A\};$$
$$A + B = \{a + b \mid a \in A,\, b \in B\};$$
$$A - B = A + (-B);$$
$$n \cdot A = \{na \mid a \in A\} \quad \text{and} \quad nA = \underbrace{A + A + \ldots + A}_{n \text{ summands}}.$$

A subset C of an Abelian group G is called symmetric if $-C = C$. Let D be a subset of G, then $D \bigcap (-D)$ and $D \bigcup (-D)$ are symmetric subsets, and

$$\langle D \rangle = \bigcup_{n=1}^{\infty} n\left(D \cup (-D)\right) = \bigcup_{n=1}^{\infty} \underbrace{\left((D \cup (-D)) + \ldots + (D \cup (-D))\right)}_{n \text{ summands}}$$

is the smallest subgroup in G containing D (it is called a subgroup generated by the subset D).

B.4. Throughout this book by a ring (unless otherwise stated) an associative ring (possibly without the unitary element) is meant, i.e. a set R with two associative operations (addition and multiplication) such that:

R is an Abelian group with respect to addition;

there are fulfilled the distributivity laws: $a \cdot (b+c) = a \cdot b + a \cdot c$ and $(b+c) \cdot a = b \cdot a + c \cdot a$ for all $a,\, b,\, c \in R$.

An element a of a ring R with the unitary element 1 is called invertible if there exists $b \in R$ such that $a \cdot b = b \cdot a = 1$.

If all non-zero elements of R are invertible, then R is called a skew field (a division ring). A commutative skew field is called a field.

An element a of a ring R is said to be nilpotent if $a^n = 0$ for some natural n.

Let $0 \neq r \in R$, $0 \neq s \in R$ and $r \cdot s = 0$, then r is called a left divisor of zero and s a right divisor of zero.

Let R be a ring (semigroup), then the set $Z = \{r \in R \mid x \cdot r = r \cdot x \text{ for all } x \in R\}$ is called a centre of the ring (semigroup).

We define the operation " $\circ$ " on a ring R putting $a \circ b = a + b - a \cdot b$. It can be easily verified that the introduced operation is associative, and $a \circ 0 = 0 \circ a = a$ for any $a \in R$.

An element $a \in R$ is called right quasi-regular (or quasi-invertible in other terminology) if $a \circ b = 0$ for some $b \in R$. A left quasi-regularity (quasi-invertibility) is defined similarly.

If $a \in R$ is both right and left quasi-regular, then it is called quasi-regular. Clearly, any nilpotent element is quasi-regular.

A ring, all of whose elements are quasi-regular, is called a quasi-regular ring.

B.5. Let R be a ring, then the set R considered with only the addition (multiplication) operation is called the additive group (multiplicative semigroup) of the ring R.

B.6. By an R-module M (unless otherwise stated) we mean a left module over a ring R, i.e. an Abelian group M with given left multiplication by elements of R such that the following conditions are satisfied:

$$r \cdot (m_1 + m_2) = r \cdot m_1 + r \cdot m_2, \quad (r_1 + r_2) \cdot m = r_1 \cdot m + r_2 \cdot m_2, \quad r_1 \cdot (r_2 m) = (r_1 r_2) \cdot m$$

for all $r_1, r_2, r \in R$ and $m_1, m_2, m \in M$ (if R is a ring with the unitary element 1, and $1 \cdot m = m$ for any $m \in M$, then M is called unitary). If R is a skew field, then a unitary R–module is called a vector space over R.

B.7. Let A and B be Abelian groups (R-modules, rings), then a mapping $\varphi : A \to B$ is called a homomorphism if $\varphi(a + b) = \varphi(a) + \varphi(b)$ for all $a, b \in A$ (and, in addition, if $\varphi(r \cdot a) = r \cdot \varphi(a)$ for all $r \in R$, $a \in A$ in the case of R-modules, and $\varphi(a \cdot b) = \varphi(a) \cdot \varphi(b)$ for all $a, b \in A$ in the case of rings). A homomorphism, being also:

a bijection, is called an isomorphism;

an injective mapping, is called a monomorphism.

If $A = B$, then a homomorphism is called an endomorphism.

Let $\varphi : A \to B$, $\psi : B \to C$ be homomorphisms of Abelian groups (rings, R-modules), then $\psi \circ \varphi : A \to C$ is a homomorphism of Abelian groups (rings, R-modules, respectively), too.

B.8. Let A and B be Abelian groups, then over $\mathrm{Hom}(A, B)$ we denote the set of all homomorphisms from A to B. For $\varphi \in \mathrm{Hom}(A, B)$ put $\ker \varphi = \{a \in A \mid \varphi(a) = 0\}$ (a kernel of the homomorphism φ). The set $\mathrm{Hom}(A, B)$ is an Abelian group with respect to addition operation " $+$ ", where $(\xi + \varphi)(a) = \xi(a) + \varphi(a)$ for any $a \in A$ and any $\xi, \varphi \in \mathrm{Hom}(A, B)$.

B.9. Let A be an Abelian group (R-module, ring) and $B \subseteq A$. Then B is called a subgroup (submodule, subring), if B is a group (R-module, ring) with respect to the existing operations.

Let R be a ring and $I \subseteq R$, then I is a called left (right) ideal if I is a subgroup of the additive group of R and $r \cdot i \in I$ ($i \cdot r \in I$) for all $i \in I$, $r \in R$.

Let S be a multiplicative semigroup and $I \subseteq S$, then I is called a left (right) ideal if $s \cdot i \in I$ ($i \cdot s \in I$) for all $i \in I$, $s \in S$.

If I is both left and right ideal of a ring (semigroup), then I is called a two-sided ideal or, briefly, an ideal of the ring (semigroup).

B.10. Let R be a ring, $R^2 \neq 0$ and R has no proper ideals (i.e. ideals different from R and $\{0\}$), then R is called simple.

A simple commutative ring is a field.

B.11. Let B, C be subgroups (submodules, ideals) of an Abelian group (module, ring) A, then denote over A/B the factor-group (factor-module, factor-ring) of the group (module, ring) A by B, i.e. the set $\{a + B \mid a \in A\}$ of cosets with the following operation of addition: $(a + B) + (c + B) = a + c + B$; and with the operation of multiplication: $r \cdot (c + B) = r \cdot c + B$ in the case of R-modules, and $(a + B) \cdot (c + B) = a \cdot c + B$ in the case of rings.

Over $\omega = \omega_B : A \to A/B$ denotes the natural homomorphism: $\omega(a) = a + B$ for all $a \in A$. It is clear that $\omega^{-1}(\omega(S)) = S + B$ for any subset $S \subseteq A$. In particular, if $S = S + B$, then $\omega^{-1}(\omega(S)) = S$.

Let $\varphi : A \to A'$ be a homomorphism of Abelian groups (modules, rings), and $B \supseteq \ker \varphi$, then the natural mapping $\xi_\varphi : A/B \to \varphi(A)$, for which $\xi_\varphi(a + B) = \varphi(a)$, is a homomorphism. Moreover, $\varphi = \xi_\varphi \circ \omega_B$ and there exists a bijection between all subgroups (submodules, subrings) of $\varphi(A)$ and those subgroups (submodules, subrings) of A which contain $\ker \varphi$. If $B = \ker \varphi$, then ξ_φ is an isomorphism.

Let $B \subseteq C$, then the natural mapping

$$\mu : (AB)/(C/B) \to A/C,$$

for which $\mu((a + B) + (C/B)) = a + C$, is an isomorphism of Abelian groups (modules, rings).

Let D be a subgroup (submodule, subring) of A, then the natural mapping

$$\psi : D/(D \bigcap B) \to (D + B)/B,$$

for which $\psi(a + (B \cap D)) = a + B$, is an isomorphism of Abelian groups (modules, rings).

THEOREM. For any Abelian group A and a non-zero element $a \in A$ there exists a homomorphism $\xi: A \to \mathbb{R}/\mathbb{Z}$ such that $\xi(a) \neq 0$.

B.12. Let $n \in \mathbb{N}$, R be a ring and $A, B \subseteq R$. Let M be an R-module, $D, E \subseteq M$, and C be a subset of either R or M, then put:

$A \cdot C = \{a \cdot c \,|\, a \in A, \, c \in C\};$

$AC = \left\{ \sum_{i=1}^{k} a_i \cdot b_i \,\middle|\, a_i \in A, \, c_i \in C, \, 1 \leq i \leq k, \, k \in \mathbb{N} \right\};$

$A^{(1)} = A$ and $A^{(n)} = A \cdot A^{(n-1)}$, for $n > 1$;

$A^1 = A$ and $A^n = AA^n$, for $n > 1$

(it is clear that $A^n = \left\{ \sum_{i=1}^{k} b_i \,\middle|\, b_i \in A^{(n)}, \, 1 \leq i \leq k, \, k \in \mathbb{N} \right\} $);

$(A \colon B)_R = \{r \in R \,|\, r \cdot B \subseteq A\};$

$(D \colon E)_R = \{r \in R \,|\, r \cdot E \subseteq D\};$

$(D \colon A)_M = \{m \in M \,|\, A \cdot m \subseteq D\}.$

For $r \in R$ and $m \in M$ instead of $(A : \{r\})_R$ and $(D : \{m\})_M$ we will write $(A : r)_R$ and $(D : m)_M$, respectively.

In this case:

if E is a subgroup of the group M, then $(E : D)_R$ and $(E : A)_M$ are subgroups of the groups $R(+)$ and M, respectively;

if E is a submodule of the R-module M, then $(E : D)_R$ is a left ideal of the ring R;

if E and D are submodules of the R-module M, then $(E : D)_R$ is an ideal of the ring R;

if E is a subgroup of the group M and A is a right ideal of the ring R, then $(E : A)_M$ is a submodule of the R-module M.

B.13. A subset S of a ring R is called nilpotent if $S^n = \{0\}$ for some natural n. Clearly, an element of a nilpotent set is a nilpotent element (see B.4).

Let R be a finite associative commutative ring without nilpotent elements, then R is (isomorphic to) a finite direct sum of finite fields.

B.14. Let M be an R-module, $S \subseteq M$ and $Q \subseteq R$. If $Q \cdot S \subseteq S$, then the subset S is called Q-stable.

B.15. A partially ordered group is a group G which is a partially ordered set (with a relation $\leq$) such that $c \cdot a \leq c \cdot d$, and that $a \cdot c \leq b \cdot c$ follows from $a \leq b$, for all $a, b, c \in G$.

Let e be the unitary element of the partially ordered group G, then the subset

$$P = P(G) = \{g \in G \,|\, g \geq e\}$$

is called a positive cone.

A subset B of a group G is a positive cone for some partial order on G if and only if B satisfies the following conditions: $B \cap B^{-1} = \{e\}$ and $B \cdot B \subseteq B$. In addition, if G is not a commutative group, then also $x \cdot B \cdot x^{-1} \subseteq B$ for any $x \in B$.

If the order on G is linear, then G is called a linearly ordered group (it is equivalent to $P \cup P^{-1} = G$ for a positive cone P).

A partially ordered group G is called archimedean if from

$$a^n \leq b \text{ for all } n \in \mathbb{Z} \text{ and some } b \in G$$

follows that a is the unitary element of G.

GOELDER THEOREM. A linearly ordered group is archimedean if and only if it is isomorphic (saving the order) to a subgroup of the additive group $\mathbb{R}(+)$ of real numbers with the natural order. In particular, all archimedean linearly ordered groups are commutative.

Let a linear order " $\leq$ " on a group G satisfy the condition:

$$x \cdot y \leq z \cdot y \text{ for any } x, y, z \in G \text{ for which, } x \leq z,$$

then G is called right-ordered.

It takes place according the following theorem (see [231,234]).

Theorem. A group G admits linear order, with respect to which it is right-ordered, in each of the following cases:

1) G is the group of automorphisms of a linearly ordered set;

2) G is a torsion-free Abelian group;

3) G is a free solvable group*;

4) G is a free nilpotent group*;

5) G is a free group.

B.16. A ring F (an R-module M) is called a free ring (free R-module) generated by set X if:

1) $X \subseteq F$ ($X \subseteq M$);

2) any subring (any submodule) of the ring F (module M) containing X coincides with the ring F (module M);

3) for any ring (R-module) A any mapping $\varphi : X \longrightarrow A$ is extendable to a ring (module) homomorphism of the ring F (R-module M) to the ring (module) A.

For any set X (ring R) there exists a free ring (free R-module) generated by X.

B.17. Let R be a ring with the unitary element and G be a monoid (i.e. a semigroup with unitary element).

Let's consider the set

$$RG = \left\{ \sum_{i=1}^{k} r_i \cdot g_i \,\middle|\, r_i \in R,\, g_i \in G,\, 1 \le i \le k,\, k \in \mathbb{N},\, g_i \ne g_j \text{ for } i \ne j \right\}.$$

Let's identify with 0 any element $\sum_{i=1}^{k} r_i \cdot g_i \in RG$ where $r_i = 0$ for $1 \le i \le k$, and convert RG into a ring by defining operations of addition and multiplication on RG as follows.

Let $\sum_{i=1}^{n} r_i \cdot g_i$, $\sum_{j=1}^{m} p_j \cdot h_j \in RG$. Put $n = m$ and $g_i = h_i$ for $1 \le i \le n$, adding, if necessary, elements of the form $0 \cdot g$ and carrying on the renumeration. Then put:

$$\sum_{i=1}^{n} r_i \cdot g_i + \sum_{j=1}^{m} p_j \cdot h_j = \sum_{i=1}^{n} (r_i + p_i) \cdot g_i \quad \text{and}$$

$$\left(\sum_{i=1}^{n} r_i \cdot g_i \right) \cdot \left(\sum_{j=1}^{m} p_j \cdot h_j \right) = \sum_{k=1}^{t} \sum_{g_i \cdot h_i = s_k} (r_i \cdot p_j) \cdot s_k,$$

where $\{ s_1, \dots, s_t \} = \{ g_i \cdot h_i \,|\, 1 \le i \le n,\, 1 \le j \le m \}$.

The ring RG is called a semigroup ring of the ring R and group G.

---

* Definitions of the nilpotent and solvable groups can be found, f.i., in [244].

B.18. Let R be a commutative ring and S be some multiplicative system, i.e. a subsemigroup of the multiplicative semigroup of the ring, every element of which not being a divisor of zero in R. Put $R_S = \{(r,s) \,|\, r \in R, s \in S\}$.

Let's identify such pairs (r_1, s_1) and (r_2, s_2) if $r_1 \cdot s_2 = r_2 \cdot s_1$, and define the operations of addition and multiplication on the set R_S, putting:

$$(r,s) + (p,q) = (r \cdot q + p \cdot s, \; s \cdot q) \quad \text{and} \quad (r,s) \cdot (p,q) = (r \cdot p, s \cdot q).$$

It is easy to check that R_S is a ring with respect to these operations. This ring is called a quotient ring of the ring R by the multiplicative system S, and its elements are written as fractions $\frac{r}{s}$, where $r \in R$, $s \in S$. It is clear that $\left\{ \frac{r \cdot s}{s} \middle| r \in R, \, s \in S \right\}$ is a subring of R_S isomorphic to the ring R.

Let R be an associative commutative ring without divisors of zero, then the ring $R_{(R \setminus \{0\})}$ is called a field of quotients of the ring R.

B.19. Let $\mathbb{H}$ be a 4-dimensional vector space over the field $\mathbb{R}$ of real numbers and $\{\bar{e}, \bar{i}, \bar{j}, \bar{k}\}$ be its basis. Define the multiplication operation on the basis in the following way:

$$\bar{e}^2 = \bar{e}, \; \bar{i}^2 = \bar{j}^2 = \bar{k}^2 = -\bar{e};$$

$$\bar{e} \cdot \bar{i} = \bar{i}, \; \bar{e} \cdot \bar{j} = \bar{j}, \; \bar{e} \cdot \bar{k} = \bar{k};$$

$$\bar{i} \cdot \bar{j} = -\bar{j} \cdot \bar{i} = \bar{k}, \; \bar{j} \cdot \bar{k} = -\bar{k} \cdot \bar{j} = \bar{i}, \; \bar{k} \cdot \bar{i} = -\bar{i} \cdot \bar{k} = \bar{j}.$$

Extending this operation onto $\mathbb{H}$ with the help of the distributivity laws as well as the properties of commutation of the real numbers with elements of the basis, we get that $\mathbb{H}$ becomes a skew field. This field is called a field of quaternions.

C. Topology

C.1. A topological space is a pair (X, τ) consisting of a set X and some family τ of subsets of the set X such that:

$\emptyset, X \in \tau$;

if $U_1, U_2 \in \tau$, then $U_1 \bigcap U_2 \in \tau$;

if $\{U_\gamma \,|\, \gamma \in \Gamma\} \subseteq \tau$, then $\bigcup_{\gamma \in \Gamma} U_\gamma \in \tau$.

Subsets of X belonging to τ are called open, and their complements in X are called closed. Thus, a union of a finite number and an intersection of any number of closed subsets is a closed.

A subset both open and closed is called open-closed. If τ contains all subsets of X, then the topological space (X, τ) is said to be discrete. If $\tau = \{\emptyset, X\}$, then the topological space (x, τ) is called anti-discrete.

If it is clear from the context what topology τ on X is considered, then we'll use also the denotation X for (X, τ).

It could be easily verified that if $(X, \leq)$ is a linearly ordered set without borders (i.e. without both the smallest and greatest elements (see A.10)), then the family of all subsets in X, each of them being a union of intervals sets (see A.9), defines the topology on X which is called an interval topology.

C.2. Let (X, τ) be a topological space. A family $\mathcal{B}$ of subsets of X is called a basis of open subsets or, briefly, a basis: if $\mathcal{B} \subseteq \tau$ and any subset $V \in \tau$ is a union of a certain family of subsets from $\mathcal{B}$.

It could be easily verified that a family $\mathcal{B}$ of subsets of a set X is a basis of a certain topology τ on X if and only if the following conditions are fulfilled:

1) $X = \bigcup_{U \in \mathcal{B}} U$;

2) for any $U_1, U_2 \in \mathcal{B}$ and any $x \in U_1 \bigcap U_2$ there exists $U \in \mathcal{B}$ such that $x \in U \subseteq U_1 \bigcap U_2$.

C.3. A subset U of a topological space (X, τ) is called a neighborhood of element $x \in X$ if there exists an open set V such that $x \in V \subseteq U$. Any subset containing a neighborhood of a point is itself a neighborhood of this point.

A subset U is open in (X, τ) if and only if U is a neighborhood of any of its elements.

An element $x \in X$ is called isolated in (X, τ) if $\{x\}$ is a neighborhood of the element x.

A family $\mathcal{B}_x$ of subsets in (X, τ) is called a basis of neighborhoods of $x \in X$ if any subset of $\mathcal{B}_x$ is a neighborhood of x and any neighborhood of the element x contains some subset from $\mathcal{B}_x$.

We say that a topological space X satisfies the first axiom of countability if any of its elements has a countable basis of neighborhoods.

A neighborhood of a subset $S \subseteq X$ is a subset U which is a neighborhood of every element $x \in S$. It is equivalent to saying that there exists an open subset V for which $S \subseteq V \subseteq U$.

C.4. A metric space is a pair (X, ρ), where X is a set and ρ is a real-valued function (called metric) defined on the set $X \times X$ and satisfying the following conditions:

(M.1) $\rho(x, y) \geq 0$ for any elements $x, y \in X$, moreover, $\rho(x, y) = 0$ if and only if $x = y$;

(M.2) $\rho(x, y) = \rho(y, x)$ for any elements $x, y \in X$;

(M.3) $\rho(x, y) \leq \rho(x, z) + \rho(z, y)$ for any elements $x, y, z \in X$.

Let τ_ρ be the family of all subsets U of a metric space (X, ρ) such that for every element $x \in U$ there exists a positive real number ε for which $\{z \in X \mid \rho(x, z) < \varepsilon\} \subseteq U$.

It is clear that (X, τ_ρ) is a topological space. We say that the topology τ_ρ is determined by the metric ρ.

The family $\mathcal{B}_x = \{U(x, \varepsilon) \mid \varepsilon > 0\}$, where $U(x, \varepsilon) = \{x \in X \mid \rho(x, y) < \varepsilon\}$ is a basis of neighborhoods of the element x in the topological space (X, τ_ρ).

C.5. Let (X, τ) be a topological space and $A \subseteq X$, then the intersection of all closed subsets containing A is the smallest closed subset containing A. It is called a closure of A and is denoted over $[A]_{(X,\tau)}$ (sometimes over $[A]_X$ and even $[A]$). A subset A is closed in topological space (X, τ) if and only if it coincides with its closure.

Clearly, $[A \cup B]_X = [A]_X \cup [B]_X$ and $[A]_X = \{x \in X \mid A \cap U \neq \emptyset$ for any neighborhood U of the element $x\}$ for any subsets $A, B \subseteq X$.

A subset A is called dense in (X, τ) if $[A]_X = X$.

Let V be an open subset and A be a dense subset in a topological space X, then it can be easily verified that $[V]_X = [V \cap A]_X$.

C.6. An element $a \in (X, \tau)$ is called an accumulation point (a limit) of a sequence $a_1, a_2, \ldots$ in (X, τ) if for any neighborhood V of a and any $n \in \mathbb{N}$ (for some $n \in \mathbb{N}$) we get that $a_i \in V$ for some $i \geq n$ (for all $i \geq n$).

Any limit of a sequence is its accumulation point.

C.7. A mapping f of a topological space X to a topological space Y is called continuous at element $x \in X$ if for any neighborhood V of the element $f(x)$ in the topological space Y there exists a neighborhood U of the element x in X such that $f(U) \subseteq V$ (it is equivalent to saying that $f^{-1}(V)$ is a neighborhood of x in X).

A mapping f is called continuous if it is continuous at every $x \in X$.

The following statements are equivalent:

f is continuous mapping from X to Y;

$f^{-1}(V)$ is an open subset in X for any open subset V in Y;

$f^{-1}(S)$ is a closed subset in X for any closed subset S in Y;

$f([A]_X) \subseteq [f(A)]_Y$ for any subset $A \subseteq X$.

Let X and Y be topological spaces, then a mapping $f : X \to Y$ is called open (closed) if $f(S)$ is open (closed, respectively) subset in Y for any open (closed, respectively) subset S in X.

If $f : X \to Y$ and $g : Y \to Z$ be continuous (open, closed) mappings of topological spaces, then $g \circ f : X \to Z$ is a continuous (open, closed respectively) mapping of topological spaces, too.

Let $f : X \to Y$ be a bijection, then the following statements are equivalent:

f is an open mapping;

f is an closed mapping;

$f^{-1} : Y \to X$ is a continuous mapping.

A continuous and open bijection $f : X \to Y$ is called a homeomorphism, and topological spaces X and Y in this case are said to be homeomorphic.

Let $f : X \to Y$ be a bijection of topological spaces X and Y, then the following statements are equivalent:

f is a homeomorphism;

f and f^{-1} are continuous mappings;

f and f^{-1} are open mappings;

f and f^{-1} are closed mappings;

$f([S]_X) = [f(S)]_Y$ for any subset $S \subseteq X$;

if $\mathcal{B}_x(X)$ be a basis of neighborhoods of $x \in X$, then $\{f(U) \mid U \in \mathcal{B}_x(X)\}$ is a basis of neighborhoods of the element $f(x) \in Y$.

A topological space X is called homogeneous if for any $x, y \in X$ there exists a homeomorphism f of the space X onto itself such that $f(x) = y$.

C.8. Let τ_1 and τ_2 be two topologies on a set X. We say that τ_1 is weaker than τ_2 (use the denotation $\tau_1 \leq \tau_2$) if $\tau_1 \subseteq \tau_2$, that is, any subset of X which is open in (X, τ_1), is open in (X, τ_2), too. If $\tau_1 \leq \tau_2$, then we say also that the topology τ_2 is stronger then τ_1 (use the denotation $\tau_2 \geq \tau_1$). It is clear that the set of all topologies on X with the relation " $\leq$ " is a partially ordered set.

For any family $\{\tau_\gamma \mid \gamma \in \Gamma\}$ of topologies on a set X there exist topologies stronger than all τ_γ (for example, the discrete topology) and there exist topologies weaker than all τ_γ (e.g., the anti-discrete topology).

If $\tau_1 \leq \tau_2$ and $\tau_1 \neq \tau_2$, then there exist an element $x \in X$ and such neighborhood V of x in topological space (X, τ_2) which is not a neighborhood of x in (X, τ_1).

For any two topologies τ_1 and τ_2 on X the following statements are equivalent:

$\tau_1 \leq \tau_2$;

any subset of X which is closed in (X, τ_1) is closed in (X, τ_2), too;

for every element $x \in X$ any its neighborhood in (X, τ_1) is its neighborhood in (X, τ_2), too;

$[S]_{(X,\tau_1)} \supseteq [S]_{(X,\tau_2)}$ for any subset $S \subseteq X$;

the identical mapping $\varepsilon : X \to X$ is a continuous mapping of (X, τ_2) onto (X, τ_1).

C.9. Let $\{\tau_\gamma \mid \gamma \in \Gamma\}$ be a non-empty family of topologies on X and $\tau = \bigcap_{\gamma \in \Gamma} \tau_\gamma$ (i.e. τ consists of all subsets of X each of which is open in (X, τ_γ), for all $\gamma \in \Gamma$). Then (X, τ) is a topological space. The topology τ is the greatest lower bound of the family of topologies $\{\tau_\gamma \mid \gamma \in \Gamma\}$ on X and is denoted over $\inf\{\tau_\gamma \mid \gamma \in \Gamma\}$.

It is clear that $\tau \leq \tau_\gamma$ for any $\gamma \in \Gamma$.

Let $\{\tau_\gamma \mid \gamma \in \Gamma\}$ be a non-empty family of topologies on X and $\{\tau'_\omega \mid \omega \in \Omega\}$ be the family of topologies on X such that $\tau_\gamma \leq \tau'_\omega$ for any $\gamma \in \Gamma$ and any $\omega \in \Omega$. Since the discrete topology belongs to $\{\tau'_\omega \mid \omega \in \Omega\}$, then this family is non-empty and, hence, we can consider the topology $\tau' = \inf\{\tau'_\omega \,|\, \omega \in \Omega\}$. The topology τ' is called the least upper bound of the family $\{\tau_\gamma \,|\, \gamma \in \Gamma\}$ and is denoted over $\sup\{\tau \,|\, \gamma \in \Gamma\}$.

The topology τ' is stronger than any topology τ_γ.

For every $x \in X$ consider the family $\mathcal{B}_x(X, \tau')$ of subsets U of X such that there exist finite sets of topologies $\tau_{\gamma_1}, \ldots, \tau_{\gamma_n}$ and neighborhoods $U_1, \ldots, U_n$ of x in the topological spaces $(X, \tau_{\gamma_1}), \ldots, (X, \tau_{\gamma_n})$, respectively, such that $U = \bigcap_{i=1}^n U_i$. The family $\mathcal{B}_x(X, \tau')$ is a basis of neighborhoods of the element x in (X, τ').

C.10. Let (X, τ) be a topological space, $Y \subseteq X$ and $\tau|_Y = \{U \bigcap Y \,|\, U \in \tau\}$. Then $(Y, \tau|_Y)$ is a topological space. The topology $\tau|_Y$ is called a topology induced on Y by the topology τ, and the space $(Y, \tau|_Y)$ is called a subspace of the space (X, τ).

Let (X, τ) be a topological space, $Y \subseteq X$ and τ' be a topology on Y. Then the following statements are equivalent:

$\tau' = \tau|_Y$;

a subset $S \subseteq Y$ is closed in (Y, τ') if and only if $S = Q \bigcap Y$, where Q is a certain closed subset in (X, τ);

$[S]_{(Y, \tau')} = Y \bigcap [S]_{(X, \tau)}$, for any subset $S \subseteq Y$;

for any element $y \in Y$ a subset $U \subseteq Y$ is a neighborhood of y in (Y, τ') if and only if $U = V \bigcap Y$, where V is a certain neighborhood of y in (X, τ).

Let Y be an open (closed) subset of the topological space (X, τ), then a subset $S \subseteq Y$ is open (closed, respectively) in the topological space $(Y, \tau|_Y)$ if and only if it is open (closed, respectively) in (X, τ).

Let Y be an open subset of a topological space (X, τ), $y \in Y$ and $\mathcal{B}_y$ be a basis of neighborhoods of y in $(Y, \tau|_Y)$, then $\mathcal{B}_y$ is a basis of neighborhoods of y in (X, τ).

Let $\varphi : (X, \tau) \to (X', \tau')$ be a continuous mapping of topological spaces, and $Y \subseteq X$. Then:

the mapping $\varphi|_Y : Y \to X'$ (see A.4) is a continuous mapping of the topological space $(Y, \tau|_Y)$ onto the topological space (X', τ');

the mapping φ is a continuous mapping of the topological space (X, τ) onto the topological space $(\varphi(X), \tau'|_{\varphi(X)})$.

C.11. Consider the following conditions for the topological space (X, τ), that are called the axioms of separation:

(T_0) for any distinct $x, y \in X$ at least one of them has a neighborhood which does not contain the second element;

(T_1) for any distinct $x, y \in X$ each of them has a neighborhood not containing the second element;

(T_2) for any distinct elements $x, y \in X$ there are neighborhoods U_x and U_y of x and y, respectively, such that $U_x \cap U_y = \emptyset$;

(T_3) for any closed subset $S \subseteq X$ and any $x \notin S$ there are neighborhoods U and V of the element x and the subset S, respectively, such that $U \cap V = \emptyset$;

(T_4) for a closed subset $S \subseteq X$ and any element $x \notin S$ there exists a continuous real-valued function f satisfying the following conditions:

$$f(x) = 0, \quad f(S) = \{1\} \quad \text{and} \quad 0 \le f(y) \le 1 \quad \text{for all} \quad y \in X;$$

(T_5) for any two closed non-intersecting subsets $A, B \subseteq X$ there are neighborhoods U and V of subsets A and B, respectively, such that $U \cap V = \emptyset$.

Clearly, $T_2 \Rightarrow T_1 \Rightarrow T_0$ and $T_4 \Rightarrow T_3$.

A topological space satisfying the condition (T_i) is called T_i-space, $i = 0, 1, 2, 3, 4, 5$. In addition, T_2-spaces are often called Hausdorff.

T_1-space is said to be regular (completely regular, normal) if it satisfies also the condition (T_3) (the conditions (T_4) or (T_5), respectively).

Any normal space is completely regular, any completely regular space is regular, and any regular space is Hausdorff.

There are examples of topological spaces showing that all the axioms of separation presented above are different. In particular, there exist completely regular spaces which are not normal.

It is clear that any finite T_1-space is discrete.

Let τ_1 and τ_2 be topologies on a set X such that $\tau_1 \le \tau_2$. If (X, τ_1) is a T_i-space, then (X, τ_2) is a T_i-space for $i = 0, 1, 2$.

All the given axioms of separation of topological space, except for the normality axiom, are valid also for arbitrary subspaces. Subspace of a normal space may not be a normal space, whereas every closed subspace of a normal space is normal, too.

C.12. Let (X, τ) be a topological space and $X = \bigcup_{\gamma \in \Gamma} S_\gamma$, where $S_\gamma \in \tau$ for $\gamma \in \Gamma$, then the family $\{S_\gamma \mid \gamma \in \Gamma\}$ is called an open cover of the topological space (X, τ).

A topological space (X, τ) is called compact if from any open cover $\{S_\gamma \mid \gamma \in \Gamma\}$ it is possible to choose a finite subfamily $\{S_{\gamma_1}, \ldots, S_{\gamma_n}\}$, which is an open cover of (X, τ), too.

A topological space (X, τ) is compact if and only if the intersection of any family of closed subsets, which is a basis of some filter in X (see A.12), is not empty.

A subset Y of a topological space (X, τ) is called compact if $(Y, \tau|_Y)$ is a compact space. It is equivalent to saying that for any family $\{S_\gamma \mid \gamma \in \Gamma\} \subseteq \tau$, for which $Y \subseteq \bigcup_{\gamma \in \Gamma} S_\gamma$, there exists a finite number of the subsets $S_{\gamma_1}, \ldots, S_{\gamma_n}$ such that $Y \subseteq \bigcup_{i=1}^{n} S_{\gamma_i}$.

A topological space is called locally compact if every element has a neighborhood which is a compact subset.

It is clear that every compact space is locally compact.

A closed subset of a compact (locally compact) space is compact (locally compact).

A compact subset of a Hausdorff topological space is closed.

Let f be a continuous mapping of a topological space (X, τ) onto a topological space (X', τ'). Then:

if S is a compact subset in (X, τ), then $f(S)$ is a compact subset in (X', τ');

if (X, τ) is compact and (X', τ') is Hausdorff, then the mapping f is closed;

if (X, τ) is compact, (X', τ') is Hausdorff and f is a bijection, then f is a homeomorphism.

Let (X, τ_1) be a compact space and (X, τ_2) be a Hausdorff space. If $\tau_2 \leq \tau_1$, then $\tau_2 = \tau_1$.

Let (X, τ) be a compact space and $\{x_i \mid i \in \mathbb{N}\}$ be a sequence of elements in X, then there exists an element $x \in X$ which is its accumulation point (see C.6).

C.13. A topological space (X, τ) is called connected if there are no open-closed subsets in it different from X and $\emptyset$. A subset Y of (X, τ) is called connected if $(Y, \tau|_Y)$ is a

1.1.3. REMARK. It is easy to verify that conditions (AC) and (AIC) are equivalent to the following condition:

Subtraction Continuity Condition (SC). The mapping $(a,b) \to a - b$ of the topological space $A \times A$ onto the topological space A is continuous, i.e. for any two elements $a, b \in A$ [an]d arbitrary neighborhood U of the element $a - b$ there exist neighborhoods V and W of [ele]ments a and b respectively such that $V - W = U$.

1.1.4. EXAMPLE. Let A be a Abelian group. It is easy to verify that the condition is true in the case of discrete topology on A. Hence, A is a topological Abelian group discrete topology (See C.1). In this manner any Abelian group could be considered [to]pological group.

EXAMPLE. Let A be an Abelian group. It is easy to verify that the condition [tr]ue in the anti-discrete topology on A (see C.1). Consequently, A is a topological [gr]oup. In this manner, any Abelian group could be considered as a topological [th]e anti-discrete topology.

[D]EFINITION. A ring R is called a topological ring if a topology is defined [R] and the additive group of the ring R is a topological group in this topology, [i]ng condition is valid:

Continuity Condition (MC). The mapping $(a,b) \to a \cdot b$ of the topological [t]he topological space R is continuous.

[REMA]RK. Condition (MC) exhibits the following:

[eleme]nts $a, b \in R$ and arbitrary neighborhood U of the element $a \cdot b$ there [V] and W of elements a and b, respectively, such that $V \cdot W \subseteq U$.

By virtue of Definition 1.1.6, the additive group of any topological [Ab]elian group. Conversely, if A is a topological Abelian group, then into the ring by the definition of zero multiplication on A, i.e. $b \in A$. In doing so, condition (MC) is fulfilled, and, hence, A is [topo]logical ring. In this manner any topological Abelian group may [ic]al ring with zero multiplication.

connected space.

Clearly, if Y is a connected subset in (X, τ), and S is an open-closed subset in (X, τ), then either $Y \subseteq S$ or $Y \cap S = \emptyset$.

Let Y be a connected subset of a topological space (X, τ) and $f : (X, \tau) \to (X', \tau')$ be a continuous mapping. Then $f(Y)$ is a connected subset in (X', τ').

Let $\{S_\gamma | \gamma \in \Gamma\}$ be a non-empty family of connected subsets of (X, τ). Then:

$[S_\gamma]_X$ is a connected subset in (X, τ) for any $\gamma \in \Gamma$;

if $\bigcap_{\gamma \in \Gamma} S_\gamma \neq \emptyset$, then $\bigcup_{\gamma \in \Gamma} S_\gamma$ is a connected subset in (X, τ).

For any element x of (X, τ) the union of all connected subsets containing x is called a connected component of the element x (denoted over C_x).

It is obvious that the subset C_x is closed and is the greatest connected subset in (X, τ) containing x. Let $f : (X, \tau) \to (X', \tau')$ be a homeomorphism of topological spaces, then for any $x \in X$ the subset $f(C_x)$ is a connected component of the element $f(x)$ in (X', τ').

A topological space (X, τ) is called totally disconnected if $C_x = \{x\}$ for any $x \in X$.

C.14. A topological space (X, τ) is called zero-dimensional if any of its elements has a basis of neighborhoods consisting of open-closed subsets. It is equivalent to saying that for every closed subset S and any $x \notin S$ there exists a continuous real-valued function f on (X, τ) such that:

$$f(X) = \{0,1\}, \quad f(S) = \{1\}, \quad f(x) = 0.$$

It is clear that any zero-dimensional Hausdorff space is totally disconnected. The inverse statement, in general is not valid. However, any Hausdorff locally compact totally disconnected space is zero-dimensional.

C.15. Let X be a set, (Y, τ) be a topological space, and $f : X \to Y$. Let $\tau' = \{f^{-1}(U) | U \in \tau\}$, then (X, τ') is a topological space and f is a continuous mapping. If, moreover, f is a surjection, then f is an open mapping of (X, τ') onto (Y, τ).

A topology τ' is called a prototype of τ with respect to the mapping f.

C.16. Let (X_1, τ_1) and (X_2, τ_2) be topological spaces, $X = X_1 \times X_2$ (see A.3). We define the mapping $f_i : X \to X_i$ as follows: $f_i((x_1, x_2)) = x_i$ for $i = 1, 2$. Let τ_i' be the prototype of the topology τ_i with respect to the mapping f_i, $i = 1, 2$, and $\tau = \sup\{\tau_1', \tau_2'\}$ (see C.9). Then (X, τ) is a topological space which is called a product of topological spaces (X_1, τ_1) and (X_2, τ_2).

A subset $U \subseteq X$ is a neighborhood of an element (x_1, x_2) in the space (X, τ) if and only if $U \supseteq U_1 \times U_2$, where U_i is a certain neighborhood of the element x_i in the space (X_i, τ_i), $i = 1, 2$.

TYCHONOFF THEOREM. A product of compact spaces is itself a compact space.

1. Main Concepts and General Information

The numerous examples taken from algebra and analysis suggest that the same may be a carrier of an algebraic structure (as an example, the ring or module struc and a topological space structure at once. As this takes place, the algebraic ope are assumed to be continuous in the appropriate topology. This permits using algebraic and topological methods simultaneously for investigation of those ob

The first five sections of the present chapter contain definitions and and g teristics of topological Abelian groups, rings and modules. Topological rin also Abelian groups, and this fact is essential to the majority of the sta 1.6 and 1.8 contain the concepts and results peculiar to the topologic

The majority of the results included in this chapter (especially in for the specialists in topological algebra, but could be useful for

It's possible to find associated results also in [24, 26, 61, 6? 110, 114-116, 118-120, 125].

§ 1.1. Main Definition

1.1.1. DEFINITION. Abelian group A is called defined on the set A and the following conditions a

Addition Continuity Condition (AC). This cor $a + b$ of the topological space $A \times A$ (see C.16)

Additive Inversion Continuity Condition ($a \to -a$ of the topological space A onto it'

1.1.2. REMARK. In the terms of respectively the following:

for any two elements $a, b \in A$ ar exist such neighborhoods V and

for any element $a \in A$ and neighborhood V of the elem

(SC)
in the
as a to

1.1.5
(SC) is t
Abelian g
group in th

1.1.6. D
onto the set *R*
and the followi

Multiplicatio
space $R \times R$ to t

1.1.7. **REMA**
for any two elem
exist neighborhoods

1.1.8. **REMARK**
ring is a topological Ab
A could be transformed
setting $a \cdot b = 0$ for any a
transformed into the topo
be considered as a topologi

1.1.9. EXAMPLE. Let R be a ring, then its additive group could be transformed into a topological Abelian group by endowing R with the discrete or anti-discrete topology (see examples 1.1.4 and 1.1.5). It is easy to verify also that the ring R satisfies condition (MC) in both topologies. In this manner any ring could be considered as a topological ring in the discrete or anti-discrete topology.

1.1.10. REMARK. As it was shown in the examples above (see Examples 1.1.4 and 1.1.5), the same Abelian group (ring) may be transformed into a topological group (topological ring) in various ways. Because of this, symbols (A, τ) will be used when it would be necessary for designation of the topological Abelian group (topological ring), where A is an Abelian group (ring) and τ is a topology considering onto A.

1.1.11. EXAMPLE. Let $\mathbb{R}$ be the field of real numbers (with canonical operations of addition and multiplication, and this field is endowed with the interval topology (see C.1) relative to the natural linear order on $\mathbb{R}$ (this is a topology in which all of the interval unions, and only they, represent open sets). The additive and multiplicative operations of the field of real numbers are continuous as the functions of two variables, and, hence, the conditions (SC) and (MC) are fulfilled. Thus, $\mathbb{R}$ is a topological ring.

1.1.12. DEFINITION. A ring R is called a normed (pseudonormed) ring if a non-negative real function ξ is specified onto R, and this function satisfies the following conditions:

Norm Ring Condition 1 (NR1): $\xi(r) = 0$ if and only if $r = 0$;

Norm Ring Condition 2 (NR2): $\xi(r_1 - r_2) \leq \xi(r_1) + \xi(r_2)$, for any $r_1, r_2 \in R$;

Norm Ring Condition 3 (NR3): $\xi(r_1 \cdot r_2) = \xi(r_1) \cdot \xi(r_2)$, (respectively, $\xi(r_1 \cdot r_2) \leq \xi(r_1) \cdot \xi(r_2)$), for any $r_1, r_2 \in R$.

The number $\xi(r)$ is called a norm (pseudonorm) of the element $r \in R$, and the function ξ is called a norm (pseudonorm) on the ring R. It is well understood that any norm on the ring R is a pseudonorm on this ring.

1.1.13. REMARK. Let ξ be a pseudonorm on a ring R, then the following conditions are true:

(1) $\xi(-r) = \xi(r)$ for any $r \in R$;

(2) $\xi(r_1) - \xi(r_2) \leq \xi(r_1 - r_2)$ for any $r_1, r_2 \in R$.

Indeed, $\xi(-r) = \xi(0 - r) \leq \xi(0) + \xi(r) = \xi(r)$, that is $\xi(-r) \leq \xi(r)$. In view of $\xi(r) = \xi(-(-r)) \leq \xi(-r)$, we get $\xi(r) = \xi(-r)$, that is (1) is proved.

Further, $\xi(r_1) = \xi(r_1 - 0) = \xi(r_1 - (r_2 - r_2)) = \xi((r_1 - r_2) - (-r_2)) \leq \xi(r_1 - r_2) + \xi(r_2)$, thus $\xi(r_1) - \xi(r_2) \leq \xi(r_1 - r_2)$, that is (2) is proved, too.

1.1.14. EXAMPLE. Let R be a pseudonormed ring, and ξ be a pseudonorm on R. Then the real-valued function $\rho(x, y) = \xi(x - y)$ satisfies the metric axioms (see C.4). Thus, R is a metric space.

Let's verify that the ring operations are continuous in the topology τ_ξ defined by this metric. Let $a, b \in R$, and let ϵ be a positive number. Consider the two following cases.

1) Let $a = b = 0$. Then put $\delta = \min\left\{\frac{\epsilon}{2}, \sqrt{\epsilon}\right\}$. Let $x, y \in R$ satisfy conditions $\rho(x, a) < \delta$ and $\rho(y, b) < \delta$, then:

$$\rho(x - y, a - b) = \rho(x - y, 0) = \xi(x - y) \leq \xi(x) + \xi(y)$$
$$= \xi(x - a) + \xi(y - b) = \rho(x, a) + \rho(y, b) < \delta + \delta \leq \epsilon$$

and $\rho(xy, ab) = \rho(xy, 0) = \xi(xy) \leq \xi(x) \cdot \xi(y) = \xi(x-a) \cdot \xi(x-b) = \rho(x, a) \cdot \rho(y, b) < \delta\delta \leq \epsilon$.

2) Let at least one element of a and b differ from zero, for instance, $a \neq 0$. At that time $\xi(a) \neq 0$ and $2\xi(a) + \xi(b) \neq 0$. Let $\delta = \min\left\{\frac{\epsilon}{2}, \xi(a), \frac{\epsilon}{2\xi(a)+\xi(b)}\right\}$. Let $x, y \in R$ be such that $\rho(x, a) < \delta$ and $\rho(y, b) < \delta$, then $\xi(x - a) < \delta$, and, hence, $\xi(x) - \xi(a) \leq \xi(x - a) < \delta$, that is $\xi(x) < \xi(a) + \delta$. Thus,

$$\rho(x - y, a - b) = \xi\big((x - a) - (y - b)\big) \leq \xi(x - a) + \xi(y - b)$$
$$= \rho(x, a) + \rho(y, b) < \delta + \delta \leq \epsilon,$$

and

$$\rho(xy, ab) = \xi(xy - ab) = \xi(xy - xb + xb - ab)$$
$$\leq \xi\big(x \cdot (y - b) + (x - a) \cdot b\big)$$
$$\leq \xi(x) \cdot \xi(y - b) + \xi(x - a) \cdot \xi(b)$$
$$< \big(\xi(a) + \delta\big) \cdot \delta + \delta \cdot \xi(b)$$
$$\leq \big(2\xi(a) + \xi(b)\big) \cdot \delta \leq \epsilon.$$

Hence, the operations of subtraction and multiplication are continuous in the ring R with the topology specified by the metric $\rho(x, y)$.

Thus, any pseudonormed ring is a topological ring.

1.1.15. EXAMPLE. The function corresponding each real number $x \in \mathbb{R}$ to its absolute value $|x|$ apparently satisfies conditions (NR1) to (NR3), and, hence, $\mathbb{R}$ is a normed ring, and consequently is a topological ring. Notice that a subset is open in this topology if and only if each of its elements belong to this subset together with some open interval containing this element. Because of this, the subsets, which are unions of the open intervals, and only they, represent open subsets in this topology. Thus, the topology, which we have just defined onto $\mathbb{R}$, coincides with the interval topology (see Example 1.1.11).

1.1.16. EXAMPLE. The function corresponding complex number $z = a + bi \in \mathbb{C}$ to its absolute value $|z| = \sqrt{a^2 + b^2}$ satisfies conditions (NR1) to (NR3) and, hence, $\mathbb{C}$ is a normed ring, i.e. it is a topological ring.

1.1.17. EXAMPLE. Let $C([0; 1], \mathbb{R})$ be the ring of all continuous real-valued functions specified on the segment $[0, 1]$. Let $\xi(f) = \max\{|f(x)| \big| x \in [0, 1]\}$ for $f \in C([0, 1], \mathbb{R})$.

Let's verify that $\xi(f)$ is a pseudonorm. It is easy to see that $\xi(f) \geq 0$. Moreover, if $\xi(f) = 0$, then $f = 0$, i.e. condition (NR1) is satisfied.

Further,

$$\xi(f - g) = \max\{|(f - g)(x)| \big| x \in [0, 1]\}$$
$$= \max\{|f(x) - g(x)| \big| x \in [0; 1]\} \leq \max\{|f(x)| + |g(x)| \big| x \in [0; 1]\}$$
$$\leq \max\{|f(x)| \big| x \in [0; 1]\} + \max\{|g(x)| \big| x \in [0; 1]\} = \xi(f) + \xi(g),$$

i.e. condition (NR2) is satisfied.

In conclusion,

$$\xi(f \cdot g) = \max\{|(f \cdot g)(x)| \big| x \in [0; 1]\}$$
$$= \max\{|f(x) \cdot g(x)| \big| x \in [0; 1]\} \leq \max\{|f(x)| \cdot |g(x)| \big| x \in [0; 1]\}$$
$$\leq \max\{|f(x)| \big| x \in [0; 1]\} \cdot \max\{|g(x)| \big| x \in [0; 1]\} = \xi(f) \cdot \xi(g),$$

i.e. condition (RN3) is satisfied, too. Thus, the ring $C([0;1], \mathbb{R})$ is pseudonormed ring, and consequently is a topological ring.

1.1.18. EXAMPLE. The function corresponding each quaternion $h = h_0 + h_1 \cdot i + h_2 \cdot j + h_3 \cdot k$ from the ring $\mathbb{H}$ (see B.19) to its absolute value $|h| = \sqrt{h_0^2 + h_1^2 + h_2^2 + h_3^2}$, as it is known, satisfies conditions (NR1) to (NR3), and, hence, the ring $\mathbb{H}$ is normed ring and consequently a topological ring.

1.1.19. EXAMPLE. Let some prime number p be fixed. For each non-zero rational number $r \in \mathbb{Q}$ there exists a unique integer k such that $r = \frac{a}{b} \cdot p^k$, where $a, b \in \mathbb{Z}$, and a, b are not divisible by p. Put $\xi_p(r) = \frac{1}{2^k}$ and $\xi_p(0) = 0$. Then the non-negative real-valued function ξ_p is specified on $\mathbb{Q}$.

Clearly, the function ξ_p satisfies condition (NR1).

Let's verify that ξ_p satisfies also conditions (NR2) and (NR3), and hence, is a norm on $\mathbb{Q}$.

Actually, let

$$r_1 = \frac{a_1}{b_1} \cdot p^{k_1}, \quad r_2 = \frac{a_2}{b_2} \cdot p^{k_2}$$

and assume that a_1, b_1, a_2 and b_2 are not divisible by p. Assume also that $k_1 \geq k_2$ for the sake of definiteness. Then

$$\xi_p(r_1) = \frac{1}{2^{k_1}}, \, \xi_p(r_2) = \frac{1}{2^{k_2}},$$

and

$$\xi_p(r_1 - r_2) = \xi_p\left(\frac{a_1}{b_1} \cdot p^{k_1} - \frac{a_2}{b_2} \cdot p^{k_2}\right) = \xi_p\left(\frac{b_2 a_1 p^{k_1 - k_2} - a_2 b_1}{b_1 b_2} \cdot p^{k_2}\right)$$

$$\leq \frac{1}{2^{k_2}} \leq \frac{1}{2^{k_1}} + \frac{1}{2^{k_2}} = \xi_p(r_1) + \xi_p(r_2).$$

Also

$$\xi_p(r_1 \cdot r_2) = \xi_p\left(\frac{a_1}{b_1} \cdot p^{k_1} \cdot \frac{a_2}{b_2} \cdot p^{k_2}\right) = \xi_p\left(\frac{a_1 a_2}{b_1 b_2} \cdot p^{k_2 + k_2}\right).$$

By virtue of the fact that none of numbers $a_1 a_2$ and $b_1 b_2$ are divisible by p, it follows that

$$\xi_p\left(\frac{a_1 a_2}{b_1 b_2} \cdot p^{k_1 + k_2}\right) = \frac{1}{2^{k_1 + k_2}}.$$

Consequently

$$\xi_p(r_1 r_2) = \frac{1}{2^{k_1+k_2}} = \frac{1}{2^{k_1}} \cdot \frac{1}{2^{k_2}} = \xi_p(r_1) \cdot \xi_p(r_2).$$

Finally, the ring $\mathbb{Q}$ is a topological ring in topology τ_p defined by the norm ξ_p. The topology τ_p is called a p-adic topology on rational numbers.

1.1.20. DEFINITION. A skew field (field) K is called a topological skew field (field) if it is a topological ring and the following condition is satisfied:

Multiplicative Inversion Continuity Condition (MIC). This condition implies that the mapping $x \to x^{-1}$ of the subspace $K \setminus \{0\}$ onto itself is continuous, i.e. that for any non-zero element $x \in K$ and any neighborhood U of x^{-1} there exist a neighborhood V of the element x such that $(V \setminus \{0\})^{-1} \subseteq U$.

1.1.21. REMARK. The multiplicative group of non-zero elements of the topological field is a topological Abelian group.

1.1.22. EXAMPLE. Let K be a skew field (field). Consider discrete or anti-discrete topology onto K. In both cases condition (MIC) is satisfied, and, hence, any skew field (field) is a topological skew field (field) in the discrete or anti-discrete topology.

1.1.23. REMARK. Let a skew field K be a normed ring. Then (it will be shown in Proposition 1.2.14) the topology specified on K by the metric $\rho(x, y) = \xi(x - y)$ (see Example 1.1.14) satisfies condition (MIC), i.e. K is a topological skew field. For instance, the field of real numbers $\mathbb{R}$ with interval topology (see Examples 1.1.15), the field of complex numbers $\mathbb{C}$ (see Example 1.1.16), and the field of rational numbers $\mathbb{Q}$ with the p-adic topology for any prime number p (see Example 1.1.19) are topological fields, and quaternion skew field $\mathbb{H}$ is the topological skew field (see Example 1.1.18).

1.1.24. DEFINITION. Let R be a topological ring. A left R-module M is called a topological left R-module if on M is specified a topology such that M is a topological Abelian group and the following condition is satisfied:

Ring Multiplication Continuity Condition (RMC). The mapping $(r, m) \to rm$ of the topological space $R \times M$ to the topological space M is continuous, i.e. for any $r \in R$ and

$m \in M$ and arbitrary neighborhood U of the element $r \cdot m$ in M there exist a neighborhood V of the element r in R and a neighborhood W of the element m in M such that $V \cdot W \subseteq U$.

1.1.25. REMARK. Any topological left R-module M is a topological Abelian group. Consequently M satisfies conditions (AC) and (AIC) (see Definition 1.1.1 and Remark 1.1.2), or M satisfies condition (SC) (see Remark 1.1.3), which is equivalent to the above-mentioned conditions.

1.1.26. REMARK. In a similar way it is possible to investigate right topological modules over a topological ring. Any topological ring R is both a topological left R-module and a topological right R-module. Hereafter, by a topological R-module is meant a topological left R-module, unless otherwise stated.

1.1.27. EXAMPLE. Let A be a topological Abelian group. It is well understood that A is a $\mathbb{Z}$-module (the discrete topology is introduced on $\mathbb{Z}$). Let's show that condition (RMC) is satisfied.

Indeed, let $b = k \cdot a$, where $k \in \mathbb{Z}$ and $a \in A$. Taking into consideration condition (AIC), we can suppose that $k > 0$, i.e.

$$b = \underbrace{a + a + \ldots + a}_{k \text{ summmmands}} .$$

Let U be a neighborhood of the element b in A. In compliance with condition (AC), there exists a neighborhood W of the element a in A such that

$$\underbrace{W + W + \ldots + W}_{k \text{ summands}} \subseteq U.$$

Since the discrete topology is introduced onto $\mathbb{Z}$, the subset $V = \{k\}$ is a neighborhood of the element k in $\mathbb{Z}$. Hence,

$$V \cdot W = \{k\} \cdot W \subseteq \underbrace{W + W + \ldots + W}_{k \text{ summands}} \subseteq U,$$

i.e. condition (RMC) is satisfied (see Definition 1.1.24).

By this means any topological Abelian group in the natural way is the topological $\mathbb{Z}$-module over the ring of integers $\mathbb{Z}$ with the discrete topology.

1.1.28. EXAMPLE. Let R be a topological ring, and let M be R-module. As it is noted in Example 1.1.5, M is a topological group in the anti-discrete topology. It is well understood that in this case condition (RMC) is satisfied, i.e. M with anti-discrete topology is a topological R-module.

1.1.29. REMARK. Let (R, τ) be a topological ring, and let M be a topological (R, τ)-module. Consider a ring topology τ' onto R such that $\tau \leq \tau'$. Let's verify that M is a topological (R, τ')-module, i.e. that condition (RMC) is satisfied for the module M and the topological ring (R, τ').

Let $r \in R$, $m \in M$, and U be a neighborhood of the element $r \cdot m$ in M. By virtue of the fact that M is a topological (R, τ)-module, there exist a neighborhood V of the element r in (R, τ) and a neighborhood W of the element m in M such that $V \cdot W \subseteq U$. Since $\tau \leq \tau'$, then V is a neighborhood of the element r in (R, τ'), and therefore, condition (RMC) is satisfied for the topological ring (R, τ').

1.1.30. DEFINITION. Let K be a topological skew field. A unitary topological K-module E is called a topological vector space over K.

1.1.31. EXAMPLE. In the natural way (see Remark 1.1.26), the additive group of a topological skew field K is a topological vector space over K.

1.1.32. EXAMPLE. The field of complex numbers $\mathbb{C}$ (see Example 1.1.16) and quaternion skew field $\mathbb{H}$ (see Example 1.1.18) with the topology specified by the norm $|\cdot|$ are vector spaces over the field of real numbers $\mathbb{R}$ endowed with the interval topology.

1.1.33. EXAMPLE. The ring $C([0;1], \mathbb{R})$ of the real-valued continuous functions on segment $[0;1]$ with the ordinary operations of addition and multiplication by the real numbers, endowed with the topology specified by pseudonorm $\xi(f) = \max\{|f(x)| \, | \, x \in [0;1]\}$ (see Example 1.1.17), is a topological vector space over the field $\mathbb{R}$ endowed with the interval topology.

1.1.34. PROPOSITION. *Let A be a topological Abelian group, $a \in A$, B and C be subsets of A. Then the following statements are true:*

(1) mappings $\varphi_a : A \to A$ and $\varphi : A \to A$, where $\varphi_a(x) = x + a$ and $\varphi(x) = -x$, are homeomorphic mappings of the topological space A onto itself;

(2) the following conditions are equivalent:

  (a) subset B is open (closed);

  (b) subset $-B$ is open (closed);

  (c) subset $B + a$ is open (closed). (Among other things a subset $U \subseteq A$ is a neighborhood of the element a if and only if $U - a$ is a neighborhood of 0.)

(3) if subset B is open, then $B + C$ is also an open subset.

PROOF. (1) It is readily apparent that mappings φ_a and φ are bijective mappings. Continuity of these mappings follows from conditions (AC) and (AIC). As for the inverse mappings, their continuity is obvious by virtue of $(\varphi_a)^{-1} = \varphi_{-a}$ and $\varphi^{-1} = \varphi$.

(2) Implication (a) $\Rightarrow$ (b) is obvious in view of fact that $-B = \varphi(B)$. By virtue of the fact that $B + a = \varphi_a(\varphi(-B))$, implication (b)$\Rightarrow$ (c) is true. Finally, implication (c) $\Rightarrow$ (a) follows from fact that $B = (\varphi_a)^{-1}(B + a)$.

(3) By virtue of the fact that $B + C = \bigcup_{c \in C}(B + c)$, it follows that $B + C$ is a union of open subsets and, hence, is open.

1.1.35. COROLLARY. The topological space of any topological Abelian group A is a homogeneous space (see C.7), i.e. for any a, $b \in A$ there exists a homeomorphism $\psi : A \to A$ such that $\psi(a) = b$.

Indeed, it is enough to consider the equality $\psi = \varphi_{b-a}$.

1.1.36. REMARK. Since the space of a topological Abelian group is homogeneous, all the local topological properties can be defined and verified for only one element. Usually, 0 is selected as such element.

1.1.37. COROLLARY. A topological Abelian group is discrete if and only if the group contains at least one isolated element.

1.1.38. PROPOSITION. *Let B and C be subsets of topological Abelian group A. Then the following statements are true:*

(1) if B and C are compact subsets then $B + C$ is a compact subset in A;

(2) if B is a closed subset and C is a compact subset then $B + C$ is a closed subset in A.

PROOF. (1) Let ψ be a mapping of topological space $A \times A$ onto topological space A, where $\psi(a, b) = a + b$. In view of condition (AC), the mapping is continuous, and, hence, its restriction onto subspace $B \times C$ of $A \times A$ is continuous, too. Thus, $B + C = \psi(B \times C)$ is a compact subset in A (as a continuous image of compact space $B \times C$, see C.16 and C.12).

(2) Let us assume the contrary, i.e. that $B + C$ is not a closed subset in A and let $x \in [B + C]_A$, but $x \notin B + C$. Then $(x - C) \cap B = \emptyset$, that is $x - C \subseteq A \backslash B$. By virtue of the fact that B is a closed subset in A, we get that $A \backslash B$ is an open subset in A and, hence, $A \backslash B$ is a neighborhood in A of any element of the type of $x - c$, where $c \in C$. In view of condition (SC) (see Remark 1.1.3), there exist neighborhoods U_c and V_c of elements x and c, respectively, in A (these neighborhoods can be chosen among open ones), such that $U_c - V_c \subseteq A \backslash B$. Thus, $\{V_c \mid c \in C\}$ is an open cover of the compact subset C, and, hence (see C.12), there exists a finite set of elements $c_1, c_2, \ldots, c_n$ from C such that

$$V = \bigcup_{i=1}^{n} V_{c_i} \supseteq C .$$

Consequently,

$$U = \bigcap_{i=1}^{n} U_{c_i}$$

is a neighborhood of the element x. Thus, $U - V \subseteq A \backslash B$. Since $V \supseteq C$, then $U - C \subseteq A \backslash B$, and, hence, $y - C \subseteq A \backslash B$ for any $y \in U$. Therefore, $(y - C) \cap B = \emptyset$, that is $y \notin B + C$. It is thereby shown that $U \cap (B + C) = \emptyset$, that is $x \notin [B + C]_A$. The contradiction to the assumption that $x \in [B + C]_A$ is obtained. Thus $B + C$ is clossed. This completes the proof.

1.1.39. COROLLARY. The sum $B + C$ of a closed subset B and a finite subset C of any topological Abelian group is its closed subset.

1.1.40. REMARK. The sum of two closed subsets of a topological Abelian group may be not closed. Indeed, the subsets $\mathbb{N} = \{1, 2, \dots\}$ and $B = \left\{\frac{1}{2} - 2, \frac{1}{3} - 3, \dots\right\}$ are closed in the topological Abelian group $\mathbb{R}$ of real numbers with the interval topology, but the subset $\mathbb{N} + B$ is not closed. Actually, $0 \notin \mathbb{N} + B$, but $0 \in [\mathbb{N} + B]_{\mathbb{R}}$ - by virtue of the fact that $\mathbb{N} + B$ contains the sequence $\left\{\frac{1}{2}, \frac{1}{3}, \dots\right\}$ converging to 0.

1.1.41. PROPOSITION. *Let B and C be subsets of a topological Abelian group A. Then the following statements are true:*

(1) $[B + C]_A \supseteq [B]_A + [C]_A$;

(2) $[-B]_A = -[B]_A$;

(3) $[B - C]_A \supseteq [B]_A - [C]_A$;

(4) if C is a compact subset,* then $[B + C]_A = [B]_A + [C]_A = [B]_A + C$ and $[B - C]_A = [B]_A - [C]_A = [B]_A - C$.

PROOF. (1) Let $x \in [B]_A + [C]_A$ and let U be a neighborhood of the element x. Then $x = b + c$, where $b \in [B]_A$ and $c \in [C]_A$, and, hence, there exist neighborhoods V and W in A of the elements b and c respectively, such that $V + W \subseteq U$. By virtue of the fact that $V \bigcap B \neq \emptyset$ and $W \bigcap C \neq \emptyset$, elements $b_1 \in V \bigcap B$ and $c_1 \in W \bigcap C$ can be found. Thus, $b_1 + c_1 \in B + C$ and $b_1 + c_1 \in V + W \subseteq U$, that is $(B + C) \bigcap U \neq \emptyset$. Consequently, $[B + C]_A \supseteq [B]_A + [C]_A$.

(2) The validity of $[-B]_A = -[B]_A$ results from the fact that the mapping $x \mapsto -x$ is a homeomorphism of the topological space A onto itself (see Proposition 1.1.34, Item 1).

(3) Inclusion $[B - C]_A \supseteq [B]_A - [C]_A$ results from items (1) and (2).

(4) Inclusions $[B + C]_A \supseteq [B]_A + [C]_A \supseteq [B]_A + C$ follow from item (1). In view of Proposition 1.1.38, $[B]_A + C$ is a closed subset of A. Thus, $[B + C]_A \subseteq \big[[B]_A + C\big]_A = [B]_A + C$, and, hence, $[B + C]_A = [B]_A + [C]_A = [B]_A + C$.

1.1.42. PROPOSITION. *Let R be a topological ring, A a topological R-module, $r \in R$, $a \in A$, and Q a subset in R, B a subset in A. Then the following statements are true:*

---

* In a general case, if to allow the topology on A be non-Hausdorff, the subset C can be not closed in A.

(1) the mapping $\varphi_r : A \to A$, *where* $\varphi_r(x) = r \cdot x$, $x \in A$, *is a continuous mapping of the topological space A into itself;*

(2) the mapping $\varphi_a : R \to A$, *where* $\varphi_a(x) = x \cdot a$, $x \in R$, *is a continuous mapping of the topological space R to the topological space A;*

(3) $[Q \cdot B]_A \supseteq [Q]_R \cdot [B]_A$;

(4) if subsets Q and B are compact, then $Q \cdot R$ is a compact subset.

PROOF. Statements (1) and (2) result from the definition of a topological module (see Definition 1.1.24). Statements (3) and (4) are proved similarly to the proofs of statement (1), Proposition 1.1.41, and statement (1), Proposition 1.1.38, respectively.

1.1.43. COROLLARY. Let A be a topological ring, $a \in A$, and let B and C be subsets in A. Then the following statements are true:

(1) the mappings $\psi_a : A \to A$ and $\psi_a' : A \to A$, where $\psi_a(x) = x \cdot a$ and $\psi_a'(x) = a \cdot x$ for $x \in A$, are continuous mappings of the topological space A into itself;

(2) $[B \cdot C]_A \supseteq [B]_A \cdot [C]_A$;

(3) if B and C are compact subsets, then $B \cdot C$ is a compact subset in A.

1.1.44. PROPOSITION. *Let A be a topological ring with the unitary element and M be a topological A-module. Let $a \in A$ be an invertible element, then mappings $\varphi_a : M \to M$ (see Proposition 1.1.42), $\psi_a : A \to A$ and $\psi_a' : A \to A$ (see Corollary 1.1.43) are homeomorphic mappings of the topological spaces M and A correspondingly onto themselves (see C.7).*

PROOF. Let B be an open subset of M, and $b' \in \psi_a(B)$. Then $b' = \psi_a(b) = a \cdot b$ for some $b \in B$.

By virtue of the fact that B is an open subset, we get that B is a neighborhood of the element b in M. From $b = a^{-1} \cdot b'$ and condition (RMC) (see Definition 1.1.24) follows the existence of a neighborhood U' of the element b' in M such that $a^{-1} \cdot U' \subseteq B$. Then $U' \subseteq a \cdot B = \varphi_a(B)$ and, hence, $\varphi_a(B)$ is neighborhood of the element b' in M, i.e. $\varphi_a(B)$ is an open subset of M. Hence, φ_a is open mapping.

In the same manner is proved that ψ_a and ψ_a' are open mappings too.

In view of the fact that all the mappinds φ_a, ψ_a and ψ'_a are bijections, the proposition is proved completely.

1.1.45. COROLLARY. *Let A be a topological ring with the unitary element, $a \in A$ be an invertible element and $x \in A$. Then the following statements are equivalent:*

(1) U is a neighborhood of the element x in A;

(2) $U \cdot a$ is a neighborhood of the element $x \cdot a$ in A;

(3) $a \cdot U$ is a neighborhood of the element $a \cdot x$ in A.

PROOF. Let's consider the homeomorphism $\psi_a : A \to A$ (see Proposition 1.1.44). Since $x \cdot a = \psi_a(x)$ and $U \cdot a = \psi_a(U)$, then $(1) \Rightarrow (2)$.

The mapping $\theta_a : A \to A$, where $\theta_a(z) = a \cdot (z \cdot a^{-1})$ for $z \in A$, is the composition of the homeomorphic mappings ψ_a and ψ'_a (see Proposition 1.1.44), and, hence, is a homeomorphism. Since $\theta_a(x \cdot a) = a \cdot x$ and $\theta_a(U \cdot a) = a \cdot U$, then $(2) \Rightarrow (3)$.

Equalities $\psi'_{a^{-1}}(a \cdot x) = x$ and $\psi'_{a^{-1}}(a \cdot U) = U$ are obtained by considering of the homeomorphism $\psi'_{a^{-1}} : A \to A$. Then from Proposition 1.1.44 follows that U is a neighborhood of the element x. Thus, $(3) \Rightarrow (1)$.

1.1.46. COROLLARY. Let a be an invertible element of a topological ring A with the unitary element. Then the following statements are equivalent:

(1) U is a neighborhood of 0 in A;

(2) $U \cdot a$ is a neighborhood of 0 in A;

(3) $a \cdot U$ is a neighborhood of 0 in A.

1.1.47. COROLLARY. Let a be an invertible element of a topological ring R with the unitary element and let $B \subseteq R$. Then the following conditions are equivalent:

(1) B is open (closed);

(2) $a \cdot B$ is open (closed);

(3) $B \cdot a$ is open (closed).

1.1.48. PROPOSITION. *Let R be a topological skew field and let $0 \neq a \in R$. If element a is an accumulation point (a limit) of a sequence of non-zero elements $a_1, a_2, \ldots \in K$ (see C.6), then the element a^{-1} is an accumulation point (a limit) of the sequence $a_1^{-1}, a_2^{-1}, \ldots$ in the skew field K.*

PROOF. Let U be a neighborhood of the element a^{-1}, and let V be a neighborhood of the element a such that $(V\backslash\{0\})^{-1} \subseteq U$ (see Definition 1.1.20).

By virtue of the fact that a is an accumulation point (a limit) of the sequence $a_1, a_2, \ldots$, we get that for any $n \in \mathbb{N}$ (there exists $n \in \mathbb{N}$) there exists $i \geq n$ (for any $i \geq n$) such that $a_i \in V$. Since $a_i^{-1} \neq 0$, then $a_i^{-1} \in (V\backslash\{0\})^{-1} \subseteq U$, i.e. a^{-1} is an accumulation point (a limit) of the sequence $a_1^{-1}, a_2^{-1}, \ldots$.

1.1.49. PROPOSITION. *Let K be a topological skew field. Then the mapping $\theta :$ $K\backslash\{0\} \to K\backslash\{0\}$, where $\theta(x) = x^{-1}$ for $x \neq 0$, is a homeomorphism of the topological subspace $K\backslash\{0\}$ onto itself.*

PROOF. In view of Definition 1.1.20, θ is a continuous mapping. Since $\theta = \theta^{-1}$, then θ is a homeomorphism.

§ 1.2. Neighborhoods of Zero

1.2.1. PROPOSITION. *Let a family $\mathcal{B}_0$ of subsets of a topological Abelian group A be a basis of neighborhoods of zero (see C.3) in A. Then the following conditions (Basis of Neighborhood conditions) are satisfied:*

(BN1) $0 \in \bigcap_{V \in \mathcal{B}_0} V$;

(BN2) for any subsets U and V from $\mathcal{B}_0$ there exists a subset $W \in \mathcal{B}_0$ such that $W \subseteq U \bigcap V$;

(BN3) for any subset $U \in \mathcal{B}_0$ there exists a subset $V \in \mathcal{B}_0$ such that $V + V \subseteq U$;

(BN4) for any subset $U \in \mathcal{B}_0$ there exists a subset $V \in \mathcal{B}_0$ such that $-V \subseteq U$;

Besides, if $a \in A$, then $\mathcal{B}_a = \{a + V \mid V \in \mathcal{B}_0\}$ is a basis of neighborhoods of the element a.

PROOF. The fulfillment of conditions (BN1) and (BN2) results from the definition of a basis of neighborhoods of an element in a topological space. The fulfillment of conditions (BN3) and (BN4) follows from the fact that $\mathcal{B}_0$ is a basis of neighborhoods of zero in A, from conditions (AC) and (AIC) (see Definition 1.1.1) and from condition (BN2) with the consideration of the fact that $0 + 0 = 0$ and $-0 = 0$.

If $a \in A$, then the mapping $\varphi_a : A \to A$ is a homeomorphism in view of Proposition 1.1.34, and, hence, $\mathcal{B}_a = \{a + V \mid V \in \mathcal{B}_0\} = \{\varphi_a(V) \mid V \in \mathcal{B}_0\}$ is a basis of neighborhoods of the element a.

1.2.2. PROPOSITION. *Let $\mathcal{B}_0$ be a basis of neighborhoods of zero of a topological ring A. In this case conditions (BN1) to (BN4) are satisfied together with the following conditions:*

(BN5) for any subset $U \in \mathcal{B}_0$ there exists a subset $V \in \mathcal{B}_0$ such that $V \cdot V \subseteq U$;

(BN6) for any subset $U \in \mathcal{B}_0$ and any element $a \in A$ there exists a subset $V \in \mathcal{B}_0$ such that $a \cdot V \subseteq U$ and $V \cdot a \subseteq U$.

PROOF. Since $\mathcal{B}_0$ is a basis of neighborhoods of zero of the additive topological group $A(+)$, the fulfillment of conditions (BN1) - (BN4) follows from Proposition 1.2.1. The fulfillment of conditions (BN5) and (BN6) results from condition (MC) (see Definition 1.1.6 and Remark 1.1.7) with regard to $0 \cdot 0 = 0$ and $0 \cdot a = a \cdot 0 = 0$ for any $a \in A$.

1.2.3. PROPOSITION. *Let R be a topological ring and let $\mathcal{B}_0$ be a basis of neighborhoods of zero of a topological R-module A. Then conditions (BN1) to (BN4) are satisfied together with the following conditions:*

(BN5') for any subset $U \in \mathcal{B}_0$ there exist a subset $V \in \mathcal{B}_0$ and a neighborhood W of zero in R such that $W \cdot V \subseteq U$;

(BN6') for any subset $U \in \mathcal{B}_0$ and any element $r \in R$ there exists a subset $V \in \mathcal{B}_0$ such that $r \cdot V \subseteq U$;

(BN6'') for any subset $U \in \mathcal{B}_0$ and any element $a \in A$ there exists a neighborhood W of zero in R such that $W \cdot a \subseteq U$.

PROOF. To prove these conditions, it is necessary to use Proposition 1.2.1, condition (RMC) (see Definition 1.1.24), and to take account of $0 \cdot a = r \cdot 0 = 0$ for any $r \in R$ and $a \in A$.

1.2.4. THEOREM. *Let $\mathcal{B}_0$ be a non-empty family of subsets of an Abelian group A, satisfying conditions (BN1) to (BN4). Then there exists a topology (and moreover, a unique topology) on the set A such that A is a topological group and $\mathcal{B}_0$ is a basis of neighborhoods of zero in this topology.*

PROOF. Let $\mathcal{B} = \{B \subseteq A \mid$ for any $b \in B$ there exists $U_b \in \mathcal{B}_0$ such that $b + U_b \subseteq B\}$. Let's show that $\mathcal{B}$ specifies some topology on A and is the family of all open subsets in this topology (see C.1) and besides, $\mathcal{B}_0$ is a basis of neighborhoods of zero in A in this topology.

It is obvious that $\emptyset$ and A belong to $\mathcal{B}$.

Let B_1 and $B_2 \in \mathcal{B}$, and let $b \in B_1 \cap B_2$. Then there exist $U_1, U_2 \in \mathcal{B}_0$ such that $b + U_1 \subseteq B_1$ and $b + U_2 \subseteq B_2$. In view of condition (BN2), there exists $U_3 \in \mathcal{B}_0$ such that $U_3 \subseteq U_1 \cap U_2$. Hence, $b + U_3 \subseteq B_1 \cap B_2$. Thus, $B_1 \cap B_2 \in \mathcal{B}$.

Let $\emptyset \neq \mathcal{A} \subseteq \mathcal{B}$, $B' = \bigcup_{B \in \mathcal{A}} B$ and $b \in B'$. Then $b \in B_0$ is true for some $B_0 \in \mathcal{A}$, and, hence, $b + U \subseteq B_0$ for some $U \in \mathcal{B}_0$. Consequently, $b + U \subseteq B'$, and, hence, $B' \in \mathcal{B}$.

Thus, on A is defined a topology and $\mathcal{B}$ is the family of all open subsets in this topology.

Let's show that for any element $a \in A$ the family $\mathcal{B}_a = \{a + U \mid U \in \mathcal{B}_0\}$ is a basis of neighborhoods of the element a in this topology. Verify first that for any $U \in \mathcal{B}_0$ the subset $a + U$ is a neighborhood of the element a.

Let $U \in \mathcal{B}_0$ and $V_a = \{x \in A \mid$ there exists $U_x \in \mathcal{B}_0$ such that $x + U_x \subseteq a + U\}$. Let is obvious that $a \in V_a$ and $V_a \subseteq a + U$. If $x \in V_a$, then $x + U_x \subseteq a + U$ and $U_x \in \mathcal{B}_0$. In view of property (BN3), there exists $U'_x \in \mathcal{B}_0$ such that $U'_x + U'_x \subseteq U_x$. By virtue of this fact, $(x + U'_x) + U'_x \subseteq x + U_x \subseteq a + U$, and, hence, $x + U'_x \subseteq V_a$. Thus, $V_a \in \mathcal{B}$, i.e. V_a is an open set, and because of this, $a + U$ is a neighborhood of the element a.

Now, let's verify that $\mathcal{B}_a$ is a basis of neighborhoods of the element a. Let W be a neighborhood of the element a in the constructed topology, then there exists $B \in \mathcal{B}$ such that $a \in B \subseteq W$. On the strength of the definition of $\mathcal{B}$, there exists $U \in \mathcal{B}_0$ such that $a + U \subseteq B \subseteq W$. Since $a + U \in \mathcal{B}_a$, then $\mathcal{B}_a$ is a basis of neighborhoods of the element a. In particular, $\mathcal{B}_0$ is a basis of neighborhoods of zero.

Now, let us show that A is topological Abelian group in the constructed topology. For it let's verify that condition (SC) is fulfilled (see Remark 1.1.3). Let $a, b \in A$, and let W be a neighborhood of the element $a - b$. Then there exists $U \in \mathcal{B}_0$ such that $(a - b) + U \subseteq W$. On the strength of conditions (BN3) and (BN2), there exists $V \in \mathcal{B}_0$ such that $V - V \subseteq U$. As it was shown above, the subsets $a + V$ and $b + V$ are neighborhoods of the elements a and b respectively and $(a + V) - (b + V) = (a - b) + (V - V) \subseteq (a - b) + U \subseteq W$. Thus, condition (SC) is fulfilled, and hence, A is a topological group with basis $\mathcal{B}_0$ of neighborhoods of zero. Denote this topological group over $(A, \mathcal{B})$.

To complete the proof of the theorem it remains to verify that if some system $\mathcal{B}'$ of the subsets of the group A defines a topology on A, and this system is the family of all open subsets in the topology, and besides, $(A, \mathcal{B}')$ is a topological group with the basis $\mathcal{B}_0$ of neighborhoods of zero, then $\mathcal{B} = \mathcal{B}'$.

Let $B \in \mathcal{B}$ and $b \in B$. Then $b + U \subseteq B$ for some $U \in \mathcal{B}_0$. Since $\mathcal{B}_0$ is a basis of neighborhoods of zero in the topological group $(A, \mathcal{B}')$ and in view of Proposition 1.2.1, $b + U$ is a neighborhood of the element b in $(A, \mathcal{B}')$. Consequently, every element $b \in B$ enters in B together with some of its neighborhood relative to the topology defined by system $\mathcal{B}'$. That means that B is an open subset of the topological group $(A, \mathcal{B}')$ (see C.3), and, hence, $B \in \mathcal{B}'$. Thus, $\mathcal{B} \subseteq \mathcal{B}'$. Now, let $B' \in \mathcal{B}'$, hence, B' is a neighborhood of any of its elements in the topology defined by system $\mathcal{B}'$. Then for every $b \in B'$ the subset $B' - b$ is a neighborhood of zero in $(A, \mathcal{B}')$, on the strength of Proposition 1.1.34. Because of this, there exists $U_b \in \mathcal{B}_0$ such that $U_b \subseteq B' - b$, that is $b + U_b \subseteq B'$. From the definition of system $\mathcal{B}$ it follows that $B' \in \mathcal{B}$. Thus, $\mathcal{B}' \subseteq \mathcal{B}$.

1.2.5. THEOREM. *Let $\mathcal{B}_0$ be a family of subsets of a ring A, satisfying conditions (BN1) to (BN6). Then there exists (a unique) topology on A in which A is a topological ring and $\mathcal{B}_0$ is a basis of neighborhoods of zero.*

PROOF. On the strength of Theorem 1.2.4, there exists a topology (and besides, a unique topology) on A, such that the additive group of the ring A is a topological group and $\mathcal{B}_0$ is a basis of neighborhoods of zero in this topology. It remains to verify that condition (MC) is satisfied (see Definition 1.1.6 and Remark 1.1.7). Let $a, b \in A$ and U be a neighborhood of the element $a \cdot b$. In view of Proposition 1.2.1, $\mathcal{B}_a$, $\mathcal{B}_b$ and $\mathcal{B}_{a \cdot b}$, where $\mathcal{B}_x = \{x + V \mid V \in \mathcal{B}_0\}$, are bases of neighborhoods respectively of elements a, b and $a \cdot b$ in the topological group $A(+)$. Hence, there exists a neighborhood $V \in \mathcal{B}_0$ such that $a \cdot b + V \subseteq U$. Using conditions (BN2), (BN3), (BN5) and (BN6), it is possible to choose neighborhoods V_1 and V_2 from $\mathcal{B}_0$ such that $a \cdot V_2 + V_1 \cdot b + V_1 \cdot V_2 \subseteq V$. Then $a + V_1$ and $b + V_2$ are neighborhoods of the elements a and b respectively. Besides, $(a + V_1) \cdot (b + V_2) \subseteq a \cdot b + a \cdot V_2 + V_1 \cdot b + V_1 \cdot V_2 \subseteq a \cdot b + V \subseteq U$, i.e. condition (MC) is satisfied.

The theorem is proved.

1.2.6. THEOREM. *Let R be a topological ring, and let $\mathcal{B}_0$ be a system of subsets of R-module A, satisfying conditions (BN1) to (BN4), (BN5'), (BN6') and (BN6''). Then there exists (a unique) topology on A in which A is a topological R-module with the basis $\mathcal{B}_0$ of neighborhoods of zero.*

PROOF. The proof of this theorem is analogous to the proof of the previous theorem.

1.2.7. EXAMPLE. Let R be a ring, and let $\mathcal{B}_0$ contain only one subset $\{0\}$. In this case $\mathcal{B}_0$ satisfies conditions (BN1) to (BN6), and, hence, $\mathcal{B}_0$ defines the topology which transforms R to a topological ring with the basis $\mathcal{B}_0$ of neighborhoods of zero. The topology, which is determined in such a manner on R, is the discrete topology (see Example 1.1.9).

1.2.8. EXAMPLE. Let R be a ring, and let $\mathcal{B}_0$ contain only one subset R, that is $\mathcal{B}_0 = \{R\}$. In this case $\mathcal{B}_0$ satisfies conditions (BN1) to (BN6). The topology, which is determined on R with the help of $\mathcal{B}_0$, is the anti-discrete topology (see Example 1.1.9).

1.2.9. EXAMPLE. Let ξ be a pseudonorm on a ring R. For every $n = 1, 2, \ldots$ let $V_n = \{x \in R | \xi(x) < \frac{1}{2^n}\}$. Then the system $\mathcal{B}_0 = \{V_n | n = 1, 2, \ldots\}$ satisfies conditions (BN1) to (BN6). Topology τ_ξ , which is defined by system $\mathcal{B}_0$ on R, is called an interval topology of the pseudonormed ring. This topology is coincident with the topology defined by metric $\rho(x, y) = \xi(x - y)$ on R (see Example 1.1.14).

1.2.10. EXAMPLE. Let I be a non-zero ideal of a ring R. The system $\mathcal{B}_0 = \{I^k | k = 1, 2, \ldots\}$ of ideals of the ring R satisfies conditions (BN1) to (BN6), and, hence, it defines a topology on R, where this system is a basis of neighborhoods of zero. This topology is called I-adic topology. It is easy to verify that I-adic topology is the discrete topology if and only if the ideal I is nilpotent (see B.13). If R is a simple non-nilpotent ring, then the anti-discrete topology is the unique I-adic topology on R. Let $I = m \cdot \mathbb{Z}$ be the ideal of the ring of integers $\mathbb{Z}$, then the I-adic topology is called an m-adic topology of the ring of integers (in the case when $m = p$ is a prime number, then this topology coincides on $\mathbb{Z}$ with the topology induced by p-adic topology of the field of rational numbers (see Example 1.1.19)).

1.2.11. EXAMPLE. Let R be a ring, $\mathcal{M}$ be a non-empty system of non-zero ideals of the ring R, and let $\mathcal{B}_0$ be the family of any finite intersections of ideals from $\mathcal{M}$. Then the family $\mathcal{B}_0$ satisfies conditions (BN1) to (BN6), and, hence, the family specifies a topology on R and represents a basis of neighborhoods of zero in this topology.

1.2.12. PROPOSITION. *Let $\mathcal{B}_0$ be a basis of neighborhoods of zero of a topological skew field K, then conditions (BN1) to (BN6) are satisfied together with the following condition:*

(BN7) for any $U \in \mathcal{B}_0$ there exists $V \in \mathcal{B}_0$ such that $((1 + V) \backslash \{0\})^{-1} \subseteq 1 + U$.

PROOF. The fulfillment of conditions (BN1) to (BN6) results from Proposition 1.2.2 by virtue of the fact that K is a topological ring.

Let $U \in \mathcal{B}_0$, then $1 + U$ is a neighborhood of the unitary element on the strength of Proposition 1.1.34. Since $1^{-1} = 1$, then condition (MIC) (see Definition 1.1.20) implies that there exists a neighborhood W of the element 1 such that $(W \backslash \{0\})^{-1} \subseteq 1 + U$. On the strength of Proposition 1.2.1, the family $\mathcal{B}_1 = \{1 + V \mid V \in \mathcal{B}_0\}$ of subsets of the skew field K is a basis of neighborhoods of 1. Consequently there exists $V \in \mathcal{B}_0$ such that $1 + V \subseteq W$. Thus, $((1+V)\backslash\{0\})^{-1} \subseteq (W\backslash\{0\})^{-1} \subseteq 1 + U$, concluding the proof.

1.2.13. THEOREM. *Let K be a skew field, and let $\mathcal{B}_0$ be a non-empty family of subsets of the skew field, satisfying conditions (BN1) to (BN7). Then there exists (a unique) topology on K in which K is a topological skew field with the basis $\mathcal{B}_0$ of neighborhoods of zero.*

PROOF. On the strength of Theorem 1.2.5, there exists (a unique) topology on K in which K is a topological ring and $\mathcal{B}_0$ is a basis of neighborhoods of zero. It remains to show that K is a topological skew field in this topology, that is that condition (MIC) is satisfied. Let $0 \neq x \in K$ and U be a neighborhood of the element x^{-1}. Since $\mathcal{B}_0$ is a basis of neighborhoods of zero, then there exists $V \in \mathcal{B}_0$ such that $x^{-1} + V \subseteq U$. In compliance with condition (BN6), there exists a neighborhood $V_1 \in \mathcal{B}_0$ such that $x^{-1} \cdot V_1 \subseteq V$. On the strength of condition (BN7), there exists a neighborhood $V_2 \in \mathcal{B}_0$ such that $((1+V_2)\backslash\{0\})^{-1} \subseteq 1 + V_1$. Finally, on the strength of condition (BN6), there exists $V_3 \in \mathcal{B}_0$ such that $V_3 \cdot x^{-1} \subseteq V_2$. Then $W = x + V_3$ is a neighborhood of element x, besides,

$$
\begin{aligned}
(W\backslash\{0\})^{-1} &= ((x + V_3)\backslash\{0\})^{-1} = \left(((1 + V_3 \cdot x^{-1}) \cdot x)\backslash\{0\} \right)^{-1} \\
&= x^{-1} \cdot ((1 + V_3 \cdot x^{-1})\backslash\{0\})^{-1} \subseteq x^{-1} \cdot ((1 + V_2)\backslash\{0\})^{-1} \\
&\subseteq x^{-1} \cdot (1 + V_1) = x^{-1} + x^{-1} \cdot V_1 \subseteq x^{-1} + V \subseteq U.
\end{aligned}
$$

This means that condition (MIC) is fulfilled.

1.2.14. PROPOSITION. *Let ξ be a norm on a skew field K. Then the family $\mathcal{B}_0 = \{V_n \mid n = 1, 2, \dots\}$, where $V_n = \{x \in K \mid \xi(x) < \dfrac{1}{2^n}\}$, satisfies conditions (BN1) to (BN7), and, hence, the family defines a topology on K in which K is a topological skew field and $\mathcal{B}_0$ is a basis of neighborhoods of zero in this topology.*

PROOF. As it was noted above, $\mathcal{B}_0$ satisfies conditions (BN1) to (BN6) (see Example 1.2.9). Let's show that condition (BN7) is satisfied, too.

Preliminarily it should be pointed out that $\xi(1) = 1$ results from $\xi(1) = \xi(1 \cdot 1) = \xi(1) \cdot \xi(1)$. Besides, $1 = \xi(1) = \xi(x \cdot x^{-1}) = \xi(x) \cdot \xi(x^{-1})$ for any non-zero element

$x \in K$, hence, $\xi(x^{-1}) = \dfrac{1}{\xi(x)}$. Let n be an arbitrary natural number. Let's show that $(1+V_{n+1})^{-1} \subseteq 1+V_n$. Indeed, for every $x \in K$ from $1 = \xi(1) = \xi(1+x-x) \leq \xi(1+x)+\xi(x)$ follows that $\xi(1 + x) \geq 1 - \xi(x)$. Since $\xi(-1) = \xi(1) = 1 > \dfrac{1}{2^{n+1}}$, then $-1 \notin V_{n+1}$, and, hence,

$$\xi\big((1 + v)^{-1} - 1\big)$$
$$= \xi\big((1 + v)^{-1} - (1 + v)^{-1} \cdot (1 + v)\big) \cdot \xi\Big((1 + v)^{-1} \cdot \big(1 - (1 + v)\big)\Big)$$
$$= \xi\big((1 + v)^{-1}\big) \cdot \xi(-v) \leq \frac{\xi(v)}{1 - \xi(v)} < \frac{\frac{1}{2^{n+1}}}{1 - \frac{1}{2}} = \frac{1}{2^n}$$

for any $v \in V_{n+1}$. Consequently, $(1 + v)^{-1} - 1 \in V_n$, that is $(1 + v)^{-1} \in 1 + V_n$. Thus, $(1 + V_{n+1})^{-1} \subseteq 1 + V_n$, i.e. condition (BN7) is fulfilled. In view of Theorem 1.2.13, K is a topological skew field with the basis $\mathcal{B}_0$ of neighborhoods of zero.

1.2.15. COROLLARY. Each of the skew fields $\mathbb{Q}, \mathbb{C}, \mathbb{H}$ is a topological skew field in the topology specified by the appropriate norm (see Examples 1.1.19, 1.1.15, 1.1.16 and 1.1.18).

1.2.16. EXAMPLE. Let R be a discrete ring, and let M be a R-module. Let $\mathcal{M}$ be a non-empty family of submodules of M, and let $\mathcal{B}_0$ be the family of all finite intersections of submodules from $\mathcal{M}$. Then the family $\mathcal{B}_0$ satisfies conditions (BN1) to (BN4), (BN5'), (BN6'), (BN6''), and, hence, the family $\mathcal{B}_0$ specifies a topology on M. This topology converts M into a topological R-module with the basis $\mathcal{B}_0$ of neighborhoods of zero.

1.2.17. EXAMPLE. Let R be a ring, I be a two-sided ideal of the ring R, and let M be a R-module. Let R be considered as a topological ring in the I-adic topology (see Example 1.2.10), then the family $\mathcal{B}_0 = \{I^n M \,|\, n = 1, 2, \dots\}$ of subsets of module M satisfies conditions (BN1) to (BN4), (BN5'), (BN6'), (BN6''), and, hence, the family $\mathcal{B}_0$ specifies a topology on M with the basis $\mathcal{B}_0$ of neighborhoods of zero. This topology is called an I-adic topology on R-module M.

1.2.18. EXAMPLE. Let R be a topological ring, and let M be a R-module. It is obvious that the family $\mathcal{B}_0 = \{M\}$, consisting of only one subset M, satisfies conditions

(BN1) to (BN4), (BN5′), (BN6′), (BN6″), and the family converts M to a topological R-module (this topology coincides with the anti-discrete topology).

1.2.19. REMARK. Let R be a ring with the discrete topology and M be a R-module. Then the family $\mathcal{B}_0 = \{0\}$, consisting of only one subset $\{0\}$, specifies the discrete topology on M in which M is a topological module over ring R endowed with the discrete topology.

If we assume a non-discrete topology on R, then a R-module M endowed with the discrete topology may be not a topological R-module. Indeed, let R be a non-discrete topological field, and M be a non-zero vector space over R, then M in the discrete topology is not a topological vector space over R. Actually, let $0 \neq m \in M$. Since $\{0\}$ is a neighborhood of zero in M, then by condition (BN6″) there exists a neighborhood U of zero in R such that $U \cdot m \subseteq \{0\}$, i.e. $u \cdot m = 0$ for every $u \in U$. But it is true only if $U = \{0\}$, which contradicts the non-discreteness of the topology on the field R.

1.2.20. THEOREM. *Let A be an Abelian group, and let Σ be a family of topologies on A such that (A, τ) is a topological group for every $\tau \in \Sigma$. Then (A, τ') is a topological group, where $\tau' = \sup\{\tau \,|\, \tau \in \Sigma\}$.*

PROOF. Let us denote the family of all neighborhoods of the element a in the topological space (A, τ) over $\mathcal{B}_a(\tau)$. Then (see C.9)

$$\mathcal{B}_a(\tau') = \left\{ U \subseteq A \,\Big|\, U = \bigcap_{i=1}^{n} U_i, U_i \in \mathcal{B}_a(\tau_i), \tau_i \in \Sigma, i = 1, 2, \ldots, n, \ n \in \mathbb{N} \right\}.$$

Let's prove that (A, τ') is a topological Abelian group, i.e. let's verify the fulfillment of condition (SC) (see Remark 1.1.3).

Let $a, b \in A$, and let $U \in \mathcal{B}_{a-b}(\tau')$. Then there exist topologies $\tau_1, \tau_2, \ldots, \tau_n \in \Sigma$ such that $U = \bigcap_{i=1}^{n} U_i$, where $U_i \in \mathcal{B}_{a-b}(\tau_i)$. On the strength of the fact that (A, τ_i) is a topological Abelian group for $i = 1, 2, \ldots, n$, we get that (A, τ_i) satisfies condition (SC), and, hence, there exist neighborhoods $V_i \in \mathcal{B}_a(\tau_i)$ and $W_i \in \mathcal{B}_b(\tau_i)$ such that $V_i - W_i \subseteq U_i$ for every $i = 1, 2, \ldots, n$. Then

$$V = \bigcap_{i=1}^{n} V_i \in \mathcal{B}_a(\tau')$$

and

$$W = \bigcap_{i=1}^{n} W_i \in \mathcal{B}_b(\tau'),$$

besides,

$$V - W = \bigcap_{i=1}^{n} V_i - \bigcap_{i=1}^{n} W_i \subseteq \bigcap_{i=1}^{n} (V_i - W_i) \subseteq \bigcap_{i=1}^{n} U_i = U.$$

This means that condition (SC) is satisfied. Therefore, the theorem is proved.

1.2.21. THEOREM. *Let R be a ring (a skew field), and let Σ be a family of topologies on R such that (R, τ) is a topological ring (a topological skew field) for every $\tau \in \Sigma$. Let $\tau' = \sup\{\tau \mid \tau \in \Sigma\}$. Then (R, τ') is a topological ring (a topological skew field).*

PROOF. On the strength of Theorem 1.2.20, (R, τ') is a topological group. Let's verify that condition (MC) is satisfied. Let $a, b \in R$ and $U \in \mathcal{B}_{a \cdot b}(\tau')$. Then there exist topologies $\tau_1, \tau_2, \ldots, \tau_n \in \Sigma$ such that $U = \bigcap_{i=1}^{n} U_i$, where $U_i \in \mathcal{B}_{a \cdot b}(\tau_i)$ for $i = 1, 2, \ldots, n$. Applying condition (MC) to each of topological rings (R, τ_i), where $i = 1, 2, \ldots, n$, we get that neighborhoods $V_i \in \mathcal{B}_a(\tau_i)$ and $W_i \in \mathcal{B}_b(\tau_i)$ can be found such that $V_i \cdot W_i \subseteq U_i$ for $i = 1, 2, \ldots, n$. Then

$$V = \bigcap_{i=1}^{n} V_i \in \mathcal{B}_a(\tau')$$

and

$$W = \bigcap_{i=1}^{n} W_i \in \mathcal{B}_b(\tau'),$$

besides,

$$V \cdot W = \left(\bigcap_{i=1}^{n} V_i \right) \cdot \left(\bigcap_{i=1}^{n} W_i \right) \subseteq \bigcap_{i=1}^{n} (V_i \cdot W_i) \subseteq \bigcap_{i=1}^{n} U_i = U,$$

that is (R, τ') satisfies condition (MC), and, hence, (R, τ') is a topological ring.

If (R, τ) is a topological skew field for any topology $\tau \in \Sigma$, then (R, τ') is a topological ring on the strength of the just proved condition. It remains to verify the fulfillment of condition (MIC) for (R, τ'). Let $0 \neq a \in R$ and $U \in \mathcal{B}_{a^{-1}}(\tau')$. Then there exist topologies $\tau_1, \tau_2, \ldots, \tau_n \in \Sigma$ such that $U = \bigcap_{i=1}^{n} U_i$, where $U_i \in \mathcal{B}_{a^{-1}}(\tau_i)$, for $i = 1, 2, \ldots, n$. Applying condition (MIC) to each topological skew field (R, τ_i), where $i = 1, 2, \ldots, n$, we get that a neighborhood $V_i \in \mathcal{B}_a(\tau_i)$ can be found such that $(V_i \backslash \{0\})^{-1} \subseteq U_i$, for $i = 1, 2, \ldots, n$. Then $V = \bigcap_{i=1}^{n} V_i \in \mathcal{B}_a(\tau')$, besides,

$$(V \backslash \{0\})^{-1} = \left((\bigcap_{i=1}^{n} V_i) \backslash \{0\} \right)^{-1} \subseteq \bigcap_{i=1}^{n} U_i = U.$$

That is (R, τ') satisfies condition (MIC), and, hence, (R, τ') is a topological skew field.

1.2.22. THEOREM. *Let R be a topological ring, and M be a R-module. Let Σ be a family of topologies on M such that (M, τ) is a topological R-module for every $\tau \in \Sigma$. Then (M, τ') is a topological R-module, where $\tau' = \sup\{\tau \mid \tau \in \Sigma\}$.*

PROOF. On the strength of Theorem 1.2.20, it is enough to verify that (M, τ') satisfies condition (RMC) (refer to Definition 1.1.24). Let $r \in R$ and $m \in M$. Let $U \in \mathcal{B}_{rm}(\tau')$, then $U = \bigcap_{i=1}^{n} U_i$, where $U_i \in \mathcal{B}_{rm}(\tau_i)$, for some $\tau_i \in \Sigma$, $i = 1, 2, \ldots, n$. Since (M, τ_i) is a topological R-module for $i = 1, 2, \ldots, n$, then there exist neighborhoods V_i of element r in R and neighborhoods W_i of element m in (M, τ_i) such that $V_i \cdot W_i \subseteq U_i$ for $i = 1, 2, \ldots, n$. Then $V = \bigcap_{i=1}^{n} V_i$ is a neighborhood of the element r in R, and $W = \bigcap_{i=1}^{n} W_i \in \mathcal{B}_m(\tau')$, besides,

$$V \cdot W = \left(\bigcap_{i=1}^{n} V_i\right) \cdot \left(\bigcap_{i=1}^{n} W_i\right) \subseteq \bigcap_{i=1}^{n}(V_i \cdot W_i) \subseteq \bigcap_{i=1}^{n} U_i = U.$$

This means that (M, τ') is a topological R-module.

As it is shown in the following example, an Abelian group A, being a topological Abelian group (a topological ring, a topological module) in each topology of some family Σ of topologies on A, may be not a topological Abelian group in the topology $\tau'' = \inf\{\tau \mid \tau \in \Sigma\}$ (see C.9).

1.2.23. EXAMPLE. Let $A = \{(a, b) \mid a, b \in \mathbb{R}\}$ be the Abelian group of pairs (a, b) of real numbers with termwise addition (i.e. $(a_1, b_1) + (a_2, b_2) = (a_1 + a_2, b_1 + b_2)$). Let $U_{n,1} = \{(a, b) \mid |a| < \frac{1}{n}, b = 0\}$ and $U_{n,2} = \{(a, b) \mid a = 0, |b| < \frac{1}{n}\}$ for every natural number n. Then the following statements are true:

(1) $(0, 0) \in U_{n,i}$ for $i = 1, 2$ and $n = 1, 2, \ldots$;

(2) if $m \geq n$, then $U_{m,i} \subseteq U_{n,i}$ for $i = 1, 2$;

(3) $U_{2n,i} + U_{2n,i} \subseteq U_{n,i}$ for $i = 1, 2$ and $n = 1, 2, \ldots$;

(4) $-U_{n,i} = U_{n,i}$ for $i = 1, 2$ and $n = 1, 2, \ldots$.

On the strength of Theorem 1.2.4, there exist topologies τ_1 and τ_2 on A such that (A, τ_1) and (A, τ_2) are topological Abelian groups with bases of neighborhoods of zero $\mathcal{B}_{0,1} = \{U_{n,1} \mid n = 1, 2, \ldots\}$ and $\mathcal{B}_{0,2} = \{U_{n,2} \mid n = 1, 2, \ldots\}$ respectively in these topologies. Consider the topology $\tau_3 = \inf\{\tau_1, \tau_2\}$ on A, and show that (A, τ_3) is not a topological Abelian group.

Note that the subset

$$U = \{(a, b) \in A \mid \mid a \mid < 1, \mid b \mid < 1, \mid a \mid \neq \mid b \mid\} \cup \{(0, 0)\}$$

is an open neighborhood of 0 in the topological space (A, τ_3). (Consider A as a two-dimensional plane with X–axis as horisontal and Y–asis as vertical one. Every point (a, b) of U enters into U together with some horizontal interval and together with some vertical interval. The centers of these intervals are situated at the point (a, b).)

Let us assume now that (A, τ_3) is a topological group. Then for the neighborhood U of zero in (A, τ_3) there exists a neighborhood V of zero such that $V + V \subseteq U$. Hence, there exists a number $n \geq 1$ such that $U_{n,1} \subseteq V$ and $U_{n,2} \subseteq V$. Now, let x be a non-zero real number such that $\mid x \mid < \frac{1}{n}$, then $(x, 0) \in U_{n,1}$ and $(0, x) \in U_{n,2}$. But in this case,

$$(x, x) = (x, 0) + (0, x) \in U_{n,1} + U_{n,2} \subseteq V + V \subseteq U,$$

which contradicts the choice of the subset U. Thus, (A, τ_3) is not a topological group.

1.2.24. REMARK. As it was shown in Example 1.2.23, topological Abelian groups (A, τ_1) and (A, τ_2) may be considered as topological $\mathbb{Z}$-modules and as topological rings with the zero multiplication. By this means, (A, τ_3) may be simultaneously considered as an example of a ring or a module, which is not topological in the topology $\tau_3 = \inf\{\tau_1, \tau_2\}$.

1.2.25. PROPOSITION. *Let S be a subset of a topological Abelian group A with a basis $\mathcal{B}_0$ of neighborhoods of zero. Then $[S]_A = \bigcap_{V \in \mathcal{B}_0}(S + V)$.*

PROOF. Let $x \in [S]_A$ and $V \in \mathcal{B}_0$. Let also $V' \in \mathcal{B}_0$ and $-V' \subseteq V$, then $(x + V') \bigcap S \neq \emptyset$, because $x \in [S]_A$. Then $x \in S - V' \subseteq S + V$. Therefore, $[S]_A \subseteq S + V$, and, hence, $[S]_A \subseteq \bigcap_{V \in \mathcal{B}_0}(S + V)$.

Now, let $y \in \bigcap_{V \in \mathcal{B}_0}(S + V)$, and let U be a neighborhood of zero in A. Let's choose a neighborhood $V \in \mathcal{B}_0$ of zero such that $-V \subseteq U$. Since $y \in S + V$, then $S \bigcap(y + U) \supseteq S \bigcap(y - V) \neq \emptyset$, that is $\bigcap_{V \in \mathcal{B}_0}(S + V) \subseteq [S]_A$.

1.2.26. COROLLARY. *Let $\mathcal{B}_0$ be a basis of neighborhoods of zero of a topological Abelian group A. Then $\bigcap_{V \in \mathcal{B}_0} V$ is a closed set.*

PROOF. In view of Proposition 1.2.25, $\bigcap_{V \in \mathcal{B}_0} V$ is a closure of a one-element subset $\{0\}$ in A.

1.2.27. COROLLARY. *Let U and V be neighborhoods of zero of topological Abelian group A such that $V + V \subseteq U$, then $[V]_A \subseteq U$.*

PROOF. It is obvious that the family $\mathcal{B}_0$ of all neighborhoods of zero in A is a basis of neighborhoods of zero of A. In the view of Proposition 1.2.25,

$$[V]_A = \bigcap_{W \in \mathcal{B}_0} (V + W) \subseteq V + V \subseteq U.$$

1.2.28. PROPOSITION. *Let A be a topological Abelian group then the following statements are true:*

(1) A has a basis of neighborhoods of zero consisting of symmetric open neighborhoods;

(2) A has a basis of neighborhoods of zero consisting of symmetric closed neighborhoods.

PROOF. Let $\mathcal{B}_0$ be a basis of neighborhoods of zero in A. Then for every $V \in \mathcal{B}_0$ there exists an open subset W_V in A such that $0 \in W_V \subseteq V$. In view of Proposition 1.1.34, the subset $-W_V$ is open, besides, $0 \in -W_V$. Hence, subset $W_V \bigcap (-W_V)$ is an open symmetric neighborhood of zero, being contained in V. Then from condition (BN2) follows that the family $\mathcal{B}_0' = \{W_V \bigcap (-W_V) \mid V \in \mathcal{B}_0\}$ is a basis of symmetric open neighborhoods of zero in A. Thus, statement (1) is proved.

Now, let $\mathcal{B}_0'' = \{[U]_A \mid U \in \mathcal{B}_0'\}$. From Proposition 1.1.41 and from the symmetry of every neighborhood $U \in \mathcal{B}_0'$ follows that $-[U]_A = [-U]_A = [U]_A$. Hence, $\mathcal{B}_0''$ is a family of symmetric closed neighborhoods of zero of A. Let's check that $\mathcal{B}_0'$ is a basis of neighborhoods of zero in A. Indeed, if $U \in \mathcal{B}_0'$, then, in view of condition (BN3), there exists a neighborhood $U' \in \mathcal{B}_0'$ such that $U' + U' \subseteq U$. Consequently, in view of Corollary 1.2.27, $[U']_A \subseteq U$, and, hence, $\mathcal{B}_0''$ is a basis of neighborhoods of zero in A.

1.2.29. COROLLARY. *Let A be a topological Abelian group and $a \in A$. Then the following statements are true:*

(1) the element a has a basis of neighborhoods consisting of open neighborhoods;

(2) the element a has a basis of neighborhoods consisting of closed neighborhoods.

PROOF. The statement results from Propositions 1.2.28 and 1.1.34.

1.2.30. COROLLARY. Let R be a topological ring, M be a topological R-module, $a \in R, m \in M, S \subseteq R, N \subseteq M$. Let also $\mathcal{B}_0(R)$ be a basis of neighborhoods of zero in R, $\mathcal{B}_0(M)$ be a basis of neighborhoods of zero in M. Then the following statements are true:

(1) R has a basis of neighborhoods of zero consisting of symmetric open neighborhoods;

(2) M has a basis of neighborhoods of zero consisting of symmetric open neighborhoods;

(3) R has a basis of neighborhoods of zero consisting of symmetric closed neighborhoods;

(4) M has a basis of neighborhoods of zero consisting of symmetric closed neighborhoods;

(5) the element a has a basis of neighborhoods consisting of open neighborhoods;

(6) the element m has a basis of neighborhoods consisting of open neighborhoods;

(7) the element a has a basis of neighborhoods consisting of closed neighborhoods;

(8) the element m has a basis of neighborhoods consisting of closed neighborhoods;

(9) $[S]_R = \bigcap_{U \in \mathcal{B}_0(R)} (S + U)$;

(10) $[N]_M = \bigcap_{V \in \mathcal{B}_0(M)} (N + V)$;

(11) the subset $\bigcap_{U \in \mathcal{B}_0(R)} U$ is closed in R;

(12) the subset $\bigcap_{V \in \mathcal{B}_0(M)} V$ is closed in M.

1.2.31. PROPOSITION. *Let a be an invertible element of a topological ring R with the unitary element, $\mathcal{B}_x(R)$ be a basis of neighborhoods of the element $x \in R$. Then $\{a \cdot U | U \in \mathcal{B}_x(R)\}$ and $\{U \cdot a | U \in \mathcal{B}_x(R)\}$ are bases of neighborhoods of the elements $a \cdot x$ and $x \cdot a$, respectively.*

PROOF. The proof results from Corollary 1.1.45.

1.2.32. COROLLARY. Let a be an invertible element of a topological ring R with the unitary element, $\mathcal{B}_0(R)$ be a basis of neighborhoods of the ring R. Then $\{a \cdot U | U \in \mathcal{B}_0(R)\}$ and $\{U \cdot a | U \in \mathcal{B}_0(R)\}$ are bases of neighborhoods of zero of the ring R.

§ 1.3. Questions of Separability

1.3.1. PROPOSITION. *Let $\{U_i \mid i = 0, \pm 1, \pm 2, \dots\}$ be a system of symmetrical neighborhoods of zero of a topological Abelian group A such that $U_i + U_i \subseteq U_{i-1}$. Then for an arbitrary element $a \in A$ and an arbitrary integer n there exists a real-valued function $f_{n,a}(x)$ on A satisfying the following conditions:*

1) $0 \leq f_{n,a}(x) \leq 1$ for all $x \in A$;

2) $f_{n,a}(a) = 1$;

3) $f_{n,a}(x) = 0$, if $x \notin a + U_n$;

4) if $x_1 - x_2 \in U_{n+m}$, then $|f_{n,a}(x_1) - f_{n,a}(x_2)| \leq \frac{1}{2^{m-1}}$.

PROOF. Let n be an arbitrary non-negative integer. For any binary-rational number $\frac{p}{2^k}$, where $p = 0, 1, \dots, 2^k$ let's construct a non-empty subset $X_{\frac{p}{2^k}} \subseteq A$ such that

$$X_{\frac{p+1}{2^k}} + U_{n+k} \subseteq X_{\frac{p}{2^k}}$$

for all $p = 0, 1, \dots, 2^k - 1$. The construction will be realized by the induction by k.

Let $k = 0$, then $p = 0, 1$. Put $X_0 = a + U_n$ and $X_1 = \{x \mid x + U_n \subseteq X_0\}$. Since $a \in X_0$ and $a \in X_1$, then the subsets X_0 and X_1 are not empty, and, besides, it is evident that $X_1 + U_n \subseteq X_0$, in particular, $X_1 \subseteq X_0$.

Suppose that for any non-negative number k, where $k < l$, and for all $p = 0, 1, \dots, 2^k$ we have constructed non-empty subsets $X_{\frac{p}{2^k}} \subseteq A$ such that

$$X_{\frac{p+1}{2^k}} + U_{n+k} \subseteq X_{\frac{p}{2^k}}$$

for all $p = 0, 1, \dots, 2^k - 1$. Now let p be any of the numbers $0, 1, 2, \dots, 2^l$. If p is even, i.e. $p = 2p'$, then from $0 \leq p \leq 2^l$ results that $0 \leq p' \leq 2^{l-1}$, then we put

$$X_{\frac{p}{2^l}} = X_{\frac{p'}{2^{l-1}}}.$$

If $p = 2p' + 1$, then from $0 \leq p \leq 2^l$ we get that $0 \leq p' < 2^{l-1}$, then put

$$X_{\frac{p}{2^l}} = \left\{ x \mid x + U_{n+l} \subseteq X_{\frac{p'}{2^{l-1}}} \right\}.$$

Now let's show that $X_{\frac{p+1}{2^l}} + U_{n+l} \subseteq X_{\frac{p}{2^l}}$ for all $p = 0, 1, \ldots, 2^l - 1$.

Indeed, if $p = 2p'$ and $0 \le p < 2^l$, then $p + 1 = 2p' + 1$, and $1 \le p + 1 \le 2^l$. According to the construction,

$$X_{\frac{p+1}{2^l}} = \left\{ x \middle| x + U_{n+l} \subseteq X_{\frac{p'}{2^{l-1}}} \right\}.$$

Then

$$X_{\frac{p+1}{2^l}} + U_{n+l} \subseteq X_{\frac{p'}{2^{l-1}}},$$

and since

$$X_{\frac{p'}{2^{l-1}}} = X_{\frac{2p'}{2^l}} = X_{\frac{p}{2^l}},$$

then we obtain the necessary inclusion

$$X_{\frac{p+1}{2^l}} + U_{n+l} \subseteq X_{\frac{p}{2^l}}.$$

If $p = 2p' + 1$ and $0 \le p < 2^l$, then $p + 1 = 2(p' + 1)$ and $0 \le p' + 1 \le 2^{l-1}$. According to the construction,

$$X_{\frac{p+1}{2^l}} = X_{\frac{p'+1}{2^{l-1}}}.$$

Since

$$X_{\frac{p'+1}{2^{l-1}}} + U_{n+l} + U_{n+l} \subseteq X_{\frac{p'+1}{2^{l-1}}} + U_{n+l-1} \subseteq X_{\frac{p'}{2^{l-1}}},$$

then

$$X_{\frac{p+1}{2^l}} + U_{n+l} = X_{\frac{p'+1}{2^{l-1}}} + U_{n+l} \subseteq \left\{ x \middle| x + U_{n+l} \subseteq X_{\frac{p'}{2^{l-1}}} \right\} = X_{\frac{2p'+1}{2^l}} = X_{\frac{p}{2^l}},$$

i.e. we have obtained the necessary inclusion in this case, too.

Thus, the sets X_r are constructed for any binary-rational number $0 \le r \le 1$.

Now let's define a real-valued function $f_{n,a}(x)$ on A as the following:

$$f_{n,a}(x) = \begin{cases} \sup\{r \mid x \in X_r\}, & \text{if } x \in X_0; \\ 0, & \text{if } x \notin X_0. \end{cases}$$

It is clear that $0 \le f_{n,a}(x) \le 1$, i.e. condition 1) is fulfilled.

Since $a \in X_1$, then $f_{n,a}(a) = 1$, i.e. condition 2) is fulfilled.

The fulfillment of condition 3) results from the construction of the set X_0 and from the definition of the function $f_{n,a}(x)$ on $A \backslash X_0$.

Now let's verify the fulfillment of condition 4). Let $x_1, x_2 \in A$, and $x_1 - x_2 \in U_{n+m}$ (for more definiteness let $f_{n,a}(x_1) \geq f_{n,a}(x_2)$). If $f_{n,a}(x_1) \leq \frac{1}{2^{m-1}}$, then, since $f_{n,a}(x_2) \geq 0$, we get that

$$|f_{n,a}(x_1) - f_{n,a}(x_2)| \leq f_{n,a}(x_1) \leq \frac{1}{2^{m-1}}.$$

If $f_{n,a}(x_1) > \frac{1}{2^{m-1}}$, then $x_1 \in X_r$ for some binary-rational number $r > \frac{1}{2^{m-1}}$. Decompose r in two summands: $r = \frac{k}{2^m} + \frac{q}{2^l}$, where $0 \leq \frac{q}{2^l} < \frac{1}{2^m}$. Then

$$x_2 \in x_1 + U_{n+m} \subseteq X_r + U_{n+m} \subseteq X_{\frac{k}{2^m}} + U_{n+m} \subseteq X_{\frac{k-1}{2^m}}.$$

Since

$$\frac{k-1}{2^m} = r - \frac{q}{2^l} - \frac{1}{2^m} > r - \frac{1}{2^m} - \frac{1}{2^m} = r - \frac{1}{2^{m-1}},$$

then

$$x_2 \in X_{\frac{k-1}{2^m}} \subseteq X_{r - \frac{1}{2^{m-1}}}.$$

Thus, $x_2 \in X_{r-2^{-(m-1)}}$ for all r, such that $r > \frac{1}{2^{m-1}}$, and $x_1 \in X_r$. Therefore,

$$f_{n,a}(x_2) \geq \sup\left\{r - \frac{1}{2^{m-1}} \Big| x_1 \in X_r\right\} = \sup\{r | x_1 \in X_r\} - \frac{1}{2^{m-1}} = f_{n,a}(x_1) - \frac{1}{2^{m-1}}.$$

Thus,

$$f_{n,a}(x_1) - f_{n,a}(x_2) < \frac{1}{2^{m-1}},$$

i.e. condition 4) is fulfilled.

1.3.2. THEOREM. *For any topological Abelian group A the following conditions (see C.11) are equivalent:*

1) A is a totally regular space;

2) A is a regular space;

3) A is a Hausdorff space;

4) $\{0\}$ is closed subset in A;

5) if $\mathcal{B}_0$ is a basis of neighborhoods of zero of A, then $\bigcap_{V \in \mathcal{B}_0} V = \{0\}$;

6) A is a T_0-space;

7) A is a T_1-space.

PROOF. It is evident that 1) $\Longrightarrow$ 2) $\Longrightarrow$ 3).

Now we show that 3) $\Longrightarrow$ 4). Indeed, let's assume the contrary, i.e. that $0 \neq a \in [\{0\}]_A$. Then, due to item 3), there exist neighborhoods U and V of the elements a and 0 correspondingly such that $U \cap V = \emptyset$. In particular, $0 \notin U$, i.e. $U \cap \{0\} = \emptyset$, that contradicts the fact that $a \in [\{0\}]_A$. Thus, 3) $\Longrightarrow$ 4).

Let $\{0\}$ be closed subset in A and $\mathcal{B}_0$ be a basis of neighborhoods of zero in A. Then, due to Proposition 1.2.25, we get $\{0\} = [\{0\}]_A = \bigcap_{V \in \mathcal{B}_0} (\{0\} + V) = \bigcap_{V \in \mathcal{B}_0} V$, i.e. 4) $\Longrightarrow$ 5).

Let $\mathcal{B}_0$ be a basis of neighborhoods of zero in A, and $\bigcap_{V \in \mathcal{B}_0} V = \{0\}$. Let $x, y \in A$ and $x \neq y$, then $x - y \neq 0$, hence, there exists a neighborhood $V_0 \in \mathcal{B}_0$ such that $x - y \notin V_0$. Therefore, $x \notin y + V_0$. Thus, A is T_0-space, i.e. 5) $\Longrightarrow$ 6).

Let A be a T_0-space and x, y be distinct elements of A. Then there exists a neighborhood of zero U in A such that $x \notin y + U$ or $y \notin x + U$. Due to Proposition 1.2.28, we can consider that U is a symmetrical neighborhood of zero. Now let's show that $y \notin x + U$ if $x \notin y + U$ and that $x \notin y + U$ if $y \notin x + U$. Indeed, assume the contrary, i.e. that $y \in x + U$. Then $x \in y - U = y + U$, that contradicts the choice of U. The similarily if $x \in y + U$. Thus, A is T_1-space, i.e. 6) $\Longrightarrow$ 7).

Finally, let A be T_1-space, $a \in A$, and B be a closed subset of A such that $a \notin B$. Since B is closed, then $A \backslash B$ is a neighborhood of a, hence, $(A \backslash B) - a$ is a neighborhood of zero in A. According to Proposition 1.2.28, A has a basis $\mathcal{B}_0$ of symmetrical neighborhoods of zero. It is easy to construct by induction a sequence $U_0, U_1, \ldots$ of neighborhoods of zero from $\mathcal{B}_0$ such that $U_0 \subseteq (A \backslash B) - a$ and $U_i + U_i \subseteq U_{i-1}$ for all $i = 1, 2, \ldots$. Put also $U_i = A$ for $i < 0$.

Then the system $\{U_i | i \in \mathbb{Z}\}$ of symmetrical neighborhoods of zero in A satisfies the conditions of Proposition 1.3.1, hence, we can define on A a real-valued function $f_{0,a}(x)$ such that:

1) $0 \leq f_{0,a}(x) \leq 1$ for all $x \in A$;

2) $f_{0,a}(a) = 1$;

3) $f_{0,a}(x) = 0$ for all $x \notin a + U_0$;

4) if $x_1 - x_2 \in U_m$, then $\left| f_{0,a}(x_1) - f_{0,a}(x_2) \right| \leq \frac{1}{2^{m-1}}$.

Due to 4) and Proposition 1.1.34, $f_{0,a}(x)$ is continuous on A. If $x \in B$, hence, $x - a \notin (A \backslash B) - a$, then $x - a \notin U_0$. Therefore, $x \notin a + U_0$, hence, due to 3), $f_{0,a}(x) = 0$. Thus, A is a totally regular space, i.e. 7) $\Longrightarrow$ 1). The theorem is proved.

1.3.3. COROLLARY. For any topological ring (module) A the statements 1)-7) of Theorem 1.3.2 are equivalent.

1.3.4. DEFINITION. A topological Abelian group (topological ring, topological module) is called a Hausdorff group (Hausdorff ring, Hausdorff module) if one of the conditions 1)-7) of Theorem 1.3.2 is fulfilled.

1.3.5. REMARK. In Remark 6.2.7 will be constructed a Hausdorff topological ring, whose space is not normal.

1.3.6. COROLLARY. *Let $\mathcal{B}_0$ be a system of subsets of Abelian group A, satisfying the conditions (BN2) - (BN4) of Proposition 1.2.1 and the condition*

(BN1') $\bigcap_{V \in \mathcal{B}_0} V = \{0\}$.

Then there exists (besides, unique) topology on A such that A be a Hausdorff topological Abelian group, and $\mathcal{B}_0$ be a basis of neighborhoods of zero.

PROOF. Since condition (BN1') supposes the fulfillment of condition (BN1), then due to Theorem 1.2.4, there exists (besides, unique) topology on A such that A be a topological group and $\mathcal{B}_0$ be a basis of neighborhoods of zero. Then the fulfillment of condition (BN1') means the fulfillment of condition 5) of the Theorem 1.3.2, hence A in the topology, defined by the system $\mathcal{B}_0$, is a Hausdorff group, that completes the proof.

1.3.7. COROLLARY. *Let $\mathcal{B}_0$ be a system of subsets of a ring A, satisfying condition (BN1') and conditions (BN2) - (BN6) (see Propositions 1.2.1 and 1.2.2). Then there exists (besides, unique) topology on A such that A is a Hausdorff topological ring, and $\mathcal{B}_0$ is a basis of neighborhoods of zero.*

PROOF is analogous to the proof of Corollary 1.3.6 (using Theorems 1.2.5 and 1.3.2).

1.3.8. COROLLARY. *Let R be a topological ring, $\mathcal{B}_0$ be a system of subsets of R-module A, satisfying condition (BN1') and conditions (BN2) - (BN4), (BN5'), (BN6') and (BN6'') (see Proposition 1.2.3). Then there exists (besides, unique) topology on A such that A is a Hausdorff topological R-module, and $\mathcal{B}_0$ is a basis of neighborhoods of zero.*

PROOF is analogous to the proof of Corollary 1.3.6 (using Theorems 1.2.6 and 1.3.2).

1.3.9. REMARK. Any discrete Abelian group is Hausdorff. Therefore, any discrete ring and any discrete module is Hausdorff, too. It is also clear that anti–discrete groups, anti–discrete rings, and anti–discrete modules are Hausdorff if and only if they are one-element sets.

1.3.10. REMARK. Any pseudonormed ring (see Definition 1.1.12) is Hausdorff. In particular, rings $\mathbb{Q}$ (with p-adic topology), $\mathbb{R}$, $\mathbb{C}$, $\mathbb{H}$, $C([0,1], \mathbb{R})$ are Hausdorff.

§ 1.4. Submodules, Subrings and Ideals

1.4.1. DEFINITION. Let (A, τ) be a topological Abelian group. A subset B of A is called a subgroup of topological group (A, τ) if B is a subgroup of A and B is endowed with the topology $\tau|_B$, induced by the topology τ (see C.10).

1.4.2. REMARK. A subgroup B of a topological Abelian group A is a topological group itself.

Indeed, let $b_1, b_2 \in B$ and U be a neighborhood of the element $b_1 - b_2$ in the topological space $(B, \tau|_B)$. Then $U = U_1 \cap B$, where U_1 is a neighborhood of the element $b_1 - b_2$ in the topological space (A, τ). Let V_1' and V_1'' be neighborhoods of elements b_1 and b_2 correspondingly in the topological space (A, τ) such that $V_1' - V_1'' \subseteq U_1$. Then $V' = V_1' \cap B$ and $V'' = V_1'' \cap B$ are neighborhoods of elements b_1 and b_2 correspondingly in the topological space B. Besides, $V' - V'' \subseteq (V_1' - V_1'') \cap B \subseteq U_1 \cap B = U$, i.e. B with the induced topology satisfies condition (SC) (see Remark 1.1.3). Hence, $(B, \tau|_B)$ is a topological group.

1.4.3. DEFINITION. Let R be a topological ring, M be a topological R-module. A subset Q of the ring R (a subset N of R-module M) is called a subring of the topological ring R (a submodule of the topological R-module M) if Q is a subring of R (if N is a submodule of R-module M) and ring Q (R-module N) is endowed with the topology induced by the topology of the ring R (R-module M).

1.4.4. REMARK. A subring Q of a topological ring R is a topological ring. A submodule N of a topological R-module M is a topological R-module.

Indeed, due to Remark 1.2.2, Q and N are topological Abelian groups. Let x, y be elements of Q and U be a neighborhood of $x \cdot y$ in the topological space Q, then $U = U_1 \cap Q$, where U_1 is a certain neighborhood of $x \cdot y$ in the topological space R. Let V_1 and W_1 be neighborhoods of the elements x and y correspondingly in the topological space R such that $V_1 \cdot W_1 \subseteq U_1$. Then $V = V_1 \cap Q$ and $W = W_1 \cap Q$ are neighborhoods of elements x and y correspondingly in the topological space Q, besides,

$$V \cdot W = (V_1 \cap Q) \cdot (W_1 \cap Q) \subseteq (V_1 \cdot W_1) \cap Q \subseteq U_1 \cap Q = U,$$

i.e. Q in the induced topology satisfies condition (MC) (see Definition 1.1.6 and Remark 1.1.7), hence, Q is a topological ring.

Analogously it shown is that N in the induced topology satisfies condition (RMC) (see Definition 1.1.24), i.e. that N is a topological R-module.

1.4.5. PROPOSITION. *Let B be a subgroup of a topological Abelian group A. Then $[B]_A$ is a subgroup of the topological group A, too.*

PROOF. Due to the Proposition 1.1.41, $[B]_A - [B]_A \subseteq [B - B]_A = [B]_A$, hence, $[B]_A$ is a subgroup of the topological group A.

1.4.6. PROPOSITION. *Let Q be a subset of a topological ring R, and N be a subset of a topological R-module M. If N is a Q-stable subset (see B.14), then $[N]_M$ is a $[Q]_R$-stable subset.*

PROOF. Since N is a Q-stable subset, then $Q \cdot N \subseteq N$. Then, due to Proposition 1.1.42, $[Q]_R \cdot [N]_M \subseteq [Q \cdot N]_M \subseteq [N]_M$, i.e. $[N]_M$ is a $[Q]_R$-stable subset.

1.4.7. PROPOSITION. *Let R be a topological ring and M be a topological R-module. Let Q be a subring of the ring R, and N be a Q-submodule of R-module M, then:*

1) $[Q]_R$ is a subring of the topological ring R;

2) $[N]_M$ is a $[Q]_R$-module.

PROOF. Due to Proposition 1.4.5, $[Q]_R$ and $[N]_M$ are subgroups of topological Abelian groups R and M correspondingly. Since Q is a Q-stable subset of the topological R-module R and N is a Q-stable subset of the topological R-module M, then, due to Proposition 1.4.6:

1) $[Q]_R$ is a $[Q]_R$-stable subset of the topological R-module R, i.e. $[Q]_R$ is a subring of R;

2) $[N]_M$ is a $[Q]_R$-stable subset of the topological R-module M, and since $[Q]_R$ is a subring of R, then $[N]_M$ is a $[Q]_R$-module.

1.4.8. COROLLARY. *Let Q be a dense subring of a topological ring R and N be a Q-submodule of a topological R-module M. Then $[N]_M$ is a submodule of the topological*

R-module M. In particular, the closure of any submodule of a topological R-module is also a topological R-module.

1.4.9. COROLLARY. Let Q be a dense subring of a topological ring R, and I be a left (right, two-sided) ideal of the ring Q. Then $[I]_R$ is a left (right, two-sided) ideal of the ring R. In particular, the closure of a left (right, two-sided) ideal of the ring R is also left (right, two-sided) ideal of the ring R.

1.4.10. COROLLARY. *Let Q be a dense left (right, two-sided) ideal of a topological ring R, and P be a left (right, two sided) ideal of the ring Q. Then $[P]_Q$ is a left (right, two sided) ideal of the ring R.*

PROOF. Since $[P]_Q = [P]_R \cap Q$, then $[P]_Q$ is an intersection of two left (right, two-sided) ideals Q and $[P]_R$ of the ring R, hence, is a left (right, two-sided) indeed, too.

1.4.11. COROLLARY. *Let $\mathcal{B}_0$ be a basis of neighborhoods of zero of a topological R-module M, then $M_0 = \bigcap_{V \in B_0} V$ is the smallest closed submodule of M.*

PROOF. Due to Proposition 1.2.25, $M_0 = [\{0\}]_M$, then, according to the Corollary 1.4.8, M_0 is a closed submodule of M. Let N be a closed submodule of M, then, from $\{0\} \subseteq N$ results that $M_0 = [\{0\}]_M \subseteq [N]_M = N$, i.e. M_0 is the smallest closed submodule of M.

1.4.12. REMARK. *Let B_0 be a basis of neighborhoods of zero of the topological ring R, then $R_0 = \bigcap_{V \in B_0} V$ is the smallest closed two-sided ideal of R.*

PROOF. The result follows from Corollary 1.4.11, considering R as a left and right topological R-module.

1.4.13. REMARK. In view of Theorem 1.3.2, it is easy to see that in a topological ring R (topological R-module M) the smallest closed ideal (smallest closed submodule) equals zero if and only if R (R-module M) is Hausdorff. Therefore, any topological ring without closed proper ideals, in particular, any topological skew field (field) is either Hausdorff or anti-discrete.

1.4.14. REMARK. Let Q be a subring of a topological ring R and N be a submodule of a topological R-module M. Let Q_0 and R_0 be the smallest closed ideals of the topological rings Q and R correspondingly, N_0 and M_0 be the smallest closed submodules of the topological R-modules N and M correspondingly. Then $Q_0 = R_0 \bigcap Q$ and $N_0 = M_0 \bigcap N$.

1.4.15. DEFINITION. A topological ring R (a topological R-module M) is called topologically simple if the ring R (R-module M) has no closed ideals (closed submodules) different from R_0 and R (from M_0 and M) (see Corollaries 1.4.12 and 1.4.11).

1.4.16. PROPOSITION. *Any dense subring of a topologically simple topological ring is topologically simple. A dense submodule of a topologically simple topological module is a topologically simple module.*

PROOF. Let Q be a dense subring of a topological ring R, and let Q_0 and R_0 be the smallest closed ideals of the rings Q and R correspondingly. Let P be a closed ideal of Q, then, due to Corollary 1.4.9, $[P]_R$ is a closed ideal of R, and since the ring R is topologically simple, then either $[P]_R = R_0$, or $[P]_R = R$. Since $P = [P]_Q = [P]_R \bigcap Q$, then either $P = R_0 \bigcap Q = Q_0$ (see Remark 1.4.14), or $P = R \bigcap Q = Q$, i.e. Q is a topologically simple ring.

The second part of the proposition is proved analogously.

Any simple endowed ring with arbitrary ring topology, in particular any topological skew field, is topologically simple ring. But there exist also topologically simple rings which are not simple. The trivial example of such rings is any ring which is not simple and is endowed with the anti–discrete topology. The following example shows that there also exist non-simple Hausdorff topologically simple rings.

1.4.17. EXAMPLE. Let $\mathbb{Q}[\theta]$ be the set of all real numbers of the type of $\sum_{i=0}^{n} q_i \cdot \theta^i$, where $q_i \in \mathbb{Q}$, n is a non-negative integer, and θ is a fixed transcendental number. Then $\mathbb{Q}[\theta]$ is a subring of the field $\mathbb{R}$ of real numbers generated by the subset $\mathbb{Q} \bigcup \{\theta\}$. If we consider an interval topology onto $\mathbb{R}$, then $\mathbb{Q}[\theta]$ is a dense subring of the topological ring $\mathbb{R}$. Since $\mathbb{R}$ is a field, it is a topologically simple ring. Then, due to Proposition 1.4.16, $\mathbb{Q}[\theta]$ is also a topologically simple ring, too. Besides, it is clear that $I_k = \theta^k \cdot \mathbb{Q}[\theta]$ is a proper ideal of the ring $\mathbb{Q}[\theta]$ for any $k = 1, 2, \ldots$. It is also clear that the topological ring $\mathbb{Q}[\theta]$ is Hausdorff as a subring of the Hausdorff ring $\mathbb{R}$.

The ring $\mathbb{Q}[\theta]$ can also be considered as an example of a topologically simple $\mathbb{Q}[\theta]$-module which is not simple.

1.4.18. PROPOSITION. *A subgroup B of a topological Abelian group A is open if and only if it is a neighborhood of at least one of its elements. Any open subgroup of the group A is closed in A.*

PROOF. Let B be an open subgroup of a topological Abelian group A, then B is a neighborhood of each of its elements. Let, besides, $\mathcal{B}_0$ be a basis of neighborhoods of zero of A, then, taking into account that B is a neighborhood of zero of A, and applying Proposition 1.2.25, we obtain that

$$[B]_A = \bigcap_{V \in \mathcal{B}_0} (V + B) \subseteq B + B = B.$$

Therefore, B is a closed subgroup of the topological group A.

Vice versa, let $b_0 \in B$ and B be a neighborhood of the element b_0, then for any element $b \in B$ (due to Proposition 1.1.34), $B = b + (B - b_0)$ is a neighborhood of the element b. Therefore, B is an open subgroup of the topological Abelian group A, being a neighborhood of each of its elements.

1.4.19. COROLLARY. If a subgroup of the additive group of a topological ring R (topological R-module M) is open, then it is also closed. In particular, any open subring, any open left (right, two-sided) ideal of the ring R or open submodule of any R-module M is closed.

1.4.20. COROLLARY. A connected topological Abelian group A (see C.13) has no open subgroups different from A.

A connected topological ring R has no open subgroups of additive group, in particular open subrings, open left (right, two sided) ideals different from the R.

A connected topological module M does not contain open subgroups of the additive group, in particular open submodules, different from M.

1.4.21. COROLLARY. *A connected topological Abelian group A is generated as a group (see B.3) by any neighborhood of each of its elements.*

PROOF. Indeed, let $a \in A$ and U be a neighborhood of a, then the subgroup $< U >$ generated by the set U is a neighborhood of the element a, too. Due to Proposition 1.4.18, $< U >$ is an open subgroup of A, and, according to Corollary 1.4.20, $< U >= A$.

1.4.22. PROPOSITION. *In a topological Abelian group A the connected component of zero $C(A)$ (see C.13) is a closed subgroup. If $a \in A$, then $C(A) + a$ is a connected component of a.*

PROOF. Let $a \in A$. The mapping $\varphi_a : A \to A$, where $\varphi_a(x) = x + a$ for $x \in A$, is, due to Proposition 1.1.34, a homeomorphic mapping of the topological space A to itself and, besides, $\varphi_a(0) = a$. Then $C(A) + a$ is a connected component of a.

Let $c \in C(A)$, then $C(A) - c$ is a connected component of $-c$, besides, $0 \in C(A) - c$ and, hence, $C(A) - c \subseteq C(A)$. Then

$$C(A) - C(A) = \bigcup_{c \in C(A)} (C(A) - c) \subseteq C(A),$$

i.e. $C(A)$ is a subgroup of the group A. Since $C(A)$ is a closed subset in A, the proposition is proved completely.

1.4.23. COROLLARY. *The connected component of zero of a topological ring R is a closed two-sided ideal, and the connected component of zero of a topological R-module M is a closed submodule.*

PROOF. Let $C(M)$ be the connected component of zero of an R-module M. Then, due to Proposition 1.4.22, $C(M)$ is a closed subgroup of the additive group of M. Let $r \in R$, then the mapping $f_r : M \to M$, where $f_r(m) = rm$ for $m \in M$, is a continuous mapping of the topological space M to itself. Then $r \cdot C(M) = f_r(C(M))$ is a connected subset in M and besides, $0 \in r \cdot C(M)$. Therefore, $r \cdot C(M) \subseteq C(M)$, i.e. $C(M)$ is a closed submodule of the module M.

Considering R as left and right topological R-modules, we obtain that the connected component $C(R)$ of the topological ring R is a closed two-sided ideal of R.

1.4.24. PROPOSITION. *Let B be a dense subgroup of a topological Abelian group A (see C.5) and U be a neighborhood of zero of the topological group B. Then $[U]_A$ is a neighborhood of zero of the topological group A.*

PROOF. Indeed, $U = V \cap B$, where V is some neighborhood of zero of the topological group A. Let V_1 be a neighborhood of zero of A such that $V_1 + V_1 \subseteq V$. To complete the proof, we'll show that $V_1 \subseteq [U]_A$.

Let $x \in V_1$ and W be a neighborhood of zero of A. Without loss of generality we can suppose that $W \subseteq V_1$. Then $x + W \subseteq V_1 + V_1 \subseteq V$ and since B is dense subgroup in A, then $(x + W) \cap B \neq \emptyset$. Therefore,

$$(x + W) \cap U = (x + W) \cap (V \cap B) = (x + W) \cap B \neq \emptyset,$$

i.e. $x \in [U]_A$.

1.4.25. COROLLARY. *Let B be a dense subgroup of a topological Abelian group A. If topological group B has a basis of neighborhoods of zero consisting of subgroups, then the topological group A has a basis of neighborhoods of zero consisting of subgroups, too.*

PROOF. The result follows from Propositions 1.4.24 and 1.4.5.

1.4.26. COROLLARY. Let Q be a dense subring of a topological ring R and let Q have a basis of neighborhoods of zero consisting of subgroups (of subrings, of left, right, two sided ideals). Then the topological ring R has a basis of neighborhoods of zero consisting of subgroups (of subrings, of left, right, two-sided ideals).

1.4.27. COROLLARY. Let N be a dense submodule of a topological R-module M and let N have a basis of neighborhoods of zero consisting of subgroups (of submodules). Then the topological R-module M has a basis of neighborhoods of zero consisting of subgroups (of submodules).

1.4.28. PROPOSITION. *Let R be a topological ring, N be a closed non-empty subset of a topological R-module M, $x \in M$ and $a \in R$. Then the subset $(N : x)_R$ is closed in R, and the subset $(N : a)_M$ is closed in M (see B.12).*

PROOF. Let $\mathcal{B}_0(M)$ be a basis of neighborhoods of zero of M, $r \in [(N : x)_R]_R$, $m \in [(N : a)_M]_M$ and $U \in \mathcal{B}_0(M)$. Due to conditions (BN6′) and (BN6″) (see Proposition 1.2.3), there exist neighborhoods of zero V_U in R and W_U in M such that $V_U \cdot x \subseteq U$ and $a \cdot W_U \subseteq U$. We can choose elements $r_U \in (N : x)_R$ and $m_U \in (N : a)_M$ such that

$r - r_U \in V_U$ and $m - m_U \in W_U$. Then $r \cdot x - r_U \cdot x = (r - r_U) \cdot x \in V_U \cdot x \subseteq U$ and $a \cdot m - a \cdot m_U = a \cdot (m - m_U) \in a \cdot W_U \subseteq U$. Since $r_U \cdot x \in N$ and $a \cdot m_U \in N$, then $r \cdot x \in r_U \cdot x + U \subseteq N + U$, and analogously $a \cdot m \in N + U$. Hence,

$$r \cdot x, a \cdot m \in \bigcap_{U \in \mathcal{B}_0(M)} (N + U) = [N]_M = N,$$

i.e. $r \in (N : x)_R$ and $m \in (N : a)_M$. Thus, $(N : x)_R$ and $(N : a)_M$ are closed in the topological ring R and in the topological module M correspondingly.

1.4.29. COROLLARY. *Let R be a topological ring, N be a closed non-empty subset of a topological R-module M, $X \subseteq M, A \subseteq R$. Then:*

1) $(N : X)_R$ is a closed subset in R, and $(N : A)_M$ is a closed subset in M;

2) if N is a subgroup of the additive group M, then $(N : X)_R$ and $(N : A)_M$ are closed subgroups of the additive groups of R and M correspondingly;

3) if N is a a subgroup of the additive group of M and A is a right ideal of R, then $(N : A)_M$ is a closed submodule of M;

4) if N is a submodule of M, then $(N : X)_R$ is a closed left ideal of R;

5) if X and N are submodules of M, then $(N : X)_R$ is a closed two-sided ideal of R.

PROOF. Since $(N : X)_R = \bigcap_{x \in X}(N : x)_R$ and $(N : A)_M = \bigcap_{a \in A}(N : a)_M$, then $(N : X)_R$ and $(N : A)_M$ are closed subsets as intersections of closed subsets (see Proposition 1.4.28). Thus, the statement 1) is proved. The statements 2) - 5) result from 1) and from the corresponding statements of B.12.

1.4.30. COROLLARY. Let R be a topological ring, $A \subseteq R$, and let X be a subset of a Hausdorff topological R-module M. Then:

1) $(0 : X)_R$ is a closed left ideal of R;

2) if X is a submodule of M, then $(0 : X)_R$ is a closed two-sided ideal of R;

3) if A is a right ideal of R, then $(0 : A)_M$ is a closed submodule of M.

1.4.31. COROLLARY. In a Hausdorff topological ring R a left (right) annulator $(0 : A)_R$ of any non-empty subset $A \subseteq R$ is a closed left (right) ideal of R.

1.4.32. PROPOSITION. *Let R be a topological ring and X be a discrete submodule of a topological R-module M. Then for any $x \in X$ the subset $(0 : x)_R$ is an open ideal of R.*

PROOF. Let $r \in (0 : x)_R$ and U be a neighborhood of zero in M such that $U \cap X = \{0\}$. Let V be a neighborhood of zero in R such that $V \cdot x \subseteq U$. Since $V \cdot x \subseteq X$, then $V \cdot x = \{0\}$, i.e. $V \subseteq (0 : x)_R$. Therefore, from Proposition 1.4.18 and B.12 follows that $(0 : x)_R$ is an open left ideal of R.

1.4.33. COROLLARY. Let I be a discrete left (right) ideal of a topological ring R. Then a left (right) annulator of any of the elements of I is an open left (right) ideal of R. In particular, a non-discrete topological ring without zero divisors does not contain discrete left, right or two-sided ideals.

1.4.34. REMARK. It follows from Proposition 1.4.32 that if R is a non-discrete topological ring and M is an R-module such that $(0 : m)_R = \{0\}$ for some $m \in M$, then M with the discrete topology is not a topological R-module (compare to Remark 1.2.19).

1.4.35. PROPOSITION. *If a topological Abelian group A contains at least one Hausdorff closed subgroup, then A is a Hausdorff group, too.*

PROOF. Let B be a closed Hausdorff subgroup of A. Then, due to Theorem 1.3.2 and Definition 1.3.4, the subset $\{0\}$ is closed in the topological group B. Since B is closed in A, then $\{0\}$ is closed in A, too, i.e. A is a Hausdorff group.

1.4.36. PROPOSITION. *In a Hausdorff topological Abelian group A any discrete subgroup B is closed.*

PROOF. Assume the contrary hypothesis, i.e. that $[B]_A \neq B$, and let $a \in [B]_A \backslash B$. Since B is a discrete subgroup, then there exists a neighborhood U of zero in A such that $U \cap B = \{0\}$. Let's choose a neighborhood V of zero in A such that $V - V \subseteq U$. Then $(a + V) \cap B \neq \emptyset$. Let $b \in (a + V) \cap B$. Then $b \neq a$, because $a \notin B$ and $b \in B$. Since A is a Hausdorff group, then there exists a neighborhood W of zero in A such that $W \subseteq V$ and $b \notin a + W$. Let's show that $(a + W) \cap B = \emptyset$.

Indeed, if to assume the contrary, i.e. that there exists some element $c \in (a + W) \cap B$, we obtain that $c \neq b$, because $b \notin a+W$. But, on the other hand, $b-c \in (a+V)-(a+W) = V - W \subseteq V - V \subseteq U$. Taking into account that $b - c \in B$, we obtain the contradiction to the last assumtion, because $0 \neq b - c \in U \cap B = \{0\}$. Thus, $(a + W) \cap B = \emptyset$.

Hence, $a \notin [B]_A$, which is a contradiction to the first hypothesis. Then $[B]_A = B$. This completes the proof.

1.4.37. PROPOSITION. *For the topological group $\mathbb{R}(+)$ (see Example 1.1.11) the following statements are true:*

1) any non-discrete subgroup is dense;

2) any closed subgroup, different from $\mathbb{R}(+)$, is discrete and is of the type of $a \cdot \mathbb{Z}$, where a is a non-negative real number.

PROOF. Let A be a non-discrete subgroup of $\mathbb{R}(+)$ and d be a non-negative number. Let $\epsilon > 0$ be an arbitrary positive number of $\mathbb{R}$, then since $A \cap (-\epsilon, \epsilon)$ is a symmetrical neighborhood of zero of A and A is a non-discrete group, then there exists $x \in \mathbb{R}$ such that $0 < x \in A \cap (-\epsilon; \epsilon)$, i.e. $0 < x < \epsilon$. Let n be a natural number such that $n \cdot x \leq d < (n+1)x$. Then $0 \leq d - n \cdot x < x < \epsilon$, hence $|x - d| < \epsilon$. Since $nx \in A$, then $A \cap (d - \epsilon; d + \epsilon) \neq \emptyset$, which means that $a \in [A]_{\mathbb{R}}$. Thus, $[A]_{\mathbb{R}}$ contains all the positive numbers from $\mathbb{R}$. Since $[A]_{\mathbb{R}}$ is a subgroup of $\mathbb{R}$, then $[A]_{\mathbb{R}}$ also contains all the negative numbers. Therefore, $[A]_{\mathbb{R}} = \mathbb{R}$, i.e. the statement 1) is proved.

Let B be a closed subgroup of $\mathbb{R}$, different from $\mathbb{R}$. Then from 1) follows that B is a discrete subgroup. If $B = \{0\}$, then, putting $a = 0$, we get that $B = \{0\} = 0 \cdot \mathbb{Z}$, i.e. in that case the statement 2) is proved. If $B \neq \{0\}$, then there exists $0 < b \in B$. Since the segment $[0; b]$ is a compact subset in $\mathbb{R}$ and B is closed in $\mathbb{R}$, then $[0; b] \cap B$ is a compact subset in a discrete space B. Hence, $[0; b] \cap B$ is a finite subset. Let $a = \min\{x \mid x \in ([0; b] \cap B)\setminus\{0\}\}$. For any positive element $x \in B$ let m be the whole integer part of $\frac{x}{a}$. Then $x - m \cdot a \in B$, and $0 \leq x - m \cdot a < a$. Therefore, due to the choice of $a \in B$, it follows that $x - m \cdot a = 0$, i.e. $x = m \cdot a$. Hence, all the positive $x \in B$ are of the type of $m \cdot a$. Since B is a group, then all negative elements of B are of the same type, too. Thus, $B = a \cdot \mathbb{Z}$, i.e. statement 2) is proved in this case, too.

1.4.38. PROPOSITION. *In a Hausdorff ring R the closure $[Q]_R$ of a commutative subring Q is a commutative subring, too.*

PROOF. Let $a, b \in [Q]_R$ and $\mathcal{B}_0$ be a basis of symmetrical neighborhoods of zero of R. For any neighborhood $U \in \mathcal{B}_0$ let's choose a neighborhood $V_U \in \mathcal{B}_0$ such that

$$a \cdot V_U + b \cdot V_U + V_U \cdot a + V_U \cdot b + V_U \cdot V_U + V_U \cdot V_U + V_U \cdot V_U + V_U \cdot V_U + V_U \cdot V_U + V_U \cdot V_U \subseteq U.$$

Since $a, b \in [Q]_R$, then there exists elements $a_U, b_U \in Q$ such that $a - a_U \in V_U$ and $b - b_U \in V_U$. Then $a \in a_U + V_U$ and $b \in b_U + V_U$. Therefore,

$$a \cdot b - b \cdot a \in (a_U + V_U) \cdot (b_U + V_U) - (b_U + V_U) \cdot (a_U + V_U)$$
$$\subseteq a_U \cdot b_U + a_U \cdot V_U + V_U \cdot b_U + V_U \cdot V_U - b_U \cdot a_U$$
$$- b_U \cdot V_U - V_U \cdot a_U - V_U \cdot V_U.$$

Taking into account that V is a symmetrical neighborhood of zero in R and $a_U b_U - b_U a_U = 0$, we obtain that

$$a \cdot b - b \cdot a \in a_U \cdot V_U + V_U \cdot b_U + b_U \cdot V_U + V_U \cdot a_U + V_U \cdot V_U + V_U \cdot V_U$$
$$\subseteq (a + V_U) \cdot V_U + V_U \cdot (b + V_U) + (b + V_U) \cdot V_U + V_U \cdot (a + V_U)$$
$$+ V_U \cdot V_U + V_U \cdot V_U \subseteq a \cdot V_U + V_U \cdot b + b \cdot V_U + V_U \cdot a + V_U \cdot V_U$$
$$+ V_U \cdot V_U + V_U \cdot V_U + V_U \cdot V_U + V_U \cdot V_U + V_U \cdot V_U \subseteq U.$$

Therefore, $a \cdot b - b \cdot a \in \bigcap_{U \in \mathcal{B}_0} U = \{0\}$ (see Theorem 1.3.2), that means that the subring $[Q]_R$ is commutative.

1.4.39. PROPOSITION. *The center $Z(R)$ of a Hausdorff topological ring R is a closed commutative subring.*

PROOF. Let $x \in R$. Let's consider the subset $Z_x(R) = \{y \mid y \in R, y \cdot x = x \cdot y\}$ and show that it is closed. Indeed, let $a \in [Z_x(R)]_R$ and $y \in R$. Let $\mathcal{B}_0(R)$ be a basis of neighborhoods of zero of R, then for any neighborhood $U \in \mathcal{B}_0(R)$ let's choose a neighborhood $V_U \in \mathcal{B}_0(R)$ such that $V_U \cdot x - x \cdot V_U \subseteq U$. Since $a \in [Z_x(R)]_R$, then there exists an element $a_U \in Z_x(R)$ such that $a - a_U \in V_U$. From $a_U \in Z_x(R)$, i.e. $a_U \cdot x - x \cdot a_U = 0$, follows that

$$a \cdot x - x \cdot a \in (a_U + V_U) \cdot x - x \cdot (a_U + V_U) \subseteq a_U \cdot x - x \cdot a_U + V_U \cdot x - x \cdot V_U \subseteq U.$$

Hence, $a \cdot x - x \cdot a \in \bigcap_{U \in \mathcal{B}_0(R)} U = \{0\}$, i.e. $a \cdot x - x \cdot a = 0$, therefore, $a \in Z_x(R)$. Thus, $Z_x(R)$ is a closed subset in R and since $Z(R) = \bigcap_{x \in R} Z_x(R)$, then the center $Z(R)$ is closed as an intersection of closed subsets.

1.4.40. DEFINITION. Let (K, τ) be a topological skew field (a topological field). A subset H of K is called a skew subfield of the topological skew field K (a subfield of the topological field K), if H is a skew subfield of K (a subfield of K) and H is endowed with the topology induced by τ.

1.4.41. PROPOSITION. *A skew subfield H of a topological skew field (a topological field) K is a topological skew field (a topological field).*

PROOF. Due to the Remark 1.4.4, H is a subring of the topological ring K. Let's verify the fulfillment of condition (MIC) (see Definition 1.1.20) for the topological ring $(H, \tau|_H)$.

Let $0 \neq x \in H$ and U' be a neighborhood of the element x^{-1} in the topological ring $(H, \tau|_H)$. Then there exists a neighborhood U of x^{-1} in (K, τ) such that $U \bigcap H = U'$. Since condition (MIC) is fulfilled in (K, τ), then it is possible to find a neighborhood V of x in (K, τ) such that $(V \backslash \{0\})^{-1} \subseteq U$. Then $V \bigcap H$ is a neighborhood of x in $(H, \tau|_H)$, besides,

$$(V \cap H \backslash \{0\})^{-1} = \left((V \backslash \{0\}) \cap (H \backslash \{0\}) \right)^{-1}$$
$$= (V \backslash \{0\})^{-1} \cap (H \backslash \{0\})^{-1} = (V \backslash \{0\})^{-1} \cap H \subseteq U \cap H = U'.$$

This completes the proof.

1.4.42. PROPOSITION. *Let K be a topological skew field, and H be its skew subfield. Then $[H]_K$ is a skew subfield of K.*

PROOF. If $[H]_K = K$, the statement is evident. Let $[H]_K \neq K$. Then K is not an anti–discrete skew field. Hence, K is a Hausdorff skew field (see Remark 1.4.13). Due to Proposition 1.4.7, $[H]_K$ is a subring of K. Let $0 \neq x \in [H]_K$. Assume that $x^{-1} \notin [H]_K$. Then $K \backslash [H]_K$ is a neighborhood of x^{-1}. Therefore, since the skew field K is Hausdorff and due to condition (MIC), there exists a neighborhood V of x in K such that $0 \notin V$ and $V^{-1} \subseteq K \backslash [H]_K$. Since $V \bigcap H \neq \emptyset$, then there exists a non-zero element $y \in V \bigcap H$. Then

$$y^{-1} \in (V)^{-1} \cap (H \backslash \{0\})^{-1} = (V)^{-1} \cap H \subseteq (K \backslash [H]_K) \cap H = \emptyset.$$

The obtained contradiction shows that $x^{-1} \in [H]_K$. Then $[H]_K$ is a skew subfield of K. This completes the proof.

1.4.43. REMARK. Let R be a topological ring and its subring Q is a skew field, then $[Q]_R$ can be not a skew field.

Indeed, it was shown in Example 1.4.17 that for a transcendental number θ the field of rational numbers $\mathbb{Q}$ is a subfield of the ring $\mathbb{Q}[\theta]$, which is not a field, but meantime,

$$[\mathbb{Q}]_{\mathbb{Q}[\theta]} = [\mathbb{Q}]_{\mathbb{R}} \cap \mathbb{Q}[\theta] = \mathbb{R} \cap \mathbb{Q}[\theta] = \mathbb{Q}[\theta].$$

1.4.44. EXAMPLE. Due to Proposition 1.4.39, the center $Z(K)$ of a Hausdorff topological skew field K is its closed subfield.

1.4.45. REMARK. Any totally disconnected Abelian group A is Hausdorff.

Indeed, due to Proposition 1.4.22, the connected component $C(A)$ of A is a closed subgroup and, since $C(A) = \{0\}$, then, due to Theorem 1.3.2, the group A is Hausdorff.

1.4.46. THEOREM. *Let A be a locally compact totally disconnected group, U be a neighborhood of zero of A. Then there exists an open compact subgroup B of A such that $B \subseteq U$.*

PROOF. Due to Remark 1.4.45, A is a Hausdorff group. Then, according to C.14, there exists an open compact neighborhood P of zero of the group A such that $P \subseteq U$. Let $Q = \{q \in A | P + q \subseteq P\}$ and $B = Q \cap (-Q)$.

Let's show that B is an open compact subgroup of A, being contained in U. Note that $0 \in Q$, and, hence, $B \neq \emptyset$. Then, it is evident that $B = -B$. Besides, since $B \subseteq Q$, then $P + B \subseteq P$. Therefore, $P + (B + B) = (P + B) + B \subseteq P + B \subseteq P$, consequently $B + B \subseteq Q$. Since $B = -B$, then $B + B = (-B) + (-B) = -(B + B) \subseteq -Q$. Thus, $B + B \subseteq Q \cap (-Q) = B$. Since $B = -B$, then B is a subgroup of A. Let $q \in Q$, then $p + q \in P$ for any element $p \in P$. Since P is an open subset, then, due to condition (AC) (see Remark 1.1.2), there exist neighborhoods U_p and V_p of the elements p and q in A correspondingly, such that $U_p + V_p \subseteq P$. According to Corollary 1.2.29, we can suggest that U_p is an open neighborhood of p. Then $\{U_p | p \in P\}$ is an open cover of the compact subset P of A. Hence, there exist elements $p_1, p_2, \ldots, p_n \in P$ such that $\bigcup_{i=1}^{n} U_{p_i} \supseteq P$. Let $V = \bigcap_{i=1}^{n} V_{p_i}$. Then

$$P + V \subseteq \bigcup_{i=1}^{n} U_{p_i} + \bigcap_{i=1}^{n} V_{p_i} \subseteq \bigcup_{i=1}^{n} (U_{p_i} + V_{p_i}) \subseteq P,$$

i.e. $V \subseteq Q$. Since V is a neighborhood of q in Q, then q is an internal point of Q. Thus, all the points of Q are internal ones, hence, Q is an open subset. Therefore, the subgroup B is open as the intersection of open subsets Q and $-Q$ (see Proposition 1.1.34). Due to Proposition 1.4.18, the subgroup B is closed, besides, $B \subseteq Q \subseteq P + Q \subseteq P$. Then B is a compact subgroup of A as a closed subset of the compact space P. Since $P \subseteq U$, then $B \subseteq U$. This completes the proof of the theorem.

1.4.47. COROLLARY. A locally compact totally disconnected Abelian group has a basis of neighborhoods of zero consisting of open compact subgroups.

1.4.48. PROPOSITION. *Let A be a closed subset and B a compact subset of a topological skew field K, and $0 \notin B$. Then the subsets $A \cdot B$ and $B \cdot A$ are closed.*

PROOF. If a skew field K is non-Hausdorff, then, due to the Remark 1.4.13, K is an anti-discrete skew field, hence $A = K$. Since $B \neq \{0\}$, then $A \cdot B = B \cdot A = K$ and the statement is proved.

Let K be a Hausdorff skew field, $x \in [A \cdot B]_K$, but $x \notin A \cdot B$. Then $(x \cdot B^{-1}) \bigcap A = \emptyset$, i.e. $x \cdot B^{-1} \subseteq K \backslash A$. Since A is a closed subset of K, then $K \backslash A$ is an open neighborhood of any element $x \cdot b^{-1}$, where $b \in B$. Due to conditions (MC) and (MIC) (see Remark 1.1.7 and Definition 1.1.20), for any $b \in B$ there exist (we can consider open) neighborhoods U_b and V_b of the elements x and b correspondingly, such that $U_b \cdot V_b^{-1} \subseteq K \backslash A$ (since the skew field K is Hausdorff, and $0 \notin B$, the neighborhood V_b can be chosen so that $0 \notin V_b$ for $b \in B$). Then $\{V_b \mid b \in B\}$ is an open cover of the compact subset B, hence we can choose a finite number of the elements $b_1, b_2, \ldots, b_n \in B$ such that $V = \bigcup_{i=1}^{n} V_{b_i} \supseteq B$. Thus, $U = \bigcap_{i=1}^{n} U_{b_i}$ is a neighborhood of x, besides, $0 \notin V$ and $U \cdot V^{-1} \subseteq K \backslash A$. Since $V \supseteq B$, then $U \cdot B^{-1} \subseteq K \backslash A$, hence, $y \cdot B^{-1} \subseteq K \backslash A$ for any $y \in U$. Therefore, $(y \cdot B^{-1}) \bigcap A = \emptyset$, i.e. $y \notin A \cdot B$. Thus, it is proved that $U \bigcap (A \cdot B) = \emptyset$, i.e. $x \notin [A \cdot B]_K$. The obtained contradiction proves that $A \cdot B$ is closed subset.

The proof that $B \cdot A$ is closed subset is analogous.

1.4.49. PROPOSITION. *Let A be a subset and B a compact subset of a topological skew field K, and $0 \notin B$. Then:*

1) $[A \cdot B]_K = [A]_K \cdot [B]_K = [A]_K \cdot B$;

2) $[B \cdot A]_K = [B]_K \cdot [A]_K = B \cdot [A]_K$.

PROOF. Due to the Corollary 1.1.43,

$$[A \cdot B]_K \supseteq [A]_K \cdot [B]_K \supseteq [A]_K \cdot B.$$

On the other hand, due to Proposition 1.4.48, $[A]_K \cdot B$ is a closed subset in K, consequently,

$$[A \cdot B]_K \subseteq \left[[A]_K \cdot B\right]_K = [A]_K \cdot B.$$

Hence,

$$[A \cdot B]_K = [A]_K \cdot [B]_K = [A]_K \cdot B,$$

i.e. the statement 1) is proved.

The statement 2) is proved analogously.

The following example shows that in Propositions 1.4.48 and 1.4.49 the condition $0 \notin B$ is indispensable.

1.4.50. EXAMPLE. Consider the field $\mathbb{R}$ of real numbers with the interval topology. It is clear that the subset $A = \mathbb{Z}$ is closed in $\mathbb{R}$.

Let's verify that the subset $B = (\mathbb{Z}\setminus\{0\})^{-1} \bigcup \{0\}$ is closed, too. Indeed, let $x \in \mathbb{R}$ and $x \notin (\mathbb{Z}\setminus\{0\})^{-1} \bigcup \{0\}$. Then $x \neq 0$. Without loss of generality we can consider that $x > 0$. Since $x \notin (\mathbb{Z}\setminus\{0\})^{-1}$, we can choose a natural number $n \in \mathbb{Z}$ such that $\frac{1}{n} < x < \frac{1}{n-1}$. Therefore, the interval $\left(\frac{1}{n}; \frac{1}{n-1}\right)$ is a neighborhood of x, besides,

$$\left(\frac{1}{n}; \frac{1}{n-1}\right) \bigcap ((\mathbb{Z}\setminus\{0\})^{-1} \cup \{0\}) = \emptyset,$$

i.e. the subset $B = (\mathbb{Z}\setminus\{0\})^{-1} \bigcup \{0\}$ is closed in $\mathbb{R}$. Thus, the subset B is compact as a closed subset of the segment $[0; 1] \subset \mathbb{R}$. It is evident that $A \cdot B = \mathbb{Q}$, hence, $A \cdot B$ is not closed in $\mathbb{R}$.

1.4.51. COROLLARY. *Let K be a topological skew field, $S \subset K$, and $0 \notin S$. Then the subset S is open in K if and only if the subset S^{-1} is open in K.*

PROOF. The statement is evident if a skew field K is anti–discrete. If K is not anti–discrete, then, due to the Remark 1.4.13, K is a Hausdorff skew field, i.e. $\{0\}$ is a closed subset and $K\backslash\{0\}$ an open subset in K.

Let S be an open subset in K. Then S is an open subset in the subspace $K\backslash\{0\}$. Due to Proposition 1.1.49, S^{-1} is an open subset in $K\backslash\{0\}$. Since $K\backslash\{0\}$ is open in K, then S^{-1} is an open subset in K (see C.10). Vice versa, if S^{-1} is an open subset in K, then $S = (S^{-1})^{-1}$ is open in K.

§ 1.5. Homomorphisms

1.5.1. DEFINITION. Let A and A' be topological Abelian groups. A mapping $\varphi : A \to A'$ is called a continuous (open) homomorphism if φ is a homomorphism of Abelian groups and a continuous (open) mapping of the topological spaces. A homomorphism of groups, which is at the same time open and continuous, is called a topological homomorphism. The subgroup $\varphi(A)$ of the topological group A' is called a continuous, open and topologically homomorphic image of the topological group A, correspondingly. If $\varphi : A \to A'$ is a continuous (open, topological) homomorphism, which is also a bijection, then φ is called a continuous (open, topological) isomorphism, and the topological group A' is called the continuous (open, topological) isomorphic image of the topological group A.

1.5.2. REMARK. Let A and A' be topological Abelian groups and $\varphi : A \to A'$ be an isomorphism of these groups. Then the following conditions are equivalent:

φ is open mapping;

$\varphi^{-1} : A' \to A$ is continuous mapping.

Therefore, a topological isomorphism of topological groups is an isomorphism of these groups, being a homeomorphism of the corresponding spaces.

1.5.3. DEFINITION. Let R and R' be topological rings, $\varphi : R \to R'$ be a homomorphic (isomorphic) mapping of the rings R to R'. We call φ a continuous, open or topological homomorphism (isomorphism) of the topological ring R to (onto) the topological ring R' if φ is correspondingly continuous, open or a topological homomorphism (isomorphism) of the additive topological group of R to (onto) the additive topological group of R'. Meantime, the subring $\varphi(R)$ of R' is called a continuous, open or topological homomorphic (isomorphic) image of R.

1.5.4. DEFINITION. Let R be a topological ring, M and M' be topological R-modules, $\varphi : M \to M'$ be a homomorphic (isomorphic) mapping of R-module M to R-module M'. We say that φ is a continuous, open or topological homomorphism (isomorphism) of the topological R-module M to (onto) the topological R-module M' if φ

is correspondingly a continuous, open or topological homomorphism (isomorphism) of the additive topological group of the topological R-module M to (onto) the additive topological group of the topological R-module M'. Meantime, the submodule $\varphi(M)$ of M' is called a continuous, open or topological homomorphic (isomorphic) image of M.

1.5.5. PROPOSITION. *Let A and A' be topological Abelian groups, and $\varphi : A \to A'$ be a homomorphic mapping of A to A'. Then:*

1) φ is a continuous homomorphism if and only if $\varphi^{-1}(U')$ is a neighborhood of zero in A for any neighborhood U' of zero in A';

2) φ is an open homomorphism if and only if $\varphi(U)$ is a neighborhood of zero in A' for any neighborhood U of zero in A;

3) φ is a topological homomorphism if and only if for any neighborhoods U and U' of zero in A and A' correspondingly, $\varphi(U)$ and $\varphi^{-1}(U')$ are neighborhoods of zero in A' and A respectively.

PROOF. 1) If φ is a continuous homomorphism, then it is, in particular, continuous at the point $0 \in A$, and $\varphi(0) = 0$. Then the necessity of the condition in the statement 1) results from C.7.

Let now $a \in A$ and V' be a neighborhood of the element $\varphi(a)$. Then, due to Proposition 1.1.34, $V' - \varphi(a)$ is a neighborhood of zero in A'. Therefore, $U = \varphi^{-1}(V' - \varphi(a))$ is the neighborhood of zero in A. Then $U + a$ is the neighborhood of a in A, besides,

$$\varphi(U + a) = \varphi(U) + \varphi(a) = \varphi\left(\varphi^{-1}(V' - \varphi(a))\right) + \varphi(a) = V' - \varphi(a) + \varphi(a) = V,$$

that means that the mapping $\varphi : A \to A'$ is continuous.

2) If $\varphi : A \to A'$ is an open mapping, and U is a neighborhood of zero in A, then, due to Proposition 1.2.28, there exists an open neighborhood V of zero in A such that $V \subseteq U$. Then $\varphi(V)$ is an open subset in A', besides, $0 \in \varphi(V) \subseteq \varphi(U)$, i.e. $\varphi(U)$ is a neighborhood of zero in A'.

Vice versa, let B be an open subset in A, and $b' \in \varphi(B)$. Then there exists an element $b \in B$ such that $\varphi(b) = b'$. Since B is a neighborhood of the element b in A, then $B - b$ is a neighborhood of zero in A. Hence, $\varphi(B - b)$ is a neighborhood of zero in A', besides,

$$\varphi(B - b) = \varphi(B) - \varphi(b) = \varphi(B) - b'.$$

Therefore, $\varphi(B) = \varphi(B-b)+b'$ is a neighborhood of b' in A'. Thus, $\varphi(B)$ is a neighborhood of each of its elements, i.e. $\varphi(B)$ is an open subset in A'.

The statement 3) results from statements 1), 2) and Definition 1.5.1.

1.5.6. PROPOSITION. *Let A be a topological Abelian group, B be a subgroup of A, $\mathcal{B}_0(A)$ be a basis of neighborhoods of zero in A and $\omega : A \to A/B$ be a natural homomorphism of the group A to its factor-group A/B (see B.11). Then the system*

$$\mathcal{B}_0(A/B) = \{\omega(U) \mid U \in \mathcal{B}_0(A)\}$$

of the subsets of factor group A/B satisfies conditions (BN1) - (BN4) (see Proposition 1.2.1), hence, specifies on A/B a topology, in which A/B is a topological group with the basis $\mathcal{B}_0(A/B)$ of neighborhoods of zero.

PROOF. It is clear that $0 = \omega(0) \in \omega(U)$ for any neighborhood $U \in \mathcal{B}_0(A)$, i.e. condition (BN1) is fulfilled.

Let $U, V \in \mathcal{B}_0(A)$. There exists $W \in \mathcal{B}_0(A)$ such that $W \subseteq U \cap V$. Then $\omega(W) \in \mathcal{B}_0(A/B)$, besides, $\omega(W) \subseteq \omega(U) \cap \omega(V)$, i.e. condition (BN2) is fulfilled.

Let $U \in \mathcal{B}_0(A)$, then there exists $V \in \mathcal{B}_0(A)$ such that $V + V \subseteq U$. Then $\omega(V) \in \mathcal{B}_0(A/B)$, besides, $\omega(V) + \omega(V) = \omega(V + V) \subseteq \omega(U)$, i.e. condition (BN3) is fulfilled.

Since $\omega(-V) = -\omega(V)$ for each $V \in \mathcal{B}_0(A)$, then condition (BN4) is fulfilled, too.

Due to Theorem 1.2.4, $\mathcal{B}_0(A/B)$ specifies on A/B a topology such that A/B is a topological group and $\mathcal{B}_0(A/B)$ is the basis of neighborhoods of zero.

1.5.7. PROPOSITION. *Let A be a topological Abelian group, B be a subgroup of A, $\mathcal{B}_0(A)$ and $\mathcal{B}_0'(A)$ be bases of neighborhoods of zero in A and $\omega : A \to A/B$ be the natural homomorphism of the group A to the group A/B. Then the topologies τ and τ', being specified on A/B according to Proposition 1.5.6 by the systems*

$$\mathcal{B}_0(A/B) = \{\omega(U) \big| U \in \mathcal{B}_0(A)\}$$

and

$$\mathcal{B}_0'(A/B) = \{\omega(U') \big| U' \in \mathcal{B}_0'(A)\}$$

correspondingly, coincide.

PROOF. Due to Proposition 1.5.6, $\mathcal{B}_0(A/B)$ and $\mathcal{B}_0'(A/B)$ specify on A/B the topologies τ and τ', in each one of them A/B is a topological group, besides, $\mathcal{B}_0(A/B)$ is a basis of neighborhoods of zero of the topological group $(A/B, \tau)$, and $\mathcal{B}_0'(A/B)$ is a basis of neighborhoods of zero of the topological group $(A/B, \tau')$.

Let $U \in \mathcal{B}_0(A)$, then there exists $U' \in \mathcal{B}_0'(A)$ such that $U' \subseteq U$. Then $\omega(U') \in \mathcal{B}_0'(A/B)$, besides, $\omega(U') \subseteq \omega(U)$, i.e. $\omega(U)$ is a neighborhood of zero of the topological group $(A/B, \tau')$.

Let $V' \in \mathcal{B}_0'(A)$, then there exists $V \in \mathcal{B}_0(A)$ such that $V \subseteq V'$. Then $\omega(V) \in \mathcal{B}_0(A/B)$, besides, $\omega(V) \subseteq \omega(V')$. Therefore, $\mathcal{B}_0(A/B)$ is a basis of neighborhoods of zero of the topological group $(A/B, \tau')$.

Due to Proposition 1.5.6 and Theorem 1.2.4, there exists a unique topology on A/B such that A/B is a topological group and $\mathcal{B}_0(A/B)$ is a basis of neighborhoods of zero. Therefore, $\tau = \tau'$.

1.5.8. DEFINITION. Let A be a topological Abelian group with a basis $\mathcal{B}_0(A)$ of neighborhoods of zero, B be its subgroup, ω be the natural homomorphism of the group A to its factor group A/B. The group A/B endowed with the topology, whose basis of neighborhoods of zero is the system

$$\mathcal{B}_0(A/B) = \big\{\omega(U)\big|U \in \mathcal{B}_0(A)\big\},$$

is called a factor-group of the topological group A by the subgroup B.

1.5.9. REMARK. It follows from Proposition 1.5.7 that the definition of the factor-group A/B of the topological Abelian group A by subgroup B does not depend on the choice of a basis of neighborhoods of zero of A.

1.5.10. PROPOSITION. *Let A/B be the factor group of a topological Abelian group A by a subgroup B. Then the natural homomorphism $\omega : A \to A/B$ is the topological homomorphism of topological group A onto the topological group A/B.*

PROOF. Let $\mathcal{B}_0(A)$ be a basis of neighborhoods of zero of A. Then, due to Definition 1.5.8,

$$\mathcal{B}_0(A/B) = \big\{\omega(U)\big|U \in \mathcal{B}_0(A)\big\}$$

is a basis of neighborhoods of zero of the topological group A/B. Let V be a neighborhood of zero of A, then there exists a neighborhood $U \in \mathcal{B}_0(A)$ such that $U \subseteq V$. Then $\omega(U) \subseteq \omega(V)$. Since $\omega(U) \in \mathcal{B}_0(A/B)$, then $\omega(V)$ is a neighborhood of zero in A/B. Therefore, due to Proposition 1.5.5, ω is an open homomorphism.

Let W be a neighborhood of zero in A/B, then there exists a neighborhood $T \in \mathcal{B}_0(A)$ such that $\omega(T) \subseteq W$. Then $\omega^{-1}(W) \supseteq T$, hence, $\omega^{-1}(W)$ is a neighborhood of zero in A. From Proposition 1.5.5 follows that ω is a continuous homomorphism. Thus, ω is topological homomorphism.

1.5.11. REMARK. The factor group A/B of a locally compact Abelian group A by a subgroup B is a locally compact group.

Indeed, let U be a compact neighborhood of zero in A, $\omega : A \to A/B$ be the natural homomorphism. Then $\omega(U)$ is a compact neighborhood of zero in the topological group A/B, i.e. A/B is a locally compact group (see Remark 1.1.36).

1.5.12. THEOREM. *Let A and A' be topological Abelian groups, $\varphi : A \to A'$ be a homomorphism from A onto A', and $B = \ker \varphi$. Let $\omega : A \to A/B$ be the natural homomorphism and $\psi : A/B \to A'$ be the natural isomorphism (see B.11), then:*

1) if φ is a continuous homomorphism, then ψ is a continuous isomorphism of the topological group A/B onto the topological group A';

2) if φ is an open homomorphism, then ψ is an open isomorphism of the topological group A/B onto the topological group A';

3) if φ is a topological homomorphism, then ψ is a topological isomorphism of the topological group A/B onto the topological group A'.

PROOF. 1) Let φ be a continuous homomorphism and U be a neighborhood of zero in A'. Then $\varphi^{-1}(U)$ is a neighborhood of zero in A, and $\omega(\varphi^{-1}(U))$ is a neighborhood of zero in A/B. From the definition of the mappings ω and ψ follows that $\psi\big(\omega(\varphi^{-1}(U))\big) = U$. Since ψ is an isomorphism, then $\omega(\varphi^{-1}(U)) = \psi^{-1}(U)$. Since $\omega(\varphi^{-1}(U))$ is a neighborhood of zero in A/B, we obtain that ψ is a continuous isomorphism.

2) Let φ be an open homomorphism and V be a neighborhood of zero in A/B. Then $\omega^{-1}(V)$ is a neighborhood of zero in A, and $\varphi(\omega^{-1}(V))$ is a neighborhood of zero in A'.

Since $\varphi(\omega^{-1}(V)) = \psi(V)$, then ψ is an open isomorphism.

The statement 3) results from 1) and 2).

1.5.13. COROLLARY. Let B be a subgroup of a topological Abelian group A and $\omega : A \to A/B$ be the natural homomorphism. The topology of the factor group A/B is the strongest of all the group topologies on A/B (see C.8) relative to which the homomorphism ω is continuous.

1.5.14. COROLLARY. *Let A be a topological Abelian group, B and C be its subgroups, $B \subseteq C$, μ be the natural isomorphism of the group $(A/B)/(C/B)$ onto the group A/C (see B.11). Then μ is a topological isomorphism of the topological group $(A/B)/(C/B)$ onto the topological group A/C.*

PROOF. Let

$$\omega_B : A \to A/B, \quad \omega_C : A \to A/C$$

and

$$\omega_{C/B} : A/B \to (A/B)/(C/B)$$

be the natural topological homomorphisms. Since

$$\omega_{C/B}\omega_B : A \to (A/B)/(C/B)$$

is a topological homomorphism (as superposition of two topological homomorphisms) and $\ker(\omega_{C/B}\omega_B) = C$, then, due to Theorem 1.5.12, the natural isomorphism

$$\psi : A/C \to (A/B)/(C/B)$$

is a topological isomorphism. It is clear that $\psi^{-1} = \mu$.

1.5.15. PROPOSITION. *Let A be a topological Abelian group, B be its subgroup, and $\omega : A \to A/B$ be the natural topological homomorphism from A to the factor-group A/B. If $C \subseteq A$ and $C + B = C$, then $\omega([C]_A) = [\omega(C)]_{A/B}$. In particular, if C is a closed subset of A and $C + B = C$, then $\omega(C)$ is a closed subset of the group A/B.*

PROOF. Let $\mathcal{B}_0(A)$ be a basis of neighborhoods of zero of A. According to the Definition 1.5.8,

$$\mathcal{B}_0(A/B) = \{\omega(U) | U \in \mathcal{B}_0(A)\}$$

is a basis of neighborhoods of zero of A/B. Due to Proposition 1.2.25,

$$\left[\omega(C)\right]_{A/B} = \bigcap_{U \in \mathcal{B}_0(A)} \left(\omega(C) + \omega(U)\right) = \bigcap_{U \in \mathcal{B}_0(A)} \omega(C + U).$$

Since

$$(C + U) + B = (C + B) + U = C + U,$$

then, due to A.6,

$$\bigcap_{U \in \mathcal{B}_0(A)} \omega(C + U) = \omega\Big(\bigcap_{U \in \mathcal{B}_0(A)} (C + U)\Big) = \omega\big([C]_A\big).$$

1.5.16. PROPOSITION. *Let B be a compact subgroup of a topological Abelian group A. Then the natural homomorphism $\omega : A \to A/B$ is closed mapping (see C.7).*

PROOF. Let S be a closed subset of the topological group A. Then, due to Proposition 1.1.38, $S + B$ is a closed subset of A. Since $(S + B) + B = S + B$, then, using Proposition 1.5.15, we obtain that

$$\omega(S) = \omega(S + B) = \omega([S + B]_A) = \left[\omega(S + B)\right]_{A/B} = \left[\omega(S)\right]_{A/B},$$

i.e. $\omega(S)$ is closed subset in A/B. Therefore, $\omega : A \to A/B$ is a closed mapping.

1.5.17. PROPOSITION. *Let A be a topological Abelian group, B be its subgroup, and $\omega : A \to A/B$ be the natural topological homomorphism of the topological group A onto the topological group A/B. Then the canonic bijective correspondence between all the subgroups of A/B and all the subgroups of A containing the subgroup B (see B.11), induced by the mapping ω, is a bijective correspondence between all closed (all open, all dense) subgroups of the topological group A/B and all closed (open, dense) subgroups of the topological group A containing the subgroup B.*

PROOF. Let C be a subgroup of A, and $C \supseteq B$. Then $C + B = C$ and, hence, $\omega^{-1}\big(\omega(C)\big) = C$.

Let subgroup C be closed in A, then $[C]_A = C$, and from Proposition 1.5.15 follows that

$$\left[\omega(C)\right]_{A/B} = \omega\big([C]_A\big) = \omega(C),$$

i.e. $\omega(C)$ is closed subgroup in A/B.

Vice versa, let $\omega(C)$ be a closed subgroup in A/B, then, since the homomorphism ω is continuous, the subgroup $C = \omega^{-1}(\omega(C))$ is closed in A.

Let C be an open subgroup of A, then, since the homomorphism ω is open, $\omega(C)$ is an open subgroup of A/B.

Vice versa, let $\omega(C)$ be an open subgroup of A/B, then, since the homomorphism ω is continuous, $C = \omega^{-1}(\omega(C))$ is an open subgroup of A.

Let C be a dense subgroup of A, then, since the homomorphism ω is continuous, the subgroup $\omega(C)$ is dense in A/C.

Vice versa, let $\omega(C)$ be a dense subgroup of A/B, then from

$$\omega([C]_A) = [\omega(C)]_{A/B} = A/B = \omega(A)$$

follows that

$$A \subseteq \omega^{-1}\left(\omega([C]_A)\right) = [C]_A + B \subseteq [C+B]_A = [C]_A.$$

This means that the subgroup C is dense in A.

1.5.18. COROLLARY. *Let A/B be the factor-group of a topological Abelian group A by its subgroup B. Then:*

1) A/B is Hausdorff if and only if B is a closed subgroup of A;

2) A/B is anti-discrete if and only if B is a dense subgroup of A;

3) A/B is discrete if and only if B is an open subgroup of A.

PROOF. Indeed, the group A/B is Hausdorff (anti-discrete, discrete) if and only if its zero subgroup is closed (dense, open) in A/B. Then, due to Proposition 1.5.17, it takes place if and only if the subgroup B is closed (dense, open) in the group A.

1.5.19. PROPOSITION. *Let (A, τ) be a topological Abelian group, B and C be subgroups of A such that $B \subseteq C$. A topology on the group C/B can be specified in two natural ways:*

1) to consider on C the topology $\tau\big|_C$ induced by τ and to specify on C/B a topology according to Definition 1.5.8 (the resulting topological group we denote over $(C, \tau\big|_C)/B)$;

2) let's denote over τ' a topology specified on A/B according to Definition 1.5.8. Then consider on C/B the topology $\tau'|_{C/B}$, induced by the topology τ', i.e. consider the subgroup $(C/B, \tau'|_{C/B})$ of the topological group $(A/B, \tau')$.

It turns out that these topologies coincide.

PROOF. Indeed, let $\omega : A \to A/B$ be the natural homomorphism, and $\omega|_C$ be its restriction on C. Then $\omega|_C$ is the natural homomorphism of the group C onto the group C/B. Since $\omega : (A, \tau) \to (A/B, \tau')$ is a continuous homomorphism, then

$$\omega|_C : (C, \tau|_C) \to (C/B, \tau'|_{C/B})$$

is continuous homomorphism, too.

Let's show that $\omega|_C$ is open mapping. Let U be a neighborhood of zero in $(C, \tau|_C)$. Then $U = V \cap C$, where V is some neighborhood of zero in (A, τ). Since $\omega : (A, \tau) \to (A/B, \tau')$ is the open homomorphism of topological groups, then $\omega(V)$ is a neighborhood of zero in $(A/B, \tau')$, hence, $\omega(V) \cap (C/B)$ is a neighborhood of zero in $(C/B, \tau'|_{C/B})$. Then from A.6 follows that

$$\left(\omega|_C\right)(U) = \left(\omega|_C\right)(V \cap C) = \omega(V \cap C) = \omega(V) \cap \omega(C) = \omega(V) \cap (C/B),$$

i.e. $\left(\omega|_C\right)(U)$ is a neighborhood of zero in $(A/B, \tau')$. Hence, $\omega|_C$ is an open homomorphism.

Thus,

$$\omega|_C : (C, \tau|_C) \to (C/B, \tau'|_{C/B})$$

is topological homomorphism. Then, due to Theorem 1.5.12, the identity mapping of the topological group $(C, \tau|_C)/B$ on the topological group $(C/B, \tau'|_{C/B})$ is a topological isomorphism. This completes the proof.

1.5.20. COROLLARY. *Let B and C be subgroups of a topological Abelian group A, and ψ be the natural isomorphism of the group $C/(B \cap C)$ to the group $(C + B)/B$ (see B.11). Then ψ is continuous isomorphism of the topological group $C/(B \cap C)$ onto the topological group $(C + B)/B$.*

PROOF. Let $\omega : A \to A/B$ be the natural topological homomorphism. Then its restriction $\omega|_C$ on C is a continuous homomorphism of the topological group C onto the topological group $(C+B)/B$ (see B.11 and Proposition 1.5.19), besides, $\ker(\omega\big|_C) = B \bigcap C$. Applying Theorem 1.5.12 to $\varphi = \omega\big|_C$, we obtain that the natural homomorphism ψ of the topological group $C/(B \bigcap C)$ onto the topological group $(C + B)/B$ is continuous.

1.5.21. PROPOSITION. *Let B be the smallest closed subgroup of a topological Abelian group A (see Corollary 1.4.11) and U be a neighborhood of zero in A. Then there exists a neighborhood of zero V in A such that:*

1) $V \subseteq U$;

2) $V + B = V$.

PROOF. Let W be a neighborhood of zero in A such that $W + W \subseteq U$. Let $V = W + B$. Then V is a neighborhood of zero in A, and, due to Corollary 1.4.11, $V = W + B \subseteq W + W \subseteq U$, i.e. 1) is proved.

Since $V + B = W + B + B = W + B = V$, then 2) is proved.

1.5.22. PROPOSITION. *Let C be a subgroup of a topological Abelian group A and B be the smallest closed subgroup in A. Then the natural isomorphism $\psi : C/(B \bigcap C) \to (C + B)/B$ is a topological isomorphism of the topological group $C/(B \bigcap C)$ to the topological group $(C + B)/B$.*

PROOF. Due to Corollary 1.5.20, the isomorphism ψ is continuous. Let W be a neighborhood of zero in $C/(C \bigcap B)$ and $\omega : C \to C/(C \bigcap B)$ be the natural homomorphism, then $\omega^{-1}(W)$ is a neighborhood of zero in C. Hence, $\omega^{-1}(W) = U \bigcap C$, where U is some neighborhood of zero in A. Due to Proposition 1.5.21, there exists a neighborhood V of zero in A such that $V \subseteq U$ and $V + B = V$. Then $V \bigcap (C + B)$ is a neighborhood of zero in $C + B$ and $\omega_B (V \bigcap (C + B))$ is a neighborhood of zero in $(C + B)/B$, where

$$\omega_B : C + B \to (C + B)/B$$

is the natural homomorphism. Taking into account A.6 and B.11, we obtain that

$$\omega_B (V \cap (C + B)) = \omega_B (V) \cap \omega_B (C + B) = \omega_B (V) \cap \omega_B (C) =$$
$$\omega_B ((V + B) \cap C) = \omega_B (V \cap C) \subseteq \omega_B (U \cap C) = \omega_B (\omega^{-1}(W)) = \psi(W).$$

Thus, $\psi(W)$ is a neighborhood of zero in $(C + B)/B$. Hence, ψ is an open isomorphism.

1.5.23. PROPOSITION. *Let B be a compact subgroup and C – a closed subgroup of a topological Abelian group A. Then the natural isomorphism*

$$\psi : C/(C \cap B) \to (C + B)/B$$

is topological isomorphism.

PROOF. Since, due to Corollary 1.5.20, the isomorphism $\psi : C/(C \cap B) \to (C + B)/B$ is a continuous mapping, then it remains to show that ψ is open mapping. Since ψ is an isomorphism, then, due to C.7, it is enough to show that ψ is closed mapping.

Let S be a closed subset of the topological group $C/(C \cap B)$. Then, due to B.11, $\psi(S) = \omega_B \left(\omega_{B \cap C}^{-1} (S) \right)$, where $\omega_B : A \to A/B$ and $\omega_{B \cap C} : C \to C/(C \cap B)$ are the natural homomorphisms. Taking into account that the mapping $\omega_{B \cap C}$ is continuous, we obtain that $\omega_{B \cap C}^{-1} (S)$ is a closed subset of the topological group C. Since $\omega_B : A \to A/B$ is a closed mapping (see Proposition 1.5.16), then $\omega_B \left(\omega_{B \cap C}^{-1} (S) \right)$ is a closed subset in A/B. Then $\omega_B \left(\omega_{B \cap C}^{-1} (S) \right) = \psi(S)$ is a closed subset of the topological group $(C + B)/B$. Thus, the mapping $\psi : C/(C \cap B) \to (C + B)/B$ is a topological isomorphism.

1.5.24. PROPOSITION. *Let C be a subgroup of a topological Abelian group A, and B be a closed subgroup of the group C. Then the topological groups C/B and $(C+[B]_A)/[B]_A$ are topologically isomorphic.*

PROOF. Due to Corollary 1.5.20, the natural isomorphism $\psi : C/([B]_A \cap C) \to (C + [B]_A)/[B]_A$ is continuous. Since the subgroup B is closed in C and $[B]_A \cap C = [B]_C = B$, then ψ is a continuous isomorphism of the topological group C/B onto the topological group $(C + [B]_A)/[B]_A$. Let's show that ψ is also an open mapping.

Let $\omega : A \to A/[B]_A$ and $\omega_B : C \to C/B = C/([B]_A \cap C)$ be the natural homomorphisms. Let T be a neighborhood of zero in the topological group C/B, then $U = \omega_B^{-1}(T)$ is a neighborhood of zero in the topological group C, besides, $U + B = U$. Let V be a neighborhood of zero in the topological group A such that $U = V \cap C$, then there exists a neighborhood W of zero in A such that $W + W \subseteq V$. Then $W + [B]_A$ is a neighborhood of zero in the topological group A and, due to Proposition 1.2.25,

$$(W + [B]_A) \cap C \subseteq (W + (B + W)) \cap C \subseteq$$
$$(V + B) \cap C \subseteq ((V \cap C) + B) = U + B = U.$$

Hence (see A.6),

$$\psi(T) = \omega\big(\omega_B^{-1}(T)\big) = \omega(U) \supseteq \omega\big((W + [B]_A) \cap C\big) = \big(\omega(W + [B]_A)\big) \cap (\omega(C)).$$

Since $\omega(C) = (C + [B]_A)/[B]_A$ and $\omega(W + [B]_A)$ is a neighborhood of zero in the topological group $A/[B]_A$, then $\omega(W + [B]_A) \cap ((C + [B]_A)/[B]_A)$ is a neighborhood of zero in the topological group $(C + [B]_A)/[B]_A$, being contained in $\psi(T)$. It completes the proof.

1.5.25.   EXAMPLE. In a general case, the natural homomorphism $\psi : C/(B \cap C) \to (C + B)/B$ can be not open. Let's consider the group $\mathbb{R}$ of real numbers with the interval topology, and its subgroups $C = \mathbb{Z}$ and $B = \theta \cdot \mathbb{Z}$, where θ is a positive irrational number. It is clear, that the discrete topologies are induced onto $\mathbb{Z}$ and $\theta \cdot \mathbb{Z}$, which is why these subgroups are closed in $\mathbb{R}$ (see Proposition 1.4.36). The subgroup $\mathbb{Z} + \theta \cdot \mathbb{Z}$ differs from $\mathbb{R}$ (the first is a countable group, and the second is uncountable) and it also contains non-zero rational and irrational numbers. Hence, it cannot be presented as $\alpha \cdot \mathbb{Z}$, where α is a non-negative real number. Then, due to Propositions 1.4.36 and 1.4.37, $\mathbb{Z} + \theta \cdot \mathbb{Z}$ is a non-discrete subgroup in $\mathbb{R}$. Since $\{0\}$ is an open subset in $\mathbb{Z}$, then $\mathbb{Z}$ is not an open subgroup in $\mathbb{Z} + \theta \cdot \mathbb{Z}$. Then $(\mathbb{Z} + \theta \cdot \mathbb{Z})/\mathbb{Z}$ is a non-discrete topological group (see Corollary 1.5.18). Since the group $\theta \cdot \mathbb{Z}$ is discrete, then from Corollary 1.5.13 follows that $(\theta \cdot \mathbb{Z})/(\mathbb{Z} \cap \theta \cdot \mathbb{Z})$ is a discrete group. Thus, ψ is not an open mapping.

1.5.26. PROPOSITION. *Let B be a subgroup of a topological Abelian group A, $\omega : A \to A/B$ be the natural topological homomorphism and U be a closed neighborhood of zero in A such that:*

1) $\omega(U)$ is a compact neighborhood of zero of the topological group A/B;

2) $U \cap B$ is a compact neighborhood of zero of the topological group B.

Let V be a closed symmetric neighborhood of zero of the topological group A such that $V + V + V + V \subseteq U$. Then V is a compact subset of the topological group A.

PROOF. Let Δ be a system of closed subsets of topological space V, which is a basis of some filter (see A.12). Since $S \subseteq V \subseteq U$ for $S \in \Delta$, then $\omega(S) \subseteq \omega(U)$. Hence, the system of closed subsets

$$\Delta' = \left\{ [\omega(S)]_{\omega(U)} \middle| S \in \Delta \right\}$$

is a basis of some filter of the compact space $\omega(U)$. Let (see C.12)

$$a' \in \bigcap_{S \in \Delta} [\omega(S)]_{\omega(U)} \subseteq \bigcap_{S \in \Delta} [\omega(S)]_{A/B}.$$

Let's choose in the topological group A a basis $\mathcal{B}_0(A)$ of neighborhoods of zero such that $W \subseteq V$ for any neighborhood $W \in \mathcal{B}_0(A)$. Since $\mathcal{B}_0(A/B) = \{\omega(W) | W \in \mathcal{B}_0(A)\}$ is a basis of neighborhoods of zero of the topological group A/B, then, due to Proposition 1.2.25, we get

$$[\omega(S)]_{A/B} = \bigcap_{W \in \mathcal{B}_0(A)} (\omega(S) + \omega(W)) = \bigcap_{W \in \mathcal{B}_0(A)} \omega(S + W),$$

hence,

$$a' \in \bigcap_{S \in \Delta} \left(\bigcap_{W \in \mathcal{B}_0(A)} \omega(S + W) \right).$$

Thus, $a' \in \omega(S + W) \subseteq \omega(V + V)$ for any $S \in \Delta$ and $W \in \mathcal{B}_0(A)$. Let an element $a \in V + V$ be such that $\omega(a) = a'$. Then for any $S \in \Delta$ and $W \in \mathcal{B}_0(A)$ there exist elements $s \in S$, $w \in W$ and $b \in B$ such that $a = s + w + b$. From $-b = s - a + w$ we obtain that $B \cap ((S - a) + W) \neq \emptyset$. Hence, $B \cap [(S - a) + W]_A \neq \emptyset$.

Since

$$[S - a + W]_A \subseteq (S - a + W) + W \subseteq V + V + V + V \subseteq U,$$

then the system

$$\left\{ B \cap [(S - a) + W]_R \middle| S \in \Delta, W \in \mathcal{B}_0(A) \right\}$$

of closed subsets of $B \cap U$ is a basis of some filter of the topological space $B \cap U$ (which is compact due to 2)). Therefore, there exists an element

$$c \in \bigcap_{S \in \Delta, W \in B_0(A)} \left(B \cap [(S-a)+W]_A \right)$$

$$\subseteq B \cap \left(\bigcap_{S \in \Delta} \left(\bigcap_{W \in B_0(A)} ((S-a)+W+W) \right) \right)$$

$$\subseteq B \cap \left(\bigcap_{S \in \Delta} \left(\bigcap_{W' \in B_0(A)} ((S-a)+W') \right) \right) = B \cap \left(\bigcap_{S \in \Delta} [S-a]_A \right).$$

Since the neighborhood V is closed in A, and each of the subsets $S \in \Delta$ is closed in the space V, then each of the subsets $S \in \Delta$ is closed in A. Therefore, according to the Proposition 1.1.41, $[S-a]_A = S - a$. Thus,

$$c \in B \cap \left(\bigcap_{S \in \Delta} (S-a) \right) \subseteq \bigcap_{S \in \Delta} (S-a) = \left(\bigcap_{S \in \Delta} S \right) - a,$$

hence, $\bigcap_{S \in \Delta} S \neq \emptyset$. Therefore, the neighborhood V is compact (see C.12).

1.5.27. COROLLARY. *Let B be a subgroup of a topological Abelian group A. If the topological groups B and A/B are locally compact, then the topological group A is locally compact, too.*

PROOF. Let S be a compact neighborhood of zero in the group B, T be a compact neighborhood of zero in the group A/B, and $\omega : A \to A/B$ be the natural topological homomorphism. Without loss of generality we can assume that the neighborhood S is closed in B. Hence, $S = S_1 \cap B$, where S_1 is some closed neighborhood of zero of the group A. Since $\omega(S_1)$ is a neighborhood of zero of the topological group A/B, then there exists a closed neighborhood of zero T_1 in A/B such that $T_1 \subseteq T \cap \omega(S_1)$. Then T_1 is a compact neighborhood of zero in A/B. Putting $U = \omega^{-1}(T_1) \cap S_1$, we obtain that U is a closed neighborhood of zero of the group A, besides:

1) $\omega(U) = \omega(\omega^{-1}(T_1) \cap S_1) = T_1 \cap \omega(S_1) = T_1$ (see A.6),

2) $U \cap B = \omega^{-1}(T_1) \cap S_1 \cap B = \omega^{-1}(T_1) \cap S$ is a closed (hence, compact) neighborhood of zero in the group B.

Let's choose a closed neighborhood V of zero in A such that $V + V + V + V \subseteq U$. Then, due to Proposition 1.5.26, V is a compact subset of the group A. Hence, A is a locally compact group.

1.5.28. COROLLARY. *Let B be a compact subgroup of a topological Abelian group A. If the topological group A/B is compact, then A is compact group, too.*

PROOF. The statement results from Proposition 1.5.26, if we put $U = V = A$.

1.5.29. PROPOSITION. *Let B be a connected subgroup of a topological Abelian group A. If the group A/B is connected, then the group A is connected, too.*

PROOF. Assume the contrary hypothesis, i.e. that S is a non-empty open-closed subset of the topological group A and $S \neq A$. Since $S \cap B$ is an open-closed subset of the connected topological group B, then either $S \cap B = \emptyset$ (in that case $B \subseteq A \backslash S$), or $S \cap B = B$ (in that case $B \subseteq S$). Hence, in any case B is contained in some open-closed subset T, different from A. Let's show that $T + B = T$.

Indeed, let $t \in T$, then $t \in (t+B) \cap T$, i.e. $(t+B) \cap T$ is a non-empty open-closed subset of the connected space $t + B$ (see Proposition 1.1.34 and C.13). Hence, $(t+B) \cap T = t+B$. Therefore, $t + B \subseteq T$ for any $t \in T$. Thus,

$$T + B = \bigcup_{t \in T}(t + B) \subseteq T.$$

Since $T + B \supseteq T$, then $T + B = T$.

Due to Propositions 1.5.10 and 1.5.15, $\omega(T)$ is a non-empty open-closed subset of the connected group A/B. Then $\omega(T) = A/B$. Since $T + B = T$, then $T = A$. This is a contradiction to the assumption. This completes the proof.

1.5.30. COROLLARY. *Let C(A) be the connected component of zero of a topological Abelian group A. Then the topological group A/C(A) is totally disconnected.*

PROOF. Let $C(A/C(A))$ be the connected component of zero of $A/C(A)$ and $\omega : A \to A/C(A)$ be the natural topological homomorphism. Applying Proposition 1.5.29 to the topological groups $\omega^{-1}\big(C(A/C(A))\big)$ and $C(A)$, and using Proposition 1.5.24, we obtain that $\omega^{-1}\big(C(A/C(A))\big)$ is a connected subgroup of the group A. Therefore, $\omega^{-1}\big(C(A/C(A))\big) = C(A)$, hence, $C(A/C(A)) = \{0\}$. Using Proposition 1.4.22, we obtain that $A/C(A)$ is totally disconnected group.

1.5.31. PROPOSITION. *Let (R, τ) be a topological ring, I be an ideal of R and $(R/I, \tau')$ be the factor-group of the additive group of the topological ring (R, τ) by the subgroup I. Then the topology τ' is a ring topology on R/I. The topological ring $(R/I, \tau')$ is called a factor-ring of the topological ring (R, τ).*

PROOF. Let $\mathcal{B}_0(R)$ be a basis of neighborhood of zero of the topological ring (R, τ), and $\omega : R \to R/I$ be the natural homomorphism (see B.11). Then, due to Proposition!1.5.6 and Definition 1.5.8, the system $\mathcal{B}_0(R/I) = \{\omega(U) \mid U \in \mathcal{B}_0(R)\}$ is a basis of neighborhoods of zero of the topological group $(R/I, \tau')$. Let's show that $\mathcal{B}_0(R/I)$ satisfies conditions (BN5) and (BN6) (see Proposition 1.2.2).

Indeed, let $V \in \mathcal{B}_0(R/I)$ and $U \in \mathcal{B}_0(R)$ be such that $V = \omega(U)$. Due to condition (BN5) for $\mathcal{B}_0(R)$ there exists a neighborhood $W \in \mathcal{B}_0(R)$ such that $W \cdot W \subseteq U$. Then $\omega(W) \in \mathcal{B}_0(R/I)$, besides, $\omega(W) \cdot \omega(W) = \omega(W \cdot W) \subseteq \omega(U) = V$, i.e. $\mathcal{B}_0(R/I)$ satisfies condition (BN5).

Let $y \in R/I$. Then let's choose any element $x \in \omega^{-1}(y)$. Due to the fulfillment of condition (BN6) for $\mathcal{B}_0(R)$, there exists a neighborhood $T \in \mathcal{B}_0(R)$ such that $x \cdot T \subseteq U$ and $T \cdot x \subseteq U$. Then $\omega(T) \in \mathcal{B}_0(R/I)$, besides, $y \cdot \omega(T) = \omega(x) \cdot \omega(T) = \omega(x \cdot T) \subseteq \omega(U) = V$ and, analogously, $\omega(T) \cdot y \subseteq V$. That means that $\mathcal{B}_0(R/I)$ satisfies condition (BN6). Due to Theorem 1.2.5, τ' is a ring topology.

1.5.32. PROPOSITION. *Let R be a topological ring, (M, τ) be a topological R-module, N be a submodule of the topological module (M, τ) and $(M/N, \tau')$ be the factor-group of the additive group of the topological module (M, τ) by the subgroup N. Then $(M/N, \tau')$ is a topological R-module. We call it a factor-module of the topological module (M, τ).*

PROOF. Let $\mathcal{B}_0(M)$ be a basis of neighborhoods of zero of the topological R-module (M, τ) and $\omega : M \to M/N$ be the natural homomorphism of the modules. Then, due to Proposition 1.5.6 and Definition 1.5.8, the system $\mathcal{B}_0(M/N) = \{\omega(U) | U \in \mathcal{B}_0(M)\}$ is a basis of neighborhoods of zero of the topological Abelian group $(M/N, \tau')$. Let's show that $\mathcal{B}_0(M/N)$ satisfies conditions (BN5'), (BN6') and (BN6'') (see Proposition 1.2.3).

Let $V \in \mathcal{B}_0(M/N)$, $U \in \mathcal{B}_0(M)$ and $V = \omega(U)$. Since $\mathcal{B}_0(M)$ satisfies condition (BN5'), there exist a neighborhood $W \in \mathcal{B}_0(M)$ and a neighborhood S of zero of the ring R such that $S \cdot W \subseteq U$. Then $\omega(W) \in \mathcal{B}_0(M/N)$, besides, $S \cdot \omega(W) = \omega(S \cdot W) \subseteq \omega(U) = V$, i.e. $\mathcal{B}_0(M/N)$ satisfies condition (BN5′).

Now let $V \in \mathcal{B}_0(M)/N)$ and $r \in R$. Let also $U \in \mathcal{B}_0(M)$ and $\omega_N(U) = V$, then, applying condition (BN6′) to $\mathcal{B}_0(M)$, we can choose a neighborhood $W \in \mathcal{B}_0(M)$ such that $r \cdot W \subseteq U$. Then $\omega(W) \in \mathcal{B}_0(M/N)$ and $r \cdot \omega(W) = \omega(r \cdot W) \subseteq \omega(U) = V$, i.e. $\mathcal{B}_0(M/N)$ satisfies condition (BN6′).

Finally, let $V \subseteq \mathcal{B}_0(M/N)$ and $m' \in M/N$. Let $U \in \mathcal{B}_0(M)$ and $\omega(U) = V$, then let's choose an element $m \in \omega^{-1}(m')$. Applying condition (BN6″) to $\mathcal{B}_0(M)$, we can choose a neighborhood S of zero in the ring R such that $S \cdot m \subseteq U$. Then

$$S \cdot m' = S \cdot \omega(m) = \omega(S \cdot m) \subseteq \omega(U) = V,$$

i.e. $\mathcal{B}_0(M/N)$ satisfies condition (BN6″). Due to Theorem 1.2.6, $(M/N, \tau')$ is a topological R-module.

1.5.33. THEOREM. *Let R be a topological ring, (M, τ) be a topological R-module. Let N also be an R-module, $f : N \to M$ be a homomorphism of modules, and τ' be a prototype of the topology τ in N with respect to the mapping f (see C.15). Then (N, τ') is a topological R-module, and $f : (N, \tau') \to (M, \tau)$ is a continuous homomorphism. If, moreover, f is a surjection and $A = \ker f$, then the modules $(N, \tau')/A$ and (M, τ) are topologically isomorphic.*

PROOF. Due to C.15, (N, τ') is a topological space, where $\Delta = \{f^{-1}(U)|U$ is an open set in $M\}$ is the system of all the open subsets, and $f : (N, \tau') \to (M, \tau)$ is a continuous mapping. Let's show that (N, τ') is a topological R-module. To do it let's verify the fulfillment of conditions (SC) and (RMC) (see Remark 1.1.3 and Definition 1.1.24).

Indeed, let $x_1, x_2 \in N$ and V be a neighborhood of the element $x_1 - x_2$ in (N, τ'). Without loss of generality we can assume that V is an open subset in (N, τ'), i.e. $V \in \Delta$. Then $V = f^{-1}(U)$, where U is some open subset in (M, τ), besides, $f(x_1 - x_2) \in f(V) = U$. Hence, U is a neighborhood of the element $f(x_1) - f(x_2) = f(x_1 - x_2)$ in (M, τ). Since (M, τ) is a topological R-module, then there exist open neighborhoods U_1 and U_2 of the elements $f(x_1)$ and $f(x_2)$ in (M, τ) correspondingly, such that $U_1 - U_2 \subseteq U$. Then $V_i = f^{-1}(U_i)$, for $i = 1, 2$, are open subsets in (N, τ') and $x_i \in V_i$. Therefore, V_i is a neighborhood of the element x_i in (N, τ') for $i = 1, 2$. Besides,

$$V_1 - V_2 = f^{-1}(U_1) - f^{-1}(U_2) = f^{-1}(U_1 - U_2) \subseteq f^{-1}(U) = V,$$

i.e. condition (SC) is fulfilled. Hence, (N, τ') is a topological group.

Now let $x \in N$, $r \in R$, and V be a neighborhood of the element $r \cdot x$ in (N, τ'). As above, we can assume that $V = f^{-1}(U)$, where U is a certain open set in (M, τ). Then $r \cdot f(x) = f(r \cdot x) \in U$, i.e. U is a neighborhood of the element $r \cdot f(x)$ in (M, τ). Since (M, τ) is a topological R-module, then there exist a neighborhood W of the element r in R and an open neighborhood U_1 of the element $f(x)$ in (M, τ), such that $W \cdot U_1 \subseteq U$. Then $f^{-1}(U_1)$ is a neighborhood of the element x in (N, τ'). Besides,

$$W \cdot f^{-1}(U_1) = f^{-1}(W \cdot U_1) \subseteq f^{-1}(U) = V,$$

i.e. condition (RMC) is fulfilled, too. Thus, (N, τ') is a topological module.

If f is a surjection, then, according to C.15, the mapping f is open. Then from Theorem 1.5.12 follows that the modules $(N, \tau')/A$ and (M, τ) are topologically isomorphic.

1.5.34. COROLLARY. *Let (A, τ) be a topological Abelian group, B be an Abelian group, $f : B \to A$ be a group homomorphism and τ' be the prototype of the topology τ in B with respect to the mapping f. Then (B, τ') is a topological group and $f : (B, \tau') \to (A, \tau)$ is a continuous homomorphism. If, moreover, f is a surjection and $C = \ker f$, then the groups $(B, \tau')/C$ and (A, τ) are topologically isomorphic.*

PROOF. The statement follows from Theorem 1.5.33, if we consider the ring of integers $\mathbb{Z}$ with the discrete topology as the ring R.

1.5.35. THEOREM. *Let (R, τ) be a topological ring and R' be a ring. Let $f : R' \to R$ be a homomorphism of rings, and τ' be the prototype of the topology τ in R' with respect to the mapping f. Then (R', τ') is a topological ring and $f : (R', \tau') \to (R, \tau)$ is a continuous homomorphism. If, moreover, f is a surjection and $Q = \ker f$, then the rings $(R', \tau')/Q$ and (R, τ) are topologically isomorphic.*

PROOF. It follows from Corollary 1.5.34 that $(R'(+), \tau')$ is an additive topological group. Besides, $f : (R'(+), \tau') \to (R(+), \tau)$ is a continuous group homomorphism.

Let's show that (R', τ') is a topological ring. To do it let's verify the fulfillment of condition (MC) (see Remark 1.1.7). Indeed, let $a', b' \in R'$ and U' be a neighborhood of the element $a' \cdot b'$. Like in Theorem 1.5.33, we can assume that $U' = f^{-1}(U)$, where U is certain open set in (R, τ). Since $f(a') \cdot f(b') = f(a' \cdot b') \in f(U') = U$, then U is a neighborhood of the element $f(a') \cdot f(b')$. Hence, there exist open neighborhoods U_1 and U_2 of the elements $f(a')$ and $f(b')$ correspondingly in (R, τ), such that $U_1 \cdot U_2 \subseteq U$. Then $f^{-1}(U_1)$ and $f^{-1}(U_2)$ are neighborhoods of the elements a' and b' correspondingly in (R', τ'). Besides,

$$f^{-1}(U_1) \cdot f^{-1}(U_2) = f^{-1}(U_1 \cdot U_2) \subseteq f^{-1}(U) = U',$$

i.e. condition (MC) is fulfilled. Hence, (R', τ') is a topological ring.

If f is a surjection and $Q = \ker f$, then, due to C.15, f is open homomorphism. Then from Theorem 1.5.12 follows that the topological rings $(R', \tau')/Q$ and (R, τ) are topologically isomorphic.

1.5.36. PROPOSITION. *Let R be a pseudonormed ring and ξ be a pseudonorm on R. Let also I be an ideal of R, closed in the interval topology τ_ξ on R (see Examples 1.2.9 and 1.1.14), and τ' be the topology on the factor-ring $(R, \tau_\xi)/I$ (see Definition 1.5.8 and Proposition 1.5.6). Then:*

1) the real-valued function ξ' on R/I for which $\xi'(r + I) = \inf\{\xi(a) | a \in (r + I)\}$, is a pseudonorm on R/I;

2) the interval topology $\tau_{\xi'}$ on the pseudonormed ring R/I defined by the pseudonorm ξ', coincides with the topology τ'.

PROOF. Since $\xi(r) \geq 0$ for all $r \in R$, then the function ξ' is correctly defined on R/I, besides, $\xi'(r + I) \geq 0$ for any element $(r + I) \in R/I$.

Since $0 \leq \xi'(I) \leq \xi(0) = 0$, then $\xi'(I) = 0$. Vice versa, let $\xi'(r + I) = 0$. Since $\xi'(r + I) = \inf\{\xi(a) \mid a \in r + I\}$, then for any positive number ϵ there exists an element $a \in r + I$ such that $\xi(a) \leq \epsilon$. Hence, $0 \in [r + I]_{(R, \tau_\xi)}$. Since the ideal I and, consequently, the subset $r + I$, are closed in (R, τ_ξ), then $[r + I]_{(R, \tau_\xi)} = r + I$. Therefore, $0 \in r + I$, then $r + I = I$. Thus, the function ξ' satisfies condition (NR1) (see Definition 1.1.12).

Let $r_1 + I$, $r_2 + I \in R/I$ and ϵ be an arbitrary positive number. From the definition of the function ξ' on R/I follows that there exist elements $a'_\epsilon \in r_1 + I$ and $a''_\epsilon \in r_2 + I$ such that $\xi(a'_\epsilon) < \xi'(r_1 + I) + \frac{\epsilon}{2}$ and $\xi(a''_\epsilon) < \xi'(r_2 + I) + \frac{\epsilon}{2}$. Then

$$\xi(a'_\epsilon - a''_\epsilon) \leq \xi(a'_\epsilon) + \xi(a''_\epsilon) < \xi'(r_1 + I) + \xi'(r_2 + I) + \epsilon.$$

Since $a'_\epsilon - a''_\epsilon \in (r_1 + I) - (r_2 + I)$, then

$$\xi'\big((r_1 + I) - (r_2 + I)\big) \leq \xi(a'_\epsilon - a''_\epsilon).$$

Therefore,

$$\xi'\big((r_1 + I) - (r_2 + I)\big) < \xi'(r_1 + I) + \xi'(r_2 + I) + \epsilon.$$

Thus,

$$\xi'((r_1 + I) - (r_2 + I)) \leq \xi'(r_1 + I) + \xi'(r_2 + I),$$

i.e. the function ξ' satisfies condition (NR2) (see Definition 1.1.12).

Let $r_1 + I$, $r_2 + I \in R/I$ and ϵ be a positive number. Let's choose a number $\delta > 0$ such that $\delta \cdot (\xi'(r_1 + I) + \xi'(r_2 + I)) + \delta^2 < \epsilon$. Due to the definition of the function ξ', there exist elements $a'_\delta \in (r_1 + I)$ and $a''_\delta \in (r_2 + I)$ such that $\xi(a'_\delta) < \xi'(r_1 + I) + \delta$ and $\xi(a''_\delta) < \xi'(r_2 + I) + \delta$. Then

$$\begin{aligned} \xi(a'_\delta \cdot a''_\delta) \leq \xi(a'_\delta) \cdot \xi(a''_\delta) &< \big(\xi'(r_1 + I) + \delta\big) \cdot \big(\xi'(r_2 + I) + \delta\big) \\ &= \xi'(r_1 + I) \cdot \xi'(r_2 + I) + \delta^2 + \delta \cdot \big(\xi'(r_1 + I) + \xi'(r_2 + I)\big) \\ &< \xi'(r_1 + I) \cdot \xi'(r_2 + I) + \epsilon. \end{aligned}$$

Since $a'_\delta \cdot a''_\delta \in (r_1 + I) \cdot (r_2 + I)$, then

$$\xi'\big((r_1 + I) \cdot (r_2 + I)\big) \leq \xi(a'_\delta \cdot a''_\delta) < \xi'(r_1 + I) \cdot \xi'(r_2 + I) + \epsilon.$$

Hence,

$$\xi'\big((r_1 + I) \cdot (r_2 + I)\big) \leq \xi'(r_1 + I) \cdot \xi'(r_2 + I),$$

i.e. the function ξ' satisfies condition (NR3) (see Definition 1.1.12).

Thus, if ξ is a pseudonorm on the ring R, then ξ' is a pseudonorm on the ring R/I, i.e. statement 1) is proved.

Let $\mathcal{B}_0(R) = \{V_n | n = 1, 2, \dots\}$ and $\mathcal{B}_0'(R/I) = \{V_n' | n = 1, 2, \dots\}$, where $V_n = \{r | r \in R, \xi(r) < \frac{1}{2^n}\}$ and $V_n' = \{r + I | r + I \in R/I, \xi'(r + I) < \frac{1}{2^n}\}$, for $n = 1, 2, \dots$. Then, due to Example 1.2.9, $\mathcal{B}_0(R)$ is a basis of neighborhoods of zero of the topological ring (R, τ_ξ), and $\mathcal{B}_0'(R/I)$ is a basis of neighborhoods of zero of the topological ring $(R/I, \tau_{\xi'})$.

Let $\omega : R \to R/I$ be the natural homomorphism. Let's show that $\omega(V_n) = V_n'$ for $n = 1, 2, \dots$. Indeed, if $r \in V_n$, then $\xi(r) < \frac{1}{2^n}$, then $\xi'(r + I) \leq \xi(r) < \frac{1}{2^n}$, i.e. $\omega(r) = r + I \in V_n'$. Hence, $\omega(V_n) \subseteq V_n'$ for $n = 1, 2, \dots$. Vice versa, if $r + I \in V_n'$, then $\xi'(r + I) < \frac{1}{2^n}$. Therefore, there exists an element $a \in r + I$ such that $\xi(a) < \frac{1}{2^n}$, i.e. $a \in V_n$. It is clear that $\omega(a) = r + I$. Thus, $\omega(V_n) = V_n'$ for $n = 1, 2, \dots$. This proves that the topologies τ' and $\tau_{\xi'}$ coincide on the ring R/I, i.e. statement 2) is proved.

1.5.37. **REMARK.** *Let A and B be topological Abelian groups, and let $C\text{--Hom}(A, B) = \{f : A \to B | f$ is continuous homomorphism of topological groups\}. Then with the operation of addition of homomorphisms $C\text{--Hom}(A, B)$ is a subgroup of the Abelian group $\mathrm{Hom}(A, B)$ (see B.8).*

PROOF. Indeed, let $f, g \in C - \mathrm{Hom}(A, B)$ and U be a neighborhood of zero of the topological group B. Then there exists a neighborhood V of zero in B such that $V - V \subseteq U$. Since the homomorphisms $f : A \to B$ and $g : A \to B$ are continuous, then, due to Proposition 1.5.5, $f^{-1}(V)$ and $g^{-1}(V)$ are neighborhoods of zero in A. Since

$$(f - g)(f^{-1}(V) \cap g^{-1}(V)) \subseteq f(f^{-1}(V) \cap g^{-1}(V)) - g(f^{-1}(V) \cap g^{-1}(V)) \subseteq$$
$$f(f^{-1}(V)) - g(g^{-1}(V)) = V - V \subseteq U,$$

then

$$f^{-1}(V) \cap g^{-1}(V) \subseteq (f - g)^{-1}(U).$$

Since $f^{-1}(V) \cap g^{-1}(V)$ is a neighborhood of zero in the group A, then $(f - g)^{-1}(U)$ is a neighborhood of zero in A, too, i.e. the homomorphism $(f - g) : A \to B$ is continuous.

Hence, $f - g \in C\text{--Hom}(A, B)$. Thus, it was shown that $C\text{--Hom}(A, B)$ is a subgroup of the group $\text{Hom}(A, B)$.

1.5.38. PROPOSITION. *Let A and B be topological Abelian groups, $\mathcal{B}_0(B)$ be a basis of symmetrical neighborhoods of zero of B (see Proposition 1.2.28), and $K(A)$ be the system of all compact subsets of the group A. Then the system*

$$\mathcal{B}_0\big(C\text{--Hom}(A, B)\big) = \big\{W(S, U)\big|\, S \in K(A),\, U \in \mathcal{B}_0(B)\big\}$$

of subsets of the group $C\text{--Hom}(A, B)$, of the type of

$$W(S, U) = \big\{f \big|\, f \in C\text{--Hom}(A, B),\, f(S) \subseteq U\big\},$$

satisfies conditions (BN1) - (BN4) (see Proposition 1.2.1), and, hence (see Proposition 1.2.4), defines on the group $C\text{--Hom}(A, B)$ a topology, in which $C\text{--Hom}(A, B)$ is a topological group and $\mathcal{B}_0(C\text{--Hom}(A, B))$ is a basis of neighborhoods of zero.

Moreover, if B is Hausdorff, then $C\text{--Hom}(A, B)$ is Hausdorff, too.

PROOF. It is clear that the condition (BN1) is fulfilled.

Let $S_1, S_2 \in K(A)$ and $U_1, U_2 \in \mathcal{B}_0(B)$. Then $S_1 \cup S_2 \in K(A)$ and there exists a neighborhood $U_3 \in \mathcal{B}_0(B)$ such that $U_3 \subseteq U_1 \bigcap U_2$. Then $f(S_1 \cup S_2) \subseteq U_3$ for each $f \in W(S_1 \cup S_2, U_3)$. Therefore, $f(S_1) \subseteq f(S_1 \cup S_2) \subseteq U_3 \subseteq U_1$. Analogously, $f(S_2) \subseteq U_2$. Thus, $f \in W(S_1, U_1) \bigcap W(S_2, U_2)$, i.e. condition (BN2) is fulfilled.

Let $S \in K(A)$ and $U \in \mathcal{B}_0(B)$. Then there exists $V \in \mathcal{B}_0(B)$ such that $V + V \subseteq U$. Further,

$$(f + g)(S) \subseteq f(S) + g(S) \subseteq V + V \subseteq U,$$

i.e. $f + g \in W(S, U)$ for any $f, g \in W(S, V)$. Hence, $W(S, V) + W(S, V) \subseteq W(S, U)$, i.e. condition (BN3) is fulfilled.

Finally, if $f \in W(S, U)$, then $(-f)(S) = -\big(f(S)\big) \subseteq -U = U$, i.e. $(-f) \in W(S, U)$. Hence, $-W(S, U) \subseteq W(S, U)$ for any $S \in K(A)$ and $U \in \mathcal{B}_0(A)$, i.e. condition (BN4) is fulfilled.

Thus, $C\text{--Hom}(A, B)$ is a topological group with the basis $\mathcal{B}_0(C\text{--Hom}(A, B))$ of neighborhoods of zero .

At last, let B be Hausdorff and

$$f \in \bigcap_{S \in K(A), U \in \mathcal{B}_0(B)} W(S, U).$$

Since $\{a\} \in K(A)$ for any $a \in A$, then $f \in \bigcap_{U \in \mathcal{B}_0(B)} W(\{a\}, U)$, i.e. $f(a) \in U$ for any $U \in \mathcal{B}_0(B)$. Since the group B is Hausdorff, then $\bigcap_{U \in \mathcal{B}_0(B)} U = \{0\}$, hence, $f = 0$. Thus, condition (BN1′) (see Corollary 1.3.6) is fulfilled. Hence, C–$\mathrm{Hom}(A, B)$ is a Hausdorff topological group.

1.5.39. EXAMPLE. Let $\mathbb{R}$ be the additive group of real numbers with the interval topology and $\mathbb{Z}$ be the subgroup of all integers. Then $\mathbb{R}$ is a Hausdorff group, and $\mathbb{Z}$ is its discrete subgroup. Due to Proposition 1.4.36, $\mathbb{Z}$ is a closed subgroup of the group $\mathbb{R}$. Then the factor-group $\mathbb{T} = \mathbb{R}/\mathbb{Z}$ is Hausdorff (see Corollary 1.5.18). Moreover, $\mathbb{T} = \omega([0; 1])$, where $\omega : \mathbb{R} \to \mathbb{T}$ is the natural homomorphism. Since $[0; 1] \subseteq \mathbb{R}$ is compact and the homomorphism ω is continuous (see Proposition 1.5.10), then $\mathbb{T}$ is a compact group.

Now let A be a topological Abelian group. The Hausdorff topological group $Ch(A) = C$–$\mathrm{Hom}(A, \mathbb{T})$ (see Proposition 1.5.38) is called the character group of the group A, and its elements, i.e. continuous homomorphisms from A to $\mathbb{T} = \mathbb{R}/\mathbb{Z}$, are called the characters of the group A.

In conclusion of this paragraph, we quote the classical theorem on the existence of a sufficiently large number of characters. The proof of it can be found in, for example, [115, 326, 333].

1.5.40. THEOREM. Let A be a locally compact Abelian group. Then for any non-zero element $a \in A$ there exists a character $f : A \to \mathbb{T}$ of the group A, such that $f(a) \neq 0$.

§ 1.6. Bounded Subsets

1.6.1. DEFINITION. Let R be a topological ring, M be a topological R-module. A subset $S \subseteq M$ is called bounded if for any neighborhood U of zero in M there exists a neighborhood V of zero in R such that $V \cdot S \subseteq U$. A topological R-module M is called bounded if M is a bounded subset of the module M.

1.6.2. DEFINITION. A subset S of the topological ring R is called bounded from left (right) if S is a bounded subset of the topological left (right) R-module $R(+)$, i.e. for any neighborhood U of zero in R there exists a neighborhood V of zero in R such that $V \cdot S \subseteq U$ (correspondingly, $S \cdot V \subseteq U$). A subset S of the topological ring R, bounded from left and from right, is called bounded. A topological ring R is called bounded from left (bounded from right, bounded) if R is a bounded from left (correspondingly, bounded from right, bounded) subset of the ring R.

1.6.3. PROPOSITION. *Let K be a skew field endowed with a non-discrete ring topology τ. A subset $S \subseteq K$ is bounded from left (right) if and only if for any neighborhood V of zero in the topological ring (K, τ) there exists a non-zero element $a \in K$ such that $a \cdot S \subseteq U$ (correspondingly, $S \cdot a \subseteq U$).*

PROOF. Let U be a neighborhood of zero in K. Suppose that a subset S is bounded from left. Then there exists a neighborhood V of zero in K such that $V \cdot S \subseteq U$. Since the topology τ is non-discrete, then there exists a non-zero element $a \in V$. Then $a \cdot S \subseteq U$.

Vice versa, let V be a neighborhood of zero in K such that $V \cdot V \subseteq U$ and k be such non-zero element from K that $k \cdot S \subseteq V$. We can choose a neighborhood W of zero in K such that $W \cdot k^{-1} \subseteq V$. Then $W \cdot S = W \cdot k^{-1} \cdot k \cdot S \subseteq V \cdot V \subseteq U$. Hence, we have proved that the subset S is bounded from left.

Analogously the statement is proved for the subset S being bounded from right.

1.6.4. PROPOSITION. *Any compact subset of a topological R-module M is bounded. In particular, any finite subset in M is bounded.*

PROOF. Let P be a compact subset and U be a neighborhood of zero of the module M. Then, due to the condition (RMC) (see Definition 1.1.24), for any element $m \in P$ there exist a neighborhood V_m of zero in R and a neighborhood W_m of m in M such that

$V_m \cdot W_m \subseteq U$. Due to Corollary 1.2.30, we can assume that W_m is an open neighborhood of the element m. Since $\{W_m | m \in P\}$ is an open cover of the compact subset P (see C.12) of the topological R-module M, then there exist elements $m_1, m_2, \dots, m_k \in P$ such that $P \subseteq \bigcup_{i=1}^{k} W_{m_i}$. Then $V = \bigcap_{i=1}^{k} V_{m_i}$ is a neighborhood of zero in R and

$$V \cdot P \subseteq V \cdot \left(\bigcup_{i=1}^{k} W_{m_i} \right) \subseteq \bigcup_{i=1}^{k} (V_{m_i} \cdot W_{m_i}) \subseteq U,$$

i.e. P is a bounded subset of the R-module M.

1.6.5. COROLLARY. *Any compact (in particular, any finite) subset Q of a topological ring R is bounded.*

PROOF. Considering R as left and right topological R-modules, we get that Q is bounded from left and from right in R.

1.6.6. EXAMPLE. Any topological R-module M endowed with the anti-discrete topology is bounded. In particular, any anti-discrete ring is bounded.

Indeed, if U is a neighborhood of zero in M, then $U = M$. Since R is a neighborhood of zero in R, then $R \cdot M \subseteq M = U$.

1.6.7. EXAMPLE. Let R be a discrete ring, then any topological R-module M is bounded. In particular, any discrete ring is bounded.

Indeed, $\{0\}$ is a neighborhood of zero in R. Therefore, $\{0\} \cdot M = \{0\} \subseteq U$ for any neighborhood U of zero in M.

1.6.8. EXAMPLE. If a topological ring R has a basis of neighborhoods of zero consisting of left (right, two-sided) ideals of the multiplicative semigroup of the ring R (see B.5), then R is a bounded from right (bounded from left, bounded) ring. In particular, any topological ring with zero multiplication is bounded.

1.6.9. PROPOSITION. *Let τ_ξ be a topology on a ring R, defined by a pseudonorm ξ and $Q \subseteq R$. If $\xi(Q) < k$ for some $0 \le k \in \mathbb{R}$, then Q is a bounded subset.*

PROOF. Due to Example 1.2.9,

$$\left\{ V_n | V_n = \left\{ a \in R \mid \xi(a) < \frac{1}{2^n} \right\}, n \in \mathbb{N} \right\}$$

is a basis of neighborhoods of zero in (R, τ_ξ). There exist $m \in \mathbb{N}$ such that $k < 2^m$. Then

$$\xi(v \cdot q) \leq \xi(v) \cdot \xi(q) \leq \frac{1}{2^{n+m}} \cdot 2^m = \frac{1}{2^n}$$

for any $n \in \mathbb{N}$, $v \in V_{n+m}$ and $q \in Q$, i.e. $V_{n+m} \cdot Q \subseteq V_n$. Analogously $Q \cdot V_{n+m} \subseteq V_n$. Therefore, Q is a bounded subset.

1.6.10. PROPOSITION. *Let an interval topology τ_ξ, defined by a pseudonormon ξ on a skew field R, be non-discrete. Then for any subset Q of the topological ring (R, τ_ξ) the following conditions are equivalent:*

1) the subset Q is bounded from left in (R, τ_ξ);

2) there exists a natural number k such that $\xi(q) \leq 2^k$ for any $q \in Q$;

3) the subset Q is bounded from right in (R, τ_ξ);

4) the subset Q is bounded in (R, τ_ξ).

PROOF. The system

$$\mathcal{B}_0(R) = \left\{ V_n \middle| V_n = \left\{ a \in R \middle| \xi(a) < \frac{1}{2^n} \right\}, n \in \mathbb{N} \right\}$$

of the subsets of the skew field R is a basis of neighborhoods of zero of the topological ring (R, τ_ξ) (see Example 1.2.9).

Let the subset Q be bounded from left and let n be a natural number such that $V_n \cdot Q \subseteq V_1$. Since the topology τ_ξ is non-discrete, then there exists a non-zero element $v \in V_n$. Then we can choose a natural number k such that $\xi(v^{-1}) \leq 2^{k+1}$. Since $v \cdot q \in V_1$ for any $q \in Q$, then $\xi(v \cdot q) < \frac{1}{2}$. Therefore,

$$\xi(q) = \xi(v^{-1} \cdot v \cdot q) \leq \xi(v^{-1}) \cdot \xi(v \cdot q) \leq 2^{k+1} \cdot \frac{1}{2} = 2^k,$$

i.e. 1)$\Longrightarrow$ 2).

Due to Proposition 1.6.9, 2) $\Longrightarrow$ 1), i.e. the conditions 1) and 2) are equivalent. Analogously is proved that the conditions 2) and 3) are equivalent, which implies the equivalence of the conditions 1) and 4).

1.6.11. EXAMPLE. Let R be a ring of linear transformations of a countable-dimensional linear space over the field $\mathbb{R}$, i.e. R is isomorphic to the ring of all finite-row matrices over the ring $\mathbb{R}$ (any row of these matrices has only a finite number of non-zero elements). For any $n \in \mathbb{N}$, we define as I_n the set of all such matrices $(a_{i,j}) \in R$ such that $a_{i,j} = 0$ for $1 \leq i \leq n$ and $j \in \mathbb{N}$.

It is clear, that I_n is a right ideal of the ring R for any $n \in \mathbb{N}$. Since $I_n \cap I_m = I_k$, where $k = \max\{n, m\}$, then the system $\mathcal{B}_0(R) = \{I_n | n \in \mathbb{N}\}$ satisfies the conditions (BN1) - (BN5) (see Propositions 1.2.1 and 1.2.2). Let's prove that the condition (BN6) is fulfilled, too. Since I_n is a right ideal, then it is enough to show that for any $n \in \mathbb{N}$ and any matrix $(a_{i,j}) \in R$ there exists $m \in \mathbb{N}$ such that $(a_{i,j}) \cdot I_m \subseteq I_n$.

Since $(a_{i,j})$ is a finite-row matrix, then we can put $m = \max\{j \mid a_{i,j} \neq 0 \text{ for some } i \leq n\}$. Then $a_{i,j} = 0$ for $1 \leq i \leq n$ and $j = m + 1, m + 2, \ldots$. It is clear that $(a_{i,j}) \cdot I_m \subseteq I_n$. Hence, $\mathcal{B}_0(R)$ satisfies the condition (BN6), and therefore, due to Proposition 1.2.5, R is a topological ring with $\mathcal{B}_0(R)$ as the basis of neighborhoods of zero.

Due to the Proposition 1.6.9, R is a topological ring, bounded from left.

Now we'll show that R is not bounded from right. To do so it is enough to verify for any $n \in \mathbb{N}$ the existence of a matrix $(a_{ij}) \in R$ such that $(a_{ij}) \cdot I_n \not\subseteq I_1$. Denote over $E_{i,j}$ a matrix having 1 at the (i, j) place, and zero at all other places. Then $E_{1,n+1} \cdot E_{n+1,1} = E_{1,1} \notin I_1$, but $E_{n+1,1} \in I_n$ for any $n \in \mathbb{N}$. Thus, R is a topological ring, bounded from left but not bounded from right.

1.6.12. PROPOSITION. *Let S be a subring of a topological ring R and A be a bounded subset of a topological R-module M. Let N be some S-submodule in M containing A, then A is a bounded subset of the topological S-module N.*

PROOF. Let U be a neighborhood of zero in N. Then $U = V \cap N$ for a certain neighborhood V of zero in M. Since A is a bounded subset in M, then there exists a neighborhood W of zero in R such that $W \cdot A \subseteq V$. Then $W \cap S$ is a neighborhood of zero in S and

$$(W \cap S) \cdot A \subseteq W \cdot A \cap S \cdot A \subseteq V \cap N = U.$$

1.6.13. COROLLARY. *Let A be a bounded from left (bounded from right, bounded) subset of a topological ring R and Q be a subring of R, which contains A. Then A is a bounded from left (bounded from right, bounded) subset of the topological ring Q.*

PROOF. The statement follows from Proposition 1.6.12, if we consider R as a left or right topological R-module and Q as a left or right its topological Q-submodule.

1.6.14. REMARK. Let R be a topological ring, S be a bounded subset of the topological R-module M. If $S_1 \subseteq S$, then S_1 is a bounded subset of the topological R-module M.

1.6.15. REMARK. Let S be a bounded from left (bounded from right, bounded) subset of a topological ring R and $S_1 \subseteq S$. Then S_1 is a bounded from left (bounded from right, bounded) subset of the topological ring R.

1.6.16. PROPOSITION. *The closure of a bounded subset P of a topological R-module M is bounded.*

PROOF. Let U be a neighborhood of zero in M and V be a closed neighborhood of zero in M such that $V \subseteq U$. Since P is a bounded subset of R-module M, then there exists a neighborhood W of zero in R such that $W \cdot P \subseteq V$. Then

$$W \cdot [P]_M \subseteq [W]_R \cdot [P]_N \subseteq [W \cdot P]_M \subseteq [V]_M = V \subseteq U,$$

i.e. $[P]_M$ is a bounded subset.

1.6.17. COROLLARY. The closure of a bounded from left (bounded from right, bounded) subset of a topological ring is a subset bounded from left (bounded from right, bounded).

1.6.18. PROPOSITION. *A topological skew field K is bounded from left (from right) if and only if it is either anti-discrete or discrete.*

PROOF. Let the topology on K be not anti-discrete and K be bounded from left. Due to Remark 1.4.13, K is a Hausdorff skew field. Let U be a neighborhood of zero in K such that it does not contain the unitary element. Then there exists a neighborhood of zero V in K such that $V \cdot K \subseteq U$.

Let $V \neq \{0\}$ and $0 \neq a \in V$, then $1 = a \cdot a^{-1} \in V \cdot K \subseteq U$. That contradicts the choice of the neighborhood U. Thus, $V = \{0\}$, i.e. K is a discrete skew field.

1.6.19. PROPOSITION. *The sum and the union of a finite number of bounded subsets of a topological R-module M are bounded, too. In particular, if any of the subsets*

$S_1, \ldots, S_n$ of a topological ring R is bounded from left (bounded from right, bounded), then the subsets $\sum_{i=1}^n S_i$ and $\bigcup_{i=1}^n S_i$ are bounded from left (bounded from right, bounded), too.

PROOF. It is enough to analyze the case of two bounded subsets. Let S_1 and S_2 be bounded subsets of the R-module M. Let U be a neighborhood of zero in M, then we can choose a neighborhood U_1 of zero in M and neighborhoods V, W_1, W_2 of zero in R such that:

1) $U_1 + U_1 \subseteq U$;

2) $W_1 \cdot S_1 \subseteq U_1$ and $W_2 \cdot S_2 \subseteq U_1$;

3) $V \subseteq W_1 \bigcap W_2$.

Then $V \cdot (S_1 + S_2) \subseteq V \cdot S_1 + V \cdot S_2 \subseteq W_1 \cdot S_1 + W_2 \cdot S_2 \subseteq U_1 + U_1 \subseteq U$ and $V \cdot (S_1 \bigcup S_2) = (V \cdot S_1) \bigcup (V \cdot S_2) \subseteq (W_1 \cdot S_1) \bigcup (W_2 \cdot S_2) \subseteq U_1 \subseteq U$, i.e. $S_1 + S_2$ and $S_1 \bigcup S_2$ are bounded subsets of the R-module M.

1.6.20. THEOREM. *Let $\mathcal{B}_0(R)$ be a basis of neighborhoods of zero of a topological ring R, then R is a union of not more then $2^{|\mathcal{B}_0(R)|}$ of its bounded subsets.*

PROOF. Let Φ be the set of all mappings of the set $\mathcal{B}_0(R)$ to itself. Then $\mid \Phi \mid = 2^{|\mathcal{B}_0(R)|}$ (see A.7). For each mapping $\varphi \in \Phi$ consider subset $S_\varphi = \{x \in R \mid \varphi(V) \cdot x \subseteq V$ and $x \cdot \varphi(V) \subseteq V$ for any neighborhood $V \in \mathcal{B}_0(R)\}$ in the ring R. Since $S_\varphi \cdot \varphi(V) \subseteq V$ and $\varphi(V) \cdot S_\varphi \subseteq V$ for any neighborhood $V \in \mathcal{B}_0(R)$, then S_φ is a bounded subset in R.

Let's verify that $R = \bigcup_{\varphi \in \Phi} S_\varphi$. Let $x \in R$. For any neighborhood $V \in \mathcal{B}_0(R)$ we can choose a neighborhood $V' \in \mathcal{B}_0(R)$ such that $x \cdot V' \subseteq V$ and $V' \cdot x \subseteq V$. Putting $\varphi_x(V) = V'$, we define the mapping $\varphi_x : \mathcal{B}_0(R) \to \mathcal{B}_0(R)$, therefore, $\varphi_x \in \Phi$. Since $x \in S_{\varphi_x}$, then $R = \bigcup_{\varphi \in \Phi} S_\varphi$.

1.6.21. PROPOSITION. *Let Q be a bounded from left subset of a topological ring R and S be a bounded subset of a topological R-module M, then $Q \cdot S$ is a bounded subset of the R-module M.*

PROOF. Let U be a neighborhood of zero in M and V be a neighborhood of zero in R such that $V \cdot S \subseteq U$. We can choose a neighborhood W of zero in R such that $W \cdot Q \subseteq V$. Then

$$W \cdot (Q \cdot S) = (W \cdot Q) \cdot S \subseteq V \cdot S \subseteq U,$$

i.e. $Q \cdot S$ is a bounded subset of the topological R-module M.

1.6.22. COROLLARY. Let each of the subsets Q_i for $i = 1, 2, \ldots, n$ of a topological ring R be bounded from left (bounded from right, bounded), then the subset $Q_1 \cdot Q_2 \cdot \ldots \cdot Q_n$ is bounded from left (bounded from right, bounded), too.

1.6.23. EXAMPLE. Consider the field of real numbers $\mathbb{R}$ with the interval topology and a countable family $\{S_i | S_i = [-i; i], i = 1, 2, \ldots\}$ of its subsets. Due to Proposition 1.6.9, each S_i is a bounded subset in $\mathbb{R}$. It is clear that

$$\mathbb{R} = \bigcup_{i=1}^{\infty} S_i = \sum_{i=1}^{\infty} S_i = \prod_{i=1}^{\infty} S_i,$$

but, due to Proposition 1.6.18, $\mathbb{R}$ is not a bounded ring. Therefore, $\mathbb{R}$ is not a bounded $\mathbb{R}$-module. Thus, we get that the condition on the finiteness of the number of the subsets in statements 1.6.19, 1.6.20 and 1.6.22 is important.

1.6.24. REMARK. Let (R, τ) be a topological ring, S be a bounded subset of a topological (R, τ)-module M. If τ' is a ring topology on R such that $\tau \leq \tau'$, then S is a bounded subset of the topological (R, τ')-module M (see Remark 1.1.29).

Indeed, let U be a neighborhood of zero in M, and V be a neighborhood of zero in (R, τ) such that $V \cdot S \subseteq U$. Since $\tau \leq \tau'$, then V is a neighborhood of zero in the topological ring (R, τ'). Therefore, S is a bounded subset of the topological (R, τ')-module M.

1.6.25. PROPOSITION. *Let R be a topological ring, M be an R-module and $S \subseteq M$. Let also Σ be a family of topologies on M such that M is a topological R-module in each one of them and S be a bounded subset. Then S is a bounded subset of the topological R-module (M, τ'), where $\tau' = \sup\{\tau | \tau \in \Sigma\}$ (see Theorem 1.2.22).*

PROOF. Let $\mathcal{B}_0(\tau)$ be a basis of neighborhoods of zero in the topological R-module (M, τ) for the topology $\tau \in \Sigma$ and U be a neighborhood of zero in the topological R-module (M, τ'). Then $U \supseteq \bigcap_{i=1}^{n} U_i$, where $U_i \in \mathcal{B}_0(\tau_i)$, $\tau_i \in \Sigma$ for $i = 1, 2, \ldots, n$. Let's choose a neighborhood V_i of zero in the topological ring R such that $V_i \cdot S \subseteq U_i$, for $i = 1, 2, \ldots, n$. Then $V - \bigcap_{i=1}^{n} V_i$ is a neighborhood of zero in the topological ring R, where $V \cdot S \subseteq V_i \cdot S \subseteq U_i$ for any $i = 1, 2, \ldots, n$. Therefore, $V \cdot S \subseteq \bigcap_{i=1}^{n} U_i \subseteq U$, i.e. S is a bounded subset of the topological R-module (M, τ').

1.6.26. PROPOSITION. *Let Σ be a family of the ring topologies on a ring R, such that a subset S of R is bounded from left (bounded from right, bounded) in the topological ring (R, τ), for each $\tau \in \Sigma$. Then S is a bounded from left (bounded from right, bounded) subset of the topological ring (R, τ'), where $\tau' = \sup\{\tau | \tau \in \Sigma\}$ (see Theorem 1.2.21).*

PROOF. Let the subset S be bounded from left in (R, τ) for any $\tau \in \Sigma$. Since $(R(+), \tau)$ is a topological (R, τ)-module, then S is a bounded subset of the topological (R, τ)-module $(R(+), \tau)$ for any $\tau \in \Sigma$. Due to Remark 1.6.24, S is a bounded subset of the topological (R, τ')-module $(R(+), \tau)$ for any topology $\tau \in \Sigma$. Due to Proposition 1.6.25, S is a bounded subset of the topological (R, τ')-module $(R(+), \tau')$, i.e. S is a bounded from left subset of the topological ring (R, τ').

The proof of the right or two-sided boundedness of the subset S is made analogously.

1.6.27. PROPOSITION. *Let R be a topological ring, M and M' be topological R-modules. Let φ be a continuous homomorphism from the module M to the module M' and a subset N is bounded in M. Then the subset $\varphi(N)$ is bounded in M'.*

PROOF. Let U' be a neighborhood of zero in the topological R-module M'. Due to Proposition 1.5.5, $\varphi^{-1}(U')$ is a neighborhood of zero in the topological R-module M. Since N is a bounded subset, then there exists a neighborhood V of zero in the ring R such that $V \cdot N \subseteq \varphi^{-1}(U')$. Then

$$V \cdot \varphi(N) = \varphi(V \cdot N) \subseteq \varphi(\varphi^{-1}(U')) = U',$$

i.e. $\varphi(N)$ is a bounded subset in M'.

1.6.28. PROPOSITION. *Let R and R' be topological rings, $\varphi : R \to R'$ be a topological homomorphism from the ring R to the ring R'. Let a subset S be bounded from left (bounded from right, bounded) in the ring R, then the subset $\varphi(S)$ is bounded from left (bounded from right, bounded) in the ring R'.*

PROOF. Let U' be a neighborhood of zero in the topological ring R'. Then, due to Proposition 1.5.5, $\varphi^{-1}(U')$ is a neighborhood of zero in the topological ring R. Since the subset S of the ring R is bounded from left, then there exists a neighborhood V of zero in R such that $V \cdot S \subseteq \varphi^{-1}(U')$. Due to Proposition 1.5.5, $\varphi(V)$ is a neighborhood of zero in the ring R'. Then

$$\varphi(V) \cdot \varphi(S) = \varphi(V \cdot S) \subseteq \varphi\big(\varphi^{-1}(U')\big) = U',$$

i.e. the subset $\varphi(S)$ is bounded from left in R'.

When the subset S of the ring R is bounded from right or bounded, the proof is analogous.

But, in general, the property to be bounded from left (bounded from right, bounded) for subsets of topological rings is not kept by continuous isomorphisms (unlike the topological homomorphisms of rings and continuous homomorphisms of modules).

Actually, let τ_1 be the discrete and τ_2 be the interval topology on the field of real numbers $\mathbb{R}$. Then $(\mathbb{R}, \tau_1)$ is a bounded ring, and $(\mathbb{R}, \tau_2)$ is not bounded (see Remark 1.6.18). It is clear that the identical mapping from the $\mathbb{R}$ to $\mathbb{R}$ is a continuous isomorphism of the topological field $(\mathbb{R}, \tau_1)$ on to the topological field $(\mathbb{R}, \tau_2)$.

1.6.29. THEOREM. *Let R be a topological ring, (M, τ) be a topological R-module, N be a certain R-module, $f : N \to M$ be a module homomorphism and τ' be a prototype in N of the topology τ with respect to the mapping f (see Proposition 1.5.33). Let S be a bounded subset in (M, τ), then $f^{-1}(S)$ is a bounded subset in the topological R-module (N, τ').*

PROOF. Let V be a certain neighborhood of zero in (N, τ'). Like in Proposition 1.5.33, we can consider that $V = f^{-1}(U)$, where U is an open subset in (M, τ). Since S is a bounded subset and U is a neighborhood of zero in (M, τ), then there exists a neighborhood of zero W in R such that $W \cdot S \subseteq U$. Then

$$W \cdot f^{-1}(S) = f^{-1}(W \cdot S) \subseteq f^{-1}(U) = V,$$

i.e. $f^{-1}(S)$ is a bounded subset in the topological module (N, τ').

1.6.30. THEOREM. *Let (R, τ) be a topological ring, R' be a ring, $f : R' \to R$ be a ring homomorphism, τ' be the prototype in R' of the topology τ with respect to the mapping f (see Proposition 1.5.38). Let a subset S be bounded from left (bounded from right, bounded) in (R, τ), then the subset $f^{-1}(S)$ is bounded from left (bounded from right, bounded) in (R', τ').*

PROOF. Let U' be a neighborhood of zero in (R', τ'). Like in Theorem 1.5.35, we can consider that $U' = f^{-1}(U)$, where U is an open subset in (R, τ). It is clear that $0 \in U$. Since the subset S is bounded from left in (R, τ) and U is a neighborhood of zero in (R, τ), then there exists an open neighborhood V of zero in (R, τ) such that $V \cdot S \subseteq U$. Then $f^{-1}(U)$ is a neighborhood of zero in (R', τ'), where

$$f^{-1}(U) \cdot f^{-1}(S) = f^{-1}(V \cdot S) \subseteq f^{-1}(U) \subseteq U',$$

i.e. the subset $f^{-1}(S)$ is bounded from left in (R', τ').

When the subset S is bounded from right or bounded, the proof is analogous.

1.6.31. THEOREM. *A topological ring R is bounded from left (from right) if and only if R has the basis of neighborhoods of zero consisting of right (left) ideals of the multiplicative semigroup of the ring R.*

PROOF. Let R be a ring bounded from left. Let's consider a system

$$\mathcal{B}_0(R) = \{V \cup (V \cdot R) | V \text{ is a neighborhood of zero in } R\}.$$

It is clear that $\mathcal{B}_0(R)$ consists of the right ideals of the multiplicative semigroup of the ring R which are neighborhoods of zero in the ring R.

Now we verify that $\mathcal{B}_0(R)$ is a basis of neighborhoods of zero in R. Indeed, for any neighborhood U of zero in R there exists a neighborhood W of zero in R such that $W \cdot R \subseteq U$ (since the ring R is bounded from left). Putting $V = W \cap U$, we get that $V \bigcup (V \cdot R) \in \mathcal{B}_0(R)$ and

$$V \cup (V \cdot R) \subseteq U \cup (W \cdot R) \subseteq U \cup U = U.$$

The boundedness from right is proved analogously.

The inverse statement is evident (see Example 1.6.8).

1.6.32. PROPOSITION. *A topological ring R possessing a basis of neighborhoods of zero consisting of subgroups of its additive group is bounded from left (from right) if and only if R has a basis of neighborhoods of zero consisting of open right (left) ideals of the ring R.*

PROOF. Let $\mathcal{B}_0(R)$ be a basis of neighborhoods of zero consisting of subgroups of the additive group of the ring R.

If R is bounded from left, then, due to Theorem 1.6.31, for any neighborhood $U \in \mathcal{B}_0(R)$ there exists a neighborhood V_U of zero in R such that $V_U \subseteq U$ and V_U is a right ideal of the multiplicative semigroup of the ring R. Let I_U be the subgroup of the additive group of the ring R generated by the set V_U. Then I_U is a right ideal of the ring R, where $V \subseteq I_U \subseteq U$, therefore, $\mathcal{B}_0'(R) = \{ I_U \mid U \in \mathcal{B}_0(R) \}$ is a basis of neighborhoods of zero of the ring R consisting of the open right ideals (see Proposition 1.4.18).

When the R ring is bounded from right, the proof is analogous.

The inverse statement is evident (see Example 1.6.8).

1.6.33. COROLLARY. *Let locally compact totally disconnected ring (R, τ) be bounded from left (right). Then it has a basis of neighborhoods of zero consisting of open compact right (left) ideals.*

PROOF. The proof results from Corollary 1.4.47 and Proposition 1.6.32.

1.6.34. THEOREM. *A topological ring R is bounded if and only if R has a basis of neighborhoods of zero consisting of ideals of the multiplicative semigroup of the ring R.*

PROOF. Let R be a bounded ring and

$$\mathcal{B}_0(R) = \big\{ V \cup (R \cdot V) \cup (V \cdot R) \cup (R \cdot V \cdot R) \big| V \text{ is a neighborhood of zero in R } \big\}.$$

It is clear that all the elements of $\mathcal{B}_0(R)$ are ideals of the multiplicative semigroup of the ring R and are neighborhoods of zero in R.

Let's verify that $\mathcal{B}_0(R)$ is a basis of neighborhoods of zero of R. Indeed, for any neighborhood U of zero in R there exist neighborhoods W_1 and W_2 of zero in R such that $R \cdot W_1 \subseteq U$ and $W_2 \cdot R \subseteq U \cap W_1$ (since the ring R is bounded). Putting $V = W_1 \cap W_2 \cap U$ we get that

$$V \cup (R \cdot V) \cup (V \cdot R) \cup (R \cdot V \cdot R) \in \mathcal{B}_0(R)$$

and

$$V \cup (R \cdot V) \cup (V \cdot R) \cup (R \cdot V \cdot R) \subseteq$$
$$U \cup (R \cdot W_1) \cup (W_2 \cdot R) \cup \big(R \cdot (W_2 \cdot R) \big) \subseteq U \cup U \cup U \cup R \cdot W_1 = U,$$

therefore, $\mathcal{B}_0(R)$ is a basis of neighborhoods of zero in R.

The inverse statement is evident (see Example 1.6.8).

1.6.35. PROPOSITION. *A topological ring R possessing a basis of neighborhoods of zero consisting of subgroups of its additive group, is bounded if and only if R has the basis of neighborhoods of zero consisting of open ideals of the ring R.*

PROOF. Proof is analogous to the proof of Proposition 1.6.32.

1.6.36. COROLLARY. *A bounded locally compact totally disconnected ring has a basis of neighborhoods of zero consisting of open compact ideals.*

PROOF. Proof results from Corollary 1.4.47 and Proposition 1.6.35.

1.6.37. COROLLARY. A compact totally disconnected ring has a basis of neighborhoods of zero consisting of open compact ideals.

1.6.38. DEFINITION. A topological R-module M is called locally bounded if it has a bounded neighborhood of zero (i.e. a neighborhood of zero which is a bounded subset of the R-module M).

1.6.39. REMARK. A topological R-module M is locally bounded if and only if it has a basis of bounded neighborhoods of zero.

Indeed, let $\mathcal{B}_0(M)$ be a basis of neighborhoods of zero and U be a bounded neighborhood of zero of the topological R-module M, then, due to Remark 1.6.15,

$$\mathcal{B}_0'(M) = \{V \cap U \mid V \in \mathcal{B}_0(M)\}$$

is a basis of bounded neighborhoods of zero in M.

1.6.40. DEFINITION. A topological ring R is called locally bounded from left (locally bounded from right) if and only if it is a locally bounded left (locally bounded right) topological R-module. A ring R is called locally bounded if it is locally bounded from left and right.

1.6.41. REMARK. Due to the Definitions 1.6.40, 1.6.38 and 1.6.2, a topological ring R is locally bounded from left (locally bounded from right, locally bounded) if and only if it has a neighborhood of zero, which is a subset, bounded from left (bounded from right, bounded). As in Remark 1.6.39, it is easy to verify that a locally bounded from left (locally bounded from right, locally bounded) ring has a basis of neighborhoods of zero consisting of bounded from left (bounded from right, bounded) subsets.

1.6.42. REMARK. A submodule N of a locally bounded topological R-module M is also a locally bounded R-module, too. A subring of a locally bounded from left (locally bounded from right, locally bounded) topological ring is a locally bounded from left (locally bounded from right, locally bounded) topological ring, too.

Indeed, let U be a bounded neighborhood of zero of the topological R-module M and V a neighborhood of zero in N, Then $V = V_1 \bigcap N$, where V_1 is a certain neighborhood of zero of the topological R–module M. We can choose a neighborhood W of zero in the topological ring R such that $W \cdot U \subseteq V_1$. Then

$$W \cdot (N \cap U) \subseteq N \cap (W \cdot U) \subseteq N \cap V_1 = V,$$

therefore, $N \bigcap U$ is a bounded neighborhood of the topological R-module N.

The statements on subrings are proved analogously.

1.6.43. REMARK. It is evident that any bounded ring is locally bounded.

1.6.44. REMARK. Any locally compact ring is locally bounded (see Corollary 1.6.5).

1.6.45. REMARK. Let an interval topology τ_ξ, defined on a skew field R by a pseudonorm ξ, be non-discrete. Then (R, τ_ξ) is a locally bounded but not bounded topological ring.

Indeed, since (R, τ_ξ) is a non-discrete Hausdorff topological ring (see Remark 1.3.10), then, due to Proposition 1.6.18, (R, τ_ξ) is not bounded.

Due to Proposition 1.6.10, $U = \{a \mid a \in R, \ \xi(a) \leq 1\}$ is a bounded subset in (R, τ_ξ). It is clear that U is a neighborhood of zero in (R, τ_ξ).

1.6.46. THEOREM. *A topological ring R is locally bounded from left (locally bounded from right, locally bounded) if and only if R has a basis $\mathcal{B}_0(R)$ of neighborhoods of zero containing a neighborhood U such that U is a subsemigroup of the multiplicative semigroup of R, and any neighborhood $V \in \mathcal{B}_0(R)$ is a right (left, two-sided) ideal of the semigroup U.*

PROOF. If R has the mentioned basis $\mathcal{B}_0(R)$ of neighborhoods of zero, then $U \in \mathcal{B}_0(R)$ and $V \cdot U \subseteq V$ (correspondingly, $U \cdot V \subseteq V$ or $V \cdot U \subseteq V$ and $U \cdot V \subseteq V$) for any neighborhood $V \in \mathcal{B}_0(R)$. That means that the neighborhood U of zero is bounded from left (bounded from right, bounded), therefore, R is a locally bounded from left (locally bounded from right, locally bounded) topological ring.

Vice versa, let R be a locally bounded from left topological ring, V be a bounded from left neighborhood of zero in R, and V_1 be a neighborhood of zero in R such that $V_1 \subseteq V$ and $V_1 \cdot V \subseteq V$.

Let's show by induction by n that $V_1^{(n)} \subseteq V$ for any $n = 1, 2, \ldots$ (denotation see in B.12). Indeed, due to the choice of neighborhood V and V_1, we get that $V_1^{(1)} = V_1 \subseteq V$. If $V_1^{(n-1)} \subseteq V$, then $V_1^{(n)} = V_1 \cdot V_1^{(n-1)} \subseteq V_1 \cdot V \subseteq V$.

Let U be the subsemigroup of the multiplicative group of the ring R, generated by the subset V_1. Then $U = \bigcup_{n=1}^{\infty} V_1^{(n)}$. Therefore, $V_1 \subseteq U \subseteq V$, thus, U is a bounded from left neighborhood of zero (see Proposition 1.6.14), which is a subsemigroup of the multiplicative semigroup of the ring R.

Consider a system $\mathcal{B}_0(R) = \{W \bigcup (W \cdot U) \mid W$ is a neighborhood of zero in R and $W \subseteq U\}$. It is clear that all the elements of $\mathcal{B}_0(R)$ are the right ideals of the semigroup U and are the neighborhoods of zero in the ring R.

Now let's verify that $\mathcal{B}_0(R)$ is a basis of neighborhoods of zero in R. Indeed, for any neighborhood S of zero in R there exists a neighborhood of zero W_1 in R such that $W_1 \cdot U \subseteq S$ (since the neighborhood U is bounded from left). Putting $W = W_1 \bigcap U \bigcap S$, we get that $W \bigcup (W \cdot U) \in \mathcal{B}_0(R)$ and

$$W \cup (W \cdot U) \subseteq S \cup (W_1 \cdot U) \subseteq S \cup S = S.$$

The cases of the local boundedness from right and of the two-sided local boundedness are considered analogously.

1.6.47. COROLLARY. A locally bounded from left (from right) ring R has a basis of neighborhoods of zero consisting of subsemigroups of the multiplicative semigroup of the ring R.

1.6.48. THEOREM. *A topological ring R possessing a basis of neighborhoods of zero consisting of subgroups of its additive group is locally bounded from left (locally bounded from right, locally bounded) if and only if R has a basis $\mathcal{B}_0(R)$ of neighborhoods of zero containing a neighborhood $A \in \mathcal{B}_0(R)$ such that A is a subring of R, and any neighborhood $V \in \mathcal{B}_0(R)$ is a right (left, two-sided) ideal of the ring A.*

PROOF. Let W be a bounded from left neighborhood of zero in R, and V be a neighborhood of zero in R such that V is a subgroup of the additive group of the ring R and $V \subseteq W$. Due to Corollary 1.6.47, there exists a neighborhood U of zero in R such that U is a subsemigroup of the multiplicative semigroup of R, and $U \subseteq V$.

Let A be the additive subgroup generated by subsemigroup U. Then A is an open subring in R and $A \subseteq V \subseteq W$. Due to the Remark 1.6.15, A is a bounded from left subset in R, and due to Corollary 1.6.13, A is a ring, bounded from left. Since the ring R has a basis of neighborhoods of zero consisting of additive subgroups, then A also has a basis of neighborhoods of zero consisting of additive subgroups, too. Due to Proposition 1.6.32, A has a basis $\mathcal{B}_0(A)$ of neighborhoods of zero consisting of the right ideals of the ring A. According to C.10, $\mathcal{B}_0(A)$ is a basis of neighborhoods of zero in R.

The case of the boundedness from right and two-sided boundedness is considered analogously.

The inverse statement results from Theorem 1.6.46.

1.6.49. COROLLARY. A locally bounded from left (from right) ring R possessing a basis of neighborhoods of zero consisting of subgroups of its additive group has a basis of neighborhoods of zero consisting of subrings of the ring R.

1.6.50. COROLLARY. *A locally compact totally disconnected ring has a basis of neighborhoods of zero consisting of its open subrings.*

PROOF. The proof results from Propositions 1.4.47 and 1.6.4 and Corollary 1.6.49.

1.6.51. PROPOSITION. *Let R be a topological ring, $\Sigma = \{\tau_1, \tau_2, \ldots, \tau_n\}$ be a finite family of topologies on an R-module M such that (M, τ_i) is a locally bounded topological R-module for each $i = 1, 2, \ldots, n$. Then (M, τ') is a locally bounded topological module, where $\tau' = \sup\{\tau_1, \tau_2, \ldots, \tau_n\}$ (see Theorem 1.2.22).*

PROOF. Let U_i be a bounded neighborhood of zero in the topological R-module (M, τ_i), $i = 1, 2, \ldots, n$. Then $U = \bigcap_{i=1}^n U_i$ is a neighborhood of zero in the topological R-module (M, τ'). Since $U \subseteq U_i$, then, due to Remark 1.6.14, U is a bounded subset of the R-module (M, τ_i), for $i = 1, 2, \ldots n$. Due to Proposition 1.6.25, U is a bounded subset of the topological R-module (M, τ'), therefore, (M, τ') is a locally bounded R-module.

1.6.52. PROPOSITION. *Let $\Sigma = \{\tau_1, \tau_2, \ldots, \tau_n\}$ be a finite family of topologies on a ring R such that (R, τ_i) is a locally bounded from left (locally bounded from right, locally bounded) topological ring for $i = 1, 2, \ldots, n$. Then (R, τ'), where $\tau' = \sup\{\tau_1, \ldots, \tau_n\}$, is a locally bounded from left (locally bounded from right, locally bounded) topological ring (see Theorem 1.2.21).*

PROOF. Let (R, τ_i) be a locally bounded from left topological ring and U_i be a bounded from left neighborhood of zero in the topological ring (R, τ_i) for $i = 1, 2, \ldots, n$. Then $U = \bigcap_{i=1}^n U_i$ is a neighborhood of zero in the topological ring (R, τ'). Since $U \subseteq U_i$, then, due to Remark 1.6.15, U is a bounded from left subset in the topological ring (R, τ_i), for $i = 1, 2, \ldots, n$. Due to Proposition 1.6.26, U is a bounded from left neighborhood of zero in the topological ring (R, τ').

The cases of the local boundedness from right and of the local boundedness of rings (R, τ_i), for $i = 1, 2, \ldots, n$, are treated analogously.

1.6.53. REMARK. Let R be a topological ring, (M, τ) be a locally bounded topological R-module, N be a certain R-module, $f : N \to M$ be a homomorphism of modules and τ' be the prototype in N of the topology τ with respect to the mapping f. Then (N, τ') is a locally bounded topological R-module (see Proposition 1.5.33).

Indeed, let U be an open bounded neighborhood of zero in (M, τ). Then $f^{-1}(U)$ is an open bounded subset in (N, τ') containing zero (see Theorem 1.6.29), therefore, (N, τ') is a locally bounded topological R-module.

1.6.54. REMARK. Let (R, τ) be a locally bounded from left (locally bounded from right, locally bounded) ring, R' be a ring, $f : R' \to R$ be a homomorphism of the rings, τ' be the prototype in R' of the topology τ with respect to the mapping f. Then (R', τ') is a locally bounded from left (locally bounded from right, locally bounded) topological ring (see Theorem 1.5.35).

Indeed, let U be an open neighborhood of zero in (R, τ), bounded from left. Then $f^{-1}(U)$ is an open bounded subset in (R', τ') containing zero (see Theorem 1.6.30), therefore, (R', τ') is locally bounded from left topological ring.

1.6.55. REMARK. Let τ be a non-discrete Hausdorff ring topology on a skew field K. Let the topological ring (K, τ) be locally bounded from left (from right) and U be a bounded from left (from right) neighborhood of zero in (K, τ). Then $\mathcal{B}_0(K) = \{a \cdot U | a \in K \backslash \{0\}\}$ (correspondingly, $\mathcal{B}_0'(K) = \{U \cdot a | a \in K \backslash \{0\}\}$) is a basis of neighborhoods of zero in the ring (K, τ).

Indeed, due to Corollary 1.1.46, all the subsets contained in $\mathcal{B}_0(K)$ (correspondingly, in $\mathcal{B}_0'(K)$) are neighborhoods of zero in (K, τ).

Let V be a neighborhood of zero in (K, τ), then, taking into account the boundedness from left (from right) of the neighborhood U, we can choose an element $a \in K \backslash \{0\}$ such that $a \cdot U \subseteq V$ (correspondingly $U \cdot a \subseteq V$) (see Proposition 1.6.3). Thus, $\mathcal{B}_0(K)$ (correspondingly, $\mathcal{B}_0(K)$) is a basis of neighborhoods of zero in (K, τ).

1.6.56. PROPOSITION. *Let τ be a Hausdorff non-discrete ring topology on a skew field K, besides the topological ring (K, τ) is locally bounded from left (from right). Then the ring (K, τ) has a countable basis of neighborhoods of zero if and only if (K, τ) contains a non-discrete countable subset.*

PROOF. Let the ring (K, τ) be locally bounded from left and U be a bounded from left neighborhood of zero in (K, τ).

Let S be a non-discrete countable subset of the ring (K, τ), then without loss of generality we can assume that $0 \in S$ and 0 is not isolated in S (see Proposition 1.1.34). Then $S \backslash \{0\}$ is a countable subset of non-zero elements in the skew field K, where $0 \in [S \backslash \{0\}]_K$.

Consider a countable family $\mathcal{B}_0(K) = \{a \cdot U | a \in S \backslash \{0\}\}$ of subsets of K. Due to Corollary 1.1.46, all the elements of $\mathcal{B}_0(K)$ are neighborhoods of zero in (K, τ).

Let's verify that $\mathcal{B}_0(K)$ is a basis of neighborhoods of zero in (K, τ). Let V be a neighborhood of zero in (K, τ). Since the neighborhood U is bounded from left, then there exists a neighborhood W of zero in (K, τ) such that $W \cdot U \subseteq V$. Since $0 \in [S \setminus \{0\}]_K$, then $W \cap (S \setminus \{0\}) \neq \emptyset$. Let $a \in W \cap (S \setminus \{0\})$. Then $a \cdot U \subseteq V$, and it is clear that $a \cdot U \in \mathcal{B}_0(K)$. The case of the local boundedness from right of the ring (K, τ) is analyzed analogously.

Vice versa, let $\mathcal{B}_0(K) = \{U_i | i = 1, 2, \dots\}$ be a countable basis of neighborhoods of zero of the ring (K, τ). Choosing non-zero elements $a_i \in V_i$, for $i = 1, 2, \dots$, we get not more then a countable subset $S' = \{a_1, a_2, \dots\}$ of the non-zero elements of the skew field K. It is clear that the subset $S = \{0\} \bigcup S'$ is non-discrete. Since the topology τ is Hausdorff, then $| S | = \infty$. Therefore, S is a countable subset.

1.6.57. REMARK. Let τ_1 and τ_2 be non-discrete ring topologies on a skew field K, and $\tau_2 \leq \tau_1$. If S is a bounded from left (bounded from right, bounded) subset of the topological ring (K, τ_1), then S is a bounded from left (bounded from right, bounded) subset of the topological ring (K, τ_2).

Indeed, let U be a neighborhood of zero of the ring (K, τ_2). Since $\tau_2 \leq \tau_1$, then U is a neighborhood of zero of the ring (K, τ_1). Since the subset S is bounded from left in the ring (K, τ_1), then there exists a non-zero element a such that $a \cdot S \subseteq U$, i.e. S is a bounded from left subset of (K, τ_2). The cases of the boundedness from right and two-sided boundedness of S are analyzed analogously.

1.6.58. THEOREM. *Let Σ be a family of topologies on a skew field K such that (K, τ) is a locally bounded from left (locally bounded from right, locally bounded) ring for all $\tau \in \Sigma$, and let $\tau' = \sup\{\tau | \tau \in \Sigma\}$ be a non-discrete topology. Then the ring (K, τ') is locally bounded from left (locally bounded from right, locally bounded) if and only if there exists a finite number of topologies $\tau_1, \tau_2, \dots, \tau_n \in \Sigma$ such that $\tau' = \sup\{\tau_1, \tau_2, \dots, \tau_n\}$.*

PROOF. Let $\tau_1, \tau_2, \dots, \tau_n$ be topologies from Σ such that $\tau' = \sup\{\tau_1, \tau_2, \dots, \tau_n\}$, as above. Since $\tau_i \in \Sigma$, then (K, τ_i) is a locally bounded from left (locally bounded from right, locally bounded) ring.

Due to Proposition 1.6.52, (K, τ') is also a locally bounded from left (locally bounded from right, locally bounded) ring.

Vice versa, let the ring (K, τ') be locally bounded from left and U be a bounded from left neighborhood of zero in (K, τ'). Then there exist topologies $\tau_1, \tau_2, \dots, \tau_n$

$\in \Sigma$ such that $U = \bigcap_{i=1}^{n} U_i$, where U_i is a neighborhood of zero in the ring (K, τ_i), $i = 1, 2, \ldots, n$. According to Remark 1.6.55, $\mathcal{B}_0(K) = \{a \cdot U \,|\, a \in K \backslash \{0\}\}$ is a basis of neighborhoods of zero in the ring (K, τ'). Due to Remark 1.6.57, U is a bounded from left neighborhood of zero in the ring (K, τ''), where $\tau'' = \sup\{\tau_1, \tau_2, \ldots, \tau_n\}$, since $\tau'' \leq \tau'$. According to Remark 1.6.55, $\mathcal{B}_0(K)$ is a basis of neighborhoods of zero in the rings (K, τ'') and (K, τ'), therefore, $\tau' = \tau''$.

The cases of the right and two-sided local boundedness are considered analogously.

1.6.59.   EXAMPLE. For a prime number p let τ_p be the p-adic topology on the field $\mathbb{Q}$ of rational numbers. It is induced by the norm ξ_p on the $\mathbb{Q}$, besides $(\mathbb{Q}, \tau_p)$ is a locally bounded field for any prime number p (see Example 1.1.19, Corollary 1.2.15 and Remark 1.6.45). Then the field $(\mathbb{Q}, \tau)$ is non-discrete, where $\tau = \sup\{\tau_p \,|\, p$ is a prime number $\}$ (see Theorem 1.2.21).

Let's show that the field $(\mathbb{Q}, \tau)$ is not locally bounded. Indeed, assume the contrary, i.e. that $(\mathbb{Q}, \tau)$ is a locally bounded field. Then, due to Theorem 1.6.58, there exists a finite set $p_1, p_2, \ldots, p_n$ of prime numbers such that $\tau = \sup\{\tau_{p_1}, \tau_{p_2}, \ldots, \tau_{p_n}\}$. Choose a prime number p_{n+1} such that $p_{n+1} \neq p_i$, $i = 1, 2, \ldots, n$. The subset $V = \{r \in \mathbb{Q} \,|\, \xi_{p+1}(r) < \frac{1}{2}\}$ is a neighborhood of zero in the topological field $(\mathbb{Q}, \tau_{p+1})$. Since $\tau_{p+1} \leq \tau$, then V is a neighborhood of zero in $(\mathbb{Q}, \tau)$. Since $\tau = \sup\{\tau_{p_1}, \tau_{p_2}, \ldots, \tau_{p_n}\}$, then there exist integers $k_1, k_2, \ldots, k_n$ such that $\bigcap_{i=1}^{n} V_{k_i} \subseteq V$, where

$$V_{k_i} = \left\{ r \in \mathbb{Q} \,\middle|\, \xi_{p_i}(r) < \frac{1}{2^{k_i}} \right\}.$$

Consider the element $r = p_1^{k_1+1} \cdot p_2^{k_2+1} \cdot \ldots \cdot p_n^{k_n+1}$ in $\mathbb{Q}$. Since

$$\xi_{p_i}\left(p_1^{k_1+1} \cdot p_2^{k_2+1} \cdot \ldots \cdot p_n^{k_n+1} \right) = \frac{1}{2^{k_i+1}} < \frac{1}{2^{k_i}},$$

then $r \in \bigcap_{i=1}^{n} V_{k_i} \subseteq V$. But $\xi_{p_{n+1}}(r) = 1$, which contradicts the choice of the subset V. Therefore, the topological field $(\mathbb{Q}, \tau)$ is not locally bounded.

Note that the family $\Sigma = \{\tau_p \,|\, p$ is a prime number$\}$ gives an example of the family of topologies on $\mathbb{Q}$ such that each one of those is induced by a norm, but the upper limit of the family cannot be induced even by a pseudonorm (otherwise, due to Remark 1.6.45, $(\mathbb{Q}, \tau)$ would be locally bounded).

1.6.60. DEFINITION. Let τ be a Hausdorff ring topology on a skew field K. A subset $S \subseteq K$ is called inversely bounded from left (inversely bounded from right, inversely bounded) in the topological ring (K, τ) if $0 \in S$ and the subset $(K \backslash S)^{-1}$ is bounded from left (bounded from right, bounded) in the topological ring (K, τ).

1.6.61. DEFINITION. Let τ be a Hausdorff ring topology on a skew field K. The topological ring (K, τ) is called locally inversely bounded from left (locally inversely bounded from right, locally inversely bounded)* if any neighborhood of zero V in (K, τ) is an inversely bounded from left (inversely bounded from right, inversely bounded) subset in (K, τ).

1.6.62. EXAMPLE. Any discrete skew field K is locally inversely bounded.

1.6.63. EXAMPLE. Any normed skew field is locally inversely bounded in the interval topology.

Indeed, let ξ be a norm on the skew field K and V be a neighborhood of zero in K. Then there exists integer $n \geq 1$ such that

$$V_n = \left\{ a \mid a \in K, \xi(a) < \frac{1}{2^n} \right\} \subseteq V.$$

If $x \in K \backslash V_n$, then $\xi(x) \geq \frac{1}{2^n}$, therefore $\xi(x^{-1}) \leq 2^n$ (since $\xi(x) \cdot \xi(x^{-1}) = \xi(x \cdot x^{-1}) = \xi(1) = 1$). Due to Proposition 1.6.10, the subset $(K \backslash V_n)^{-1}$ is bounded. Since $(K \backslash V_n)^{-1} \supseteq (K \backslash V)^{-1}$, then $(K \backslash V)^{-1}$ is a bounded subset. Thus, the normed skew field K is locally inversely bounded.

1.6.64. PROPOSITION. *Let τ be a ring topology on a skew field K and the topological ring (K, τ) be locally inversely bounded from left (locally inversely bounded from right, locally inversely bounded). Then the topological ring (K, τ) is locally bounded from left (locally bounded from right, locally bounded).*

PROOF. We can consider that τ is a non-discrete Hausdorff topology, therefore there exists a neighborhood U of zero in (K, τ) such that $1 \notin U$. Let's choose a neighborhood

---

* Sometimes the locally inversely bounded ring topologies on a skew field are called in the literature V-topologies.

V of zero in (K, τ) such that $V \cdot V \subseteq U$. Since the ring (K, τ) is locally inversely bounded from left, then the subset $(K \backslash V)^{-1}$ is bounded from left.

Let's show that V is a bounded from left neighborhood of zero in (K, τ). Let W be a neighborhood of zero in (K, τ). According to Proposition 1.6.3, there exists a non-zero element $a \in K$ such that $a \cdot (R \backslash V)^{-1} \subseteq W$.

Let's verify that $a \cdot V \subseteq W$. Indeed, if we assume that $0 \neq x \in V$ and $a \cdot x \notin W$, then $x \notin (R \backslash V)^{-1}$, and hence, $x^{-1} \notin R \backslash V$, i.e. $x^{-1} \in V$. Therefore, $1 = x \cdot x^{-1} \in V \cdot V \subseteq U$, which contradicts the choice of the neighborhood U. Thus, $a \cdot V \subseteq W$, which means that the neighborhood V is bounded from left (see Proposition 1.6.3). The cases of the right and two-sided local inversely boundedness of the ring are considered analogously.

1.6.65. PROPOSITION. *Let τ be a Hausdorff ring topology on a skew field K and the ring (K, τ) be locally inversely bounded from left (right). Then (K, τ) is a topological skew field.*

PROOF. Let $\mathcal{B}_0(K)$ be the basis of all neighborhoods of zero in (K, τ) and W be a neighborhood of zero in (K, τ) such that $(1 + W) \bigcap W = \emptyset$. Since the topological ring (K, τ) is locally inversely bounded from left, then the subset $(K \backslash W)^{-1}$ is bounded from left. Taking into account that $1 + W \subseteq K \backslash W$, we get that the subset $(1 + W)^{-1}$ is bounded from left.

Let U be a neighborhood of zero in (K, τ), then we can choose a symmetrical neighborhood of zero V in (K, τ) such that $V \cdot (1 + W)^{-1} \subseteq U$ and $V \subseteq W$. Then $V \cdot (1 + V)^{-1} \subseteq V \cdot (1 + W)^{-1} \subseteq U$. If $v \in V$, then

$$(1 + v)^{-1} - 1 = (1 + v)^{-1} - (1 + v) \cdot (1 + v)^{-1} = (1 - 1 - v) \cdot (1 + v)^{-1}$$
$$= -v \cdot (1 + v)^{-1} \in V \cdot (1 + V)^{-1} \subseteq U.$$

Thus, $(1 + V)^{-1} \subseteq 1 + U$. It means that the basis $\mathcal{B}_0(K)$ satisfies the condition (BN7) (see Proposition 1.2.12). Due to Theorem 1.2.13, (K, τ) is a topological skew field.

1.6.66. PROPOSITION. *Let $\mathbb{R}$ be the additive group of real numbers endowed with the interval topology, $\mathbb{Z}$ be the subgroup of integers,*

$$\mathcal{B}_0(\mathbb{R}) = \left\{ V_n \mid V_n = \{x \in \mathbb{R}, \mid x \mid < \frac{1}{2^n}\}, n \in \mathbb{N} \right\}$$

be a basis of neighborhoods of zero in $\mathbb{R}$ and $\omega : \mathbb{R} \to \mathbb{T} = \mathbb{R}/\mathbb{Z}$ be the natural topological homomorphism. Let $t \in \mathbb{T}$ and k be a non-negative number such that $2^i \cdot t \in \omega(V_2)$ for $i = 0, 1, \dots, k$, then $t \in \omega(V_{k+2})$.

PROOF. We'll prove the proposition by the induction by k. If $k = 0$, then, according to the condition above, $t \in \omega(V_2)$. Suppose that from $2^i \cdot t \in \omega(V_2)$ for $i = 0, 1, \dots, k$ results that $t \in \omega(V_{k+2})$. Let $2^i \cdot t \in \omega(V_2)$ for $i = 0, 1, 2, \dots, k, k+1$. Then, due to the inductive supposition, $t \in \omega(V_{k+2})$, i.e. there exists an element $a \in V_{k+2}$ such that $t = \omega(a)$. Since $2^{k+1} \cdot t \in \omega(V_2)$, then there exists an element $b \in V_2$ such that $2^{k+1} \cdot t = \omega(b)$. Because of

$$\omega(2^{k+1} \cdot a - b) = 2^{k+1} \cdot \omega(a) - \omega(b) = 2^{k+1} \cdot t - 2^{k+1} \cdot t = 0,$$

we get that $2^{k+1} \cdot a - b$ is an integer. From

$$\mid 2^{k+1} \cdot a - b \mid \leq \mid 2^{k+1} \cdot a \mid + \mid b \mid = 2^{k+1} \cdot \mid a \mid + \mid b \mid \leq \frac{2^{k+1}}{2^{k+2}} + \frac{1}{4} = \frac{3}{4} < 1$$

follows that $2^{k+1} \cdot a - b = 0$, i.e $a = \frac{b}{2^{k+1}}$. Therefore,

$$\mid a \mid = \frac{\mid b \mid}{2^{k+1}} < \frac{1}{2^2 \cdot 2^{k+1}} = \frac{1}{2^{k+3}},$$

i.e. $a \in V_{k+3}$ thus, $t = \omega(a) \in \omega(V_{k+3})$. The proposition is completely proved.

1.6.67. THEOREM. *Let R be a locally compact ring, $C_0(R)$ be the connected component of zero of R and A be a subgroup of the additive group of the ring R such that A is a bounded from left (right) subset in R. Then $C_0(R) \cdot A = \{0\}$ (respectively, $A \cdot C_0(R) = \{0\}$).*

PROOF. Assume the contrary, i.e. that $c_0 \cdot a_0 \neq 0$ for certain elements $c_0 \in C_0(R)$ and $a_0 \in A$. Let $\mathbb{R}(+)$ be the additive group of real numbers endowed with the interval topology. Consider the basis $\mathcal{B}_0(\mathbb{R}) = \{V_n \mid n = 1, 2, \dots\}$ of neighborhoods of zero in $\mathbb{R}$, where $V_n = \{x \in \mathbb{R} \mid \mid x \mid < \frac{1}{2^n}\}$ for $n = 1, 2, \dots$ (see Examples 1.1.15 and 1.2.9). Let $\omega : \mathbb{R} \to \mathbb{T} = \mathbb{R}/\mathbb{Z}$ be the natural homomorphism, then $\mathcal{B}_0(\mathbb{T}) = \{\omega(V_n) \mid n = 1, 2, \dots\}$ is a basis of neighborhoods of zero of the Hausdorff topological group $\mathbb{T}$. Due to Theorem 1.5.40, there exists a character $f_0 : R \to \mathbb{T}$ of the additive group of R such that $f_0(c_0 \cdot a_0) \neq 0$.

We consider the subset $I_0 = \{x \in R \mid f_0(x \cdot a) = 0 \mid \text{for any } a \in A\}$ in R. Let's show that I_0 is an open subgroup of the additive group R.

Indeed, since $0 \in I_0$, then $I_0 \neq \emptyset$. Let $x_1, x_2 \in I_0$. Since f_0 is a homomorphism of the additive group $R(+)$ to the group $\mathbb{T}(+)$, then $f_0((x_1 - x_2) \cdot a) = f_0(x_1 \cdot a - x_2 \cdot a) = f_0(x_1 \cdot a) - f_0(x_2 \cdot a) = 0$ for any $a \in A$. Therefore, I_0 is subgroup of the group $R(+)$. Since the homomorphism $f_0 : R \to \mathbb{T}$ is continuous, we can choose a neighborhood of zero U in the topological group $R(+)$ such that $f_0(U) \subseteq \omega(V_2)$. Since A is a bounded from left subset of the topological ring R, then there exists a neighborhood V of zero in R such that $V \cdot A \subseteq U$.

Let's verify that $V \subseteq I_0$ (in that case we'll prove the openness of the subgroup I_0 (see Proposition 1.4.18)). Indeed, let $x \in V$. Then for a non-negative integer n we get that $2^n \cdot x \cdot A = x \cdot 2^n \cdot A \subseteq x \cdot A \subseteq V \cdot A \subseteq U$, therefore, $f_0(2^n \cdot x \cdot a) \in f_0(U) \subseteq \omega(V_2)$ for any $a \in A$. Since $f_0(2^n \cdot x \cdot a) = 2^n \cdot f_0(x \cdot a)$, then $2^n \cdot f_0(x \cdot a) \in \omega(V_2)$ for $n = 0, 1, \ldots$. Due to Proposition 1.6.66, $f_0(x \cdot a) \in \omega(V_{n+2})$ for all $n = 0, 1, \ldots$ Therefore,

$$f_0(x \cdot a) \in \bigcap_{n=0}^{\infty} \omega(V_{n+2}) = \{0\}.$$

Thus, $f_0(x \cdot a) = 0$ for any $a \in A$, i.e. $x \in I_0$. Then $V \subseteq I_0$.

Thus, I_0 is an open-closed subgroup of the topological group R (see Proposition 1.4.18). Then $C_0(R) \subseteq I_0$ (see C.13). Therefore, for any elements $c \in C_0(R)$ and $a \in A$ we get that $f_0(c \cdot a) = 0$. But it contradicts the choice of the character f_0, since $f_0(c_0 \cdot a_0) \neq 0$. Thus, $C_0(R) \cdot A = \{0\}$.

In the case of the right boundedness of the subgroup A the proof is analogous.

1.6.68. COROLLARY. A locally compact bounded from left (right) ring with 1 is totally disconnected. In particular, any compact ring with 1 is totally disconnected.

1.6.69. PROPOSITION. *Let S be a subset of a Hausdorff non-discrete locally compact skew field K. Then the following statements are equivalent:*

1) S is compact subset;

2) S is closed, bounded from the left (right) subset.

PROOF. The implication 1) $\Longrightarrow$ 2) results from Corollary 1.6.5 and C.12.

Let's prove the implication 2) $\Longrightarrow$ 1). Indeed, let U be a compact neighborhood of zero in K, and the subset S be closed and bounded from left. Due to Proposition 1.6.3, there exists a non-zero element $a \in K$ such that $a \cdot S \subset U$. Then the subset $a \cdot S$ is compact as a closed subset (see Proposition 1.1.44) of the compact subset U. Due to Proposition 1.1.44, the subset $S = a^{-1} \cdot (a \cdot S)$ is compact, too.

1.6.70. THEOREM. *Let τ be a non-discrete ring topology on the ring $\mathbb{Z}$ of integers such that the topological ring $(\mathbb{Z}, \tau)$ is locally bounded. Then in $(\mathbb{Z}, \tau)$ exists a basis of neighborhoods of zero consisting of ideals of the ring $\mathbb{Z}$.*

PROOF. It is enough to show that any neighborhood of zero in $(\mathbb{Z}, \tau)$ contains an open ideal of the ring $(\mathbb{Z}, \tau)$. Let W be a neighborhood of zero in $(\mathbb{Z}, \tau)$. We can choose in $(\mathbb{Z}, \tau)$ a closed symmetric neighborhood W_1 of zero such that $W_1 \subseteq W$. Since $(\mathbb{Z}, \tau)$ is a locally bounded ring, we can consider that W_1 is a bounded subset. We also can choose symmetric neighborhoods V_1 and V_2 of zero in the ring $(\mathbb{Z}, \tau)$ such that $V_1 + V_1 \subseteq W_1$ and $W_1 \cdot V_2 \subseteq V_1$ (the choice of the neighborhood V_2 is possible, since W_1 is bounded). Since the topology τ is non-discrete, then V_2 contains a certain natural number n.

Let's verify by induction by s that $\sum_{i=1}^{s} W_1 \cdot n^i \subseteq W_1$. Indeed, for $s = 1$ we have $W_1 \cdot n \subseteq W_1 \cdot V_2 \subseteq V_1 \subseteq W_1$. Suppose that $\sum_{i=1}^{k} W_1 \cdot n^i \subseteq W_1$ for all $k < s$. Then we get that

$$\sum_{i=1}^{s} W_1 \cdot n^i = W_1 \cdot n + \left(\sum_{i=1}^{s-1} W_1 \cdot n^i \right) \cdot n \subseteq$$
$$W_1 \cdot n + W_1 \cdot n \subseteq W_1 \cdot V_2 + W_1 \cdot V_2 \subseteq V_1 + V_1 \subseteq W_1.$$

We can choose a symmetric neighborhood U of zero in the topological ring $(\mathbb{Z}, \tau)$ such that $U \cdot t \subseteq W_1$ for all $t = 0, 1, \ldots, n - 1$, and can also choose $m \in \mathbb{Z}$ such that $0 < m \in U$ (the choice of m is possible because the topology τ is non-discrete).

Let's show that $n \cdot m \cdot k \in W_1$ for any $k \in \mathbb{Z}$. Indeed, since the neighborhood W_1 is symmetric, we can consider only the case when $k > 0$. Then $k = \sum_{i=1}^{q} r_i \cdot n^i$, where $0 \leq r_i \leq n - 1$, therefore,

$$n \cdot m \cdot k = \sum_{i=1}^{q+1} m \cdot r_i \cdot n^i \in \sum_{i=1}^{q+1} (U \cdot r_i) \cdot n^i \subseteq \sum_{i=1}^{q+1} W_1 \cdot n^i \subseteq W_1.$$

Thus, $(n \cdot m) \cdot \mathbb{Z} \subseteq W_1$.

Since $I = \left[(n \cdot m) \cdot \mathbb{Z}\right]_{(\mathbb{Z},\tau)}$ is a non-zero closed ideal of the ring $(\mathbb{Z}, \tau)$, then $\mathbb{Z}/I$ is a finite Hausdorff topological ring (see Corollary 1.5.18). Hence, $\mathbb{Z}/I$ is discrete ring (see C.11). Due to Corollary 1.5.18, the ideal I is open, besides,

$$I = \left[(n \cdot m) \cdot \mathbb{Z}\right]_{(\mathbb{Z},\tau)} \subseteq [W_1]_{(\mathbb{Z},\tau)} = W_1 \subseteq W.$$

This completes the proof of the theorem.

§ 1.7 Minimal Topologies

1.7.1. DEFINITION. Let (A, τ) be a topological Abelian group, $\Delta = \Delta(A)$ be a certain class of group topologies on A. The topology τ we call weakable[†] in the class Δ if in Δ exists a topology τ' such that $\tau' < \tau$. If in Δ does not exist any topology weaker than τ, we call the topology τ non-weakable in the class Δ.

1.7.2. EXAMPLE. The anti-discrete topology τ_0 on an Abelian group A is non-weakable in any class of group topologies on A. Any topology on A, different from the anti-discrete, is weakable in any class of group topologies on A, containing the anti-discrete topology (see C.8).

1.7.3. EXAMPLE. Let (A, τ) be a compact Abelian group and Δ be a certain class of Hausdorff group topologies, then the topology τ is non-weakable in the class Δ (see C.12).

1.7.4. EXAMPLE. If a class Δ of group topologies on an Abelian group A contains at least one non-discrete topology, then the discrete topology on the group A is weakable in Δ.

1.7.5. PROPOSITION. *For any Hausdorff topological ring (R, τ_0) the following statements are equivalent:*

1) the topology τ_0 is non-weakable in the class Δ of all Hausdorff ring topologies on R;

2) any continuous ring isomorphism f from the topological ring (R, τ_0) onto a Hausdorff topological ring (Q, τ) is a topological isomorphism.

PROOF. Let the topology τ_0 be non-weakable in the class Δ. Let $f : (R, \tau_0) \to (Q, \tau)$ be a continuous ring isomorphism, then the prototype τ' on R of the topology τ with respect to the mapping f is a ring topology on R (see Theorem 1.5.35). Since f is an isomorphism and τ is Hausdorff topology, then the topology τ' is Hausdorff, i.e. $\tau' \in \Delta$.

Let's show that $\tau' \leq \tau_0$. Indeed, if S' is an open subset of the topological ring (R, τ'), then $S' = f^{-1}(S)$, where S is a certain open subset of the topological ring (Q, τ). Since the mapping f is continuous, then S' is an open subset of the topological ring (R, τ_0). Hence, $\tau' \leq \tau_0$.

[†] In general it is not supposed that $\tau \in \Delta$.

Since τ_0 is non-weakable in the class Δ, then $\tau' = \tau_0$. Then for any subset A open in (R, τ_0) there exists a subset B open in (Q, τ) such that $A = f^{-1}(B)$, i.e. $f(A) = f(f^{-1}(B)) = B$ is an open subset in (Q, τ). It means that f is open mapping. Thus, 1) $\Longrightarrow$ 2).

If the topology τ_0 is weakable in the class Δ, then there exists a topology $\tau \in \Delta$ such that $\tau < \tau_0$. Then the identical isomorphism $\epsilon : (R, \tau_0) \to (R, \tau)$ is continuous and it is not a topological isomorphism, i.e. 2) $\Longrightarrow$ 1).

1.7.6. PROPOSITION. *Let R be a topological ring and (M, τ_0) be a Hausdorff topological R-module. Then the following statements are equivalent:*

1) the topology τ_0 is non-weakable in the class of all Hausdorff module topologies on M;

2) any continuous isomorphism f from the topological R-module (M, τ_0) onto a Hausdorff topological R-module (N, τ) is a topological isomorphism.

PROOF. Proof is analogous to the proof of Proposition 1.7.5, using Proposition 1.5.33.

1.7.7. COROLLARY. For any Hausdorff topological Abelian group (A, τ_0) the following statements are equivalent:

1) the topology τ_0 is non-weakable in the class of all Hausdorff group topologies on A;

2) any continuous isomorphism from the topological group (A, τ_0) onto a Hausdorff topological Abelian group (B, τ) is a topological isomorphism.

1.7.8. THEOREM. *Let R be a field, $\Delta = \Delta(R)$ be the class of all Hausdorff field topologies on R. If τ is a Hausdorff ring topology on R such that $\tau \notin \Delta$, then the topology τ is weakable in the class Δ.*

PROOF. Let $\mathcal{B}_0(R)$ be a basis of symmetric neighborhoods of zero of the topological ring (R, τ). Since the topology τ is Hausdorff, then without lost of generality we can consider that $1 \notin V$ for any neighborhood $V \in \mathcal{B}_0(R)$.

Let's show that the family

$$\mathcal{B}_0'(R) = \left\{ V \cdot (1 + V)^{-1} \mid V \in \mathcal{B}_0(R) \right\}$$

of the subsets of R satisfies the conditions (BN1'), (BN2) - (BN7) (see Corollary 1.3.7 and Theorem 1.2.13).

It is evident that $0 \in V \cdot (1 + V)^{-1}$ for any neighborhood $V \in \mathcal{B}_0(R)$, i.e

$$0 \in \bigcap_{V \in \mathcal{B}_0(R)} V \cdot (1 + V)^{-1}.$$

Assume that

$$0 \neq r \in \bigcap_{V \in \mathcal{B}_0(R)} V \cdot (1 + V)^{-1}.$$

Since the topology τ is Hausdorff, then there exists a neighborhood $V_0 \in \mathcal{B}_0(R)$ such that $r \notin V_0$. We can choose neighborhoods $V_1, W \in \mathcal{B}_0(R)$ such that $V_1 - V_1 \subseteq V_0$, $W \subseteq V_1$, and $r \cdot W \subseteq V_1$. Since

$$r \in \bigcap_{V \in \mathcal{B}_0(R)} V \cdot (1 + V)^{-1},$$

then $r \in W \cdot (1 + W)^{-1}$. Therefore, there exist elements $x, y \in W$ such that $r = x \cdot (1 + y)^{-1}$. Then $r \cdot (1 + y) = x$, hence,

$$r = x - r \cdot y \in W - r \cdot W \subseteq V_1 - V_1 \subseteq V_0,$$

i.e. $r \in V_0$, which contradicts the choice of the neighborhood V_0. Thus,

$$\bigcap_{V \in \mathcal{B}_0(R)} V \cdot (1 + V)^{-1} = \{0\},$$

i.e. the system $\mathcal{B}_0'(R)$ satisfies the condition (BN1′).

It is clear, that if $V_1, V_2, V_3 \in \mathcal{B}_0(R)$ and $V_3 \subseteq V_1 \cap V_2$, then $V_3 \cdot (1 + V_3)^{-1} \subseteq (V_1 \cdot (1 + V_1)^{-1}) \cap (V_2 \cdot (1 + V_2)^{-1})$, i.e. the system $\mathcal{B}_0'(R)$ satisfies the condition (BN2).

Let $V \cdot (1+V)^{-1} \in \mathcal{B}_0'(R)$. Since $V \in \mathcal{B}_0(R)$, then it is possible to choose a neighborhood $W \in \mathcal{B}_0(R)$ such that $W + W + W \cdot W + W \cdot W \subseteq V$. If $x, y, z, t \in W$, then

$$x \cdot (1 + y)^{-1} + z \cdot (1 + t)^{-1} = (x \cdot (1 + t) + z \cdot (1 + y)) \cdot (1 + y)^{-1} \cdot (1 + t)^{-1}$$
$$= (x + z + xt + xy) \cdot (1 + y + t + yt)^{-1}$$
$$\in (W + W + W \cdot W + W \cdot W) \cdot (1 + W + W + W \cdot W)^{-1} \subseteq V \cdot (1 + V)^{-1},$$

i.e.,

$$W \cdot (1 + W)^{-1} + W \cdot (1 + W)^{-1} \subseteq V \cdot (1 + V)^{-1}.$$

Thus, $\mathcal{B}'_0(R)$ satisfies the condition (BN3).

The fulfillment of the condition (BN4) is evident, since the neighborhoods of $\mathcal{B}_0(R)$ are symmetric.

Let $V \cdot (1+V)^{-1} \in \mathcal{B}'_0(R)$, $W \in \mathcal{B}_0(R)$ and $W + W + W \cdot W \subseteq V$. Then $W \cdot W \subseteq V$ and

$$x \cdot (1+y)^{-1} \cdot z \cdot (1+t)^{-1} = xz \cdot (1+y+t+yt)^{-1}$$
$$\in W \cdot W \cdot (1+W+W+W \cdot W)^{-1} \subseteq V \cdot (1+V)^{-1}$$

for any $x, y, z, t \in W$, i.e., $V \cdot (1+V)^{-1} \supseteq W \cdot (1+W)^{-1} \cdot W \cdot (1+W)^{-1}$. Thus, $\mathcal{B}'_0(R)$ satisfies the condition (BN5).

Let $V \cdot (1+W)^{-1} \in \mathcal{B}'_0(R), a \in R, \ W \in \mathcal{B}_0(R), \ W \subseteq V$ and $a \cdot W \subseteq V$. Then $a \cdot (W \cdot (1+W)^{-1}) = (a \cdot W) \cdot (1+W)^{-1} \subseteq V \cdot (1+V)^{-1}$, i.e., the condition (BN6) is fulfilled.

Let $V \cdot (1+V)^{-1} \in \mathcal{B}'_0(R)$ and $W \in \mathcal{B}_0(R)$, where $W + W \subseteq V$. Then $W \subseteq V$. If $x, y \in W$ and $1 + x \cdot (1+y)^{-1} \neq 0$, then

$$\left(1 + x \cdot (1+y)^{-1}\right)^{-1} = (1+y) \cdot (1+x+y)^{-1} = 1 - x \cdot (1+x+y)^{-1}$$
$$\in 1 - W \cdot (1+W+W)^{-1} \subseteq 1 + V \cdot (1+V)^{-1},$$

i.e. $\left(1 + W \cdot (1+W)^{-1}\right)^1 \subseteq 1 + V \cdot (1+V)^{-1}$ and, hence, the condition (BN7) is fulfilled.

Due to Theorem 1.2.13 and Corollary 1.3.7, the system $\mathcal{B}'_0(R)$ defines on R a Hausdorff topology τ' in which R is a topological field, and $\mathcal{B}'_0(R)$ is a basis of neighborhoods of zero. Hence, $\tau' \in \Delta(R)$.

Since $V \subseteq V \cdot (1+V)^{-1}$ for any neighborhood $V \in \mathcal{B}_0(R)$, then $\tau' \leq \tau$, and besides, $\tau \notin \Delta(R)$ therefore, $\tau' < \tau$. This completes the proof of the theorem.

1.7.9. PROPOSITION. *Let R be a skew field, τ be a Hausdorff ring topology on R such that the topological ring (R, τ) contains a non-discrete set S and let $\Delta = \Delta(R)$ be a class of all such ring topologies on R, that any one of them has a basis of neighborhoods of zero, whose cardinality does not exceed the cardinality of S. Then either $\tau \in \Delta$, or the topology τ is weakable in the class Δ.*

PROOF. It is enough to define on R a topology $\tau' \in \Delta$ such that $\tau' \leq \tau$. Without loss of generality we can suppose that S contains the zero element and it is not isolated in S (see Proposition 1.1.34).

It is easy to construct by induction a descending countable sequence

$$V_0 \supseteq V_1 \supseteq \ldots \supseteq V_n \supseteq V_{n+1} \supseteq \ldots$$

of the symmetrical neighborhoods of zero of the topological ring (R, τ) in the following way:

a) as V_0 we choose any symmetrical neighborhood of zero in (R, τ) that does not contain the unitary element of the skew field R (such choice is possible since the topological ring (R, τ) is Hausdorff);

b) if the neighborhoods V_i are already constructed for all $0 \leq i < n$, we choose as V_n a symmetric neighborhood of zero such that $V_n + V_n \subseteq V_{n-1}$ and $V_n \cdot V_n \subseteq V_{n-1}$.

Put $S' = \bigcup_{k=1}^{\infty} (S \backslash \{0\})^{(k)}$ (see B.12). Then S' is the subset of R consisting of non-zero elements and its cardinality does not exceed the cardinality of the set S. Let

$$\mathcal{B}_0'(R) = \{a \cdot V_n \cdot b \,|\, a, b \in S', n = 1, 2, \ldots \}.$$

Due to Corollary 1.1.46, $a \cdot V_n \cdot b$ is a neighborhood of zero in (R, τ) for any $a, b \in S', n = 1, 2, \ldots$. The cardinality of the set $\mathcal{B}_0'(R)$ does not exceed the cardinality of S.

Let's show that $\mathcal{B}_0'(R)$ satisfies the conditions (BN1'), (BN2) - (BN6) (see Proposition 1.2.1 and Corollary 1.3.7).

It is clear that $0 \in a \cdot V_n \cdot b$ for $a, b \in S'$ and $n \in \mathbb{N}$. Assume that

$$0 \neq x \in \bigcap_{n=1}^{\infty} \bigcap_{a,b \in S'} a \cdot V_n \cdot b.$$

Let V be a neighborhood of zero in (R, τ) such that $V \subseteq V_2$ and $x^{-1} \cdot V \subseteq V_2$. Since the zero element is not isolated in S, then there exists a non-zero element $a_0 \in S \cap V$. Then $a_0 \in V_2$, $x^{-1} \cdot a_0 \in V_2$ and $x \in a_0 \cdot V_2 \cdot a_0$. Therefore,

$$1 = x^{-1} \cdot x \in x^{-1} \cdot (a_0 \cdot V_2 \cdot a_0) = (x^{-1} \cdot a_0) \cdot V_2 \cdot a_0$$

$$\subseteq V_2 \cdot V_2 \cdot V_2 \subseteq V_1 \cdot V_2 \subseteq V_1 \cdot V_1 \subseteq V_0,$$

which contradicts the choice of the neighborhood V_0. Thus,

$$\{0\} = \bigcap_{n=1}^{\infty} \bigcap_{a,b \in S'} a \cdot V_n \cdot b,$$

i.e. $\mathcal{B}_0'(R)$ satisfies the condition (BN1').

Let $a'' \cdot V_n \cdot b'$, $a'' \cdot V_m \cdot b'' \in \mathcal{B}_0'(R)$ and $k = \max\{n, m\}$. Since the elements a', b', a'', b'' belong to S', then they are invertible. Therefore, due to Corollary 1.1.46, the subsets $a' \cdot V_{k+2}$, $a'' \cdot V_{k+2}$, $V_{k+2} \cdot b'$ and $V_{k+2} \cdot b''$ are neighborhoods of zero in (R, τ). Since the zero element is not isolated in S, then there exists nonzero elements $a, b \in S$ such that $a \in (a' \cdot V_{k+2}) \cap (a'' \cdot V_{k+2})$ and $b \in (V_{k+2} \cdot b') \cap (V_{k+2} \cdot b'')$. Then $a \cdot V_{k+2} \cdot b \in \mathcal{B}_0'(R)$, where

$$a \cdot V_{k+2} \cdot b \subseteq a' \cdot V_{k+2} \cdot V_{k+2} \cdot V_{k+2} \cdot b' \subseteq a' \cdot V_k \cdot b' \subseteq a' \cdot V_n \cdot b'.$$

Analogously,

$$a \cdot V_{k+2} \cdot b \subseteq a'' \cdot V_{k+2} \cdot V_{k+2} \cdot V_{k+2} \cdot b'' \subseteq a'' \cdot V_k \cdot b'' \subseteq a'' \cdot V_m \cdot b''.$$

Therefore,

$$a \cdot V_{k+2} \cdot b \subseteq (a' \cdot V_n \cdot b') \cap (a'' \cdot V_m \cdot b''),$$

i.e. $\mathcal{B}_0'(R)$ satisfies the condition (BN2).

Let $a \cdot V_n \cdot b \in \mathcal{B}_0'(R)$. Then $a \cdot V_{n+1} \cdot b \in \mathcal{B}_0'(R)$, where

$$a \cdot V_{n+1} \cdot b + a \cdot V_{n+1} \cdot b = a \cdot (V_{n+1} + V_{n+1}) \cdot b \subseteq a \cdot V_n \cdot b,$$

i.e. $\mathcal{B}_0'(R)$ satisfies the condition (BN3).

The fulfillment of the condition (BN4) is evident, since

$$-(a \cdot V_n \cdot b) = a \cdot (-V_n) \cdot b = a \cdot V_n \cdot b$$

(the neighborhood V_n is symmetric).

Let $a \cdot V_n \cdot b \in \mathcal{B}_0'(R)$. We can choose a neighborhood V of zero in (R, τ) such that $V \subseteq V_{n+3}$, $a \cdot V \subseteq V_{n+3}$ and $V \cdot b \subseteq V_{n+3}$. Since the zero element is not isolated in S, then there exists a non-zero element $c \in V \cap S$. Then $c \in V_{n+3}$, $a \cdot c \in V_{n+3}$ and $c \cdot b \in V_{n+3}$. Since $a \cdot c$, $c \cdot b \in S'$, then $a \cdot c \cdot V_{n+3} \cdot c \cdot b \in \mathcal{B}_0'(R)$, besides

$$\left(a \cdot c \cdot V_{n+3} \cdot c \cdot b\right) \cdot \left(a \cdot c \cdot V_{n+3} \cdot c \cdot b\right) = a \cdot \left(c \cdot V_{n+3} \cdot c \cdot b \cdot a \cdot c \cdot V_{n+3} \cdot c\right) \cdot b$$
$$\subseteq a \cdot \left(V_{n+3} \cdot V_{n+3} \cdot V_{n+3} \cdot V_{n+3} \cdot V_{n+3} \cdot V_{n+3}\right) \cdot b \subseteq a \cdot V_n \cdot b,$$

i.e. $\mathcal{B}_0'(A)$ satisfies the condition (BN5).

Let $a \cdot V_n \cdot b \in \mathcal{B}_0'(R)$ and $r \in R$. Due to Corollary 1.1.46, $a \cdot V_{n+1}$ and $V_{n+1} \cdot b$ are the neighborhoods of zero in (R, τ). Therefore, we can choose a neighborhood V of zero in (R, τ) such that $r \cdot V \subseteq a \cdot V_{n+1}$ and $V \cdot r \subseteq V_{n+1} \cdot b$. Since the zero element is not isolated in S, there exists a non-zero element $c \in V \cap S$. Then $r \cdot c \in a \cdot V_{n+1}$, $c \cdot r \in V_{n+1} \cdot b$, and $c \cdot V_{n+1} \cdot b \in \mathcal{B}_0'(R)$ and $a \cdot V_{n+1} \cdot c \in \mathcal{B}_0'(R)$, besides

$$r \cdot (c \cdot V_{n+1} \cdot b) = r \cdot c \cdot V_{n+1} \cdot b \subseteq a \cdot V_{n+1} \cdot V_{n+1} \cdot b \subseteq a \cdot V_n \cdot b$$

and

$$(a \cdot V_{n+1} \cdot c) \cdot r = a \cdot V_{n+1} \cdot c \cdot r \subseteq a \cdot V_{n+1} \cdot V_{n+1} \cdot b \subseteq a \cdot V_n \cdot b.$$

Hence, the condition (BN6) is fulfilled for $\mathcal{B}_0'(R)$.

According to Corollary 1.3.7, the system $\mathcal{B}_0'(R)$ defines on R Hausdorff ring topology τ', besides, $\mathcal{B}_0'(R)$ is basis of neighborhoods of zero of the topological ring (R, τ'). Due to Corollary 1.1.46, any subset of the type $a \cdot V_n \cdot b \in \mathcal{B}_0'(R)$ is a neighborhood of zero of the topological ring (R, τ), hence $\tau' \leq \tau$. Since the cardinality of $\mathcal{B}_0'(R)$ does not exceed the cardinality of the set S, then $\tau' \in \Delta$. This completes the proof of the proposition.

1.7.10. THEOREM. *Any Hausdorff non-discrete locally inversely bounded from right (left) ring topology τ on a skew field R (see Definition 1.6.61) is non-weakable in the class of all Hausdorff ring topologies on R.*

PROOF. Denote over $\Delta = \Delta(R)$ the class of all Hausdorff ring topologies on R. Assume the contrary, i.e. that there exists a topology $\tau' \in \Delta$ such that $\tau' < \tau$. Let $\mathcal{B}_0(R, \tau)$ and $\mathcal{B}_0(R, \tau')$ be bases of neighborhoods of zero of (R, τ) and (R, τ') correspondingly. Since $\tau' < \tau$, then there exists a neighborhood $V_0 \in \mathcal{B}_0(R, \tau)$ such that $U \not\subseteq V_0$ for any neighborhood $U \in \mathcal{B}_0(R, \tau')$. Since the topology τ' is Hausdorff, then there exists a neighborhood

$U_0 \in \mathcal{B}_0(R, \tau')$ such that $1 \notin U_0$. We can choose a neighborhood $U_1 \in \mathcal{B}_0(R, \tau')$ such that $U_1 \cdot U_1 \subseteq U_0$. Since $\tau' < \tau$, then U_1 is a neighborhood of zero in (R, τ). Due to Propositions 1.6.3, there exists a non-zero element $a \in R$ such that $(R \backslash V_0)^{-1} \cdot a \subseteq U_1$. Let $U_2 \in \mathcal{B}_0(R, \tau')$ and $a^{-1} \cdot U_2 \subseteq U_1$. Since $U_2 \nsubseteq V_0$, then there exists an element $b \in (R \backslash V_0) \cap U_2$. It is clear that $b \neq 0$. Then $1 = a^{-1} \cdot b \cdot b^{-1} \cdot a \in (a^{-1} \cdot U_2) \cdot (R \backslash V_0)^{-1} \cdot a \subseteq U_1 \cdot U_1 \subseteq U_0$. This contradicts the choice of the neighborhood U_0 and completes the proof of the theorem.

1.7.11. COROLLARY. *Let an interval topology τ_ξ on a skew field R, defined by a norm ξ (see Proposition 1.2.14), be non-discrete. Then the topology τ_ξ is non-weakable in the class of all Hausdorff ring topologies on R.*

PROOF. Proof results from Theorem 1.7.10, since τ_ξ is locally inversely bounded topology (see Example 1.6.63).

1.7.12. EXAMPLE. Let p and q be different prime numbers, τ_p and τ_q be correspondingly p-adic and q-adic topologies on the field $\mathbb{Q}$ of rational numbers (see Example 1.1.19). The topologies τ_p and τ_q are non-discrete and Hausdorff. Both topologies are induced by norms (see Example 1.1.19). Due to Example 1.6.63, $(\mathbb{Q}, \tau_p)$ and $(\mathbb{Q}, \tau_q)$ are locally inversely bounded fields. In the topology $\tau = \sup\{\tau_p, \tau_q\}$ the field $\mathbb{Q}$ is non-discrete locally bounded topological field (see Theorem 1.2.21 and Proposition 1.6.52). But the topological field $(\mathbb{Q}, \tau)$ is not locally inversely bounded, because from the incomparability of the topologies τ_p and τ_q follows that $\tau_p < \tau$ and $\tau_q < \tau$, i.e. the topology τ is weakable in the class of all ring topologies on $\mathbb{Q}$ (see Theorem 1.7.10). Due to Corollary 1.7.12, the topology τ cannot be defined as an interval for any definition of a norm on $\mathbb{Q}$.

Thus, this example demonstrates that:

1) there exist locally bounded fields that are not locally inversely bounded;

2) the least upper bound (even of the finite number) of locally inversely bounded field topologies can be not locally inversely bounded;

3) the least upper bound (even of the finite number) of field topologies, induced by a norm, can be not induced by a norm.

1.7.13. PROPOSITION. *Let B be a dense subgroup of a topological Abelian group*

(A, τ); $\Delta(A)$ and $\Delta(B)$ *be families of group topologies on A and B correspondingly, such that* $\tau'|_B \in \Delta(B)$ *for any topology* $\tau' \in \Delta(A)$. *If the topology* $\tau|_B$ *is non-weakable in the class* $\Delta(B)$, *then the topology* τ *is non-weakable in the class* $\Delta(A)$.

PROOF. Assume the contrary, i.e. that the topology τ is weakable in the class $\Delta(A)$. Then there exists a topology $\tau' \in \Delta(A)$ such that $\tau' < \tau$. Let $\mathcal{B}_0(A, \tau')$ and $\mathcal{B}_0(A, \tau)$ be bases of neighborhoods of zero of the topological groups (A, τ') and (A, τ) correspondingly. Then there exists a neighborhood $U \in \mathcal{B}_0(A, \tau)$ such that $U' \not\subseteq U$ for any neighborhood $U' \in \mathcal{B}_0(A, \tau')$. We can choose a neighborhood $V \in \mathcal{B}_0(A, \tau)$ such that $V - V \subseteq U$.

Since $\tau'|_B \in \Delta(B)$ and $\tau' < \tau$, then $\tau'|_B \leq \tau|_B$. Taking into account that the topology $\tau|_B$ is non-weakable in the class $\Delta(B)$, we get that $\tau'|_B = \tau|_B$. Therefore, there exists a neighborhood $V' \in \mathcal{B}_0(A, \tau)$ such that $V' \cap B \subseteq V \cap B \subseteq V$. Let W' be a neighborhood from $\mathcal{B}_0(A, \tau')$ such that $W' + W' \subseteq V'$. Since $\tau' < \tau$, then W' is a neighborhood of zero in the topological group (A, τ), thus, there exists a neighborhood $W \in \mathcal{B}_0(A, \tau)$ such that $W \subseteq W' \cap V$. If $x \in W'$, then, taking into account the density of the subgroup $B \subseteq (A, \tau)$, we get that $(x + W) \cap B \neq \emptyset$. Let $y \in (x + W) \cap B$. Then

$$y \in (x + W) \cap B \subseteq (W' + W') \cap B \subseteq (W' + W') \cap B \subseteq V' \cap B \subseteq V.$$

Therefore, $x \in y - W \subseteq V - V \subseteq U$. Thus, $W' \subseteq U$, which contradicts the choice of the neighborhood U. This completes the proof of the proposition.

1.7.14. COROLLARY. Let Q be a dense subring of a topological ring (R, τ); $\Delta(R)$ and $\Delta(Q)$ be such families of the ring topologies on R and Q correspondingly that $\tau'|_Q \in \Delta(Q)$ for any topology $\tau' \in \Delta(R)$. If the topology $\tau|_Q$ is non-weakable in the class $\Delta(Q)$, then the topology τ is non-weakable in the class $\Delta(R)$.

1.7.15. COROLLARY. Let R be a topological ring, N be a dense submodule of a topological R-module (M, τ); $\Delta(M)$ and $\Delta(N)$ be such families of module topologies on M and N correspondingly that $\tau'|_N \in \Delta(N)$ for any topology $\tau' \in \Delta(M)$. If the topology $\tau|_N$ is non-weakable in the class $\Delta(N)$, then the topology τ is non-weakable in the class $\Delta(M)$.

1.7.16. PROPOSITION. *Let* (A, τ) *be a topological Abelian group and B be its dense subgroup. Let* τ_0' *be a group topology on B such that* $\tau_0' \leq \tau|_B$ *(see C.8) and* $\mathcal{B}_0'(B, \tau_0')$ *be a* *basis of symmetric neighborhoods of zero of the topological group* (B, τ_0'), *then the system*

$$\mathcal{B}_0'(A) = \left\{ [U]_{(A,\tau)} \mid U \in \mathcal{B}_0'(B, \tau') \right\}$$

of the subsets of A *is a basis of neighborhoods of zero for a certain group topology* τ' *on the group* A, *besides,* $\tau' \leq \tau$ *and* $\tau'\big|_B = \tau_0'$.

PROOF. Let's verify that the system $\mathcal{B}_0'(A)$ of the subsets of the group A satisfies the conditions (BN1) - (BN4) (see Proposition 1.2.1).

It is evident that

$$0 \in \bigcap_{U \in \mathcal{B}_0'(B, \tau_0')} [U]_{(A,\tau)},$$

i.e., the condition (BN1) is fulfilled.

Let $U_1, U_2, U_3 \in \mathcal{B}_0'(B, \tau_0')$ and $U_3 \subseteq U_1 \cap U_2$. Then

$$[U_3]_{(A,\tau)} \subseteq [U_1]_{(A,\tau)} \cap [U_2]_{(A,\tau)},$$

i.e. $\mathcal{B}_0'(A)$ satisfies the condition (BN2).

If $U \in \mathcal{B}_0'(B, \tau_0')$, then, choosing $V \in \mathcal{B}_0'(B, \tau_0')$ such that $V + V \subseteq U$, and taking into account Proposition 1.1.41, we get that

$$[V]_{(A,\tau)} + [V]_{(A,\tau)} \subseteq [V + V]_{(A,\tau)} \subseteq [U]_{(A,\tau)},$$

i.e. the condition (BN3) is fulfilled.

From the equality $[U]_{(A,\tau)} = [-U]_{(A,\tau)}$ (see Proposition 1.1.41) follows that the condition (BN4) is also fulfilled, too.

From Theorem 1.2.4 follows that on A exists a topology τ' such that (A, τ') is a topological group, and $\mathcal{B}_0'(A)$ is a basis of neighborhoods of zero in (A, τ').

Since $\tau_0' \leq \tau\big|_B$, then any subset $U \in \mathcal{B}_0'(B, \tau_0')$ is a neighborhood of zero of the group $(B, \tau\big|_B)$. According to Proposition 1.4.24, $[U]_{(A,\tau)}$ is a neighborhood of zero of the group (A, τ), for any neighborhood $U \in \mathcal{B}_0'(B, \tau_0')$. Hence, $\tau' \leq \tau$.

Let's show that $\tau'\big|_B = \tau_0'$. Indeed, let V be a neighborhood of zero of the group $(B, \tau'\big|_B)$ and W be a neighborhood of zero of the group (A, τ') such that $V = W \cap B$. Due to the construction of the topology τ', there exists a neighborhood $U \in \mathcal{B}_0'(B, \tau_0')$ such that

$[U]_{(A,\tau)} \subseteq W$. Hence, $U \subseteq W$. Then $U = U \cap B \subseteq W \cap B = V$, i.e. V is a neighborhood of zero of the group (B, τ'_0), hence, $\tau'\big|_B \leq \tau'_0$.

Let $V \in \mathcal{B}_0(B, \tau'_0)$, $U \in \mathcal{B}_0(B, \tau'_0)$ and $U + U \subseteq V$. Then, due to Corollary 1.2.27, $[U]_{(B,\tau'_0)} \subseteq V$. Taking into account that $\tau'_0 \leq \tau\big|_B$, we get that

$$[U]_{(A,\tau)} \cap B = [U]_{(B,\tau|_B)} \subseteq [U]_{(B,\tau'_0)} \subseteq V.$$

Since $[U]_{(A,\tau)}$ is a neighborhood of zero of the group (A, τ'), then V is a neighborhood of zero of the group $(B, \tau'\big|_B)$, i.e. $\tau'_0 \leq \tau'\big|_B$.

1.7.17. PROPOSITION. *Let (R, τ) be a topological ring, Q be a dense subring of (R, τ), τ'_0 be a ring topology on Q such that $\tau'_0 \leq \tau\big|_Q$, and $\mathcal{B}'_0(Q, \tau'_0)$ be a basis of symmetric neighborhoods of zero of the topological ring (Q, τ'_0). Then the system*

$$\mathcal{B}'_0(R) = \big\{ [U]_{(R,\tau)} \big| U \in \mathcal{B}'_0(Q, \tau'_0) \big\}$$

of the subsets of the ring R is a basis of neighborhoods of zero for a certain ring topology τ' on R, besides $\tau' \leq \tau$ and $\tau'\big|_Q = \tau'_0$.

PROOF. Due to Proposition 1.7.16, $\mathcal{B}'_0(R)$ is a basis of neighborhoods of zero for a certain group topology τ' on R, besides $\tau' \leq \tau$ and $\tau'\big|_Q = \tau'_0$. Let's show that τ' is a ring topology. For it we verify that the system $\mathcal{B}'_0(R)$ satisfies the conditions (BN5) - (BN7) (see Proposition 1.2.2 and Theorem 1.2.5).

Let $U \in \mathcal{B}'_0(Q, \tau'_0)$. Applying the condition (BN5) to the basis $\mathcal{B}'_0(Q, \tau'_0)$, we can choose $V \in \mathcal{B}'_0(Q, \tau'_0)$ such that $V \cdot V \subseteq U$. Then $[V]_{(R,\tau)} \in \mathcal{B}'_0(R)$, besides,

$$[V]_{(R,\tau)} \cdot [V]_{(R,\tau)} \subseteq [V \cdot V]_{(R,\tau)} \subseteq [U]_{(R,\tau)},$$

i.e. $\mathcal{B}'_0(R)$ satisfies the condition (BN5).

Let $U \in \mathcal{B}'_0(Q, \tau'_0)$ and $a \in R$. We can find U_1, $V \in \mathcal{B}'_0(Q, \tau'_0)$ such that $U_1 + U_1 \subseteq U$, $V \subseteq U_1$ and $V \cdot V \subseteq U_1$. Since $\tau'_0 \leq \tau\big|_Q$, then V is a neighborhood of zero of the dense subring $(Q, \tau\big|_Q)$ of the ring (R, τ). Due to Proposition 1.4.24, $[V]_{(R,\tau)}$ is a neighborhood of zero of the topological ring (R, τ). Since $(Q, \tau\big|_Q)$ is a dense subring in (R, τ), then

$(a + [V]_{(R,\tau)}) \cap Q \neq \emptyset$. Let $b \in (a + [V]_{(R,\tau)}) \cap Q$. Let also $W \in \mathcal{B}_0'(Q, \tau_0')$ be such that $W \subseteq V$, $b \cdot W \subseteq V$ and $W \cdot b \subseteq V$. Then $[W]_{(R,\tau)} \in \mathcal{B}_0'(R)$, besides,

$$\begin{aligned} a \cdot [W]_{(R,\tau)} &\subseteq (b + [V]_{(R,\tau)}) \cdot [W]_{(R,\tau)} \subseteq b \cdot [W]_{(R,\tau)} + [V]_{(R,\tau)} \cdot [W]_{(R,\tau)} \\ &\subseteq [b \cdot W]_{(R,\tau)} + [V \cdot W]_{(R,\tau)} \subseteq [V]_{(R,\tau)} + [V \cdot V]_{(R,\tau)} \\ &\subseteq [U_1]_{(R,\tau)} + [U_1]_{(R,\tau)} \subseteq [U_1 + U_1]_{(R,\tau)} \subseteq [U]_{(R,\tau)}. \end{aligned}$$

Analogously is proved that $[W]_{(R,\tau)} \cdot a \subseteq [U]_{(R,\tau)}$. Thus, $\mathcal{B}_0'(R)$ satisfies the condition (BN6).

1.7.18. PROPOSITION. *Let R be a topological ring, (M, τ) be a topological R-module and N be a dense submodule of the topological R-module M. Let r_0' be such module topology on N that $\tau_0' \leq \tau\big|_N$, and let $\mathcal{B}_0(N, \tau_0')$ be a basis of symmetric neighborhoods of zero of the topological R-module (N, τ_0'). Then the system*

$$\mathcal{B}_0'(M) = \{[U]_{(M,\tau)} \mid U \in \mathcal{B}_0'(N, \tau_0')\}$$

of the subsets of R-module M is a basis of neighborhoods of zero for a certain module topology τ' on M, besides, $\tau' \leq \tau$ and $\tau'\big|_N = \tau_0'$.

PROOF. Due to Proposition 1.7.16, $\mathcal{B}_0'(M)$ is a basis of neighborhoods of zero for a certain group topology τ' on M, besides, $\tau' \leq \tau$ and $\tau'\big|_N = \tau_0'$.

Let's show that τ' is a module topology. For it we verify that the system $\mathcal{B}_0'(M)$ satisfies the conditions (BN5'), (BN6') and (BN6'') (see Proposition 1.2.3 and Theorem 1.2.6).

Let $[U]_{(M,\tau)} \in \mathcal{B}_0'(M)$, where $U \in \mathcal{B}_0'(N, \tau_0')$. Applying the condition (BN5') to the basis $\mathcal{B}_0'(N, \tau_0')$, we can find neighborhoods V of zero of the ring R and $W \in \mathcal{B}_0'(N, \tau_0')$ such that $V \cdot W \subseteq U$. Then $[W]_{(M,\tau)} \in \mathcal{B}_0'(M)$, besides,

$$V \cdot [W]_{(M,\tau)} \subseteq [V \cdot W]_{(M,\tau)} \subseteq [U]_{(M,\tau)},$$

i.e. $\mathcal{B}_0'(M)$ satisfies the condition (BN5').

Let $[U]_{(M,\tau)} \in \mathcal{B}_0'(M)$, where $U \in \mathcal{B}_0'(N, \tau_0')$ and $a \in R$. Applying the condition (BN6') to the basis $\mathcal{B}_0'(N, \tau_0')$, we can find a neighborhood $W \in \mathcal{B}_0'(N, \tau_0')$ such that $a \cdot W \subseteq U$. Then $[W]_{(M,\tau)} \in \mathcal{B}_0'(M)$, besides,

$$a \cdot [W]_{(M,\tau)} \subseteq [a \cdot W]_{(M,\tau)} \subseteq [U]_{(M,\tau)},$$

i.e. $\mathcal{B}_0'(M)$ satisfies the condition (BN6').

Let $[U]_{(M,\tau)} \in \mathcal{B}_0'(M)$, where $U \in \mathcal{B}_0'(N, \tau_0')$ and $x \in M$. We can choose a neighborhood U_1 from $\mathcal{B}_0'(N, \tau_0')$ such that $U_1 + U_1 \subseteq U$. There exist neighborhoods V of zero in R and $U_2 \in \mathcal{B}_0'(N, \tau_0')$ such that $V \cdot U_2 \subseteq U_1$. Since $[U_2]_{(M,\tau)}$ is a neighborhood of zero of the topological group (M, τ') and $\tau' \le \tau$, then $[U_2]_{(M,\tau)}$ is a neighborhood of zero of the topological group (M, τ). Since the submodule $(N, \tau\big|_N)$ is dense in (M, τ), we get that $(x + [U_2]_{(M,\tau)}) \cap N \ne \emptyset$. Let $y \in (x + [U_2]_{(M,\tau)}) \cap N$. Then $y \in N$ and $x \in y + [U_2]_{(M,\tau)}$. We can choose a neighborhood V_1 of zero in R such that $V_1 \subseteq V$ and $V_1 \cdot y \subseteq U_1$. Then

$$
\begin{aligned}
V_1 \cdot x &\subseteq V_1 \cdot \left(y + [U_2]_{(M,\tau)}\right) \subseteq V_1 \cdot y + V_1 \cdot [U_2]_{(M,\tau)} \\
&\subseteq U_1 + \left[V_1 \cdot U_2\right]_{(M,\tau)} \subseteq [U_1]_{(M,\tau)} + \left[V \cdot U_2\right]_{(M,\tau)} \\
&\subseteq [U_1]_{(M,\tau)} + [U_1]_{(M,\tau)} \subseteq [U_1 + U_1]_{(M,\tau)} \subseteq [U]_{(M,\tau)},
\end{aligned}
$$

i.e. $\mathcal{B}_0'(M)$ satisfies the condition (BN6'').

1.7.19. THEOREM. *Let Q be a subring of a non-zero ring R; $\Delta(R)$ and $\Delta(Q)$ be the classes of all Hausdorff ring topologies on R and Q correspondingly. Let τ be a topology from $\Delta(R)$ such that:*

1) (R, τ) is a topologically simple ring (see Definition 1.4.15);

2) $(Q, \tau\big|_Q)$ is a dense subring of (R, τ).

Then the topology τ is non-weakable in the class $\Delta(R)$ if and only if the topology $\tau\big|_Q$ is non-weakable in the class $\Delta(Q)$.

PROOF. Since the topology $\tau\big|_Q$ is a Hausdorff ring topology on Q, i.e. $\tau\big|_Q \in \Delta(Q)$, and since $\tau\big|_Q$ is non-weakable in the class $\Delta(Q)$, then from Corollary 1.7.14 follows that the topology τ is non-weakable in the class $\Delta(R)$.

Vice versa, let the topology τ be non-weakable in the class $\Delta(R)$. Assume that τ_0' is a Hausdorff ring topology on Q and $\tau_0' \le \tau\big|_Q$. Due to Proposition 1.7.17, there exists a ring topology τ' on R such that $\tau' \le \tau$ and $\tau'\big|_Q = \tau_0'$.

Let's show that τ' is a Hausdorff topology. Denote over R_0 the smallest closed ideal of the topological ring (R, τ'), and over Q_0 the smallest closed ideal of the topological ring $(Q, \tau_0') = (Q, \tau'\big|_Q)$. Since $\tau' \le \tau$, then R_0 is a closed ideal of the topological ring (R, τ).

Since the ring (R, τ) is topologically simple, then we get that either $R_0 = \{0\}$, or $R_0 = R$. Assume that $R_0 = R$. Since the ring (Q, τ_0') is Hausdorff, then from Remark 1.4.14 follows that $Q = R \cap Q = R_0 \cap Q = Q_0 = \{0\}$. Therefore, since the ring (R, τ) is Hausdorff, we get that $\{0\} = [\{0\}]_{(R,\tau)} = [Q]_{(R,\tau)}$, which contradicts the density of the subring Q in (R, τ). Thus, $R_0 = \{0\}$. Due to Remark 1.4.13, (R, τ') is Hausdorff ring, i.e. $\tau' \in \Delta(R)$. Since $\tau' \leq \tau$ and the topology τ is non-weakable in the class $\Delta(R)$, then $\tau' = \tau$. Hence, $\tau_0' = \tau'|_Q = \tau|_Q$, i.e. the topology $\tau|_Q$ is non-weakable in the class $\Delta(Q)$.

1.7.20. THEOREM. Let R be a topological ring, N be a submodule of a non-zero R-module M; $\Delta(M)$ and $\Delta(N)$ be the classes of all Hausdorff module topologies on M and N correspondingly. Let τ be a topology from $\Delta(M)$ such that:

1) (M, τ) is topologically simple R-module (see Definition 1.4.15);

2) $(N, \tau|_N)$ is dense submodule of the module (M, τ).

Then the topology τ is non-weakable in the class $\Delta(M)$ if and only if the topology $\tau|_N$ is non-weakable in the class $\Delta(N)$.

PROOF. The proof is analogous to the proof of Theorem 1.7.19, using Proposition 1.7.18, Remark 1.4.13 and Definition 1.4.15.

1.7.21. COROLLARY. Let B be a subgroup of a non-zero Abelian group A; $\Delta(A)$ and $\Delta(B)$ be the classes of all Hausdorff group topologies on A and B correspondingly. Let τ be a topology from the $\Delta(A)$ such that:

1) the topological group (A, τ) does not contain proper closed subgroups;

2) $(B, \tau|_B)$ is dense subgroup of the group (A, τ).

Then the topology τ is non-weakable in the class $\Delta(A)$ if and only if the topology $\tau|_B$ is non-weakable in the class $\Delta(B)$.

§ 1.8. Topological Divisors of Zero and
Topological Nilpotency

1.8.1. DEFINITION. An element a of a topological ring R is called a left (right) topological divisor of zero* if there exists a subset $S \subset R$ such that:

1) $0 \notin [S]_R$;

2) $0 \in [a \cdot S]_R$ (correspondingly, $0 \in [S \cdot a]_R$).

An element a is called a topological divisor of zero if it is a left and right topological divisor of zero, i.e. there exist subsets $S_1 \subset R$ and $S_2 \subset R$ such that $0 \notin [S_1]_R$ and $0 \notin [S_2]_R$, but as well $0 \in [a \cdot S_1]_R$ and $0 \in [S_2 \cdot a]_R$.

1.8.2. REMARK. In a Hausdorff topological ring R any left (right) divisor of zero is a left (right) topological divisor of zero.

Indeed, if a is a left divisor of zero in R and $0 \neq b$ is such that $a \cdot b = 0$, then, due to Theorem 1.3.2 and Proposition 1.1.34, the subset $\{b\}$ is closed in R and $0 \notin \{b\} = [\{b\}]_R$. It is evident that $0 \in [a \cdot \{b\}]_R = \{0\}$, i.e. a is a left topological divisor of zero in R.

1.8.3. EXAMPLE. A topological ring R with anti-discrete topology has neither left nor right topological divisors of zero, though it can even be a ring with zero multiplication (see Remark 1.1.8).

Indeed, if $0 \notin [S]_R$, then $[S]_R = \emptyset$ (since the topology of the ring R is anti-discrete). Then $[a \cdot S]_R = \emptyset$.

Thus, this example demonstrates that the condition of being Hausdorff for the ring R in Remark 1.8.2 is important.

1.8.4. REMARK. In a discrete ring R only the left (right) divisors of zero are left (right) topological divisors of zero.

The statement results from Remark 1.8.2, since any discrete ring is Hausdorff. Vice versa, if a is a left (right) topological divisor of zero, then there exists a subset $S \subset R$ such that $0 \notin [S]_R = S$ and $0 \in [a \cdot S]_R = a \cdot S$ ($0 \in [S \cdot a]_R = S \cdot a$ correspondingly). Then

---

* Sometimes the topological divisors of zero are called generalized divisors of zero (e.g. see [146, chapter II, § 10]).

$S \neq \emptyset$. There exists an element $b \in S_1$ such that $a \cdot b = 0$ (correspondingly, $b \cdot a = 0$), i.e. a is a left (right) divisor of zero in R.

1.8.5. EXAMPLE. Let p and q be different prime numbers. Consider the system $\mathcal{B}_0(\mathbb{Z}) = \{V_k | k = 1, 2, \ldots\}$ of the subsets of the ring $\mathbb{Z}$ of integers, where V_k is the ideal in $\mathbb{Z}$ generated by $p \cdot q^k$, i.e. $V_k = p \cdot q^k \cdot \mathbb{Z}$ for $k = 1, 2, \ldots$. Since V_k is ideal in $\mathbb{Z}$ for $k = 1, 2, \ldots$, then the system $\mathcal{B}_0(\mathbb{Z})$ satisfies the conditions (BN1), (BN3) - (BN6).

Moreover, from the evident inclusion $V_{k+1} \subseteq V_k$ follows that $\mathcal{B}_0(\mathbb{Z})$ satisfies the condition (BN2).

Since the only integer divisible by all natural powers of the number q is 0, then the system $\mathcal{B}_0(Z)$ satifies the condition (BN1').

Due to Corollary 1.3.7, the system $\mathcal{B}_0(\mathbb{Z})$ defines on $\mathbb{Z}$ the Hausdorff ring topology $\tau_{p \cdot q}$, besides, $\mathcal{B}_0(\mathbb{Z})$ is a basis of neighborhoods of zero of the topological ring $(\mathbb{Z}, \tau_{pq})$. The subset $S = \{q^k \mid k = 1, 2, \ldots\}$ does not contain the numbers, divisible by p, hence $S \cap V_1 = \emptyset$, i.e. $0 \notin [S]_{(\mathbb{Z}, \tau_{pq})}$. Since $pq^k \in V_k$, then $(p \cdot S) \cap V_k \neq \emptyset$ for any $k = 1, 2, \ldots$, i.e. $0 \in [p \cdot S]_{(\mathbb{Z}, \tau_{pq})}$. Therefore, p is a topological divisor of zero in the ring $(\mathbb{Z}, \tau_{pq})$. Since the ring $\mathbb{Z}$ does not contain non-zero divisors of zero, then this example demonstrates that in a non-discrete Hausdorff ring the conceptions of the divisor of zero and topological divisor of zero, in general, are different.

1.8.6. EXAMPLE. Consider in the topological ring $C([0; 1], \mathbb{R})$ of real-valued continuous functions on the segment $[0;1]$ (see Example 1.1.17) the element f such that $f(x) = x$ for any $x \in [0; 1]$. This element is not a divisor of zero in the ring $C([0; 1], \mathbb{R})$. Let's show that f is a topological divisor of zero.

For any $n = 1, 2, \ldots$ consider in $C([0; 1], \mathbb{R})$ the element f_n, where

$$f_n(x) = \begin{cases} \sin(\pi \cdot n \cdot x), & \text{if } 0 \leq x \leq \frac{1}{n}; \\ 0, & \text{if } \frac{1}{n} < x \leq 1. \end{cases}$$

Let $S = \{f_n \mid n = 1, 2, \ldots\}$. Since

$$\xi(f_n) = \max\{|f_n(x)|| \in [0; 1]\} = \max\{|f_n(x)|| x \in [0; \frac{1}{n}]\} = 1$$

for any $n = 1, 2, \ldots$, then the zero element of the ring $C([0;1], \mathbb{R})$ does not belong to $[S]_{C([0;1],\mathbb{R})}$. On the other hand, the element

$$\xi(f \cdot f_n) = \max\{|(f \cdot f_n)(x)| \, 0 \le x \le 1\} = \max\{|x| \cdot |\sin(\pi \cdot n \cdot x)| \, | \, 0 \le x \le \frac{1}{n}\} \le \frac{1}{n},$$

hence, $0 \in [f \cdot S]_{C([0,1],\mathbb{R})}$, i.e. f is a topological divisor of zero.

1.8.7. PROPOSITION. *Let R be a Hausdorff topological ring, $a \in R$. The following statements are equivalent:*

1) a is a left (right) topological divisor of zero in R;

2) there exists a neighborhood U of zero in R such that for any neighborhood V of zero in R can be found an element b from $R \backslash U$ such that $a \cdot b \in V$ (correspondingly, $b \cdot a \in V$);

3) the mapping $f_a : R \to a \cdot R$, for which $f_a(x) = a \cdot x$ for any $x \in R$ (correspondingly, the mapping $g_a : R \to R \cdot a$, for which $g_a(x) = x \cdot a$ for any $x \in R$), is not a topological isomorphism of the additive topological group $R(+)$ onto its subgroup $(a \cdot R)(+)$ (correspondingly, onto the subgroup $(R \cdot a)(+))$.

PROOF. Let a be a left topological divisor of zero in R and $S \subseteq R$. If $0 \notin [S]_R$ and $0 \in [a \cdot S]_R$, then $U = R \backslash [S]_R$ is a neighborhood of zero in R, besides,

$$R \backslash U = R \backslash (R \backslash [S]_R) = [S]_R \supseteq S.$$

Let V be a neighborhood of zero in R, then from $0 \in [a \cdot S]_R$ follows that $V \bigcap (a \cdot S) \ne \emptyset$, i.e. there exists an element $b \in S$ such that $a \cdot b \in V$. It is clear that $b \in R \backslash U$. Thus, 1)$\Longrightarrow$2).

Let U be a neighborhood of zero in R, whose existence is defined in the condition 2). Without loss of generality we can consider that U is an open neighborhood. If the mapping f_a is not a bijection, then 2) $\Longrightarrow$ 3).

Consider the case when f_a is a bijection. Then

$$f_a(R \backslash U) = f_a(R) \backslash f_a(U) = (a \cdot R) \backslash f_a(U).$$

Let W be a neighborhood of zero in $a \cdot R$. Then $W = (a \cdot R) \bigcap V$, where V is a certain neighborhood of zero in R. From the condition 2) follows that there exists an element

$b \in R \backslash U$ such that $a \cdot b \in V$. Since $a \cdot b \in a \cdot R$ and $a \cdot b = f_a(b)$, then $f_a(b) \in f_a(R \backslash U) = (a \cdot R) \backslash f_a(U)$ and $f_a(b) \in W$, hence, $((a \cdot R) \backslash f_a(U)) \bigcap W \neq \emptyset$. Therefore, $0 \in [(a \cdot R) \backslash f_a(U)]_R$. Since $0 \notin (a \cdot R) \backslash f_a(U)$, then $(a \cdot R) \backslash f_a(U)$ is not closed subset in $a \cdot R$, hence the subset $f_a(U)$ is not open in $a \cdot R$, i.e. f_a is not a topological isomorphism. Thus, 2)$\Longrightarrow$3) in that case, too.

If f_a is not a topological isomorphism, since $f_a : R \to a \cdot R$ is a continuous surjective group homomorphism (see Corollary 1.1.43), then either $\ker f_a \neq \{0\}$, or f_a is a bijection, but not open mapping.

If $\ker f_a \neq \{0\}$, then there exists a subset $Q \subset R$ such that $Q \neq \{0\}$ and $a \cdot x = f_a(x) = 0$, for any $x \in Q$, i.e. a is a left divisor of zero in R. Due to Remark 1.8.2, a is a left topological divisor of zero in R, i.e. in that case 3)$\Longrightarrow$1).

If f_a is a bijection that is not open mapping, then, due to C.7, there exists a closed subset $Q \subset R$ such that $f_a(Q) = a \cdot Q$ is not closed subset in $a \cdot R$. Let $b \in R$ be such that $a \cdot b \in [a \cdot Q]_{a \cdot R}$ and $a \cdot b \notin a \cdot Q$. Then it is clear that $b \notin Q$. Consider the subset $S = Q - b$. Then $0 \notin Q - b = [Q - b]_R = [S]_R$. Since

$$[a \cdot S]_R \supseteq [a \cdot S]_{a \cdot R} = [a \cdot Q - a \cdot b]_{a \cdot R} = [a \cdot Q]_{a \cdot R} - a \cdot b \ni 0,$$

then a is a left topological divisor of zero in R. Thus, in that case 3) $\Longrightarrow$ 1), too.

The equality of the right analogs of the statements 1) - 3) is proved analogously.

1.8.8. REMARK. Let R be a subring of a topological ring Q and an element $a \in R$ be a left (right) topological divisor of zero in R. Then a is a left (right) topological divisor of zero in the ring Q.

Indeed, let $S \subset R$, $0 \notin [S]_R$ and $0 \in [a \cdot S]_R$ (correspondingly, $0 \in [S \cdot a]_R$). Then $0 \notin [S]_Q$, since $[S]_Q \bigcap R = [S]_R$. Since $[a \cdot S]_R \subseteq [a \cdot S]_Q$ (correspondingly, $[S \cdot a]_R \subseteq [S \cdot a]_Q$), then $0 \in [a \cdot S]_Q$ (correspondingly, $0 \in [S \cdot a]_Q$), i.e. a is a left (right) topological divisor of zero in the topological ring Q.

1.8.9. PROPOSITION. *Let R be a subring of a topological ring Q with the unitary element. If an element $a \in R$ is invertible from left (right) in Q, then a is not a left (right) topological divisor of zero in the topological ring R.*

PROOF. Let $b \in Q$ and $b \cdot a = 1$. Then for any subset $S \subset Q$ from $0 \in [a \cdot S]_Q$ follows that $0 \in b \cdot [a \cdot S]_Q \subseteq [b \cdot a \cdot S]_Q = [S]_Q$, i.e. a is not a left topological divisor of zero in the ring Q. Due to Remark 1.8.8, a is not a left topological divisor of zero in the ring R either.

If a is invertible from right, the statement is proved analogously.

1.8.10. PROPOSITION. *Let a be a left (right) topological divisor of zero in a topological ring R. Then for any $b \in R$ the element $b \cdot a$ is a left topological divisor of zero in R (correspondingly, the element $a \cdot b$ is a right topological divisor of zero in R).*

PROOF. Let $S \subset R$, $0 \notin [S]_R$ and $0 \in [a \cdot S]_R$. Then $0 \in b \cdot [a \cdot S]_R \subseteq [b \cdot a \cdot S]_R$, i.e. $b \cdot a$ is a left topological divisor of zero in R.

Analogously is considered the case when a is a right topological divisor of zero in R.

1.8.11. THEOREM. *Let R be a Hausdorff compact ring and an element $a \in R$ be not a left (right) of divisor zero in R. Then a is not a left (right) topological divisor of zero in R.*

PROOF. Suppose that a is not a left divisor of zero. Then the mapping $f_a: R \longrightarrow a \cdot R$, where $f_a(r) = a \cdot r$, is a continuous bijection. By C.12, f_a is a homeomorphism, and therefore, f_a is an isomorphism of topological groups $R(+)$ and $a \cdot R(+)$. Then, by Theorem 1.8.7, a is not a left topological divisor of zero in R.

Analogously we consider the case when a is not a right divisor of zero.

1.8.12. PROPOSITION. *Let R be a topological ring and $a, b \in R$. If $a \cdot b$ is a left (right) topological divisor of zero in R, then either a or b is a left (right) topological divisor of zero.*

PROOF. Let $a \cdot b$ be a left topological divisor of zero in R and $S \subset R$ be such that $0 \notin [S]_R$, and $0 \in [(a \cdot b) \cdot S]_R$. If $0 \in [b \cdot S]_R$, then b is a left topological divisor of zero. If $0 \notin [b \cdot S]_R$, then, taking into account that $0 \in [(a \cdot b) \cdot S]_R = [a \cdot (b \cdot S)]_R$, we get that a is a left topological divisor of zero.

Analogously is considered the case when $a \cdot b$ is a right topological divisor of zero in R.

1.8.13. COROLLARY. *Let $l(R)$ be the system of all left, and $r(R)$ be the system of all right topological divisors of zero in the topological ring R. Then the subsets $R \backslash l(R)$*

and $R\backslash r(R)$ are the subsemigroups of the multiplicative semigroup of the ring R.

1.8.14. PROPOSITION. *Let R be a commutative ring, $\Delta = \Delta(R)$ be class of all Hausdorff ring topologies on R. Let $\tau \in \Delta$ and the topology τ be non-weakable in Δ. Let $a \in R$ and $a \in [a^2 \cdot R]_{(R,\tau)}$, then a is a topological divisor of zero in (R,τ) if and only if a is a divisor of zero.*

PROOF. Let a be a divisor of zero, then, since the topology τ is Hausdorff, from Remark 1.8.2 follows that a is a topological divisor of zero in (R,τ).

Vice versa, let a be a topological divisor of zero in (R,τ) and $a \in [a^2 \cdot R]_{(R,\tau)}$. Assume the contrary, i.e. that a is not a divisor of zero in R.

Consider a basis $\mathcal{B}_0(R,\tau)$ of symmetric neighborhoods of zero of the topological ring (R,τ). For any neighborhood $V \in \mathcal{B}_0(R,\tau)$ put

$$W_V = \left\{ z \in R \big| (z \cdot (a + V)) \cap V \neq \emptyset \right\}.$$

It is clear that $0 \in W_V$ for any neighborhood $V \in \mathcal{B}_0(R,\tau)$. Thus, the system

$$\mathcal{B}_0'(R) = \left\{ W_V \mid V \in \mathcal{B}_0(R,\tau) \right\}$$

of the subsets of the ring R satisfies the condition (BN1).

Let's show that $\mathcal{B}_0'(R)$ also satisfies the conditions (BN1') and (BN2) - (BN6) (see Propositions 1.2.1, 1.2.2. and Corollary 1.3.7).

Let $0 \neq b \in R$. Since a is not a divisor of zero in R, then $a \cdot b \neq 0$, hence, there exists a neighborhood $V_0 \subseteq \mathcal{B}_0(R,\tau)$ such that $a \cdot b \notin V_0$ (since the topology τ is Hausdorff). We can choose a neighborhood $V_1 \in \mathcal{B}_0(R,\tau)$ such that $V_1 - b \cdot V_1 \subseteq V_0$. Then $a \cdot b \notin V_1 - b \cdot V_1$ and hence, $(b \cdot (a + V_1)) \cap V_1 = (a \cdot b + b \cdot V_1) \cap V_1 = \emptyset$. Therefore, $b \notin W_{V_1}$ and moreover, $b \notin \bigcap_{V \in \mathcal{B}_0(R,\tau)} W_V$, i.e $\mathcal{B}_0(R,\tau)$ satisfies the condition (BN1').

Let $V_0, V_1, V_2 \in \mathcal{B}_0(R,\tau)$ and $V_0 \subseteq V_1 \cap V_2$, then $(z \cdot (a + V_0)) \cap V_0 \neq \emptyset$ for any element $z \in W_{V_0}$. Since $V_0 \subseteq V_1$ and $V_0 \subseteq V_2$, then $(z \cdot (a + V_1)) \cap V_1 \neq \emptyset$ and $(z \cdot (a + V_2)) \cap V_2 \neq \emptyset$, i.e. $z \in W_{V_1}$ and $z \in W_{V_2}$. Thus, $W_{V_0} \subseteq W_{V_1} \cap W_{V_2}$, i.e. $\mathcal{B}_0'(R)$ satisfies the condition (BN2).

Let $V_0 \in \mathcal{B}_0(R,\tau)$. We can choose a neighborhood $V_1 \in \mathcal{B}_0(R,\tau)$ such that $V_1 + V_1 \subseteq V_0$. Since $a \in [a^2 \cdot R]_{(R,\tau)}$, then there exists an element $r \in R$ such that $a^2 \cdot r \in a + V_1$. Let

$V_2 \in \mathcal{B}_0(R, \tau)$ be such that $r \cdot a \cdot V_2 + ra \cdot V_2 + r \cdot V_2 \cdot V_2 + ra \cdot V_2 + r \cdot V_2 \cdot V_2 \subseteq V_1$ and $z_1, z_2 \in W_{V_2}$. Then there exist elements $x_1, x_2 \in V_2$, for which $z_1 \cdot (a + x_1) \in V_2$ and $z_2 \cdot (a + x_2) \in V_2$. Then

$$r \cdot (a + x_1) \cdot (a + x_2) = r \cdot a^2 + r \cdot a \cdot x_1 + r \cdot a \cdot x_2 + r \cdot x_1 \cdot x_2$$
$$\in r \cdot a^2 + r \cdot a \cdot V_2 + r \cdot a \cdot V_2 + r \cdot V_2 \cdot V_2 \subseteq r \cdot a^2 + V_1 \subseteq a + V_1 + V_1 \subseteq V_0 + a.$$

Since

$$(z_1 + z_2) \cdot r \cdot (a + x_1) \cdot (a + x_2)$$
$$= r \cdot \left(z_1 \cdot (a + x_1) \cdot (a + x_2) + z_2 \cdot (a + x_1) \cdot (a + x_2) \right)$$
$$\in r \cdot \left(V_2 \cdot (a + V_2) + V_2 \cdot (a + V_2) \right)$$
$$\subseteq r \cdot a \cdot V_2 + r \cdot V_2 \cdot V_2 + r \cdot a \cdot V_2 + r \cdot V_2 \cdot V_2 \subseteq V_1 \subseteq V_0$$

and

$$z_1 \cdot z_2 \cdot r \cdot (a + x_1) \cdot (a + x_2) = r \cdot \left(z_1 \cdot (a + x_1) \cdot z_2 \cdot (a + x_2) \right)$$
$$\in r \cdot V_2 \cdot V_2 \subseteq V_1 \subseteq V_0,$$

then

$$\left((z_1 + z_2) \cdot (a + V_0) \right) \cap V_0 \ni (z_1 + z_2) \cdot r \cdot (a + x_1) \cdot (a + x_2)$$

and

$$\left((z_1 \cdot z_2) \cdot (a + V_0) \right) \cap V_0 \ni (z_1 \cdot z_2) \cdot r \cdot (a + x_1) \cdot (a + x_2).$$

Therefore,

$$\left((z_1 + z_2) \cdot (a + V_0) \right) \cap V_0 \neq \emptyset \quad \text{and} \quad z_1 \cdot z_2 \cdot (a + V_0) \cap V_0 \neq \emptyset,$$

i.e. $z_1 + z_2 \in W_{V_0}$ and $z_1 \cdot z_2 \in W_{V_0}$. Thus, $W_{V_2} + W_{V_2} \subseteq W_{V_0}$ and $W_{V_2} \cdot W_{V_2} \subseteq W_{V_0}$, i.e. $\mathcal{B}'_0(R)$ satisfies the conditions (BN3) and (BN5).

Since the neighborhood $V \in \mathcal{B}_0(R, \tau)$ is symmetric, then the subset W_V is symmetric, too, i.e. $\mathcal{B}'_0(R)$ satisfies the condition (BN4).

Finally, let $V_0 \in \mathcal{B}_0(R, \tau)$ and $b \in R$. We can choose a neighborhood $V_1 \in \mathcal{B}_0(R, \tau)$ such that $V_1 \subseteq V_0$ and $b \cdot V_1 \subseteq V_0$. Let $z \in W_{V_1}$, then $z \cdot (a + x) \in V_1$ for a certain element $x \in V_1$. Then $(z \cdot b) \cdot (a + x) = b \cdot (z \cdot (a + x)) \in b \cdot V_1 \subseteq V_0$. Since $x \in V_1 \subseteq V_0$, then $((z \cdot b) \cdot (a + V_0)) \bigcap V_0 \ni z \cdot b \cdot (a + x)$, i.e. $((z \cdot b) \cdot (a + V_0)) \bigcap V_0 \neq \emptyset$, therefore, $z \cdot b \in W_{V_0}$. Thus, $W_{V_1} \cdot b \subseteq W_{V_0}$, i.e. $\mathcal{B}_0(R)$ satisfies the condition (BN6).

Due to Corollary 1.3.7, $\mathcal{B}_0'(R)$ defines on R a Hausdorff ring topology τ.

Let V, $U_1 \in \mathcal{B}_0(R, \tau)$ and $U_1 \cdot a \subseteq V$. If $z \in U_1$, then $z \cdot (a + 0) = z \cdot a \in U_1 \cdot a \subseteq V$, therefore, $(z \cdot (a + V)) \bigcap V \ni z \cdot a$, i.e. $z \in W_V$, hence $U_1 \subseteq W_V$. Therefore, $\tau' \leq \tau$.

Further, since a is a topological divisor of zero in (R, τ), then, due to Proposition 1.8.7, there exists a neighborhood U_2 of zero in R such that for any $V \in \mathcal{B}_0(R, \tau)$ we get $a \cdot b \in V$, where b is a certain element from $R \backslash U_2$, i.e. $b \notin U_2$. Then $b \cdot (a + V) \ni b \cdot a$, thus, $b \in W_V$. Since $b \notin U_2$, then $W_V \nsubseteq U_2$ for all $V \in \mathcal{B}_0(R, \tau)$. Hence, $\tau' < \tau$, which contradicts the non-weakability of the topology τ in the class Δ. Thus, a is a divisor of zero.

1.8.15. THEOREM. *Let R be a commutative ring, $R^2 \neq \{0\}$ and $\Delta = \Delta(R)$ be the class of all Hausdorff ring topologies on R. If $\tau \in \Delta$ and the topology τ on the ring R is non-weakable in Δ and the topological ring (R, τ) is topologically simple, then it does not contain non-zero topological divisors of zero.*

PROOF. Let $0 \neq a \in R$. Let's show that a is not a topological divisor of zero. Due to Proposition 1.8.14, it is enough to show that a is not a divisor of zero. Assume the contrary, i.e. that a is a divisor of zero. Then the annulator $(0 : a)_R$ of the element a in R (see B.12) differs from $\{0\}$. Due to Corollary 1.4.31, $(0 : a)_R$ is a closed ideal of the ring (R, τ). Since R is a topologically simple ring, then $(0 : a)_R = R$ (see Definition 1.4.15), i.e. $a \cdot R = \{0\}$, hence the annulator $(0 : R)_R$ of the ring R differs from $\{0\}$. From Corollary 1.4.31 follows that $(0 : R)_R$ is a closed ideal of the ring (R, τ). Then $(0 : R)_R = R$, i.e. $R^2 = \{0\}$, which contradicts the condition of the theorem. The proof is completed.

1.8.16. DEFINITION. A subset S of a topological ring R is called topologically nilpotent if for any neighborhood U of zero in R there exists a natural number n_0 such that $S^{(n)} \subseteq U$ for all $n \geq n_0$ (see B.12). An element $a \in R$ is called topologically nilpotent if the one-element set $\{a\}$ of the ring R is topologically nilpotent, i.e. if for any neighborhood U of zero in R there exists a natural number n_0 such that $a^n \in U$ for all $n \geq n_0$.

1.8.17. REMARK. Since for any subsets S_1 and S of a ring R from the inclusion $S_1 \subseteq S$ follows that $S_1^{(n)} \subseteq S^{(n)}$ for any $n \in \mathbb{N}$, then any subset of a topologically nilpotent subset of a topological ring is a topologically nilpotent subset, too. In particular, any element of a topologically nilpotent subset is topologically nilpotent.

1.8.18. PROPOSITION. *Let ξ be a pseudonorm on a ring R and $0 < \epsilon < 1$. Then:*

1) the subset $S_\epsilon = \{x \in R \mid \xi(x) \leq \epsilon\}$ is topologically nilpotent;

2) if $\xi(a) < 1$, then a is topologically nilpotent element.

PROOF. Consider in R the basis $\mathcal{B}_0(R) = \{V_k \mid k = 1, 2, \dots\}$ of neighborhoods of zero, where $V_k = \{x \in R \mid \xi(x) < \frac{1}{2^k}\}$ for $k = 1, 2, \dots$. Let U be a neighborhood of zero in R and m be a natural number such that $V_m \subseteq U$. Since $0 < \epsilon < 1$, then we can choose a natural number n_0, for which $\epsilon^{n_0} < \frac{1}{2^m}$. If $n \geq n_0$ and $a \in S_\epsilon^{(n)}$, then $a = a_1 \cdot a_2 \cdot \dots \cdot a_n$, where $a_i \in S_\epsilon$, i.e. $\xi(a_i) \leq \epsilon$ for $i = 1, 2, \dots, n$. Therefore,

$$\xi(a) \leq \xi(a_1) \cdot \xi(a_2) \cdot \dots \cdot \xi(a_n) \leq \epsilon^n \leq \epsilon^{n_0} < \frac{1}{2^m},$$

i.e. $a \in V_m \subseteq U$. Thus, $S_\epsilon^{(n)} \subseteq U$ for all $n \geq n_0$. This means that S_ϵ is topologically nilpotent subset in R, i.e. we have proved the statement 1).

The statement 2) follows from 1).

1.8.19. REMARK. The subset $S = \{x \in R \mid \xi(x) < 1\}$ of a pseudonormed ring (R, ξ) can be not topologically nilpotent.

Indeed, let $\mathbb{R}$ be the field of real numbers with the pseudonorm $\xi(x) = \mid x \mid$ (see Example 1.1.15) and $S = \{x \in \mathbb{R} \mid \mid x \mid < 1\}$. It is clear that $S^{(n)} = S$ for $n = 1, 2, \dots$, i.e. the subset S is not topologically nilpotent.

1.8.20. DEFINITION. A subset S of a topological ring R is called topologically Σ-nilpotent if for any neighborhood U of zero in R there exists a natural number n_0 such that $S^n \subseteq U$ for all $n \geq n_0$ (see B.12). An element $a \in R$ is called topologically Σ-nilpotent if the one-element subset $\{a\}$ of the ring R is topologically Σ-nilpotent, i.e. if for any neighborhood U of zero in R there exists a natural number n_0 such that $k \cdot a^n \in U$ for all $n \geq n_0$ and $k = 1, 2, \ldots$.

1.8.21. REMARK. Since $S^{(n)} \subseteq S^n$ for any subset $S \subseteq R$ and any natural number n, then any topologically Σ-nilpotent subset (any topologically Σ-nilpotent element) of a topological ring R is a topologically nilpotent subset (topologically nilpotent element), too.

1.8.22. REMARK. Since from the inclusion $S_1 \subseteq S$ follows that $S_1^n \subseteq S^n$ for any $n \in \mathbb{N}$, then for a topological ring any subset of a topologically Σ-nilpotent subset is a topologically Σ-nilpotent subset. In particular, any element of a topologically Σ-nilpotent subset is a topologically Σ-nilpotent element.

1.8.23. REMARK. Let a topological ring R possess a basis of neighborhoods of zero consisting of additive subgroups, then the concepts of a topologically nilpotent and a topologically Σ-nilpotent subset (element) coincide.

Indeed, let $\mathcal{B}_0(R)$ be a basis of neighborhoods of zero consisting of additive subgroups of the ring R and a subset $S \subseteq R$ be topologically nilpotent. Let's verify that S is topologically Σ-nilpotent. Let $U \in \mathcal{B}_0(R)$ and n_0 be a natural number such that $S^{(n)} \subseteq U$ for $n \geq n_0$. Since U is a subgroup of the additive group $R(+)$, then $S^n \subseteq U$ for all $n \geq n_0$, i.e. S is a topologically Σ-nilpotent subset.

1.8.24. REMARK. Since discrete ring possesses the basis of neighborhoods of zero that consists of a single subgroup $\{0\}$, then the concepts of topological Σ-nilpotency and topological nilpotency of a subset (an element) for discrete rings coincide with the concept of nilpotency of a subset (an element).

1.8.25. EXAMPLE. Due to Proposition 1.8.18, in the topological field $\mathbb{R}$ of real numbers endowed with the interval topology the subset $S = \{x \in \mathbb{R} \mid \mid x \mid < \frac{1}{2}\}$ is topologically nilpotent. But S is not topologically Σ-nilpotent.

Indeed, $S^{(n)}$ contains the interval $(-\frac{1}{2^n}; \frac{1}{2^n})$, hence, the subgroup S^n, generated by the subset $S^{(n)}$ is a neighborhood of zero in $\mathbb{R}$. Therefore, S^n is an open-closed subgroup in $\mathbb{R}$ (see Proposition 1.4.18). Since the field $\mathbb{R}$ is connected, then $S^n = \mathbb{R}$ (see Corollary 1.4.20), i.e. the subset S is not topologically Σ-nilpotent.

It is also easy to verify that any of the non-zero elements from S is not topologically Σ-nilpotent.

1.8.26. REMARK. As we have seen in Example 1.8.25, a topological field can contain non-zero topologically nilpotent elements, though, due to the Proposition 1.8.9, it cannot have non-zero topological divisors of zero. Then, unlike the case of discrete rings, a topologically nilpotent element can be not a topological divisor of zero.

1.8.27. PROPOSITION. *Let K be a skew field endowed with Hausdorff ring topology and a be a non-zero topologically nilpotent element in K. Then the element a^{-1} is not topologically nilpotent.*

PROOF. Assume a contrary, i.e. that a^{-1} is a topologically nilpotent element. Since the skew field K is Hausdorff, there exists a neighborhood U of zero in K such that $1 \notin U$. We can choose a neighborhood V of zero in K such that $V \cdot V \subseteq U$. It is clear that $1 \notin V$. Since the elements a and a^{-1} are topologically nilpotent, then there exist natural numbers n_1 and n_2 such that $a^n \in V$ for $n \geq n_1$ and $a^{-n} \in V$ for $n \geq n_2$. Putting $n_0 = \max\{n_1, n_2\}$, we get that $1 = a^{n_0} \cdot a^{-n_0} \in V \cdot V \subseteq U$, which contradicts the choice of the neighborhood U.

1.8.28. COROLLARY. Let K be a skew field endowed with Hausdorff ring topology and a be a non-zero topologically Σ-nilpotent element in K. Then the element a^{-1} is not topologically nilpotent, hence, not topologically Σ-nilpotent.

1.8.29. DEFINITION. Let R be a topological ring with the unitary element. An element $a \in R$ is called a neutral element if it is invertible and none of the elements a and a^{-1} is topologically nilpotent.

1.8.30. REMARK. From Definition 1.8.29 follows that if a is a neutral element, then a^{-1} is a neutral element, too.

1.8.31. PROPOSITION. *In a normed skew field K the elements whose norm is equal to 1, and only they, are neutral.*

PROOF. Let ξ be a norm on the skew field K and $a \in K$. If $\xi(a) = 1$, then $\xi(a^n) =$

$(\xi(a))^n = 1$ and

$$\xi(a^{-n}) = \left(\xi(a^{-1})\right)^n = \frac{1}{\xi(a)^n} = 1$$

for any natural number n. Therefore, none of the elements a and a^{-1} is topologically nilpotent.

Vice versa, let $\xi(a) \neq 1$. From $\xi(a) < 1$ and from Remark 1.8.17 follows that a is a topologically nilpotent element. From $\xi(a) > 1$ follows that $\xi(a^{-1}) = \frac{1}{\xi(a)} < 1$, i.e. a^{-1} is a topologically nilpotent element. Hence, a is not a neutral element.

1.8.32. PROPOSITION. *Let R be a topological ring with the unitary element and a be an invertible element in R. Then:*

1) an element $x \in R$ is topologically nilpotent if and only if the element $a \cdot x \cdot a^{-1}$ is topologically nilpotent;

2) an element $x \in R$ is neutral if and only if the element $a \cdot x \cdot a^{-1}$ is neutral.

PROOF. 1) Let x be a topologically nilpotent element and U be a neighborhood of zero in R. We can choose a neighborhood V of zero in R such that $a \cdot V \cdot a^{-1} \subseteq U$. Let n_0 be a natural number such that $x^n \in V$ for $n \geq n_0$. Then $(a \cdot x \cdot a^{-1})^n = a \cdot x^n \cdot a^{-1} \in a \cdot V \cdot a^{-1} \subseteq U$ for $n \geq n_0$, i.e. $a \cdot x \cdot a^{-1}$ is a topologically nilpotent element.

Vice versa, let $a \cdot x \cdot a^{-1}$ be a topologically nilpotent element. Then, as it was shown above, the element $x = a^{-1} \cdot (a \cdot x \cdot a^{-1}) \cdot a = a^{-1} \cdot (a \cdot x \cdot a^{-1}) \cdot (a^{-1})^{-1}$ is topologically nilpotent.

2) Let x be a neutral element, i.e. x is invertible and the elements x and x^{-1} are not topologically nilpotent. Then the element $a \cdot x \cdot a^{-1}$ is invertible in R and, due to 1), none of the elements $a \cdot x \cdot a^{-1}$ or $(a \cdot x \cdot a^{-1})^{-1} = a \cdot x^{-1} \cdot a^{-1}$ is topologically nilpotent, i.e. $a \cdot x \cdot a^{-1}$ is a neutral element.

Vice versa, if $a \cdot x \cdot a^{-1}$ is a neutral element, then it is invertible, i.e. $a \cdot x \cdot a^{-1} \cdot b = 1$ for a certain element $b \in R$. Then $x \cdot a^{-1} \cdot b = a^{-1}$ and, hence, $x \cdot (a^{-1} \cdot b \cdot a) = 1$, i.e. the element x is invertible. As it was shown above, since the element $a \cdot x \cdot a^{-1}$ is neutral, then the element $x = a^{-1} \cdot (a \cdot x \cdot a^{-1}) a$ is neutral, too.

1.8.33.   PROPOSITION. *Let S be a topologically nilpotent (topologically Σ-nilpotent) subset of a topological ring R. Then the subset $[S]_R$ is topologically nilpo-*

tent (topologically Σ-nilpotent).

PROOF. Let S be a topologically nilpotent subset and U be a neighborhood of zero in R. Due to Corollary 1.2.30, there exists a closed neighborhood V of zero in R such that $V \subseteq U$.

Let n_0 be a natural number such that $S^{(n)} \subseteq V$ for all $n \geq n_0$. Then

$$\left([S]_R\right)^{(n)} \subseteq \left[S^{(n)}\right]_R \subseteq [V]_R = V \subseteq U$$

(see Corollary 1.1.42), hence, $[S]_R$ is a topologically nilpotent subset.

If the subset S is topologically Σ-nilpotent, the proof could be analogous, but we should state before that $([S]_R)^n \subseteq [S^n]_R$. Indeed, it is:

$$\left([S]_R\right)^n = \bigcup_{k=1}^{\infty} \underbrace{\left(\left([S]_R\right)^{(n)} + \left([S]_R\right)^{(n)} + \ldots + \left([S]_R\right)^{(n)}\right)}_{k \text{ summadns}}$$

$$\subseteq \bigcup_{k=1}^{\infty} \underbrace{\left([S^{(n)}]_R + [S^{(n)}]_R + \ldots + [S^{(n)}]_R\right)}_{k \text{ summands}}$$

$$\subseteq \bigcup_{k=1}^{\infty} \underbrace{[S^{(n)} + S^{(n)} + \ldots + S^{(n)}]_R}_{k \text{ summands}} \subseteq [S^n]_R.$$

1.8.34. PROPOSITION. *Let f be a continuous homomorphism of a topological ring R onto a topological ring R' and let S be a topologically nilpotent (topologically Σ-nilpotent) subset of R. Then $f(S)$ is a topologically nilpotent (topologically Σ-nilpotent) subset of the topological ring R'.*

PROOF. Let U' be a neighborhood of zero in R', then $f^{-1}(U')$ is a neighborhood of zero in R. There exists a natural number n_0 such that $S^{(n)} \subseteq f^{-1}(U')$ (correspondingly, $S^n \subseteq f^{-1}(U')$) for all $n \geq n_0$. Then $\left(f(S)\right)^n = f(S^n) \subseteq f\left(f^{-1}(U')\right) = U'$ (correspondingly, $\left(f(S)\right)^n = f(S^n) \subseteq f\left(f^{-1}(U')\right) = U'$) for all $n \geq n_0$, i.e. the subset $f(S)$ is topologically nilpotent (topologically Σ-nilpotent) in R'.

1.8.35. PROPOSITION. *Let S be a bounded from left (right) subset of a topological ring R, then the following statements are equivalent:*

1) S is topologically nilpotent subset;

2) $S^{(k)}$ is topologically nilpotent subset for any natural number k;

3) there exists a natural number k_0 such that $S^{(k_0)}$ is a topologically nilpotent subset.

PROOF. It is evident, that 1)$\Longrightarrow$ 2)$\Longrightarrow$3).

Let's show that 3)$\Longrightarrow$1). Let $S^{(k_0)}$ be a topologically nilpotent subset in R and U be a neighborhood of zero in R. Due to Corollary 1.6.22, the subsets $S^{(2)}, S^{(3)}, \ldots, S^{(k_0-1)}$ are bounded from left. Therefore, we can choose a neighborhood V of zero in R such that $V \cdot S^{(i)} \subseteq U$ for $i = 1, 2, \ldots, k_0 - 1$. Since the subset $S^{(k_0)}$ is topologically nilpotent, there exists a natural number n_0 such that $(S^{(k_0)})^{(n)} \subseteq V$ for all $n \geq n_0$. Let $m \geq n_0 \cdot k_0$. Then $m = k_0 \cdot q + r$, where $q \geq n_0$ and $0 \leq r < k_0$. Thus,

$$S^{(m)} = S^{(k_0 \cdot q + r)} = \left(S^{(k_0)}\right)^{(q)} \cdot S^{(r)} \subseteq V \cdot S^{(r)} \subseteq U,$$

i.e. S is a topologically nilpotent subset. Therefore, 3)$\Longrightarrow$1).

1.8.36. COROLLARY. *Let a be an element of a topological ring R. Then the following conditions are equivalent:*

1) a is a topologically nilpotent element;

2) a^k is a topologically nilpotent element for any natural number k;

3) a^{k_0} is a topologically nilpotent element for a certain natural number k_0.

PROOF. The statement results from Proposition 1.8.35 and from the boundedness of the one-element subset $\{a\}$ (see Corollary 1.6.5).

1.8.37. PROPOSITION. *Let T be the subset of all topologically nilpotent elements of a topological ring R. Then the following statements are equivalent:*

1) T is an open subset;

2) there exists an open neighborhood U of zero in R consisting of topologically nilpotent elements.

PROOF. It is evident that 1)$\Longrightarrow$2).

Let's show that 2)$\Longrightarrow$1). Let $t \in T$ and n_0 be a natural number such that $t^{n_0} \in U$. We can choose a neighborhood V of the element t such that

$$V^{(n_0)} = \underbrace{V \cdot V \cdot \ldots \cdot V}_{n_0 \text{ times}} \subseteq U.$$

Then v^{n_0} is a topologically nilpotent element, for any $v \in V$. Due to Corollary 1.8.36, any element from V is topologically nilpotent, i.e. $V \subseteq T$, which means that the subset T is open. Thus, 2)$\Longrightarrow$1).

1.8.38. PROPOSITION. *Let R be a locally bounded topological ring with the unitary element. If in R exists an invertible topologically nilpotent element t, then R possesses a topologically nilpotent neighborhood of zero.*

PROOF. Due to Remark 1.6.15 and Corollary 1.6.47, in R exists a bounded neighborhood U of zero which is a subsemigroup of the multiplicative semigroup of R, i.e. $U \cdot U \subseteq U$. Since $U \cdot t$ is a neighborhood of zero in R (see Corollary 1.1.46), then we can choose a neighborhood V of zero in R such that $V \cdot U \subseteq U \cdot t$. Since t is a topologically nilpotent element, then there exists a natural number k such that $t^k \in V$. Since the element t^k is also invertible, too, then $t^k \cdot U$ is a neighborhood of zero in R.

Producing the induction by n, we'll show that $(t^k \cdot U)^{(n)} \subseteq U \cdot t^n$ for any $n = 1, 2, \ldots$. Indeed, if $n = 1$, then $t^k \cdot U \subseteq V \cdot U \subseteq U \cdot t$. If $(t^k \cdot U)^{(n-1)} \subseteq U \cdot t^{n-1}$, then

$$\left(t^k \cdot U\right)^{(n)} = \left(t^k \cdot U\right) \cdot \left(t^k \cdot U\right)^{(n-1)} \subseteq t^k \cdot U \cdot U \cdot t^{n-1}$$
$$\subseteq \left(t^k \cdot U\right) \cdot t^{n-1} \subseteq \left(U \cdot t\right) \cdot t^{n-1} = U \cdot t^n.$$

Let's verify that the neighborhood $t^k \cdot U$ is topologically nilpotent. Let W be a neighborhood of zero in R and W_1 be a neighborhood of zero in R such that $U \cdot W_1 \subseteq W$, and let n_0 be a natural number such that $t^n \in W_1$ for all $n \geq n_0$. Then

$$\left(t^k \cdot U\right)^{(n)} \subseteq U \cdot t^n \subseteq U \cdot W_1 \subseteq W$$

for $n \geq n_0$. This completes the proof.

1.8.39. PROPOSITION. *Let R be a skew field endowed with locally bounded connected ring topology. Then R contains a non-zero topologically nilpotent element.*

PROOF. Let U be a bounded neighborhood of zero in R, and V be a neighborhood of zero in R such that $V + V \subseteq U$. Since the space R is connected, then it is non-discrete. Due to Proposition 1.6.3, there exists a non-zero element $t \in R$ such that $t \cdot U \subseteq V$. Thus, $t \cdot U + t \cdot U \subseteq U$.

Let's show that t is topologically nilpotent element. Let W be a neighborhood of zero in R and a be an element from R such that $U \cdot a \subseteq W$ (see Proposition 1.6.3). Since the topology on R is connected, there exists a natural number n such that

$$a^{-1} \in \underbrace{U + U + \ldots + U}_{2^k \text{ summands}}$$

for all $k \geq n$ (see Corollary 1.4.21). Then, applying k times the same procedure, we get:

$$t^k \cdot a^{-1} \in t^k \cdot \underbrace{(U + U + \ldots + U)}_{2^k \text{ summands}} \subseteq$$
$$t^{k-1} \cdot \underbrace{\big((t \cdot U + t \cdot U) + (t \cdot U + t \cdot U) + \ldots + (t \cdot U + t \cdot U)\big)}_{2^{k-1} \text{ summands of type } (t \cdot U + t \cdot U)} \subseteq$$
$$t^{k-1} \cdot \underbrace{(U + U + \ldots + U)}_{2^{k-1} \text{ summands}} \subseteq \ldots \subseteq t \cdot U + t \cdot U \subseteq U$$

for all $k \geq n$. Thus, $t^k = t^k \cdot a^{-1} \cdot a \in U \cdot a \subseteq W$ for all $k \geq n$, which means that the element t is topologically nilpotent.

1.8.40. COROLLARY. *Let R be a skew field endowed with a locally bounded connected ring topology. Then R possesses a topologically nilpotent neighborhood of zero.*

PROOF. The proof results from Propositions 1.8.38 and 1.8.39.

1.8.41. THEOREM. *Let R be a non-discrete locally compact Hausdorff ring. If R does not contain any non-zero divisors of zero, then R contains a non–zero topologically nilpotent element.*

PROOF. Since a compact subset of a topological ring is bounded, then, by Theorem 1.6.46, there exists a basis $\mathcal{B}_0$ of neighborhoods of zero in R such that it contains a neighborhood U_0 which is a subsemigroup of the multiplicative semigroup $R(\cdot)$, and every neighborhood $V \in \mathcal{B}_0$ is an ideal of the semigroup U_0. We may suppose that U_0 is compact and $U_0 \neq R$.

We show at the beginning that there exists a non–zero element $a_0 \in U_0$ such that $0 \in \big[\{a^k \mid k \in \mathbb{N}\}\big]_R$.

Assume the contrary, i.e. that $0 \notin \left[\{a^k \mid k \in \mathbb{N}\}\right]_R$ for any non–zero element $a \in U_0$. Then by transfinite induction for every transfinite number α we define a non–zero element b_α and a compact subset B_α in such way that:

1) B_α is a subsemigroup of the semigroup $R(\cdot)$ for each α;

2) $0 \notin B_\alpha$ for each α;

3) $b_\alpha \in B_\alpha$ for each α;

4) if $\gamma > \alpha$, then $B_\gamma \subseteq b_\alpha^k \cdot B_\alpha$ for each $k \in \mathbb{N}$.

Let U_1 be a neighborhood from $\mathcal{B}_0$ such that $[U_1]_R \neq U_0$, then we take as b_1 any non–zero element from U_1 and $B_1 = \left[\{b_1^k \mid k \in \mathbb{N}\}\right]_R$. Then $B_1 \subseteq [U_1]_R \subseteq U_0$. Thus, B_1 is a compact subsemigroup of the multiplicative semigroup $R(\cdot)$. It is clear that for $\alpha, \gamma < 2$ the conditions 1) – 4) are fulfilled.

Let now β be a transfinite number. Suppose that for every $\rho < \beta$ the element b_ρ and the subset B_ρ are defined in such way that the condition 1) – 4) are fulfilled for any $\alpha, \gamma < \beta$.

Since $\bigcap_{i=1}^n (b_{\alpha_i}^{k_i} \cdot B_{\alpha_i}) \supseteq b_\alpha^k \cdot B_\alpha$, where $k = \max\{k_1, \dots, k_n\}$ and $\alpha = \max\{\alpha_1, \dots, \alpha_k\}$, for all $k_1, \dots, k_n \in \mathbb{N}$ and $\alpha_1, \dots, \alpha_n < \beta$, we get that the family

$$\{b_\rho^k \cdot B_\rho \mid \rho < \beta, \ k \in \mathbb{N}\}$$

is a basis of some filter of the compact space U_0.

Since all $b_\rho^k \cdot B_\rho$ are compact subsets, then we have that all $b_\rho^k \cdot B_\rho$ are closed in R, and, hence, are closed in U_0. Then

$$\bigcap_{k=1}^\infty \bigcap_{\rho < \beta} b_\rho^k \cdot B_\rho \neq \emptyset.$$

We take as b_β any element from $\bigcap_{k=1}^\infty \bigcap_{\rho < \beta} b_\rho^k \cdot B_\rho$. Since $0 \notin B_1$ and $b_1 \cdot B_1 \subseteq B_1$, then we have that $b_\beta \neq 0$. Put $B_\beta = \left[\{b_\beta^k \mid k \in \mathbb{N}\}\right]_R$. It is easy to see that the conditions 1) – 4) are fulfilled for all $\alpha, \gamma \leq \beta$.

Continuing this procedure we define for every transfinite α an element b_α and a compact subset B_α in such way that they satisfy 1) – 4). Then there exists a transfinite number α_0 (for example, a transfinite number with cardinality greater than the cardinality of the set U_0) such that $B_{\alpha_0+1} = B_{\alpha_0}$, and therefore, $b_{\alpha_0} \in B_{\alpha_0+1} \subseteq b_{\alpha_0} \cdot B_{\alpha_0}$. Hence, $b_{\alpha_0} = b_{\alpha_0} \cdot b$ for some $b \in B_{\alpha_0}$.

Since $B_{\alpha_0} \subseteq [U_1]_R \neq U_0$ and $[U_1]_R$ is an ideal of the semigroup U_0, then $b \cdot c \neq c$ for some $c \in U$. Then

$$b_{\alpha_0} \cdot (b \cdot c - c) = b_{\alpha_0} \cdot b \cdot c - b_{\alpha_0} \cdot c = b_{\alpha_0} \cdot c - b_{\alpha_0} \cdot c = 0.$$

Thus, we've obtained a contradiction with the fact that R does not contain non-zero divisors of zero. Therefore,

$$0 \in \left[\left\{ a_0^k \mid k \in \mathbb{N} \right\} \right]_R$$

for some non-zero element $a_0 \in U_0$.

We show now that a_0 is a topologically nilpotent element. Indeed, let V be a neighborhood of zero from $\mathcal{B}_0$. Since

$$0 \in \left[\left\{ a_0^k \mid k \in \mathbb{N} \right\} \right]_R,$$

then we have that $a_0^k \in V$ for some $k_0 \in \mathbb{N}$.

Then $a_0^k = a_0^{k_0} \cdot a_0^{k-k_0} \in V \cdot U_0 \subseteq V$ for all $k \geq k_0$, i.e. a_0 is a non-zero topologically nilpotent element.

This completes the proof of the theorem.

1.8.42. COROLLARY. *Let R be a skew field endowed with a locally compact ring topology, then R contains a topologically nilpotent neighborhood of zero.*

PROOF. If the topology on R is non-discrete, then the statement results from Remark 1.6.44, Theorem 1.8.41 and Proposition 1.8.38. If the topology is discrete, then $\{0\}$ is a topologicaly nilpotent neighborhood of zero.

1.8.43. COROLLARY. *Let R be a skew field endowed with a locally compact ring topology, then the set T of all topologically nilpotent elements of R is an open subset in R.*

PROOF. Due to Corollary 1.8.42, R contains a topologically nilpotent neighborhood W of zero, and according to Remark 1.8.17, any element from W is topologically nilpotent. Then, due to Proposition 1.8.37, T is open subset in R.

1.8.44. THEOREM. *Let R be a skew field endowed with a ring topology τ. If the topological ring (R, τ) contains a topologically nilpotent neighborhood of zero, then (R, τ) is a topological skew field.*

PROOF. The statement is evident if the topology τ is discrete or anti-discrete (see Example 1.1.22). If the topology τ is different from anti-discrete, then it is Hausdorff (see Remark 1.4.13). Let W be a topologically nilpotent neighborhood of zero in (R, τ).

We fix up an arbitrary basis $\mathcal{B}_0(R, \tau)$ of symmetrical neighborhoods of zero in (R, τ) and show that $\mathcal{B}_0(R, \tau)$ satisfies the condition (BN7) (see Proposition 1.2.12). Let $U \in \mathcal{B}_0(R, \tau)$. There exists such neighborhood U_1 from $\mathcal{B}_0(R, \tau)$, that $U_1 + U_1 + U_1 \subseteq U \bigcap W$ and $-1 \notin U_1$. Since the neighborhood W is topologically nilpotent, then there exists a natural number n_0 such that $W^{(n)} \subseteq U_1$ for all $n \geq n_0$. Since the topology τ is non-discrete, then there exists $0 \neq w \in W$, besides, $w^{n_0-1} \cdot W$ is a neighborhood of zero in (R, τ) (see Corollary 1.1.45). From $w^{n_0-1} \cdot W \subseteq W^{(n_0)}$ follows that $W^{(n_0)}$ is a neighborhood of zero in (R, τ), hence, there exists such $V \in \mathcal{B}_0(R, \tau)$, that $V \subseteq W^{(n_0)}$. Since $V \subseteq W^{(n_0)} \subseteq U_1$, then $-1 \notin V$.

Let's show that $(1+V)^{-1} \subseteq 1+U$. Let $v \in V$. Producing the induction by k we'll verify that $\sum_{i=1}^{k} v^i \in U_1 + U_1$ for all $k = 1, 2, \ldots$. In fact, if $k = 1$, then $v \in V \subseteq U_1 \subseteq U_1 + U_1$. If $\sum_{i=1}^{k-1} v^i \in U_1 + U_1$, then

$$\sum_{i=1}^{k} v^i = v + v \cdot \sum_{i=1}^{k-1} v^i \in U_1 + v \cdot (U_1 + U_1) \subseteq U_1 + v \cdot W$$

$$\subseteq U_1 + W^{(n_0)} \cdot W = U_1 + W^{(n_0+1)} \subseteq U_1 + U_1.$$

Thus, $\sum_{i=1}^{k} v^i \in U_1 + U_1$ for all $k = 1, 2, \ldots$.

We can choose a neighborhood U_2 of zero in (R, τ) such that $(1-v)^{-1} \cdot U_2 \subseteq U_1$. Since $V \subseteq W^{(n_0)} \subseteq U_1 \subseteq W$, then, due to Remark 1.8.17, the element v is topologically nilpotent, therefore there exists a natural number m_0 such that $v^n \in U_2$ for all $n \geq m_0$. Hence,

$$(1-v)^{-1} - \left(1 + \sum_{i=1}^{m_0} v^i\right) = (1-v)^{-1} \cdot (1 - (1-v)) \cdot \left(1 + \sum_{i=1}^{m_0} v^i\right)$$

$$= (1+v)^{-1} \cdot \left(1 - 1 - \sum_{i=1}^{m_0} v^i + v + \sum_{i=1}^{m_0} v^{i+1}\right)$$

$$= (1+v)^{-1} \cdot v^{m_0+1} \in (1+v)^{-1} \cdot U_2 \subseteq U_1.$$

Thus,

$$(1-v)^{-1} \in 1 + \sum_{i=1}^{m_0} v^i + U_1 \subseteq 1 + U_1 + U_1 + U_1 \subseteq 1 + U.$$

Hence, $(1-V)^{-1} \subseteq 1+U$. Since V is symmetrical neighborhood, then $(1+V)^{-1} \subseteq 1+U$, i.e. $\mathcal{B}_0(R,\tau)$ satisfies the condition (BN7).

Due to Proposition 1.2.2, $\mathcal{B}_0(R,\tau)$ satisfies the conditions (BN1) - (BN6). Therefore, due to Theorem 1.2.13, $\mathcal{B}_0(R,\tau)$ defines on R the topology τ' in which R is a topological skew field, and $\mathcal{B}_0(R,\tau)$ is a basis of neighborhoods of zero. Due to Theorem 1.2.5, τ is the only ring topology on R in which $\mathcal{B}_0(R,\tau)$ is a basis of neighborhoods of zero. Thus, $\tau = \tau'$, hence, (R,τ) is a topological skew field.

1.8.45. COROLLARY. *Let τ be a connected locally bounded ring topology on a skew field R, then (R,τ) is a topological skew field.*

PROOF. The proof results from Corollary 1.8.40 and Theorem 1.8.44.

1.8.46. COROLLARY. *Let τ be a locally compact ring topology on a skew field R, then (R,τ) is a topological skew field.*

PROOF. The proof results from Corollary 1.8.42 and Theorem 1.8.44.

1.8.47. PROPOSITION. *Let R be a topological ring with the unitary element and U be a neighborhood of zero in R, bounded from left (right). Let an element $t \in R$ be invertible and topologically nilpotent, then $\{t^n \cdot U \mid n = 1,2,\dots\}$ (correspondingly, $\{U \cdot t^n \mid n = 1,2,\dots\}$) is a basis of neighborhoods of zero in R.*

PROOF. Let the neighborhood U be bounded from left. Since t^n is an invertible element for any $n \in \mathbb{N}$, then, due to Corollary 1.1.46, $t^n \cdot U$ is a neighborhood of zero in R for $n = 1,2,\dots$.

Let V be a neighborhood of zero in R, then we can choose a neighborhood W of zero in R and a natural number k such that $W \cdot U \subseteq V$ and $t^k \in W$. Then $t^k \cdot U \subseteq W \cdot U \subseteq V$, i.e. $\{t^n \cdot U \mid n = 1,2,\dots\}$ is a basis of neighborhoods of zero in R.

When U is the bounded from right neighborhood of zero, the proof is analogous.

1.8.48. COROLLARY. *A locally compact topological skew field possesses a countable basis of neighborhoods of zero.*

PROOF. If the skew field is non-discrete, then the statement results from Theorem 1.8.41 and Proposition 1.8.47. For the discrete skew fields the statement is evident.

1.8.49. PROPOSITION. *Let K be a Hausdorff non-discrete locally compact skew fileld, and a sequence $a_1, a_2, \ldots$ of non-zero elements of K be such that none of the elements of K is its accumulation point. Then the zero element of K is a limit of the sequence $a_1^{-1}, a_2^{-1}, \ldots$ in K (see C.6).*

PROOF. Assume the contrary, i.e. that 0 is not a limit of the sequence $a_1^{-1}, a_2^{-1}, \ldots$. According to the condition, 0 is not an accumulation point for the sequence $a_1, a_2, \ldots$ either. Hence, we can choose a neighborhood U of zero in K such that:

1) $[U]_K$ is a compact subset;

2) $1 \notin [U]_K$;

3) $a_i \notin U$, for $i = 1, 2, \ldots$.

4) $\{i \mid a_i^{-1} \notin U\}$ is infinite.

We can assume that $a_i^{-1} \notin U$ for each $i \in \mathbb{N}$ (on the contrary we will consider the sequence $\{a_{i_k} \mid a_{i_k}^{-1} \notin U, \ k \in \mathbb{N}\}$ instead of the sequence $\{a_i \mid i \in \mathbb{N}\}$).

Due to Corollary 1.8.42 and Remark 1.8.17 and since the neighborhood $[U]_K$ is bounded (see Proposition 1.6.4), there exists a topologically nilpotent neighborhood W of zero in K such that $W \cdot [U]_K \subseteq U$ and $[U]_K \cdot W \subseteq U$.

If $0 \neq t \in W$, then t is a topologically nilpotent element, besides, $t \cdot U \subseteq U$ and $U \cdot t \subseteq U$. Therefore $t^{n+1} \cdot U \subseteq t^n \cdot U$ and $U \cdot t^{n+1} \subseteq U \cdot t^n$ for any integer n. Due to Corollary 1.1.45, $t^n \cdot U$ and $U \cdot t^n$ are open neighborhoods of zero in K for $n = 1, 2, \ldots$.

Let's prove that

$$K = \bigcup_{n=-\infty}^{0} t^n \cdot U = \bigcup_{n=-\infty}^{0} U \cdot t^n.$$

Indeed, let $x \in K$. There exists a neighborhood V of zero in K such that $x \cdot V \subseteq U$ and $V \cdot x \subseteq U$. Then, choosing a natural number n_0 such that $t^{n_0} \in V$, we get that $x \cdot t^{n_0} \in x \cdot V \subseteq U$ and $t^{n_0} \cdot x \in V \cdot x \subseteq U$. Then $x \in U \cdot t^{-n_0}$ and $x \in t^{-n_0} \cdot U$.

Thus, for any $i = 1, 2, \ldots$ there exist negative integers r_i and s_i such that $a_i \in t^{r_i} \cdot U$, $a_i \notin t^{r_i+1} \cdot U$ and $a_i^{-1} \in U \cdot i^{s_i}$, $a_i^{-1} \notin U \cdot t^{s_i+1}$ (since $a_i, a_i^{-1} \notin U$). Then $t^{-r_i} \cdot a_i \in [U]_K \backslash t \cdot U$ and $a_i^{-1} t^{-s_i} \in [U]_K \backslash U \cdot t$.

Let's show that for any negative integer k the set $\{i \mid r_i \geq k\}$ is finite. Indeed, if we assume that for a negative integer k_0 the set $\{i | r_i \geq k_0\}$ is infinite, then the set

$$\{a_i \mid r_i \geq k_0\} = \{a_i \mid a_i \in t^{k_0} \cdot U\} \subseteq t^{k_0} \cdot [U]_K$$

is infinite, too. Therefore, a certain infinite sub-sequence of the sequence $a_1, a_2 \ldots$ is contained in the compact subset $t^{k_0} \cdot [U]_K$. Thus, for this sub-sequence and, hence, for the sequence $a_1, a_2, \ldots$ there exists an accumulation point in K (see C.12), which contradicts the condition of the proposition. Therefore, $\{i \mid r_i \geq k\}$ is a finite set for any negative number k.

Hence, from $a_1, a_2, \ldots$ can be extracted the sub-sequence $a_{i_1}, a_{i_2}, \ldots$ such that $r_{i_j} > r_{i_{j+1}}$ for $j = 1, 2, \ldots$. Moreover, without loss of generality we can consider that the sequence $a_1, a_2, \ldots$ itself is such that $r_i > r_{i+1}$ for $i = 1, 2, \ldots$.

From the definition of the numbers r_i and s_i follows that $t^{-r_i} \cdot a_i \in [U]_K \backslash (t \cdot U)$ and $a_i^{-1} \cdot t^{-s_i} \in [U]_K \backslash (U \cdot t)$. Since $t \cdot U$ and $U \cdot t$ are open subsets (see Proposition 1.1.44), then the subsets $[U]_K \backslash (t \cdot U)$ and $[U]_K \backslash (U \cdot t)$ are closed, hence, are compact. Thus, there exists an element $a \in K$ such that it is an accumulation point for the sequence $t^{-r_1} \cdot a_1, t^{-r_2} \cdot a_2, \ldots$ (see C.12). Moreover, $a \neq 0$, because of $t^{-r_i} \cdot a_i \notin U \cdot t$ for $i \in \mathbb{N}$.

Since K possesses a countable basis of neighborhoods of zero (see Corollary 1.8.48), then from the sequence $t^{-r_1} \cdot a_1, t^{-r_2} \cdot a_2, \ldots$ can be extracted a sub-sequence $t^{-r_{i_1}} \cdot a_{i_1}$, $t^{-r_{i_2}} \cdot a_{i_2}, \ldots$, for which a is a limit (see C.6). Without loss of generality we can consider that the sequence $t^{-r_1} \cdot a_1, t^{-r_2} \cdot a_2, \ldots$ itself coincides with this sub-sequence.

Since the elements $a_i^{-1} \cdot t^{-s_i}$ belong to the compact subset $[U]_K \backslash (U \cdot t)$, then, as above, we get that for the sequence $a_1^{-1} \cdot t^{-s_1}, a_2^{-1} \cdot t^{-s_2}, \ldots$ there exists an element $0 \neq b \in K$, which is its accumulation point. Then $a \cdot b \neq 0$, therefore we can choose neighborhoods U_0, U_{ab}, U_a and U_b of the elements $0, ab, a$ and b correspondingly, such that $U_0 \cap U_{ab} = \emptyset$ and $U_a \cdot U_b \subseteq U_{ab}$.

Since the element t is topologically nilpotent, then there exists a natural number n_0 such that $t^n \in V_0$ for all $n \geq n_0$. Since a is a limit of the sequence $t^{-r_1} \cdot a_1, t^{-r_2} \cdot a_2, \ldots$, then there exists a natural number i_0 such that $-r_{i_0} \geq n_0$ and $t^{-r_i} \cdot a_i \in U_a$ for all $i \geq i_0$. But b is an accumulation point for the sequence $a_1^{-1} \cdot t^{-s_1}, a_2^{-1} \cdot t^{-s_2}, \ldots$, therefore there exists a natural number $i_1 \geq i_0$ such that $a_{i_1}^{-1} \cdot t^{-s_{i_1}} \in U_b$. Then $r_{i_1} < r_{i_0}$, hence,

$-r_{i_1} > -r_{i_0} > n_0$. Since $-s_{i_1} > 0$, then $-r_{i_1} - s_{i_1} > n_0$, hence, $t^{-r_{i_1}-s_{i_1}} \in U_0$. Therefore

$$t^{-r_{i_1}} \cdot a_{i_1} \cdot a_{i_1}^{-1} \cdot t^{-s_{i_1}} = t^{-r_{i_1}-s_{i_1}} \in U_0.$$

On the other hand,

$$t^{-r_{i_1}} \cdot a_{i_1} \cdot a_{i_1}^{-1} \cdot t^{-s_{i_1}} \in U_a \cdot U_b \subseteq U_{ab},$$

i.e. $U_0 \bigcap U_{ab} \neq \emptyset$, which contradicts the choice of the neighborhoods U_0 and U_{ab}. Thus, the assumption that 0 is not a limit for the sequence $a_1^{-1}, a_2^{-1}, \ldots$ is wrong. This completes the proof of the proposition.

1.8.50. THEOREM. *Let K be a Hausdorff non-discrete locally compact skew field, T be the subset of all topologically nilpotent elements in K, N be the subset of all neutral elements in K and $P = \{x \mid 0 \neq x \in K, x^{-1} \in T\}$. Then:*

1) $K = T \bigcup N \bigcup P$ and $T \bigcap N = N \bigcap P = T \bigcap P = \emptyset$;

2) T and P are open subsets in K, the subsets N and $T \bigcup N$ are closed in K;

3) N and $T \bigcup N$ are compact (hence, bounded) subsets in K.

PROOF. 1) It is clear that $K = T \bigcup N \bigcup P$. Due to Proposition 1.8.27, $T \bigcap P = \emptyset$. According to the definition of a neutral element (see Definition 1.8.29), $T \bigcap N = N \bigcap P = \emptyset$.

2) According to Corollary 1.8.43, T is an open subset in K. Since K is Hausdorff, then $\{0\}$ is a closed subset in K, therefore $T \backslash \{0\}$ is an open subset in K. Due to Corollary 1.4.51, $P = (T \backslash \{0\})^{-1}$ is an open subset in K. Then $N = K \backslash (T \bigcup P)$ and $T \bigcup N = K \backslash P$ are closed subsets in K.

3) At first let's show that the subset $T \bigcup N$ is compact. Assume the contrary, i.e. that $T \bigcup N$ is not compact. Since T is a neighborhood of zero in K, then there exists an open neighborhood $U \subseteq T$ of zero in K whose closure $[U]_K$ is compact. Since the skew field K is non-discrete, then $\{0\} \neq U$. Let $0 \neq t \in U \subseteq T$. Then the subset $\bigcup_{n=1}^{k} [U]_K \cdot t^{-n}$ is compact for any $k \in \mathbb{N}$. Due to the statement 2), $T \bigcup N$ is closed in K, which, by our assumption, is not compact. Therefore, $T \bigcup N \nsubseteq \bigcup_{n=1}^{k} [U]_K \cdot t^{-n}$ for any $k \in \mathbb{N}$. Hence, for any $k \in \mathbb{N}$ we can choose an element $x_k \in T \bigcup N$ such that $x_k \notin \bigcup_{n=1}^{k} [U]_K \cdot t^{-n}$. Then $x_k \cdot t^n \notin [U]_K$ for any $k, n \in \mathbb{N}$, $k \geq n$. Since $0 \neq x_k \in T \bigcup N = K \backslash P$, then $x_k^{-1} \notin T$

for any $k \in \mathbb{N}$. But T is a neighborhood of zero, then the element 0 is not a limit for the sequence $x_1^{-1}, x_2^{-1}, \ldots$ (see C.6).

Due to Proposition 1.8.49, the sequence $x_1, x_2, \ldots$ has an accumulation point $x \in K$. Let W be a neighborhood of zero in K such that $x \cdot W \subseteq U$ and n_0 be a natural number such that $t^{n_0} \in W$. Then $x \cdot t^{n_0} \in x \cdot W \subseteq U$. Since U is an open subset, then U is neighborhood of the element $x \cdot t^{n_0}$. Hence, there exists a neighborhood S of the element x in K such that $S \cdot t^{n_0} \subseteq U$. Since x is an accumulation point of the sequence $x_1, x_2, \ldots,$ then $x_{k_0} \in S$ for a certain $k_0 \geq n_0$. Then $x_{k_0} \cdot t^{n_0} \in S \cdot t^{n_0} \subseteq U$, which contradicts the choice of the elements x_k. Thus, our assumption is wrong, therefore the subset $T \bigcup N$ is compact.

Since N is closed in K, then N is closed in $T \bigcup N$, therefore, N is compact subset.

2. Pseudonormalizability of Topological Rings and Modules

In the previous chapter we considered the examples of ring and module topologies defined by a real-valued norm or pseudonorm. Therefore, appears the interest to the problem when the ring or module topology can be defined with the help of a norm or a pseudonorm. Results of this kind in the theory of topological rings and modules are analogous to the metrical theorems in the general topology.

In § 2.1 is presented the criterion of pseudonormalizability of a topological ring. Its corollary is the criterion of pseudonormalizability of a topological field. The main result of this Para is the proof of the criterion of pseudonormalizability of a topological ring (Theorem 2.1.4). The variants of this criterion for topological rings, satisfying some additional conditions, are presented, too (see also [25, 96, 249, 283, 462]).

In § 2.2. is presented the criterion of normalizability of a topological skew field. Also the normalizability of any locally compact field is proved. Some of the results included in this Para could be found also in [129, 206, 354].

The results of the last paragraph generalize the criterion of normalizability of a topological vector space over the field of real numbers to the case of arbitrary topological modules (in particular, vector spaces). Some of the results included in this paragraph could be found also in [440].

Only Hausdorff non-discrete topological rings, skew fields, modules and vector spaces are considered in this chapter.

§ 2.1. Criterion of Pseudonormalizability of Topological Rings

2.1.1. DEFINITION. A topological ring (R, τ) is called pseudonormalized (normalized) if on R can be defined a pseudonorm (correspondingly, norm) ξ such that the interval topology τ_ξ on R, defined by the pseudonorm ξ (see 1.2.9), coincides with the initial topology τ.

2.1.2. REMARK. Consider the following conditions on a system $\{U_i \mid i \in I\}$ of subsets of a ring R, where $I = \mathbb{N}$ (the set of the natural numbers) or $I = \mathbb{Z}$ (the set of the integers):

(PNR1) $U_i \cdot U_i \subseteq U_{i+j}$ for $i, j \in I$;

(PNR2) $U_{i+1} + U_{i+1} \subseteq U_i$ for $i \in I$;

(PNR3) $\bigcup_{i \in I} U_i = R$;

(PNR4) for any element $a \in R$ there exists $n \in \mathbb{N}$ such that $a \cdot U_{i+n} \subseteq U_i$ and $U_{i+n} \cdot a \subseteq U_i$ for all $i \in I$.

2.1.3. PROPOSITION. *Let* $\{U_i \mid i \in \mathbb{Z}\}$ *be a system of symmetric neighborhoods of zero of a topological Abelian group* A *such that the conditions (PNR2) and (PNR3) are fulfilled. Then there exists a non-negative real-valued function* f *on* A *satisfying the following conditions:*

a) $f(x) \leq \frac{1}{2^{k-4}}$ *, if* $x \in U_k$;

b) $x \in U_k$, *if* $f(x) < \frac{1}{2^{k-3}}$;

c) $f(x + y) \leq f(x) + f(y)$ *for any* $x, y \in A$.

PROOF. Due to Proposition 1.3.1, for any element $a \in A$ and an arbitrary integer n, there exists a real-valued function $f_{n,a}$ on A, satisfying the following conditions:

1) $0 \leq f_{n,a}(x) \leq 1$ for all $x \in A$;

2) $f_{n,a}(a) = 1$;

3) $f_{n,a}(x) = 0$, if $x \notin a + U_n$;

4) if $x_1 - x_2 \in U_{n+m}$, then $\mid f_{n,a}(x_1) - f_{n,a}(x_2) \mid < \frac{1}{2^{m-1}}$.

For any $x \in A$ put

$$f(x) = \sup\left\{ \frac{1}{2^{n-3}} \cdot \left| f_{n,a}(z) - f_{n,a}(z+x) \right| \middle| a, z \in A, n \in \mathbb{Z} \right\}.$$

It is clear that f is a non-negative real-valued function.

Let's show that f satisfies the conditions $a) - c)$.

Let $x \in U_k$. Since $x = (z + x) - z$, then, due to the property 4),

$$\mid f_{n,a}(z) - f_{n,a}(z + x) \mid \leq \frac{1}{2^{k-n-1}}$$

for all $a, z \in A$ and $n \in \mathbb{Z}$. Therefore,

$$\frac{1}{2^{n-3}} \cdot \left| f_{n,a}(z) - f_{n,a}(z+x) \right| \leq \frac{1}{2^{k-4}},$$

hence, $f(x) \leq \frac{1}{2^{k-4}}$, i.e. f satisfies the condition a).

Since $A = \bigcup_{i=-\infty}^{+\infty} U_i$, then $f(x) < \infty$ for all $x \in A$, i.e. the function f is well defined on A.

Now let $f(x) < \frac{1}{2^{k-3}}$. We'll show that $x \in U_k$. In fact, if we assume that $x \notin U_k$, then, due to the condition 3), $f_{k,0}(x) = 0$. Then, due to the condition 2), $f_{k,0}(0) = 1$, hence,

$$f(x) \geq \frac{1}{2^{k-3}} \cdot \left| f_{k,0}(0) - f_{k,0}(x) \right| = \frac{1}{2^{k-3}},$$

which contradicts the inequality $f(x) < \frac{1}{2^{k-3}}$. Thus, $x \in U_k$, i.e. f satisfies the condition b).

Let $x, y \in A$, and ϵ be an arbitrary positive number. Then there exist elements $a_0, z_0 \in A$ and an integer n_0 such that

$$f(x + y) \leq \frac{\left| f_{n_0,a_0}(z_0) - f_{n_0,a_0}(z_0 + x + y) \right|}{2^{n_0 - 3}} + \epsilon$$

$$\leq \frac{\left| f_{n_0,a_0}(z_0) - f_{n_0,a_0}(z_0 + x) \right| + \left| f_{n_0,a_0}(z_0 + x) - f_{n_0,a_0}(z_0 + x + y) \right|}{2^{n_0 - 3}} + \epsilon$$

$$\leq f(x) + f(y) + \epsilon.$$

Therefore, $f(x + y) \leq f(x) + f(y)$, i.e. f satisfies the condition c).

2.1.4. THEOREM. *Let (R, τ) be a topological ring. Then the following conditions are equivalent:*

1) (R, τ) is a pseudonormalized ring;

2) (R, τ) is a union of a countable number of bounded subsets (see Definition 1.6.2) and possesses a basis of neighborhoods of zero $\{U_i | i \in \mathbb{N}\}$, which satisfies the conditions (PNR1) and (PNR2) (see 2.1.2);

3) (R, τ) possesses a basis of neighborhoods of zero $\{V_i | i \in \mathbb{N}\}$, which satisfies the conditions (PNR1), (PNR2) and (PNR4) (see 2.1.2);

4) (R, τ) possesses a basis of neighborhoods of zero $\{W_i | i \in \mathbb{Z}\}$, which satisfies the conditions (PNR1) - (PNR3) (see 2.1.2).

PROOF. The proof we'll realize according to the scheme: 1) $\Longrightarrow$ 2) $\Longrightarrow$ 3) $\Longrightarrow$ 4) $\Longrightarrow$ 1).

2.1.5. *PROOF.* The proof of the implication 1)$\Longrightarrow$2).

Let (R, τ) be a pseudonormalized ring, ξ be a pseudonorm on R and $\tau_\xi = \tau$ (see Definition 2.1.1). Let $i \in \mathbb{N}$, then put $U_i = \{x \in R \mid \xi(x) < \frac{1}{2^i}\}$ and $Q_i = \{x \in R \mid \xi(x) < 2^i\}$. Due to Example 1.2.9, $\{U_i \mid i \in \mathbb{N}\}$ is a basis of neighborhoods of zero in (R, τ). Due to the properties of pseudonorm (see Remark 1.1.13 and Definition 1.1.12), the conditions (PNR1) and (PNR2) are fulfilled. According to Proposition 1.6.9, Q_i is a bounded subset, $i \in \mathbb{N}$. It is clear that $R = \bigcup_{i \in \mathbb{N}} Q_i$.

2.1.6. *PROOF.* Proof of the implication 2)$\Longrightarrow$3).

Let $R = \bigcup_{i \in \mathbb{N}} S_i$, where S_i is a bounded subset in R for $i \in \mathbb{N}$. Due to Proposition 1.6.19, the subset $Q_k = \bigcup_{i=1}^{k} S_i$ is bounded, for $k \in \mathbb{N}$. Then $R = \bigcup_{k \in \mathbb{N}} Q_k$, where Q_k is a bounded subset in R and $Q_k \subseteq Q_{k+1}$ for $k \in \mathbb{N}$.

Construct by induction the following basis $\{\tilde{U}_n \mid n \in \mathbb{N}\}$ of neighborhoods of zero.

Put $\tilde{U}_1 = U_2$. Suppose that for all $k \leq n$ the neighborhoods of zero $\tilde{U}_k = \tilde{U}_{i_k}$ are already defined, besides $i_k \geq 2k$. Let i_{n+1} be a natural number such that: $i_{n+1} \geq i_n + 2$, $Q_{n+1} \cdot U_{i_{n+1}} \subseteq \tilde{U}_n$ and $U_{i_{n+1}} \cdot Q_{n+1} \subseteq \tilde{U}_n$ (the possibility of the choice of the number i_{n+1} is provided by the boundedness of the subset Q_{n+1} and by the inclusion $U_{i+1} \subseteq U_i$, which results from the condition (PNR2)). Put $\tilde{U}_{n+1} = U_{i_{n+1}}$. Then $i_{n+1} \geq i_n + 2 \geq 2n + 2 = 2(n+1)$.

It is clear that $\{\tilde{U}_n \mid n \in \mathbb{N}\}$ is a basis of neighborhoods of zero in R, besides, from $i_n \geq 2n$ follows that $\tilde{U}_n \subseteq U_{2n}$. From the construction of the sets $\tilde{U}_n$ follows that

$$Q_k \cdot \tilde{U}_{n+1} \subseteq Q_{n+1} \cdot \tilde{U}_{n+1} = Q_{n+1} \cdot \tilde{U}_{i_{n+1}} \subseteq \tilde{U}_n,$$

and $\tilde{U}_{n+1} \cdot Q_k \subseteq \tilde{U}_n$ for all $n \geq k$ and $k, n \in \mathbb{N}$.

Finally note that

$$\tilde{U}_{n+1} = U_{i_{n+1}} \subseteq U_{i_n} = \tilde{U}_n$$

and

$$\tilde{U}_n \cdot \tilde{U}_m = U_{i_n} \cdot U_{i_m} \subseteq U_{i_n + i_m} \subseteq U_{i_m} = \tilde{U}_m$$

for any $n, m \in \mathbb{N}$.

For $k \in \mathbb{N}$, put

$$\tilde{V}_k = \bigcup_{l_1 + l_2 + \ldots + l_n = k} (\tilde{U}_{l_1} \cdot \tilde{U}_{l_2} \cdot \ldots \cdot \tilde{U}_{l_n}),$$

the union is taken by all $n, l_1, \ldots, l_n \in \mathbb{N}$ such that $\sum_{i=1}^{n} l_i = k$.

Now we find out some properties of the subsets $\tilde{V}_k$ for $k \in \mathbb{N}$.

A) $\tilde{U}_k \subseteq \tilde{V}_k \subseteq U_{2k}$ for $k \in \mathbb{N}$, i.e. $\{\tilde{V}_k \mid k \in \mathbb{N}\}$ is a basis of neighbourhoods of zero in the topological ring (R, τ).

Indeed, the inclusion $\tilde{U}_k \subseteq \tilde{V}_k$ for $k \in \mathbb{N}$ is evident. Let $l_1, \ldots, l_n \in \mathbb{N}$ and $\sum_{i=1}^{n} l_i = k$. Further, since $\tilde{U}_{l_i} \subseteq U_{2l_i}$, then, using the condition (PNR1), we get that

$$\tilde{U}_{l_1} \cdot \tilde{U}_{l_2} \cdot \ldots \cdot \tilde{U}_{l_n} \subseteq U_{2l_1} \cdot U_{2l_2} \cdot \ldots U_{2l_n} \subseteq U_{2k}.$$

Therefore, $\tilde{V}_k \subseteq U_{2k}$.

B) $\tilde{V}_k \cdot \tilde{V}_l \subseteq \tilde{V}_{k+l}$ for $k, l \in \mathbb{N}$.

Indeed, let $x \in \tilde{V}_k$ and $y \in \tilde{V}_l$. Then there exist natural numbers n and m and sets $\{l_1, l_2, \ldots, l_n\}$ and $\{l'_1, l'_2, \ldots, l'_m\}$ of natural numbers such that $\sum_{i=1}^{n} l_i = k$, $\sum_{j=1}^{m} l'_j = l$ and

$$x \in \tilde{U}_{l_1} \cdot \tilde{U}_{l_2} \cdot \ldots \cdot \tilde{U}_{l_n},$$
$$y \in \tilde{U}_{l'_1} \cdot \tilde{U}_{l'_1} \cdot \ldots \cdot \tilde{U}_{l'_m}.$$

Then

$$x \cdot y \in \tilde{U}_{l_1} \cdot \tilde{U}_{l_2} \cdot \ldots \cdot \tilde{U}_{l_n} \cdot \tilde{U}_{l'_1} \cdot \tilde{U}_{l'_2} \cdot \ldots \cdot \tilde{U}_{l'_m},$$

besides,

$$l_1 + l_2 + \ldots + l_n + l'_1 + l'_2 + \ldots + l'_m = k + l.$$

Therefore, $x \cdot y \in \tilde{V}_{k+l}$, hence, $\tilde{V}_k \cdot \tilde{V}_l \subseteq \tilde{V}_{k+l}$.

C) $\tilde{V}_{k+1} \subseteq \tilde{V}_k$.

Indeed, let $x \in \tilde{V}_{k+1}$. Then there exists a set of natural numbers $l_1, l_2, \ldots, l_n$ such that $x \in \tilde{U}_{l_1} \cdot \tilde{U}_{l_2} \cdot \ldots \cdot \tilde{U}_{l_n}$ and $\sum_{i=1}^{n} l_i = k + 1$.

If $n = 1$, then

$$x \in \tilde{U}_{l_1} = \tilde{U}_{k+1} \subseteq \tilde{U}_k \subseteq \tilde{V}_k.$$

If $n > 1$ and $l_1 = 1$, then

$$x \in \tilde{U}_{l_1} \cdot \tilde{U}_{l_2} \cdot \ldots \cdot \tilde{U}_{l_n} \subseteq \tilde{U}_{l_2} \cdot \ldots \cdot \tilde{U}_{l_n},$$

besides,

$$\sum_{i=2}^{n} l_i = \left(\sum_{i=1}^{n} l_i\right) - 1 = k + 1 - 1 = k,$$

i.e. $x \in \tilde{V}_k$. Therefore, $\tilde{V}_{k+1} \subseteq \tilde{V}_k$.

If $n > 1$ and $l_1 > 1$, then

$$x \in \tilde{U}_{l_1} \cdot \tilde{U}_{l_2} \cdot \ldots \cdot \tilde{U}_{l_n} \subseteq \tilde{U}_{l_1-1} \cdot \tilde{U}_{l_2} \cdot \ldots \cdot \tilde{U}_{l_n},$$

besides,

$$(l_1 - 1) + \sum_{j=2}^{n} l_j = \left(\sum_{j=1}^{n} l_j\right) - 1 = k + 1 - 1 = k,$$

i.e. $x \in \tilde{V}_k$. Thus, $\tilde{V}_{k+1} \subseteq \tilde{V}_k$.

D) $Q_k \cdot \tilde{V}_{i_{k+1}+l} \subseteq \tilde{V}_l$ and $\tilde{V}_{i_{k+1}+l} \cdot Q_k \subseteq \tilde{V}_k$ for any $k, l \in \mathbb{N}$ (for definition of the numbers i_j see the beginning of this item in the construction of the basis $\{\tilde{U}_n \mid n \in \mathbb{N}\}$ of neighborhoods of zero).

Let $l \leq k$. Since $U_{i+1} \subseteq U_i$, then, taking into account the property A) of the basis $\{\tilde{V}_k \mid k \in \mathbb{N}\}$ of neighborhoods, we get that

$$Q_k \cdot \tilde{V}_{i_{k+1}} \subseteq Q_{k+1} \cdot U_{i_{k+1}} \subseteq \tilde{U}_k.$$

Then

$$Q_k \cdot \tilde{V}_{i_{k+1}+l} \subseteq Q_k \cdot \tilde{V}_{i_{k+1}} \subseteq \tilde{U}_k \subseteq \tilde{V}_k \subseteq \tilde{V}_l$$

for $l \leq k$.

If $l \geq k$, then the necessary inclusion we'll prove by the induction by l.

We have seen already that this inclusion is fulfilled if $l = k$. Suppose that the inclusion $Q_k \cdot \tilde{V}_{i_{k+1}+l} \subseteq \tilde{V}_l$ is fulfilled for $k \leq l \leq q$ and let's show that $Q_k \cdot \tilde{V}_{i_{k+1}+q+1} \subseteq \tilde{V}_{q+1}$. To do it, knowing the construction of the neighborhood $\tilde{V}_{i_{k+1}+q+1}$, it is enough to state that if $\{n_1, \ldots, n_m\} \subseteq \mathbb{N}$ and

$$\sum_{i=1}^{m} n_i = i_{k+1} + q + 1,$$

then

$$Q_k \cdot \tilde{U}_{n_1} \cdot \tilde{U}_{n_2} \cdot \ldots \cdot \tilde{U}_{n_m} \subseteq \tilde{V}_{q+1}.$$

Let's consider the following three cases.

I) $m \geq 2$ and $\sum_{j=1}^{m-1} n_j > i_{k+1}$.

Since $n_m \geq 1$ and

$$\sum_{j=1}^{m-1} n_j = \sum_{j=1}^{m} n_j - n_m = (i_{k+1} + q + 1) - n_m,$$

then

$$i_{k+1} < \sum_{j=1}^{m-1} n_j \leq i_{k+1} + q$$

and

$$n_m = i_{k+1} + q + 1 - \sum_{j=1}^{m-1} n_j.$$

Hence,

$$0 < \left(\sum_{j=1}^{m-1} n_j \right) - i_{k+1} \leq q,$$

therefore,

$$Q_k \cdot \tilde{U}_{n_1} \cdot \tilde{U}_{n_2} \cdot \ldots \cdot \tilde{U}_{n_{m-1}} \cdot \tilde{U}_{n_m}$$

$$\subseteq Q_k \cdot V_{n_1 + \ldots + n_{m-1}} \cdot \tilde{V}_{i_{k+1} + q + 1 - (n_1 + \ldots + n_{m-1})}$$

$$\subseteq V_{n_1 + \ldots + n_{m-1} - i_{k+1}} \cdot V_{i_{k+1} + q + 1 - (n_1 + \ldots + n_{m-1})}$$

$$\subseteq \tilde{V}_{(n_1 + \ldots + n_{m-1}) - i_{k+1} + i_{k+1} + q + 1 - (n_1 + \ldots + n_{m-1})} = \tilde{V}_{q+1}$$

(see property B) for $\tilde{V}_k$.

II) $m \geq 2$ and $\sum_{j=1}^{m-1} n_j = i_{k+1}$.

Then $n_m = q + 1$, hence,

$$Q_k \cdot \tilde{U}_{n_1} \cdot \ldots \cdot \tilde{U}_{n_{m-1}} \cdot \tilde{U}_{n_m} \subseteq Q_k \cdot \tilde{V}_{i_{k+1}} \cdot \tilde{V}_{q+1}$$

$$\subseteq \tilde{U}_k \cdot \tilde{V}_{q+1} \subseteq \tilde{V}_k \cdot \tilde{V}_{q+1} \subseteq \tilde{V}_{k+q+1} \subseteq \tilde{V}_{q+1}.$$

III) $m = 1$ or $m \geq 2$ and $\sum_{j=1}^{m-1} n_j < i_{k+1}$.

At first we'll verify that in this case $n_m \geq q + 2$. Indeed, if $m \geq 2$ and $\sum_{j=1}^{m-1} n_j < i_{k+1}$, then

$$n_m = \sum_{j=1}^{m} n_j - \sum_{j=1}^{m-1} n_j = i_{k+1} + q + 1 - \sum_{j=1}^{m-1} n_j > i_{k+1} + q + 1 - i_{k+1} = q + 1,$$

i.e. $n_m \geq q + 2$.

If $m = 1$, then $n_m = n_1 = i_{k+1} + q + 1 \geq q + 2$.

Note that

$$\tilde{U}_{n_1} \cdot \tilde{U}_{n_2} \cdot \ldots \cdot \tilde{U}_{n_m} \subseteq \tilde{U}_{q+2}.$$

Indeed, if $m = 1$, then $\tilde{U}_{n_m} \subseteq \tilde{U}_{q+2}$.

If $m \geq 2$, then

$$\tilde{U}_{n_1} \cdot \tilde{U}_{n_2} \cdot \ldots \cdot \tilde{U}_{n_{m-1}} \cdot \tilde{U}_{n_m}$$

$$\subseteq \tilde{U}_{n_1} \cdot \tilde{U}_{n_2} \cdot \ldots \cdot \tilde{U}_{n_{m-1}} \cdot \tilde{U}_{q+2} \subseteq \tilde{U}_{n_1 + n_2 + \ldots + n_{m-1} + q + 2} \subseteq \tilde{U}_{q+2}.$$

Thus,

$$\tilde{U}_{q+2} \supseteq \tilde{U}_{n_1} \cdot \tilde{U}_{n_2} \cdot \ldots \cdot \tilde{U}_{n_m},$$

hence,

$$Q_k \cdot \tilde{U}_{n_1} \cdot \tilde{U}_{n_2} \cdot \ldots \cdot \tilde{U}_{n_m} \subseteq Q_k \cdot \tilde{U}_{q+2} \subseteq Q_{q+2} \cdot \tilde{U}_{q+2} \subseteq \tilde{U}_{q+1} \subseteq \tilde{V}_{q+1}.$$

Therefore, the inclusion $Q_k \cdot \tilde{V}_{i_{k+1}+l} \subseteq \tilde{V}_l$ is fulfilled for all $k, l \in \mathbb{N}$.

Analogously is proved that $\tilde{V}_{i_{k+1}+l} \cdot Q_k \subseteq \tilde{V}_l$ for all $k, l \in \mathbb{N}$.

Now we are ready to construct the basis $\{V_k | k \in \mathbb{N}\}$. Put

$$V_k = \bigcup_{n=1}^{\infty} \bigcup_{\{m_1, \ldots, m_n\}} \left(\sum_{j=1}^{n} m_j \tilde{V}_j \right)$$

(for definition of nA see B.3), where $m_1, m_2, \ldots, m_n \in \mathbb{N} \bigcup \{0\}$ and $\sum_{j=1}^{n} \frac{m_j}{2^j} < \frac{1}{2^k}$ for $n \in \mathbb{N}$.

Let's show that $\{V_k | k \in \mathbb{N}\}$ is a basis of neighborhoods of zero in the topological ring (R, τ), satisfying the conditions (PNR1), (PNR2) and (PNR4).

Let $x \in V_k$ and $y \in V_l$. Then there exist natural numbers n and p and sets $\{m_1, m_2, \ldots, m_n\}$ and $\{m_1', m_2', \ldots, m_p'\}$ of non-negative integers such that:

$$x \in \sum_{j=1}^{n} m_j \tilde{V}_j \quad \text{and} \quad \sum_{j=1}^{n} \frac{m_j}{2^j} < \frac{1}{2^k};$$

$$y \in \sum_{s=1}^{p} m_s' \tilde{V}_s \quad \text{and} \quad \sum_{s=1}^{p} \frac{m_s'}{2^s} < \frac{1}{2^l}.$$

Therefore,

$$x \cdot y \in \left(\sum_{j=1}^{n} m_j \tilde{V}_j \right) \cdot \left(\sum_{s=1}^{p} m'_s \tilde{V}_s \right) = \sum_{j,s} (m_j \cdot m'_s)(\tilde{V}_j \cdot \tilde{V}_s)$$

$$\subseteq \sum_{j=1}^{n} \sum_{s=1}^{p} (m_j \cdot m'_s) \tilde{V}_{j+s} \subseteq \sum_{i=2}^{n+p} \left(\sum_{\{j,s|j+s=i\}} m_j \cdot m'_s \right) \tilde{V}_i,$$

besides,

$$\sum_{i=2}^{n+p} 2^{-i} \cdot \sum_{\{j,s|j+s=i\}} \cdot m_j \cdot m'_s = \sum_{i=2}^{n+p} \sum_{\{j,s|j+s=i\}} \frac{m_j \cdot m'_s}{2^{j+s}}$$

$$= \left(\sum_{j=1}^{n} \frac{m_j}{2^j} \right) \cdot \left(\sum_{s=1}^{p} \frac{m'_s}{2^s} \right) < \frac{1}{2^k} \cdot \frac{1}{2^l} = \frac{1}{2^{k+l}}.$$

Thus, $x \cdot y \in V_{k+l}$, hence $V_k \cdot V_l \subseteq V_{k+l}$, i.e. the condition (PNR1) is fulfilled.

Let $x, y \in V_{k+1}$, then there exist natural numbers n and p and sets

$$\{m_1, m_2, \dots, m_n\} \text{ and } \{m'_1, m'_2, \dots, m'_p\}$$

of non-negative numbers such that:

$$x \in m_1 \tilde{V}_1 + m_2 \tilde{V}_2 + \dots + m_n \tilde{V}_n,$$

$$y \in m'_1 \tilde{V}_1 + m'_2 \tilde{V}_2 + \dots + m'_p \tilde{V}_p;$$

$$\frac{m_1}{2} + \frac{m_2}{2^2} + \dots + \frac{m_n}{2^n} < \frac{1}{2^{k+1}};$$

$$\frac{m_1'}{2} + \frac{m'_2}{2^2} + \dots + \frac{m'_p}{2^p} < \frac{1}{2^{k+1}}.$$

Then

$$x + y \in m_1 \tilde{V}_1 + m_2 \tilde{V}_2 + \dots + m_n \tilde{V}_n + m'_1 \tilde{V}_1 + m'_2 \tilde{V}_2 + \dots + m'_p \tilde{V}_p = \sum_{r=1}^{t} (m_r + m'_r) \tilde{V}_r$$

(we consider that $m_{n+1} = m_{n+2} = \dots = m_t = 0$ if $t = p > n$ and $m'_{p+1} = m'_{p+2} = \dots = m'_t = 0$ if $t = n > p$). Besides,

$$\sum_{r=1}^{t} \frac{m_r + m_r'}{2^r} = \sum_{r=1}^{t} \frac{m_r}{2^r} + \sum_{r=1}^{t} \frac{m_r'}{2^r}$$

$$= \sum_{j=1}^{n} \frac{m_j}{2^j} + \sum_{s=1}^{p} \frac{m_s'}{2^s}$$

$$< \frac{1}{2^{k+1}} + \frac{1}{2^{k+1}} = \frac{1}{2^k}.$$

Therefore, $x + y \in V_k$, hence $V_{k+1} + V_{k+1} \subseteq V_k$, i.e. the condition (PNR2) is fulfilled.

Now let $x \in R$. Since $R = \bigcup_{k=1}^{\infty} Q_k$, then $x \in Q_k$ for a certain $k \in \mathbb{N}$. Let's show that $x \cdot \tilde{V}_{i_{k+1}+l} \subseteq V_l$ for any $l \in \mathbb{N}$. Since the property D) is valid for the neighborhoods $\tilde{V}_k$, the inclusion $x \cdot \tilde{V}_{i_{k+1}+l} \subseteq \tilde{V}_l$ is true for any $l \in \mathbb{N}$. Let $y \in V_{i_{k+1}+l}$. Then there exist $n \in \mathbb{N}$ and a set $\{m_1, m_2, \ldots, m_n\}$ of non-negative integers such that $y \in \sum_{j=1}^{n} m_j \tilde{V}_j$ and $\sum_{j=1}^{n} \frac{m_j}{2^j} < \frac{1}{2^{i_{k+1}+l}}$. Then

$$x \cdot y \in x \cdot \sum_{j=1}^{n} m_j \tilde{V}_j = \sum_{j=1}^{n} m_j (x \cdot \tilde{V}_j) \subseteq \sum_{j=1}^{n} m_j (Q_k \cdot \tilde{V}_j).$$

Note that from the inequality $\sum_{j=1}^{n} \frac{m_j}{2^j} < \frac{1}{2^{i_{k+1}+l}}$ follows that $m_j = 0$ for $j \le i_{k+1} + l$. If $j > i_{k+1} + l$, then $j - i_{k+1} > 0$. Thus,

$$Q_k \cdot \tilde{V}_j = Q_k \cdot \tilde{V}_{i_{k+1}+j-i_{k+1}} \subseteq Q_k \cdot \tilde{V}_{j-i_{k+1}}$$

(see the property D) for the neighborhoods $\tilde{V}_k$), hence,

$$x \cdot y \in \sum_{j=1}^{n} m_j \tilde{V}_{j-i_{k+1}} = \sum_{j=i_{k+1}+l}^{n} m_j \tilde{V}_{j-i_{k+1}},$$

besides,

$$\sum_{j=i_{k+1}+l}^{n} \frac{m_j}{2^{j-i_{k+1}}} = \sum_{j=1}^{n} \frac{m_j}{2^j} \cdot 2^{i_{k+1}} < \frac{2^{i_{k+1}}}{2^{i_{k+1}+l}} = \frac{1}{2^l}.$$

Thus, $x \cdot y \in V_l$, i.e. $x \cdot V_{i_{k+1}+l} \subseteq V_l$.

Analogously is proved that $V_{i_{k+1}+l} \cdot x \subseteq V_l$, i.e. the condition (PNR4) is fulfilled.

To complete the proof it is enough to verify that $\{V_k \mid k \in \mathbb{N}\}$ is a basis of neighborhoods of zero in the topological ring (R, τ).

Since $\frac{1}{2^{k+1}} < \frac{1}{2^k}$, then $\tilde{V}_{k+1} \subseteq V_k$, and since $\tilde{V}_{k+1}$ is a neighborhood of zero in the topological ring (R, τ) (see the property A) for the subset $\tilde{V}_i$), then V_k is a neighborhood of zero in (R, τ) for any $k \in \mathbb{N}$.

Let's show now that $V_k \subseteq U_{2k}$ for $k \in \mathbb{N}$. At first, due to the condition (PNR2), by the induction by m we easily get that $2^m U_n \subseteq U_{n-m}$ for all $m, n \in \mathbb{N}$ and $m < n$. Besides, by the induction by k it is easy to verify that from the condition (PNR2) follows that

$$U_{n+1} + U_{n+2} + \ldots + U_{n+k} \subseteq U_n$$

for any natural numbers n and k.

Now let $x \in V_k$. Then there exist natural number p and set $\{m_1, m_2, \ldots, m_p\}$ of non-negative integers such that $x \in \sum_{j=1}^{p} m_j \tilde{V}_j$ and $\sum_{j=1}^{p} \frac{m_j}{2^j} < \frac{1}{2^k}$. Then $m_j = 0$ for $j < k$ and $m_j < 2^{j-k}$ for $j > k$. Therefore, taking into account the property A) of the neighborhoods $\tilde{V}_k$, we get that

$$x \in \sum_{j=1}^{p} m_j \tilde{V}_j \subseteq \sum_{j=1}^{p} m_j U_{2j} \subseteq \sum_{j=k+1}^{p} 2^{j-k} U_{2j} \subseteq \sum_{j=k+1}^{p} U_{j+k}$$
$$= U_{2k+1} + U_{2k+2} + \ldots + U_{2k+p} \subseteq U_{2k}.$$

Thus, $V_k \subseteq U_{2k}$. Therefore, $\{V_k \mid k \in \mathbb{N}\}$ is a basis of neighborhoods of zero in R, which satisfies the conditions (PNR1), (PNR2), i.e. the implication 2) $\Longrightarrow$ 3) is proved.

2.1.7. *PROOF*. Proof of the implication 3)$\Longrightarrow$4).

Let $\{V_i \mid i \in \mathbb{N}\}$ be a basis of neighborhoods of zero of the topological ring (R, τ), satisfying the conditions (PNR1), (PNR2) and (PNR4). Put $W_n = V_n$, for $n > 0$ and $W_n = \{x \in R \mid V_i \cdot x \cdot V_j \subseteq V_{i+j+n}$ if $i+j+n > 0$ and $V_k \cdot x \bigcup x \cdot U_k \subseteq V_{k+n}$, if $k+n > 0\}$, for $n \leq 0$.

Since $W_n = V_n$ for $n > 0$ and $V_1 \subseteq W_n$ for $n \leq 0$, then $\{W_n \mid n \in \mathbb{Z}\}$ is a basis of neighborhoods of zero in R.

Let's show that $\{W_n \mid n \in \mathbb{Z}\}$ satisfies the conditions (PNR1) - (PNR3).

Let n_1 and n_2 be arbitrary integers. Let's verify that $W_{n_1} \cdot W_{n_2} \subseteq W_{n_1+n_2}$. To do it we consider the following cases.

CASE I. $n_1 > 0$ and $n_2 > 0$. Then

$$W_{n_1} \cdot W_{n_2} = V_{n_1} \cdot V_{n_2} \subseteq V_{n_1+n_2} = W_{n_1+n_2}.$$

CASE II. $n_1 > 0$ and $n_2 \leq 0$. If $n_1 + n_2 > 0$ then

$$W_{n_1} \cdot W_{n_2} = V_{n_1} \cdot W_{n_2} \subseteq V_{n_1+n_2} = W_{n_1+n_2}.$$

If $n_1 + n_2 \leq 0$, then for any numbers $i, j \in \mathbb{N}$, that satisfy the condition $i+j+n_1+n_2 > 0$, we get

$$V_j \cdot W_{n_1} \cdot W_{n_2} \cdot V_j = V_i \cdot V_{n_1} \cdot W_{n_2} \cdot V_j \subseteq V_{i+n_1} \cdot W_{n_2} \cdot V_j \subseteq V_{i+n_1+n_2+j}.$$

Besides, if $i + n_1 + n_2 > 0$, then

$$V_i \cdot W_{n_1} \cdot W_{n_2} = V_i \cdot V_{n_1} \cdot W_{n_2} \subseteq V_{i+n_1} \cdot W_{n_2} \subseteq V_{i+n_1+n_2}$$

and

$$W_{n_1} \cdot W_{n_2} \cdot V_i = V_{n_1} \cdot W_{n_2} \cdot V_i \subseteq V_{i+n_1+n_2}.$$

Due to the definition of the neighborhoods W_n for $n \leq 0$, we get that

$$W_{n_1} \cdot W_{n_2} \subseteq W_{n_1+n_2}.$$

CASE III. when $n_1 \leq 0$ and $n_2 > 0$, is considered analogously to the CASE II.

CASE IV. $n_1 \leq 0$ and $n_2 \leq 0$. If $i + j + n_1 + n_2 > 0$, then either $i + n_1 > 0$ or $j + n_2 > 0$. To determine, let $i + n_1 > 0$. Then

$$V_i \cdot W_{n_1} \cdot W_{n_2} \cdot V_j \subseteq V_{i+n_1} \cdot W_{n_2} \cdot V_j \subseteq V_{i+n_1+j+n_2}.$$

Besides, if $i + n_1 + n_2 > 0$, then $i + n_1 > 0$ and $i + n_2 > 0$, therefore,

$$V_i \cdot W_{n_1} \cdot W_{n_2} \subseteq V_{i+n_2} \cdot W_{n_2} \subseteq V_{i+n_1+n_2}$$

and

$$W_{n_1} \cdot W_{n_2} \cdot V_i \subseteq W_{n_1} \cdot V_{i+n_2} \subseteq V_{i+n_1+n_2}.$$

From the definition of the neighborhoods W_n for $n \leq 0$ follows that

$$W_{n_1} \cdot W_{n_2} \subseteq W_{n_1+n_2}$$

in this case, too. Thus, $\{W_n \mid n \in \mathbb{Z}\}$ satisfies the condition (PNR1).

Let n be an integer and $x, y \in W_{n+1}$. Let's verify that $x + y \in W_n$. To do it we also consider several cases.

1) $n + 1 > 1$. Then $n > 0$ and

$$x + y \in W_{n+1} + W_{n+1} = V_{n+1} + V_{n+1} \subseteq V_n = W_n.$$

2) $n + 1 = 1$, i.e. $x, y \in W_1$. Then

$$V_i \cdot (x + y) \cdot V_j \subseteq V_i \cdot x \cdot V_j + V_i \cdot y \cdot V_j \subseteq V_{i+j+1} + V_{i+j+1} \subseteq V_{i+j}$$

for any $i > 0$ and $j > 0$. Besides,

$$(x + y) \cdot V_i \subseteq x \cdot V_i + y \cdot V_i \subseteq V_{i+1} + V_{i+1} \subseteq V_i,$$

and, analogously, $V_i \cdot (x + y) \subseteq V_i$ for any $i > 0$. From the definition of the neighborhood W_0 follows that $x + y \in W_0$.

3) $n + 1 \leq 0$. If $i + j + n > 0$, then $i + j + n + 1 > 0$, therefore,

$$V_i \cdot (x + y) \cdot V_j \subseteq V_i \cdot x \cdot V_j + V_i \cdot y \cdot V_j \subseteq V_{i+j+n+1} + V_{i+j+n+1} \subseteq V_{i+j+n}.$$

Besides, if $i + n > 0$, then $i + n + 1 > 0$, therefore,

$$V_i \cdot (x + y) \subseteq V_i \cdot x + V_i \cdot y \subseteq V_{i+n+1} + V_{i+n+1} \subseteq V_{i+n}$$

and analogously $(x + y) \cdot V_i \subseteq V_{i+n}$. From the definition of the neighborhoods W_n for $n < 0$ follows that $x + y \in W_n$.

Thus, $W_{n+1} + W_{n+1} \subseteq W_n$, i.e. the condition (PNR2) is fulfilled.

Finally, let $x \in R$. According to the condition (PNR4), there exists an integer $n > 0$ such that $V_{i+n} \cdot x \subseteq V_i$ and $x \cdot V_{i+n} \subseteq V_i$ for all $i > 0$. Let i and j be such natural numbers that $i + j - 2n > 0$. Then either $i > n$ or $j > n$. To determine, let $i > n$. Then

$$V_i \cdot x \cdot V_j \subseteq V_{i-n} \cdot V_j \subseteq V_{i+j-n} \subseteq V_{i+j-2n}.$$

Besides, if $i - 2n > 0$, then $V_i \cdot x \subseteq V_{i-n} \subseteq V_{i-2n}$ and analogously $x \cdot V_i \subseteq V_{i-2n}$. From the definition of the neighborhood W_{-2n} follows that $x \in W_{-2n}$.

Thus, $R = \bigcup_{n \in \mathbb{Z}} W_n$, i.e. the condition (PNR3) is fulfilled.

Therefore, $\{W_n \mid n \in \mathbb{Z}\}$ is a basis of neighborhoods of zero of the topological ring (R, τ), which satisfies the conditions (PNR1) - (PNR3), i.e. the implication 3)$\Longrightarrow$4) is true.

2.1.8. *PROOF.* Proof of the implication 4)$\Longrightarrow$1).

Without loss of generality we can consider that the neighborhoods W_i are symmetrical, since $\{W_i \bigcap (-W_i) \mid i \in \mathbb{Z}\}$ is a countable basis of symmetrical neighborhoods of zero in (R, τ) that also satisfies the conditions (PNR1) - (PNR3). From the condition (PNR2) follows that $W_i \subseteq W_{i-1}$ for all $i \in \mathbb{Z}$.

Due to Proposition 2.1.3, on R exists a non-negative real-valued function f, satisfying the conditions a) - c). For any $x \in R$ put $\xi(x) = \frac{1}{2}(f(x) + f(-x))$. It is clear that ξ is a non-negative real-valued function on R and $\xi(-x) = \xi(x)$.

Besides, let's show that ξ has the properties a) and b) from Proposition 2.1.3.

Indeed, if $x \in W_k$, then, since the neighborhood W_k is symmetrical, we get that $-x \in W_k$. Since the function f has the property a), then

$$\xi(x) = \frac{1}{2}(f(x) + f(-x)) \leq \frac{1}{2}\left(\frac{1}{2^{k-4}} + \frac{1}{2^{k-4}}\right) = \frac{1}{2^{k-4}}.$$

If $\xi(x) < \frac{1}{2^{k-3}}$, then at least one of the non-negative numbers $f(x)$ and $f(-x)$ is less than $\frac{1}{2^{k-3}}$. Due to the property b) of the function f, and since the neighborhood W_k is symmetrical, we get that $-x, x \in W_k$, i.e. the property b) is fulfilled.

Let's show that ξ is a pseudonorm on the ring R, i.e. we'll verify the fulfillment of the conditions (NR1) - (NR3) (see Definition 1.1.12).

From the properties a) and b) of the function ξ follows that $\xi(x) = 0$ if and only if $x \in \bigcap_{i \in \mathbb{Z}} W_i = \{0\}$, i.e. ξ satisfies the condition (NR1).

Let $x, y \in R$. Due to the condition c) of the function f, we get

$$\xi(x - y) = \frac{1}{2}(f(x - y) + f(y - x)) \leq \frac{1}{2}(f(x) + f(-x) + f(y) + f(-y)) = \xi(x) + \xi(y),$$

i.e. ξ satisfies the condition (NR2).

If both x and y are non-zero elements, then, due to the condition (PNR3) of the ring (R, τ) (the ring is Hausdorff) and the inclusions $W_i \subseteq W_{i-1}$ for $i \in \mathbb{Z}$, there exist integers i_1 and i_2 such that $x \in W_{i_1}$ and $y \in W_{i_2}$, but $x \notin W_{i_1+1}$ and $y \notin W_{i_2+1}$. Due to the property b) of the function ξ, we get

$$\xi(x) \geq \frac{1}{2^{i_1+1-3}} = \frac{1}{2^{i_1-2}}$$

and $\xi(y) \geq \frac{1}{2^{i_2-2}}$. Since

$$x \cdot y \in W_{i_1} \cdot W_{i_2} \subseteq W_{i_1+i_2}$$

by the condition (PNR2), then, due to the property a) of the function ξ, we get that

$$\xi(x \cdot y) \leq \frac{1}{2^{i_1+i_2-4}} = \frac{1}{2^{i_1-2}} \cdot \frac{1}{2^{i_2-2}} \leq \xi(x) \cdot \xi(y).$$

If $x = 0$ or $y = 0$, then

$$\xi(x \cdot y) = \xi(0) = 0 \leq \xi(x) \cdot \xi(y).$$

Thus, the ξ function satisfies the condition (NR3).

Hence, ξ is a pseudonorm on the ring R.

Let's show that $\tau_\xi = \tau$. Indeed, let

$$B_0(R, \tau_\xi) = \{U_k \mid k \in \mathbb{N}\}$$

be a basis of neighborhoods of zero of the topological ring (R, τ_ξ), where

$$U_k = \left\{ x \mid x \in R, \xi(x) < \frac{1}{2^k} \right\}$$

for $k \in \mathbb{N}$. Since ξ has the property a), then from $x \in W_k$ follows that

$$\xi(x) \leq \frac{1}{2^{k-4}} < \frac{1}{2^{k-5}}.$$

Therefore, $x \in U_{k-5}$, i.e. $W_k \subseteq U_{k-5}$ for all $k > 5$. Hence, $\tau_\xi \leq \tau$.

Let now $x \in U_k$. Then

$$\xi(x) < \frac{1}{2^k} = \frac{1}{2^{(k+3)-3}},$$

and, due to the property b) of the function ξ, we have $x \in W_{k+3}$. Therefore, $U_k \subseteq W_{k+3}$ for all $k > 1$, i.e. $\tau_\xi \geq \tau$.

Thus, $\tau_\xi = \tau$, which completes the proof.

2.1.9. COROLLARY. *A connected topological ring R (see C.13) is pseudonormalized if and only if it possesses a countable basis of neighborhoods of zero, which satisfies the conditions (PNR1) and (PNR2).*

PROOF. If the ring R is pseudonormalized, then the existence of a basis of neighborhoods of zero, mentioned in the condition, results from Theorem 2.1.4.

Vice versa, let R possess a basis $\{U_i \mid i \in \mathbb{N}\}$ of neighborhoods of zero, which satisfies the conditions (PNR1) and (PNR2). Without loss of generality we can consider that U_i are symmetrical neighborhoods, since the neighborhoods $U_i \bigcap (-U_i)$ also satisfy the conditions (PNR1) and (PNR2).

Due to the condition (PNR1), U_1 is a bounded subset. Since the additive topological Abelian group $R(+)$ is connected, it is generated by the neighborhood U_1 of zero (see Corollary 1.4.21), i.e. $R = \bigcup_{n=1}^{\infty} S_n$, where

$$S_n = nS = \underbrace{U_1 + U_1 + \ldots + U_1}_{n \text{ summands}}.$$

Due to Proposition 1.6.19, each of the subsets S_n is bounded. Due to Theorem 2.1.4, the topological ring R is pseudonormalized.

2.1.10. THEOREM. *Let R be a topological ring and $U \cdot V$ be a neighborhood of zero for any neighborhoods U and V of zero. Then R is pseudonormalized if and only if it possesses a countable basis of neighborhoods of zero, which satisfies the conditions (PNR1) and (PNR2).*

PROOF. If R is pseudonormalized, then the existence of corresponding basis of neighborhoods of zero results from Theorem 2.1.4.

Vice versa, let $\mathcal{B}_0(R) = \{U_i | i \in \mathbb{N}\}$ be a basis of neighborhoods of zero in R, which satisfies the conditions (PNR1) and (PNR2). Due to Theorem 2.1.4, it is enough to show that the ring R is a union of a countable number of its bounded subsets.

Indeed, for any $n \in \mathbb{N}$ put

$$S_n = \{x \in R \mid x \cdot U_n \subseteq U_1 \text{ and } U_n \cdot x \subseteq U_1\}.$$

It is clear that $R = \bigcup_{n \in \mathbb{N}} S_n$.

Now we verify that for any $k \in \mathbb{N}$ the subsets S_k are bounded in R. Let W be a neighborhood of zero in R. There exist $n \in \mathbb{N}$ such that $U_n \subseteq W$. Then

$$S_k \cdot (U_k \cdot U_n) = (S_k \cdot U_k) \cdot U_n \subseteq U_1 \cdot U_n \subseteq U_n \subseteq W$$

and

$$(U_n \cdot U_k) \cdot S_k = U_n \cdot (U_k \cdot S_k) \subseteq U_n \cdot U_1 \subseteq U_n \subseteq W.$$

Since $U_k \cdot U_n$ and $U_n \cdot U_k$ are neighborhoods of zero in R, then S_n is a bounded subset.

2.1.11. THEOREM. *Let R be a topological ring with the unitary element. Let in R exist an invertible topologically nilpotent element d (see Definition 1.8.16) and a bounded neighborhood U of zero such that $d \cdot U = U \cdot d$. Then R is pseudonormalized ring.*

PROOF. Let V_1 be a neighborhood of zero in R for which $V_1 + V_1 \subseteq U$. Since the neighborhood U is bounded, we can choose a neighborhood V_2 of zero in R such that $V_2 \cdot U \subseteq V_1$ and $U \cdot V_2 \cdot U \subseteq U$. Since d is a topologically nilpotent element, we can choose a natural number m_0 such that $d^{m_0} \subseteq V_2$. Put $U_i = d^{m_0 \cdot (i+1)} \cdot U$ for $i \in \mathbb{N}$.

Let's show that $\{U_i \mid i \in \mathbb{N}\}$ is a basis of neighborhoods of zero in R, which satisfies the conditions (PNR1) and (PNR2).

Indeed, since $d \cdot U = U \cdot d$, hence, $d^k \cdot U = U \cdot d^k$ for any $k \in \mathbb{N}$. Then

$$U_i \cdot U_j = (d^{m_0 \cdot (i+1)} \cdot U) \cdot (d^{m_0 \cdot (j+1)} \cdot U) = d^{m_0 \cdot (i+j+1)} \cdot (U \cdot d^{m_0} \cdot U)$$

$$\subseteq d^{m_0 \cdot (i+j+1)} \cdot (U \cdot V_2 \cdot U) \subseteq d^{m_0 \cdot (i+j+1)} \cdot U = U_{i+j}$$

for any $i, j \in \mathbb{N}$, i.e. the condition (PNR1) is fulfilled.

Further,

$$V_i + V_i = d^{m_0 \cdot (i+1)} \cdot U + d^{m_0 \cdot (i+1)} \cdot U = d^{m_0 \cdot i} \cdot (d^{m_0} \cdot U + d^{m_0} \cdot U) \subseteq$$

$$d^{m_0 \cdot i} \cdot \cdot (V_2 \cdot U + V_2 \cdot U) \subseteq d^{m_0 \cdot i} \cdot (V_1 + V_1) \subseteq d^{m_0 \cdot i} \cdot U = V_{i-1}$$

for all $i = 2, 3, \dots$, i.e. the condition (PNR2) is fulfilled.

Let V and W be neighborhoods of zero in R, then, choosing a natural number k_0 such that $d^{k_0} \in V$, we get that $V \cdot W \supseteq d^{k_0} \cdot W$. Since the element d^{k_0} is invertible, then, due to Corollary 1.1.46, $V \cdot W$ is a neighborhood of zero in R.

Due to Theorem 2.1.10, the ring R is pseudonormalized.

2.1.12. COROLLARY. A topological field is pseudonormalized if and only if it is locally bounded (see Definition 1.6.40) and possesses a non-zero topologically nilpotent element.

2.1.13. COROLLARY. *A non-discrete locally compact skew field is pseudonormalized.**

---

* A stronger result is also true: any locally compact skew field is normalized.

PROOF. Let T be the set of all the topologically nilpotent elements of a locally compact skew field K. Due to Corollary 1.8.43, T is a neighborhood of zero in K, and, due to Theorem 1.8.50 and Remark 1.6.14, T is a bounded subset. Finally, due to Proposition 1.8.32, $d \cdot T \cdot d^{-1} = T$ for any non-zero element $d \in T$, hence, $d \cdot T = T \cdot d$. Now the statement easily results from Theorem 2.1.11.

2.1.14. THEOREM. *Let R be a topological ring with a countable basis of neighborhoods of zero consisting of subgroups of the additive group $R(+)$. For the ring R to be pseudonormalized, it is enough that R be a union of a countable number of its bounded subsets and possesses a bounded topologically nilpotent neighborhood of zero (see Definition 1.8.16).*

PROOF. If the ring R is pseudonormalized and ξ is a pseudonorm on R, then the set $S = \{x \in R \mid \xi(x) < \frac{1}{2}\}$ is a topologically nilpotent bounded neighborhood of zero in R (see Example 1.2.9 and Proposition 1.8.18). Besides, due to Theorem 2.1.4, the ring R is a union of the countable number of its bounded subsets.

Vice versa, let $\{U_k \mid k \in \mathbb{N}\}$ be a countable basis of neighborhoods of zero of the topological ring R that are subgroups of the group $R(+)$, and U be a bounded topologically nilpotent neighborhood of zero in R. Due to Theorem 1.6.48, the topological ring R possesses a countable basis $\{U'_k \mid k \in \mathbb{N}\}$ of neighborhoods of zero such that U'_1 is a subring of the ring R and U'_k is an ideal of the ring U'_1, for $k \in \mathbb{N}$. Choose a natural number k_0 such that $U'_{k_0} \subseteq U$. Putting $V_i = \bigcap_{j=0}^{i} U'_{k_0+j}$, we get that $\{V_i \mid i \in \mathbb{N}\}$ is a countable basis of neighborhoods of zero in R, besides:

a) V_i is a bounded topologically nilpotent subring of the ring R, for $i \in \mathbb{N}$;

b) $V_{i+1} \subseteq V_i$, for $i \in \mathbb{N}$;

c) V_i is an ideal of the ring V_1, for $i \in \mathbb{N}$.

Let

$$\tilde{V}_k = \bigcup_{i_1+\ldots+i_n=k} V_{i_1} \cdot V_{i_2} \cdot \ldots \cdot V_{i_n}$$

(the union is made by all $n, i_1, \ldots i_n \in \mathbb{N}$ such that $\sum_{j=1}^{n} i_j = k$).

As in 2.1.6, it is verified that $\tilde{V}_k \cdot \tilde{V}_l \subseteq \tilde{V}_{k+l}$, for all $k, l \in \mathbb{N}$. Since $V_k \subseteq \tilde{V}_k$, then $\tilde{V}_k$ is a neighborhood of zero of the ring R, for $k \in \mathbb{N}$.

Note that $\tilde{V}_{k+1} \subseteq \tilde{V}_k$ for any $k \in \mathbb{N}$. Indeed, let $x \in \tilde{V}_{k+1}$. Then there exist a natural number n and a set $\{i_1, i_2, \dots, i_n\}$ of the natural numbers such that $\sum_{j=1}^{n} i_j = k + 1$ and

$$x \in V_{i_1} \cdot V_{i_2} \cdot \ldots \cdot V_{i_n}.$$

If among the numbers $i_1, i_2, \dots, i_n$ exists at least one larger then 1 (let, for example, $i_p > 1$), then, taking into account the inclusion $V_{i_p} \subseteq V_{i_p-1}$, we get that

$$x \in V_{i_1} \cdot \ldots \cdot V_{i_{p-1}} \cdot V_{i_p-1} \cdot V_{i_{p+1}} \cdot \ldots \cdot V_{i_n},$$

besides,

$$i_1 + \dots + i_{p-1} + (i_p - 1) + i_{p+1} + \dots + i_n = \left(\sum_{j=1}^{n} l_j\right) - 1 = k.$$

If $i_1 = i_2 = \dots = i_n = 1$, then $x \in \underbrace{V_1 \cdot V_1 \cdot \ldots \cdot V_1}_{k+1 \text{ times}} \subseteq \underbrace{V_1 \cdot V_1 \cdot \ldots \cdot V_1}_{k \text{ times}}$. Therefore, $x \in \tilde{V}_k$, i.e. $\tilde{V}_{k+1} \subseteq \tilde{V}_k$.

Let's verify that for any $m \in \mathbb{N}$ there exists $r \in \mathbb{N}$ such that $\tilde{V}_r \subseteq V_m$, i.e. $\{\tilde{V}_k \mid k \in \mathbb{N}\}$ is a basis of neighborhoods of zero in R. Indeed, since the neighborhood V_1 is topologically nilpotent, we can choose a natural number q_0 such that $V_1^{(q)} \subseteq V_m$ (see B.12) for all $q \geq q_0$. Put $r = m \cdot q_0$ and let's show that $\tilde{V}_r \subseteq V_m$. Indeed, let $x \in \tilde{V}_r = \tilde{V}_{mq_0}$. There exist a natural number n and a set $\{i_1, i_2, \dots, i_n\}$ of natural numbers such that $\sum_{j=1}^{n} i_j = m \cdot q_0$ and $x \in V_{i_1} \cdot V_{i_2} \cdot \ldots \cdot V_{i_n}$. If $n \geq q_0$, then from $V_{i_j} \subseteq V_1$ for $j \in \mathbb{N}$ we get that

$$x \in V_{i_1} \cdot V_{i_2} \cdot \ldots \cdot V_{i_n} \subseteq V_1^{(n)} \subseteq V_m.$$

If $n < q_0$, then $i_s = \max\{i_j \mid j \in \mathbb{N}\} > m$. Since V_{i_s} is an ideal of the subring V_1 and since $V_{i_j} \subseteq V_1$ for $1 \leq j \leq n$, then

$$x \in V_{i_1} \cdot V_{i_2} \cdot \ldots \cdot V_{i_n} \subseteq V_1^{(s-1)} \cdot V_{i_s} \cdot V_1^{(n-s)} \subseteq V_{i_s} \subseteq V_m.$$

Therefore, $\tilde{V}_{m \cdot q_0} \subseteq V_m$.

Since V_m is a subgroup of the additive group of R, then $W_{m \cdot q_0} \subseteq V_m$, where W_k is a subgroup of the additive group of the ring R, generated by the neighborhood $\tilde{V}_k$, for $k \in \mathbb{N}$. Then $\{W_k \mid k \in \mathbb{N}\}$ is a basis of neighborhoods of zero of R.

Let's show that this basis satisfies the conditions (PNR1) and (PNR2).

Indeed, let $k, l \in \mathbb{N}$, $x \in W_k$ and $y \in W_l$. Since the neighborhoods $\tilde{V}_k$ and $\tilde{V}_l$ are symmetrical, then there exist elements $x_i \in \tilde{V}_k$ and $y_j \in \tilde{V}_l$, for $i = 1, 2, \ldots, n$ and $j = 1, 2, \ldots, m$ such that $x = x_1 + x_2 + \ldots + x_n$ and $y = y_1 + y_2 + \ldots + y_m$. Then $\tilde{V}_k \cdot \tilde{V}_l \subseteq \tilde{V}_{k+l}$ we get that

$$x \cdot y = \sum_{i=1}^{n} \sum_{j=1}^{m} x_i \cdot y_j \in \underbrace{\tilde{V}_{k+l} + \tilde{V}_{k+l} + \ldots + \tilde{V}_{k+l}}_{nm \text{ summands}} \subseteq W_{k+l}.$$

Thus, $W_k \cdot W_l \subseteq W_{k+l}$, i.e. the condition (PNR1) is fulfilled.

Further, from $\tilde{V}_{k+l} \subseteq \tilde{V}_k$ follows that $W_{k+1} \subseteq W_k$. Therefore, $W_{k+1} + W_{k+l} = W_{k+1} \subseteq W_k$, for any $k \in \mathbb{N}$, i.e. the condition (PNR2) is fulfilled.

Since, according to the condition of the theorem, R is a union of a countable number of its bounded subsets, then, due to Theorem 2.1.4, R is a pseudonormalized ring.

§ 2.2. Normalizability of Topological Skew Fields

2.2.1. PROPOSITION. *Let K be a topological skew field, K^* be the multiplicative group of non-zero elements of the skew field K, T be the set of all topologically nilpotent elements of the skew field K, N be the set of all neutral elements (see Definitions 1.8.16 and 1.8.29) of K. If $(T \bigcup N) \cdot T \subseteq T$, then:*

1) N is a normal subgroup of the group K^*;

2) $(T \bigcup N) \cdot (T \bigcup N) \subseteq (T \bigcup N)$, i.e. $T \bigcup N$ is a subsemigroup of the group K^*;

3) $x \cdot (T \bigcup N) \cdot x^{-1} \subseteq T \bigcup N$ for any element $x \in K^*$.

PROOF. It is clear that $1 \in N$. From Remark 1.8.30 follows that $N^{-1} = N$.

Let's show that $N \cdot N \subseteq N$. Indeed, let $a, b \in N$. If $a \cdot b \in T$, then from $a^{-1} \in N$ follows that $b = a^{-1}(a \cdot b) \in N \cdot T \subseteq (T \bigcup N) \cdot t \subseteq T$, which contradicts the neutrality of the element b. If $(a \cdot b)^{-1} \in T$, then $a^{-1} = b \cdot (b^{-1} a^{-1}) = b \cdot (a \cdot b)^{-1} \in N \cdot T \subseteq (T \bigcup N) \cdot T \subseteq T$, which contradicts the neutrality of the element a. Hence, $a \cdot b \in N$, i.e. $N \cdot N \subseteq N$. Since $N^{-1} = N$, then N is a subgroup of the group K^*. Due to Proposition 1.8.32, we get that N is a normal subgroup in K^*, i.e. we have proved the statement 1).

Note that $T \cdot N \subseteq T$. Indeed, if $t \in T$ and $a \in N$, then $a^{-1} \in N$, and from Proposition 1.8.32 follows that $a^{-1} \cdot t \cdot a \in T$, hence,

$$t \cdot a \in a \cdot T \subseteq N \cdot T \subseteq (T \cup N) \cdot T \subseteq T.$$

Therefore,

$$(T \cup N) \cdot (T \cup N) \subseteq \big((T \cup N) \cdot T\big) \cup \big((T \cup N) \cdot N\big)$$
$$\subseteq T \cup (T \cdot N) \cup (N \cdot N) \subseteq T \cup N,$$

i.e. we have proved the statement 2).

Since, due to Proposition 1.8.32, $x \cdot T \cdot x^{-1} = T$ and $x \cdot N \cdot x^{-1} = N$, then $x \cdot (T \bigcup N) \cdot x^{-1} = T \bigcup N$ for any $0 \neq x \in K$, i.e. the statement 3) is proved, too.

2.2.2. PROPOSITION. *Let K be a topological skew field, K^* be the multiplicative group of non-zero elements of K, T be the set of all topologically nilpotent elements of K, and N be the set of all neutral elements of K. If T is an open bounded from right subset* of K and $(T \bigcup N) \cdot T \subseteq T$, then on K can be defined a non-negative real-valued function f that satisfies the following conditions:

a) $f(x) = 0$ if and only if $x = 0$;

b) $f(x \cdot y) = f(x) \cdot f(y)$ for any elements $x, y \in K$;

c) $f(x) = 1$ if and only if $x \in N$;

d) $f(x^{-1}) = \frac{1}{f(x)}$ for any element $0 \neq x \in K$;

e) $f(x) < 1$ if and only if $x \in T$;

f) if V is a neighborhood of zero in K and an element $0 \neq c \in K$ is such that $T \cdot c \subseteq V$, then $x \in V$ for any element $x \in K$, for which $f(x) < f(c)$;

g) there exists a natural number n such that $f(a + b) \leq 2n \cdot \max\{f(a), f(b)\}$ for any $a, b \in K$.

PROOF. Since, due to Proposition 2.2.1 (statement 1), N is a normal subgroup of the group K^*, then we can consider the factor-group $G = K^*/N$. The element $e \cdot N = N$ is the unitary element of the group G.

Let's show that the subset

$$S = \left\{ x \cdot N \,\middle|\, x \in (T \cup N) \backslash \{0\} \right\}$$

of the group G satisfies the conditions that define a positive cone of a partially ordered group G (see B.15). Since, due Proposition 1.8.27, $(T \backslash \{0\}) \bigcup (T \backslash \{0\})^{-1} = \emptyset$, and $N^{-1} = N$, then $\left((T \bigcup N) \backslash \{0\} \right) \bigcap \left((T \bigcup N) \backslash \{0\} \right)^{-1} = N$, i.e. $S^{-1} \bigcap S = N$. The fulfillment of other conditions results from the evident inclusion $N \subseteq (T \bigcup N) \backslash \{0\}$ and from Proposition 2.2.1 (statements 2 and 3).

Since $K^* = (T \backslash \{0\}) \bigcup N \bigcup (T \backslash \{0\})^{-1}$, then $G = S \bigcup S^{-1}$ and, hence, S is a positive cone for a certain linear order of the group G, besides, $x \cdot N \leq y \cdot N$ if and only if $y \cdot x^{-1} \in (T \bigcup N) \backslash \{0\}$. In particular, the element $x \cdot N \in G$ is strongly positive if and only if $x \in (T \bigcup N) \backslash \{0\}$ and $x \notin N$, which is equivalent to $x \in T \backslash \{0\}$.

Let's show that the linear order on G, defined by the semigroup S, is Archimedean (see B.15). Let $a \cdot N$ and $b \cdot N$ be strictly positive elements of the group G. Then $a \in T \backslash \{0\}$ and $b \in T \backslash \{0\}$. Since, by the condition, T is a neighborhood of zero in K, then there exists a neighborhood U of zero in K such that $U \cdot b^{-1} \subseteq T$. Since a is a topologically nilpotent element, we can choose a natural number m such that $a^m \in U$. Then $a^m \cdot b^{-1} \in U \cdot b^{-1} \subseteq T$, besides, $a^m \cdot b^{-1} \neq 0$, i.e. $a^m \cdot b^{-1} \in T\backslash\{0\}$. Hence, $(a \cdot N)^m > b \cdot N$. Thus, the linearly ordered group G is Archimedean.

According to the Goelder theorem (see B.15), there exists an isomorphic mapping φ of the linearly ordered group G onto a subgroup A of the additive group $\mathbb{R}(+)$ of real numbers with the natural order.

Let $\omega_N : K^* \to G = K^*/N$ be the natural homomorphism (see B.11), then $\psi = \omega_N \cdot \varphi$ is a homomorphic mapping of the group K^* onto A, besides, $\ker \psi = N$.

Putting $f(x) = 2^{-\psi(x)}$ for $x \in K^*$ and $f(0) = 0$, we define a non-negative real-valued function f on K.

Let's we verify that f satisfies the conditions a) - g) of the proposition.

Indeed, according to the definition of the function f, we have $f(0) = 0$ and $f(x) > 0$ when $x \neq 0$, i.e. the condition a) is fulfilled.

Let x and y be non-zero elements from K, then $x \cdot y \neq 0$ and

$$f(x \cdot y) = 2^{-\psi(x \cdot y)} = 2^{-\varphi(\omega_N(x \cdot y))} = 2^{-\varphi(\omega_N(x) \cdot \omega_N(y))}$$
$$= 2^{-(\varphi(\omega_N(x)) + \varphi(\omega_N(y)))} = 2^{-\psi(x)} \cdot 2^{-\psi(y)} = f(x) \cdot f(y).$$

If at least one of the elements x or y is equal to zero, then the equality $f(x \cdot y) = f(x) \cdot f(y)$ is evident. Thus, the condition b) is fulfilled.

The equality $f(x) = 1$ is equivalent to $\psi(x) = 0$, which is equavalent (taking into account that φ is an isomorphism) to $x \in \ker \omega_N = N$, i.e. the condition c) is fulfilled.

Since $1 \in N$, then, by the condition b), $f(1) = 1$, hence, for any element $0 \neq x \in K$, using the condition b), we get that

$$1 = f(1) = f(x \cdot x^{-1}) = f(x) \cdot f(x^{-1}),$$

therefore, $f(x^{-1}) = \frac{1}{f(x)}$, i.e. the condition d) is fulfilled.

By the definition of the function f, the inequality $f(x) < 1$ means that either $x = 0$ or $\psi(x) > 0$. Since φ is an isomorphism of the linearly ordered groups $G = K^*/N$ and A, then $\psi(x) > 0$ is equivalent to the strict positivity of the element $x \cdot N$ in the group $G = K^*/N$, i.e. $x \in T\backslash\{0\}$. Thus, we have verified the fulfillment of the condition e).

Let V be a neighborhood of zero in K and $0 \neq c \in K$ is such element that $T \cdot c \subseteq V$. Let $x \in K$ and $f(x) < f(c)$, then

$$f(x \cdot c^{-1}) = f(x) \cdot f(c^{-1}) = \frac{f(x)}{f(c)} < 1,$$

hence, due to the condition e), $x \cdot c^{-1} \in T$. Therefore, $x \in T \cdot c \subseteq V$, i.e. the condition f) is fulfilled.

Let's verify the fulfillment of the condition g). At first we'll show that there exists a natural number n such that $f(1 + d) \leq n \cdot (1 + f(d))$ for any $d \in K$.

Assume the contrary, i.e. that for any natural number k there exists an element $d_k \in K$ such that $f(1 + d_k) > k \cdot (1 + f(d_k))$. Then

$$\frac{1}{f(1 + d_k)} + \frac{f(d_k)}{f(1 + d_k)} = \frac{1 + f(d_k)}{f(1 + d_k)} < \frac{1}{k},$$

hence,

$$\lim_{k \to +\infty} \left(\frac{1}{f(1 + d_k)} + \frac{f(d_k)}{f(1 + d_k)} \right) = 0.$$

Since $\frac{1}{f(1+d_k)} > 0$ and $\frac{f(d_k)}{f(1+d_k)} \geq 0$ for any $k \in \mathbb{N}$, then

$$\lim_{k \to +\infty} \frac{1}{f(1 + d_k)} = \lim_{k \to +\infty} \frac{f(d_k)}{f(1 + d_k)} = 0.$$

Let U and V be neighborhoods of zero in K such that $1 \notin U$ and $V + V \subseteq U$. Since, by the condition, T is a bounded from right subset in K, then, due to Proposition 1.6.3, $T \cdot c \subseteq V$ for a certain non-zero element c. Since $f(c) > 0$, then there exists a natural number k_0 such that

$$f\left((1 + d_{k_0})^{-1}\right) = \frac{1}{f(1 + d_{k_0})} < f(c)$$

and

$$f\left(d_{k_0} \cdot (1 + d_{k_0})^{-1}\right) = \frac{f(d_{k_0})}{f(1 + d_{k_0})} < f(c).$$

Thus, due to the condition f), we have $(1 + d_{k_0})^{-1} \in V$ and $d_{k_0} \cdot (1 + d_{k_0})^{-1} \in V$, therefore,

$$1 = (1 + d_{k_0}) \cdot (1 + d_{k_0})^{-1} = (1 + d_{k_0})^{-1} + d_{k_0} \cdot (1 + d_{k_0})^{-1} \in V + V \subseteq U,$$

which contradicts the choice of the neighborhood U of zero in K. Thus, there exists a natural number n such that $f(1 + d) \leq n \cdot (1 + f(d))$ for any $d \in K$.

Now let $a, b \in K$. If at least one of the elements a or b (let, for example, a) be different from zero, then

$$f(a+b) = f\big(a \cdot (1 + a^{-1} \cdot b)\big) = f(a) \cdot f(1 + a^{-1} \cdot b) \leq n \cdot f(a) \cdot \big(1 + f(a^{-1} \cdot b)\big)$$
$$= n \cdot f(a) \cdot \left(1 + \frac{f(b)}{f(a)}\right) = n \cdot \big(f(a) + f(b)\big).$$

Since $f(a) \geq 0$ and $f(b) \geq 0$, then $f(a) + f(b) \leq 2\max\{f(a), f(b)\}$, hence, $f(a+b) \leq 2n \cdot \max\{f(a), f(b)\}$, i.e. in this case the condition g) is fulfilled.

The fulfillment of the condition g) when $a = b = 0$, is evident. Hence, the condition g) is fulfilled.

2.2.3. THEOREM. *For normalizability of a topological skew field (K, τ) the fulfillment of the following conditions is sufficient and necessary:*

(NSF1). The set T of all the topologically nilpotent elements of a skew field (K, τ) is an open and bounded from right subset in (K, τ);

(NSF2). Let N be the set of neutral elements of a skew field (K, τ), then $(T \bigcup N) \cdot T \subseteq T$.

NECESSITY. Let the skew field (K, τ) be normalized, and ξ be a norm on K and $\tau_\xi = \tau$ (see Definition 2.1.1). Then, due to Proposition 1.8.18 and Example 1.2.9, the subset $V_0 = \{x \in K \,|\, \xi(x) < 1\}$ is a neighborhood of zero in (K, τ) consisting of topologically nilpotent elements. From Proposition 1.8.37 follows that T is an open subset in (K, τ). The boundedness of the subset T results from Proposition 1.6.10. Thus, the condition (NSF1) is fulfilled.

Let an element $a \in K$ be topologically nilpotent, and an element $b \in K$ be topologically nilpotent or neutral. Then $\xi(a) < 1$. Due to Propositions 1.8.18 and 1.8.31, $\xi(b) \leq 1$. Therefore, $\xi(b \cdot a) = \xi(b) \cdot \xi(a) < 1$, i.e. $b \cdot a$ is a topologically nilpotent element, hence, the condition (NSF2) is fulfilled.

SUFFICIENCY. Define on the skew field K a non-negative real-valued function f that satisfies the conditions a) - g) (see Proposition 2.2.2). Since $\lim_{\alpha \to 0}(2m)^\alpha = 1$ for any natural number m, then for the number n, mentioned in the condition g) of Proposition 2.2.2, there exists a positive number α such that $(2n)^\alpha \leq 2$.

Put $\xi(x) = \left(f(x)\right)^{\alpha}$ for $x \in K$. Let's show that so defined function ξ is a norm on K. To do it we'll verify the fulfillment of the conditions (NR1) - (NR3) (see Definition 1.1.12).

Indeed, it is clear that ξ is a non-negative real-valued function, besides, $\xi(1) = \left(f(1)\right)^{\alpha} = 1^{\alpha} = 1$. The fulfillment of the condition (NR1) is also evident (see Proposition 2.2.2, item a).

Besides (see Proposition 2.2.2, condition b),

$$\xi(x \cdot y) = \left(f(x \cdot y)\right)^{\alpha} = \left(f(x) \cdot f(y)\right)^{\alpha} = \left(f(x)\right)^{\alpha} \cdot \left(f(y)\right)^{\alpha} = \xi(x) \cdot \xi(y)$$

for any $x, y \in K$, i.e. the condition (NR3) is fulfilled.

Let's show that the function ξ satisfies the condition (NR2).

Below some properties of the function ξ which will be used in the proof of the condition (NR2) are listed.

If $x, y \in K$, then, due to Proposition 2.2.2 (condition g),

$$\xi(x + y) = \left(f(x + y)\right)^{\alpha} \le \left(2n \cdot \max\{f(x), f(y)\}\right)^{\alpha}$$
$$= (2n)^{\alpha} \cdot \max\left\{\left(f(x)\right)^{\alpha}, \left(f(y)\right)^{\alpha}\right\} \le 2 \cdot \max\left\{\left(f(x)\right)^{\alpha}, \left(f(y)\right)^{\alpha}\right\} = 2 \cdot \max\{\xi(x), \xi(y)\}.$$

By the induction by k we get that

$$\xi(a_1 + a_2 + \ldots + a_{2^k}) \le 2^k \cdot \max\left\{\xi(a_i) | i = 1, 2, \ldots, 2^k\right\}$$

for any elements $a_1, a_2, \ldots, a_{2^k} \in K$, $k \in \mathbb{N}$. In particular, since $\xi(1) = 1 = 2^0$, then for $a_1 = a_2 = \ldots = a_{2^k} = 1$ we get that $\xi(2^k \cdot 1) \le 2^k$ for $k = 1, 2, \ldots$.

Let's show that $\xi(n \cdot 1) \le 2n$ for any natural number n. Let $k = k(n)$ be a non-negative integer such that $2^k \le n < 2^{k+1}$.

The proof of the inequality $\xi(n \cdot 1) \le 2n$ we'll realize by the induction by k. If the number n is such that $k = 0$, then $n = 1$ and in that case $\xi(n \cdot 1) = \xi(1) = 1 < 2$. Suppose that for all $0 \le k < m$ the inequality $\xi(n \cdot 1) \le 2n$ is true for any n, for which $2^k \le n < 2^{k+1}$. Now let n be such that $2^m \le n < 2^{m+1}$. Since $n \cdot 1 = (n - 2^m) \cdot 1 + 2^m \cdot 1$, then

$$\xi(n \cdot 1) = \xi\left((n - 2^m) \cdot 1 + 2^m \cdot 1\right) \le 2 \max\left\{\xi\left((n - 2^m) \cdot 1\right), \xi\left(2^m \cdot 1\right)\right\}.$$

Now we consider the following two cases.

1) $\xi\big((n-2^m)\cdot 1\big) \leq \xi(2^m\cdot 1)$. Then $\max\{\xi\big((n-2^m)\cdot 1\big),\, \xi(2^m\cdot 1)\} = \xi(2^m\cdot 1)$, hence, $\xi(n\cdot 1) \leq 2\cdot 2^m \leq 2n$;

2) $\xi\big((n-2^m)\cdot 1\big) > \xi(2^m\cdot 1)$. Then

$$\max\{\xi\big((n-2^m)\cdot 1\big),\, \xi(2^m\cdot 1)\} = \xi\big((n-2^m)\cdot 1\big),$$

hence $\xi(n\cdot 1) \leq 2\cdot \xi\big((n-2^m)\cdot 1\big)$. Since $2^m \leq n < 2^{m+1}$, then $0 \leq n-2^m < 2^m$.

If $n-2^m > 0$, then there exists a natural number $q < m$ such that $2^q \leq n-2^m < 2^{q+1}$. Then, due to the inequality $n < 2^{m+1}$, we have $2n < 2^{m+2}$ and, hence, $2n - 2^{m+2} < 0$. Therefore, taking into account the inductive supposition, we get that

$$\xi(n\cdot 1) \leq 2\cdot \xi\big((n-2^m)\cdot 1\big) \leq 2\cdot 2\cdot (n-2^m) = 2n + 2n - 2^{m+2} < 2n.$$

If $n - 2^m = 0$, then $n = 2^m$ and as we have shown above,

$$\xi(n\cdot 1) = \xi(2^m\cdot 1) \leq 2^m = n < 2n.$$

Thus, $\xi(n\cdot 1) \leq 2n$ for any natural number n.

Now we'll prove that $\xi(1+a) \leq 1 + \xi(a)$ for any element $a \in K$. Since the elements 1 and a of K commutate, then, using the binomial formula, we get that

$$(1+a)^{2^n-1} = \sum_{i=0}^{2^n-1} C_{2^n-1}^i \cdot a^i,$$

and the number of the summands in this sum equals 2^n. Then, as it was shown,

$$\xi\left(\sum_{i=0}^{2^n-1} C_{2^n-1}^i \cdot a^i\right) \leq 2^n \cdot \max\{\xi(C_{2^n-1}^i \cdot a^i)\,|\,i = 0, 1, 2, \dots, 2^n - 1\}.$$

Knowing that

$$\xi(C_{2^n-1}^i \cdot a^i) = \xi\big((C_{2^n-1}^i \cdot 1)\cdot a^i\big) = \xi\big(C_{2^n-1}^i \cdot 1\big)\cdot \xi(a^i) \leq 2\cdot C_{2^n-1}^i \cdot (\xi(a))^i,$$

we get that

$$\xi\big((1+a)^{2^n-1}\big) = \xi\left(\sum_{i=0}^{2n-1} C_{2^n-1}^i \cdot a^i\right) \leq 2^n \cdot \max\{\xi(C_{2^n-1}^i \cdot a^i)\,|\,0 \leq i \leq 2^n - 1\}$$

$$\leq 2^{n+1} \cdot \max\{C_{2^n-1}^i \cdot (\xi(a))^i\,|\,i = 0, 1, \dots, 2^n - 1\} \leq 2^{n+1}\big(1 + \xi(a)\big)^{2^n-1}.$$

Then

$$\left(\xi(1+a)\right)^{2^n-1} = \xi\left((1+a)^{2^n-1}\right) \le 2^{n+1} \cdot \left(1+\xi(a)\right)^{2^n-1},$$

therefore,

$$\xi(1+a) \le 2^{\frac{n+1}{2^n-1}} \cdot \left(1+\xi(a)\right)$$

for any $n \in \mathbb{N}$. As this inequality approaches the limit as $n \to +\infty$, knowing that $\lim_{n\to+\infty} 2^{\frac{n+1}{2^n-1}} = 1$, we get that $\xi(1+a) \le 1+\xi(a)$.

Now it is easy to see that $\xi(a+b) \le \xi(a) + \xi(b)$ for any element $a, b \in K$. Indeed, from $1 = \xi(1) = \xi(x \cdot x^{-1}) = \xi(x) \cdot \xi(x^{-1})$ follows that $\xi(x^{-1}) = \frac{1}{\xi(x)}$ for any non-zero element $x \in K$. Therefore, if at least one of the elements a or b (let it be, for example, a) differs from zero, then

$$\xi(a+b) = \xi\left(a \cdot (1 + a^{-1} \cdot b)\right) = \xi(a) \cdot \xi(1 + a^{-1} \cdot b) \le \xi(a) \cdot \left(1 + \xi(a^{-1} \cdot b)\right)$$
$$= \xi(a) \cdot \left(1 + \xi(a^{-1}) \cdot \xi(b)\right) = \xi(a) \cdot \left(1 + \frac{\xi(b)}{\xi(a)}\right) = \xi(a) + \xi(b).$$

If $a = b = 0$, then $\xi(a+b) = \xi(0) = 0 = 0 + 0 = \xi(a) + \xi(b)$.

Finally, we'll show that ξ satisfies the condition (NR2). At first we note that from $1 = \xi((-1) \cdot (-1)) = \xi((-1))^2$ and $\xi(-1) > 0$ follows that $\xi(-1) = 1$. Therefore,

$$\xi(a-b) = \xi\left(a + (-1) \cdot b\right) \le \xi(a) + \xi((-1) \cdot b) = \xi(a) + \xi(-1) \cdot \xi(b) = \xi(a) + \xi(b)$$

for any elements $a, b \in K$, i.e. ξ satisfies the condition (NR2).

Thus, ξ is a norm on K, and to complete the proof of the theorem we must only verify that the interval topology τ_ξ, defined on K by the norm ξ, coincides with the initial topology τ on the skew field K.

Preliminarily note the following properties of the norm ξ:

1) $\xi(x) < 1$ if and only if $x \in T$, i.e. $T = \{x \in K \mid \xi(x) < 1\}$;

2) if V is a neighborhood of zero in K and an element $0 \ne c \in K$ is such that $T \cdot c \subseteq V$, then $x \in V$ for any $x \in K$, where $\xi(x) < \xi(c)$.

These properties result from the corresponding properties of the function f (see Proposition 2.2.1) and from the equality $\xi(x) = \left(f(x)\right)^\alpha$ for any $x \in K$ (see the definition of the function ξ).

Let V be a neighborhood of zero in (K, τ). Due to Proposition 1.6.3 and since the subset T is bounded in (K, τ), we get that there exists an element $0 \neq c \in K$ such that $T \cdot c \subseteq V$. Since $c \neq 0$, then $\xi(c) > 0$, hence there exists a natural number n for which $\frac{1}{2^n} < \xi(c)$. Then

$$V_n = \left\{ x \in K \big| \xi(x) < \frac{1}{2^n} \right\}$$

is a neighborhood of zero of the topological skew field (K, τ_ξ) (see Example 1.2.9), besides, if $x \in V_n$, then $\xi(x) < \frac{1}{2^n} < \xi(c)$, hence, $x \in V$. Therefore, $V_n \subseteq V$, i.e. V is a neighborhood of zero in (K, τ_ξ). Thus, $\tau \leq \tau_\xi$.

Let U be a neighborhood of zero in (K, τ_ξ) and k be a natural number such that

$$V_k = \left\{ x \in K \big| \xi(x) < \frac{1}{2^k} \right\} \subseteq U.$$

Since the topological skew field (K, τ) is non-discrete and T is an open subset in (K, τ), then there exists an element $0 \neq t \in T$. Since $\xi(t) < 1$, then $\xi(t^m) = (\xi(t))^m < \frac{1}{2^k}$ for a proper natural number m, besides, $t^m \neq 0$. Then $t^m \cdot T$ is a neighborhood of zero in (K, τ) (see Corollary 1.1.46), besides, for any $x \in T$ we get that $\xi(t^m \cdot x) = \xi(t^m) \cdot \xi(x) < \frac{1}{2^k}$, i.e. $t^m \cdot T \subseteq V_k \subseteq U$. Hence, U is a neighborhood of zero in (K, τ).

Thus, $\tau_\xi \leq \tau$ and, hence, $\tau = \tau_\xi$, which completes the proof of the theorem.

2.2.4. THEOREM. *For normalizability of a topological field (K, τ) the fulfillment of the following conditions is sufficient and necessary:*

(NF1) the set T of all topologically nilpotent elements of the field K is an open subset in (K, τ);

(NF2) the set $T \bigcup N$ of all topologically nilpotent or neutral elements of the field K is a bounded subset in (K, τ).

PROOF. The necessity of the condition (NF1) results from Theorem 2.2.3. Let ξ be a norm on (K, τ) and let $\tau_\xi = \tau$. Then, as we have shown in the proof of Theorem 2.2.3 (item "necessity"), $T = \{x \in K \big| \xi(x) < 1\}$. Since from Proposition 1.8.31 follows that $N = \{x \in K \big| \xi(x) = 1\}$, then $T \bigcup N = \{x \in K \big| \xi(x) \leq 1\}$. Now the condition (NF2) easily results from Proposition 1.6.10.

To prove the sufficiency, we'll verify that (K, τ) satisfies the conditions (NSF1) and (NSF2) (see Theorem 2.2.3).

At first we note that from the condition (NF2), due to Remark 1.6.15, follows that T is a bounded subset in (K, τ). Therefore, knowing the condition (NF1), we conclude that the condition (NSF1) is fulfilled.

Let $a \in T$, $b \in T \bigcup N$ and U be a neighborhood of zero in (K, τ). Knowing that the subset $T \bigcup N$ is bounded and the element a is topologically nilpotent, we can choose a neighborhood of V zero in (K, τ) and a natural number n_0 such that $(T \bigcup N) \cdot V \subseteq U$ and $a^n \in V$ for all $n \geq n_0$. From Corollary 1.8.36 and from the definition of a neutral element follows that if $b \in T \bigcup N$, then $b^k \in T \bigcup N$ for any natural number k. Therefore, for $n \geq n_0$ we get that $(b \cdot a)^n = b^n \cdot a^n \in (T \bigcup N) \cdot V \subseteq U$. Hence, the element $b \cdot a$ is topologically nilpotent, i.e. (K, τ) also satisfies the condition (NSF2).

Due to Theorem 2.2.3, (K, τ) is a normalized topological field.

2.2.5. COROLLARY. *Any locally compact field is normalized.*[*]

PROOF. The proof results from Theorems 1.8.50 and 2.2.4.

2.2.6. THEOREM. *A topological field (K, τ) is normalized if and only if it is locally inversely bounded (see Definition 1.6.61) and satisfies the condition (NF1).*

PROOF. The necessity results from Theorem 2.2.4 and Example 1.6.63.

Let (K, τ) be locally inversely bounded. Since the set T of all topologicaly nilpotent elements is an open subset in (K, τ), then T is a neighborhood of zero in (K, τ). Then

$$(T \cup N) \backslash \{0\} = (T \backslash \{0\}) \cup N = (K \backslash T)^{-1}$$

and, hence, $T \bigcup N$ is a bounded subset in (K, τ), i.e. (K, τ) satisfies the condition (NF2). Due to Theorem 2.2.4, (K, τ) is a normalized topological field.

---

[*] This statement is true for any locally compact skew field (see also the note for Corollary 2.1.13).

§ 2.3. Pseudonormalizability of Topological Modules and
Vector Spaces

2.3.1. DEFINITION. Let R be a ring and M be an R-module. Let ξ be a norm (pseudonorm) on a ring R, then we say that an R-module M is normalized (correspondingly, pseudonormalized) over (R, ξ), if on M is defined a non-negative real-valued function η which satisfies the following conditions:

(NM1) $\eta(x) = 0$ if and only if $x = 0$;

(NM2) $\eta(x - y) \leq \eta(x) + \eta(y)$ for any two elements $x, y \in M$;

(NM3) $\eta(a \cdot x) = \xi(a) \cdot \eta(x)$ (correspondingly, $\eta(a \cdot x) \leq \xi(a) \cdot \eta(x)$) for any elements $a \in R$ and $x \in M$.

The number $\eta(x)$ is called a norm (pseudonorm) of the element $x \in M$, and the function η is called a norm (pseudonorm) on R-module M (compare to Definition 1.1.12).

2.3.2. REMARK. As in Remark 1.1.13, we see that if η is a pseudonorm on a module M, then:

1) $\eta(-x) = \eta(x)$ for any element $x \in M$;

2) $\eta(x) - \eta(y) \leq \eta(x - y)$ for any elements $x, y \in M$;

3) $\eta(x + y) \leq \eta(x) + \eta(y)$ for any elements $x, y \in M$.

2.3.3. PROPOSITION. *Let R be a ring with the unitary element, ξ be a pseudonorm on R and η be a pseudonorm on an R-module M. If an element $a \in R$ is invertible, and $\xi(a) \cdot \xi(a^{-1}) = 1$, then $\eta(a \cdot m) = \xi(a) \cdot \eta(m)$ for any $m \in M$.*

PROOF. Assume the contrary, i.e. that $\eta(a \cdot m_0) \neq \xi(a) \cdot \eta(m_0)$ for a certain element $m_0 \in M$. Then $\eta(a \cdot m_0) < \xi(a) \cdot \eta(m_0)$. Therefore,

$$\eta(m_0) = \eta(1 \cdot m_0) = \eta(a^{-1} \cdot a \cdot m_0) \leq \xi(a^{-1}) \cdot \eta(a \cdot m_0) < \xi(a^{-1}) \cdot \xi(a) \cdot \eta(m_0) = \eta(m_0).$$

The contradiction received shows that $\eta(a \cdot m) = \xi(a) \cdot \eta(m)$ for any $m \in M$.

2.3.4. COROLLARY. Let ξ be a norm on a skew field K and η be a pseudonorm on a vector space E over (K, ξ). Then $\eta(a \cdot x) = \xi(a) \cdot \eta(x)$ for any $a \in K$ and any $x \in E$, i.e. η is a norm on E.

2.3.5. REMARK. Let ξ be a pseudonorm on a ring R and η be a pseudonorm on an R-module M. Let

$$V_k = \left\{ a \in R \mid \xi(a) < \frac{1}{2^k} \right\} \quad \text{and} \quad U_k = \left\{ x \in M \mid \eta(x) < \frac{1}{2^k} \right\}$$

for every $k \in \mathbb{Z}$. Then, using the conditions (NM1) - (NM3) and Remark 2.3.2, we see that the following conditions are fulfilled:

1) $\bigcap_{k=1}^{\infty} U_k = \{0\}$;

2) $U_m \subseteq U_n$, if $n \le m$;

3) $U_{k+1} + U_{k+1} \subseteq U_k$ for any $k \in \mathbb{Z}$;

4) $-U_k = U_k$ for any $k \in \mathbb{Z}$;

5) $V_n \cdot U_m \subseteq U_{n+m}$ for $n, m \in \mathbb{Z}$;

6) $a \cdot U_{n+k} \subseteq U_k$, if $\xi(a) \le 2^n$ and $n, k \in \mathbb{Z}$;

7) $V_{n+k} \cdot x \subseteq U_k$, if $\eta(x) \le 2^n$ and $n, k \in \mathbb{Z}$.

Since $\mathcal{B}_0(R) = \{V_k \mid k \in \mathbb{N}\}$ is a basis of neighborhoods of zero of the topological ring (R, τ_ξ) (see Example 1.2.9), then, due to Corollary 1.3.8, from the conditions 1) - 7) of this remark results that the system $\mathcal{B}_0(M) = \{U_k \mid k \in \mathbb{N}\}$ of the subsets of R-module M is a basis of neighborhoods of zero in a certain Hausdorff topology τ_η, where (M, τ_η) is a topological (R, τ_ξ)-module. The topology τ_η, as in the case of pseudonormed rings, we also call an interval topology, defined by the pseudonorm η on M.

2.3.6. DEFINITION. Let ξ be a norm (pseudonorm) on a ring R. A topological (R, τ_ξ)-module (M, τ) is called normalized (pseudonormalized) if on M can be defined a norm (pseudonorm) η such that the interval topology τ_η on M, defined by the norm (pseudonorm) η (see Remark 2.3.5), coincides with the initial topology τ on the module M.

2.3.7. THEOREM. *Let ξ be a pseudonorm on a ring R and τ_ξ be an interval topology defined on R by pseudonorm ξ. A topological (R, τ_ξ)-module (M, τ) is pseudonormalized if and only if it posesses a countable basis $\mathcal{B}_0(M) = \{U_i \mid i \in \mathbb{Z}\}$ of neighborhoods of zero that satisfies the following conditions:*

(PNM1) $U_{i+1} + U_{i+1} \subseteq U_i$ for $i \in \mathbb{Z}$;

(PNM2) $\bigcup_{i=-\infty}^{+\infty} U_i = M$;

(PNM3) if $a \in R$ and $\xi(a) < \frac{1}{2^n}$, then $a \cdot U_i \subseteq U_{i+n}$, for $n, i \in \mathbb{Z}$.

PROOF. The necessity of the conditions (PNM1) - (PNM3) was noted in Remark 2.3.5.

Beginning the proof of the sufficiency, make the following four notes:

1) without loss of generality we can consider that all the neighborhoods from $\mathcal{B}_0(M, \tau)$ are symmetrical subsets, since the family $\{U_i \bigcap (-U_i) \mid i \in \mathbb{N}\}$ forms a basis of neighborhoods of zero in (M, τ), which satisfies the conditions (PNM1) - (PNM3);

2) by the induction by m it is easy to conclude from the condition (PNM1) that

$$2^m U_i = \underbrace{U_i + U_i + \ldots + U_i}_{2^m \text{ summands}} \subseteq U_{i-m} \quad (\text{see B.3})$$

for any integer i and any natural number m;

3) from the condition (PNM1) also results that $U_i \subseteq U_j$, if $j \leq i$;

4) for any integers m and k, from the condition $\mid k \mid \leq \frac{1}{2^m}$ results that $kU_i \subseteq U_{i+m}$ for any integer i.

Indeed, knowing that the neighborhoods U_i are symmetrical (see item 1)), it is enough to consider the case when $k > 0$. It is evident that $m \leq 0$. If $m = 0$, then from the inequality $\mid k \mid \leq \frac{1}{2^m} = 1$ results that $k = 1$, hence, $kU_i = U_i = U_{i+m}$. If $m < 0$, then $-m$ is a natural number, hence, we get that

$$kU_i = \underbrace{U_i + U_i + \ldots + U_i}_{k \text{ summands}} \subseteq \underbrace{U_i + U_i + \ldots + U_i}_{2^{-m} \text{ summands}} \subseteq U_{i-(-m)} = U_{i+m}$$

(see item 2)).

Let f be a non-negative real-valued function on M such that:

a) $f(x) \leq \frac{1}{2^{k-4}}$, if $x \in U_k$;

b) $x \in U_k$, if $f(x) < \frac{1}{2^{k-3}}$;

c) $f(x + y) \leq f(x) + f(y)$ for any $x, y \in M$.

The existence of the function f is guaranteed by Proposition 2.1.3. For any $x \in M$ put: $\eta(x) = \inf\{\sum_{i=1}^{n} \xi(a_i) \cdot f(x_i) \mid n \in \mathbb{N}, a_i \in R \bigcup \mathbb{Z}, 0 \neq x_i \in M$ for $1 \leq i \leq n, x = \sum_{i=1}^{n} a_i \cdot x_i\}$ (if $a \in \mathbb{Z}$, then as $\xi(a)$ we consider $\mid a \mid$). Since $x = 1 \cdot x$, where $1 \in \mathbb{Z}$, then

$$0 \leq \eta(x) \leq \xi(1) \cdot f(x) = \mid 1 \mid \cdot f(x) = f(x)$$

for any $x \in M$, hence, the function η is defined on M.

For any natural number l put $W_l = \{x \in M | \eta(x) < \frac{1}{2^l}\}$.

Let's show that $\{W_l | l \in \mathbb{N}\}$ is a basis of neighborhoods of zero of the topological (R, τ_ξ)-module (M, τ). To do it we'll verify that $U_{l+5} \subseteq W_l \subseteq U_{l-3}$ for any $l \in \mathbb{N}$.

Let $x \in U_{l+5}$. Then, due to the property a) of the function f, we get that $f(x) \leq \frac{1}{2^{l+1}} < \frac{1}{2^l}$. Then from $\eta(x) \leq f(x) < \frac{1}{2^l}$ follows that $x \in W_l$, i.e. $U_{l+5} \subseteq W_l$. Let $x \in W_l$, i.e. $\eta(x) < \frac{1}{2^l}$. Then there exist a natural number p and elements $a_i \in R \bigcup \mathbb{Z}$ and $0 \neq x_i \in M$, for $i = 1, 2, \ldots, p$, such that $x = \sum_{i=1}^{p} a_i \cdot x_i$ and $\sum_{i=1}^{p} \xi(a_i) \cdot f(x_i) < \frac{1}{2^l}$. For any $i = 1, 2, \ldots, p$ let k_i be an integer such that

$$\frac{1}{2^{k_i+1}} \leq \xi(a_i) < \frac{1}{2^{k_i}}.$$

Since $x_i \neq 0$ for $i = 1, 2, \ldots, p$ and $\bigcap_{t \in \mathbb{Z}} U_t = \{0\}$, besides, $U_t \subseteq U_{t-1}$ for $t \in \mathbb{Z}$, then for any $i = 1, 2, \ldots, p$ there exists an integer $n_i = \max\{t | x_i \in U_t\}$. Then $x_i \notin U_{n_i+1}$ and, due to the properties a) and b) of the function f, we get

$$\frac{1}{2^{n_i-2}} \leq f(x_i) \leq \frac{1}{2^{n_i-4}},$$

for $i = 1, 2, \ldots, p$. Due to the condition (PNM3) and the item 4) of Remark 2.3.5, from $x_i \in U_{n_i}$ and $\xi(a_i) < \frac{1}{2^{k_i}}$ results that $a_i \cdot x_i \in U_{n_i+k_i}$, hence,

$$f(a_i \cdot x_i) \leq \frac{1}{2^{n_i+k_i-4}}, \quad \text{for} \quad i = 1, 2, \ldots, p.$$

Then

$$f(x) \leq \sum_{i=1}^{p} f(a_i \cdot x_i) \leq \sum_{i=1}^{p} \frac{1}{2^{n_i+k_i-4}}$$

$$= 2 \cdot \sum_{i=1}^{p} \frac{1}{2^{k_i+1}} \cdot \frac{1}{2^{n_i-4}} < 2 \cdot \sum_{i=1}^{p} \xi(a_i) \cdot 2^2 \cdot f(x_i) < \frac{1}{2^{l-3}},$$

therefore, $x \in U_{l-3}$.

Thus, we have shown that $\{W_l | l \in \mathbb{N}\}$ is a basis of neighborhoods of zero of (M, τ).

Let's prove now that the function η is a pseudonorm on the R-module M, i.e. verify the fulfillment of the conditions (NM1) - (NM3) (see Definition 2.3.1).

Since $0 = 0 \cdot x$ for any $x \in M$, then $0 \le \eta(0) \le \xi(0) \cdot f(x) = 0$, i.e. $\eta(0) = 0$.

Vice versa, if $\eta(x) = 0$, then $\eta(x) < \frac{1}{2^l}$ for any $l \in \mathbb{N}$, i.e. $x \in \bigcap_{l \in \mathbb{N}} W_l = \{0\}$. Knowing that $\{W_l \mid l \in \mathbb{N}\}$ (as we have just shown) is a basis of neighborhoods of zero of (M, τ) and that the topological space (M, τ) is Hausdorff, we have that $\eta(x) = 0$ if and only if $x = 0$, i.e. we have verified the fulfillment of the condition (NM1).

Let $x, y \in M$ and ϵ be an arbitrary positive number. There exist elements $a_1, a_2, \dots, a_p$, $b_1, b_2, \dots, b_q \in R \bigcup \mathbb{Z}$ and $x_1, x_2, \dots, x_p$, $y_1, y_2, \dots, y_q \in M$ such that:

$$\eta(x) + \epsilon > \sum_{i=1}^{p} \xi(a_i) \cdot f(x_i) \quad \text{and} \quad x = \sum_{i=1}^{p} a_i \cdot x_i;$$

$$\eta(y) + \epsilon > \sum_{j=1}^{q} \xi(b_j) \cdot f(y_j) \quad \text{and} \quad y = \sum_{j=1}^{q} b_j \cdot y_j.$$

Then

$$x - y = \sum_{i=1}^{p} a_i \cdot x_i + \sum_{j=1}^{p} (-b_j) \cdot y_j,$$

besides,

$$\sum_{i=1}^{p} \xi(a_i) \cdot f(x_i) + \sum_{j=1}^{q} \xi(-b_j) \cdot f(y_j)$$

$$= \sum_{i=1}^{p} \xi(a_i) \cdot f(x_i) + \sum_{j=1}^{q} \xi(b_j) \cdot f(y_j) < \eta(x) + \eta(y) + 2\epsilon,$$

therefore, $\eta(x - y) < \eta(x) + \eta(y) + 2\epsilon$.

Since the choice of $\epsilon > 0$ was arbitrary, we conclude that $\eta(x - y) \le \eta(x) + \eta(y)$, i.e. the fulfillment of the condition (NM2) is verified.

Let $a \in R$ and $x \in M$. Let ϵ be an arbitrary positive number, then there exist subsets $\{a_1, a_2, \dots, a_p\}$ in $R \bigcup \mathbb{Z}$ and $\{x_1, x_2, \dots, x_p\}$ in M such that

$$\eta(x) + \epsilon > \sum_{i=1}^{p} \epsilon(a_i) \cdot f(x_i) \quad \text{and} \quad x = \sum_{i=1}^{p} a_i \cdot x_i.$$

Then $a \cdot x = \sum_{i=1}^{p} (a \cdot a_i) \cdot x_i$, besides,

$$\sum_{i=1}^{p} \xi(a \cdot a_i) \cdot f(x_i) \le \sum_{i=1}^{p} (\xi(a) \cdot \xi(a_i) \cdot f(x_i)) = \xi(a) \cdot \left(\sum_{i=1}^{p} \xi(a_i) \cdot f(x_i) \right)$$

$$< \xi(a) \cdot (\eta(x) + \epsilon) = \xi(a) \cdot \eta(x) + \xi(a) \cdot \epsilon.$$

Hence, $\eta(a \cdot x) \le \xi(a) \cdot \eta(x) + \xi(a) \cdot \epsilon$. Since the choice of $\epsilon > 0$ was arbitrary, then $\eta(a \cdot x) \le \xi(a) \cdot \eta(x)$, i.e. the condition (NM3) is fulfilled.

Thus, η is a pseudonorm on M and the interval topology τ_η coincides with the initial topology τ, i.e. the pseudonormalizability of the topological (R, τ_ξ)-module (M, τ) is proved.

2.3.8. COROLLARY. *Let ξ be a norm on a skew field K. A topological vector space (E, τ) over the topological skew field (K, τ_ξ) is normalized if and only if it possesses a countable basis $\mathcal{B}_0(E, \tau) = \{U_i \mid i \in \mathbb{Z}\}$ of neighborhoods of zero that satisfies the conditions (PNM1) - (PNM3).*

PROOF. The proof results from Theorem 2.3.7 and Corollary 2.3.4.

2.3.9. DEFINITION. Let ξ be a norm on a skew field K and E a vector space over K. A subset S of E is called superconvex if for any finite subsets $\{a_1, a_2, \dots, a_n\}$ of K and $\{x_1, x_2, \dots, x_n\}$ of S from $\sum_{i=1}^{n} \xi(a_i) \le 1$ follows $\sum_{i=1}^{n} a_i \cdot x_i \in S$.

2.3.10. REMARK. Superconvexity of a subset $S \subseteq E$ means that $\sum_{i=1}^{n}(a_i \cdot S) \subseteq S$ for any finite subset $\{a_1, a_2, \dots, a_n\} \subseteq K$ of elements that satisfy the condition $\sum_{i=1}^{n} \xi(a_i) \le 1$. In particular, $a \cdot S \subseteq S$ for any $a \in K$, where $\xi(a) \le 1$.

Thus, knowing that $\xi(1) = \xi(-1) = 1$, we get that $-S = (-1) \cdot S \subseteq S$, hence, $S = -(-S) \subseteq -S$. Therefore, $-S = S$, i.e. any superconvex subset $S \subseteq E$ is symmetrical.

2.3.11. THEOREM. *Let ξ be a norm on a skew field K such that (K, τ_ξ) is non-discrete. A topological vector space (E, τ) over the topological skew field (K, τ_ξ) is normalized if and only if it posesses a neighborhood of zero that is a super convex and bounded subset in (E, τ) (see Definitions 2.3.9 and 1.6.1).*

PROOF. Let (E, τ) be normalized and η be a norm on E such that $\tau_\eta = \tau$. Then $U = \{x \in E | \eta(x) < 1\}$ is a neighborhood of zero in (E, τ). Let's show that the subset U is bounded and superconvex.

Let W be a neighborhood of zero in (E, τ). Since $\tau_\eta = \tau$, then there exists a natural number n such that

$$W_n = \left\{ x \in E \mid \eta(x) < \frac{1}{2^n} \right\} \subseteq W.$$

Then

$$V_n = \left\{ a \in K \mid \xi(a) < \frac{1}{2^n} \right\}$$

is a neighborhood of the topological skew field (K, τ_ξ). Let $a \in V_n$, then $\xi(a) < \frac{1}{2^n}$ and, hence,

$$\eta(a \cdot x) = \xi(a) \cdot \eta(x) < \xi(a) < \frac{1}{2^n}$$

for any $x \in U$, i.e. $a \cdot x \in W_n$ for any $x \in U$. Thus, $V_n \cdot U \subseteq W_n \subseteq W$, i.e. U is a bounded subset in (E, τ).

Let $\{x_1, x_2, \dots, x_k\} \subseteq U$. Then $\eta(x_i) < 1$ for $1 \le i \le n$. If $\{a_1, a_2, \dots, a_k\} \subseteq K$ and $\sum_{i=1}^{k} \xi(a_i) \le 1$, then

$$\eta\left(\sum_{i=1}^{k} a_i \cdot x_i\right) \le \sum_{i=1}^{k} \eta(a_i \cdot x_i) = \sum_{i=1}^{k} \xi(a_i) \cdot \eta(x_i) \le \sum_{i=1}^{k} \xi(a_i) \le 1.$$

Therefore, $\sum_{i=1}^{k} a_i \cdot x_i \in U$, which confirms the superconvexity of the neighborhood U.

Vice versa, let U be a neighborhood of zero in (E, τ) such that it is a bounded and superconvex subset. Let's show that the topological vector space is normalized.*

Let $V_k = \{a \in K \mid \xi(a) < \frac{1}{2^k}\}$ for $k \in \mathbb{N}$, then the family $\{V_k \mid k \in \mathbb{N}\}$ is a basis of neighborhoods of zero in (K, τ_ξ).

For any $n \in \mathbb{Z}$ put:

$U_n = \{x \in E \mid$ there exist $\{m_1, \dots, m_t\} \subseteq \mathbb{N} \bigcup \{0\}$ such that $\sum_{i=1}^{t} m_i \cdot 2^{-i} < 2^{-n}$ and $x \in \sum_{i=1}^{t} m_i \cdot V_i \cdot U\}$.

Let's verify that $\{U_n \mid n \in \mathbb{Z}\}$ is a basis of neighborhoods of zero in (E, τ) that satisfies the conditions (PNM1) - (PNM3) of the Theorem 2.3.7.

Let $n \in \mathbb{N}$, and $0 \ne r \in K$ be such that $\xi(r) < 2^{-n}$, then $r \cdot U \subseteq U_n$. Due to Proposition 1.1.44, $r \cdot U$ is a neighborhood of zero in (E, τ) and, hence, U_n is a neighborhood of zero in (E, τ) for any $n \in \mathbb{Z}$.

---

* The proof can be also realized this way: for any element $x \in E$ we put $\eta(x) = \inf\{\sum_i \xi(a_i) \mid x \in \sum_i (a_i \cdot U)$, the sums are finite $\}$ and verify that η is a norm on E, besides, $\tau_\eta = \tau$.

Let now W be a neighborhood of zero in (E, τ). Then, since the neighborhood U is bounded in (E, τ), there exists $n \in \mathbb{N}$ such that $V_n \cdot U \subseteq W$. Let $0 \neq d \in V_n$, i.e. $0 < \xi(d) < \frac{1}{2^n}$. We can choose $m \in \mathbb{Z}$ such that $\xi(d^{-1}) < 2^m$.

Let's show that $U_{n \cdot m} \subseteq W$. Indeed, let $x \in U_{n \cdot m}$. Then $x = \sum_{k=1}^{l} m_k \cdot v_k \cdot u_k$, where $m_k \in \mathbb{N} \bigcup \{0\}$, $v_k \in V_k$, $u_k \in U$, for $k = 1, 2, \dots, l$ and $\sum_{k=1}^{l} \frac{m_k}{2^k} < \frac{1}{2^{n \cdot m}}$. Then

$$x = d \cdot \sum_{k=1}^{n^l} (m_k \cdot d^{-1} \cdot v_k) \cdot u_k.$$

Since

$$\sum_{k=1}^{l} \xi(m_k \cdot d^{-1} \cdot v_k) = \sum_{k=1}^{l} m_k \cdot \xi(d^{-1}) \cdot \xi(v_k)$$

$$< \sum_{k=1}^{l} m_k \cdot 2^m \cdot \frac{1}{2^k} = 2^m \cdot \sum_{k=1}^{l} \frac{m_k}{2^k} < 2^m \cdot \frac{1}{2^{n \cdot m}} < 1,$$

then, since the neighborhood U is superconvex, we get

$$\sum_{k=1}^{l} (m_k \cdot d^{-1} \cdot v_k) \cdot u_k \in U$$

and, hence, $x \in d \cdot U \subseteq V_n \cdot U \subseteq W$, i.e. $U_{n \cdot m} \subseteq W$.

Thus, $\{U_n | n \in \mathbb{Z}\}$ is a basis of neighborhoods of zero in (E, τ).

Now we'll verify the fulfillment of the conditions (PNM1) - (PNM3) for the family $\{U_n \mid n \in \mathbb{N}\}$.

Let $i \in \mathbb{Z}$ and $x, y \in U_{i+1}$. Then $x = \sum_{k=1}^{l_1} m_k \cdot a_k \cdot x_k$ and $y = \sum_{s=1}^{l_2} n_s \cdot b_s \cdot y_s$, where: $a_k \in V_k$, $b_s \in V_s$, $x_k, y_s \in U$ for $1 \leq k \leq l_1$ and $1 \leq s \leq l_2$, as well as $\sum_{k=1}^{l_1} \frac{m_k}{2^k} < \frac{1}{2^{i+1}}$ and $\sum_{s=1}^{l_2} \frac{n_s}{2^s} < \frac{1}{2^{i+1}}$. Then

$$x + y = \left(\sum_{k=1}^{l_1} m_k \cdot a_k \cdot x_k \right) + \left(\sum_{s=1}^{l_2} n_s \cdot b_s \cdot y_s \right)$$

and

$$\left(\sum_{k=1}^{l_1} \frac{m_k}{2^k} \right) + \left(\sum_{s=1}^{l_2} \frac{n_s}{2^s} \right) < \frac{1}{2^{i+1}} + \frac{1}{2^{i+1}} = \frac{1}{2^i},$$

i.e. $x + y \in U_i$. Thus, $U_{i+1} + U_{i+1} \subseteq U_i$, i.e. the condition (PNM1) is fulfilled.

Let $x \in E$. Since (K, τ) is non-discrete and U is a neighborhood of zero in (E, τ), then there exists a non-zero element a in K such that $a \cdot x \in U$, i.e. $x \in a^{-1} \cdot U$. We can choose $l \in \mathbb{Z}$ such that $\xi(a^{-1}) < 2^{-l}$. From the definition of the neighborhood U_l follows that $x \in U_l$, i.e. $E = \bigcup_{n \in \mathbb{Z}} U_n$. Thus, the condition (PNM2) is fulfilled.

At last, let $a \in K$ and $\xi(a) < \frac{1}{2^m}$. Let $x \in U_n$, then $x = \sum_{k=1}^{l} m_k \cdot a_k \cdot x_k$, where: $m_k \in \mathbb{N} \cup \{0\}$, $a_k \in V_k$ (i.e. $\xi(a_k) < \frac{1}{2^k}$), $x_k \in U$ for $k = 1, 2, \dots, l$ and $\sum_{k=1}^{l} \frac{m_k}{2^k} < \frac{1}{2^n}$. Then $a \cdot x = \sum_{k=1}^{l} m_k \cdot (a \cdot a_k) \cdot x_k$, besides,

$$\xi(a \cdot a_k) = \xi(a) \cdot \xi(a_k) < \frac{1}{2^m} \cdot \frac{1}{2^k} = \frac{1}{2^{m+k}},$$

i.e. $a \cdot a_k \in V_{k+m}$ for $k = 1, 2, \dots, l$ and

$$\sum_{k=f}^{l} \frac{m_k}{2^{m+k}} = \frac{1}{2^m} \cdot \sum_{k=1}^{l} \frac{m_k}{2^k} < \frac{1}{2^m} \cdot \frac{1}{2^n} = \frac{1}{2^{m+n}}.$$

Since $x_k \in U$ for $1 \le k \le l$, then $a \cdot x \in U_{n+m}$. Therefore, $a \cdot U_n \subseteq U_{n+m}$, i.e. the condition (PNM3) is fulfilled.

From Corollary 2.3.8 follows that the topological vector space (E, τ) is normalized.

2.3.12. DEFINITION. Let ξ be a norm on a skew field K. A subset S of a vector space E over K is called absolutely convex if $a \cdot S + b \cdot S \subseteq S$ for any elements $a, b \in K$, where $\xi(a) + \xi(b) \le 1$ (compare to Definition 2.3.9).

2.3.13. REMARK. It is evident that any superconvex subset is absolutely convex. It is found out that if a norm ξ on a skew field K satisfies the condition:

(NR4) for any two elements $a, b \in K$ there exists an element $c \in K$ such that $\xi(c) = \xi(a) + \xi(b)$;

then any absolutely convex subset S of the vector space E over K is also superconvex.

Indeed, first of all by the induction by n we easily verify that for any elements $a_1, a_2, \dots, a_n \in K$ there exists an element $d \in K$ such that $\xi(d) = \sum_{i=1}^{n} \xi(a_i)$.

Now let S be an absolutely convex subset in E. Let's show that the subset S is superconvex, i.e. let's verify that for any natural number $n \ge 2$ and any subsets $\{a_1, a_2, \dots, a_n\} \subseteq K$ and $\{x_1, x_2, \dots, x_n\} \subseteq S$ from $\sum_{i=1}^{n} \xi(a_i) \le 1$ follows $\sum_{i=1}^{n} a_i \cdot x_i \in S$.

If $n = 2$ then the statement results from the absolute convexity of the subset S.

Suppose that the statement is true for all $n < k$, and let $\{a_1, a_2, \ldots, a_k\} \subseteq K$, $\{x_1, x_2, \ldots, x_k\} \subseteq S$ and $\sum_{i=1}^{k} \xi(a_i) \leq 1$.

Now we'll show that $y = \sum_{i=1}^{k} a_i \cdot x_i \in S$.

If $a_1 = a_2 = \ldots = a_{k-1} = 0$ then $y = a_k \cdot x_k$ and $\xi(a_k) \leq 1$. Therefore, $y \in 0 \cdot S + a_k \cdot S \subseteq S$ (due to an absolute convexity of the subset S).

If some of the elements $a_1, a_2, \ldots, a_{k-1}$ are non-zero, then $\sum_{i=1}^{k-1} \xi(a_i) > 0$. Then in K exists an element $d \neq 0$ such that $\xi(d) = \sum_{i=0}^{k-1} \xi(a_i)$. We get that

$$y = \sum_{i=1}^{k} a_i \cdot x_i = d \cdot \left(\sum_{i=1}^{k-1} (d^{-1} \cdot a_i) \cdot x_i \right) + a_k \cdot x_k,$$

besides,

$$\sum_{i=1}^{k-1} \xi(d^{-1} \cdot a_i) = \sum_{i=1}^{k-1} \frac{1}{\xi(d)} \cdot \xi(a_i) = \frac{1}{\xi(d)} \cdot \left(\sum_{i=1}^{k-1} \xi(a_i) \right) = \frac{1}{\xi(d)} \cdot \xi(d) = 1.$$

Hence, $z = \sum_{i=1}^{k-1} (d^{-1} \cdot a_i) \cdot x_i \in S$, knowing the inductive supposition. Then from the absolute convexity of the subset S follows that $y = d \cdot z + a_k \cdot x_k \in S$, since

$$\xi(d) + \xi(a_k) = \sum_{i=1}^{k-1} \xi(a_i) + \xi(a_k) \leq 1.$$

2.3.14. THEOREM. *Let ξ be a norm on a skew field K that satisfies the condition (NR4) (see Remark 2.3.13). A topological vector space (E, τ) over (K, τ_ξ) is normalized if and only if it possesses a neighborhood of zero that is a bounded and absolutely convex subset in (E, τ).*

PROOF. The proof easily results from Theorem 2.3.11 and Remark 2.3.13.

2.3.15. REMARK. The natural norms on the skew fields $\mathbb{R}$, $\mathbb{C}$ and $\mathbb{H}$ (see Examples 1.1.15, 1.1.16 and 1.1.18) satisfy the condition (NR4), since in the case of any of these three skew fields the element $|a| + |b|$ also belongs to this skew field, besides, $|(|a| + |b|)| = |a| + |b|$.

2.3.16. THEOREM. *A topological vector space E over the skew field $K = \mathbb{R}$, or $\mathbb{C}$, or $\mathbb{H}$ is normalized if and only if it possesses a neighborhood of zero that is a bounded and absolutely convex subset in E.*

PROOF. The proof results from Theorem 2.3.14, using Remark 2.3.15.

3. Completion of Topological Rings and Modules

One of the most important constructions in the theory of topological groups, rings and modules is the construction of completion. In this chapter we construct the completion of the topological groups, rings and modules with the help of Cauchy sequences, whose elements are indexed by a directed set. We prove some properties of the completion of topological groups, rings and modules; in particular, reveal the connection between compactness and pre–compactness of the subsets in the topological groups. We receive a criterion of invertibility of elements and the sufficient conditions for quasi-invertibility of elements in the completion of a topological ring.

The results of § § 1–4) are well known for the specialists in topological algebra, but could be useful for the beginners. The results of § 5 could be found also in [16].

§ 3.1. Directions, Cauchy Sequences and Filters

3.1.1. DEFINITION. A direction, or, in other words, a directed set, is a partially ordered set $(\Gamma, \leq)$ that satisfies the following condition:

– for any two elements $\gamma_1, \gamma_2 \in \Gamma$ there exists the third element $\gamma_3 \in \Gamma$ such that $\gamma_1 \leq \gamma_3$ and $\gamma_2 \leq \gamma_3$. If it is clear from the context what partial order is meant, we'll shortly denote the direction over Γ.*

3.1.2. REMARK. Let Γ be a direction, and $\gamma_1, \gamma_2, \ldots, \gamma_n \in \Gamma$, then there exists an element $\gamma \in \Gamma$ such that $\gamma_i \leq \gamma$ for all $i = 1, 2, \ldots, n$.

3.1.3. EXAMPLE.

a) Let $(\Gamma, \leq)$ be a linearly ordered set. Then Γ is a direction.

b) Let X be a set, $\mathcal{F} = \{A_\gamma | \gamma \in \Gamma\}$ be a filter of the set X (we consider that $A_{\gamma_1} \neq A_{\gamma_2}$ when $\gamma_1 \neq \gamma_2$). We define on Γ the following relation of partial order:

$\alpha \leq \beta$, for $\alpha, \beta \in \Gamma$ if and only if $A_\beta \subseteq A_\alpha$.

---

* The term "direction" also has another sense [131,213], as a sequence of elements of a directed set (see Definition 3.1.9.).

It is easy to check that it is really a partial order, and the condition of direction is fulfilled, because the filter was determined. We'll call Γ endowed with this order as a direction, defined by the filter $\mathcal{F}$.

c) The system of all the neighborhoods $\mathcal{B}_x = \{V_\gamma | \gamma \in \Gamma\}$ of a point x of a topological space X forms a filter, hence, we can consider the direction $(\Gamma, \leq)$, defined by the filter $\mathcal{B}_x$.

3.1.4. DEFINITION. By a subdirection Γ' of the direction Γ we call a subset $\Gamma' \subseteq \Gamma$ with the relation of partial order on Γ' induced from Γ such that for any element $\gamma \in \Gamma$ there exists an element $\gamma' \in \Gamma'$, for which $\gamma \leq \gamma'$.*

3.1.5. REMARK. It is easy to note that a subdirection Γ' of a direction Γ is a direction, too.

3.1.6. PROPOSITION. *Let* $\mathcal{B}_x = \{U_\gamma | \gamma \in \Gamma\}$ *be the system of all the neighborhoods of the point* x *of the topological space* X, Γ' *be a subset of* Γ *such that* $\mathcal{B}'_x = \{U_\gamma | \gamma \in \Gamma'\}$ *is a basis of neighborhoods of the point* x *in* X. *Then* Γ' *is a subdirection in the direction* Γ, *defined by the filter* $\mathcal{B}_x$.

PROOF. Let $\gamma \in \Gamma$. Since $\mathcal{B}'_x$ is a basis of neighborhoods of the point x in X, then there exists an element $\gamma_1 \in \Gamma'$ such that $U_{\gamma_1} \subseteq U_\gamma$. Then, according to the definition of the order, defined on the set by the system $\mathcal{B}_x$, it follows that $\gamma_1 \geq \gamma$.

3.1.7. DEFINITION. Two partially ordered sets $(\Gamma, \leq)$ and $(\Delta, \leq)$ are similar if there exists a bijection $f : \Gamma \to \Delta$ (called similarity) such that $\gamma_1 \leq \gamma_2$ if and only if $f(\gamma_1) \leq f(\gamma_2)$ for all elements $\gamma_1, \gamma_2 \in \Gamma$.

3.1.8. PROPOSITION. *Let* Γ *be a direction that does not contain the largest element and* $|\Gamma| = \aleph_0$ *(see A.7). Then* Γ *contains a subdirection that is a countable linearly ordered set, similar to the set of natural numbers* $\mathbb{N}$ *with the natural order relation.*

PROOF. We can consider set Γ to be enumerated by the set of natural numbers $\mathbb{N}$, i.e. $\Gamma = \{\gamma_n | n \in \mathbb{N}\}$. Let us construct by induction a subset $\{n_i | i \in \mathbb{N}\} \subseteq \mathbb{N}$ such that the

---

* The notion of subdirection is equivalent to the notion of cofinity of the subset.

following condition takes place:

$$\gamma_{n_i} < \gamma_{n_{i+1}} \quad \text{and} \quad \gamma_{n_i} \geq \gamma_i. \tag{*}$$

For $n_1 = 1$ it is evident that the condition $(*)$ is fulfilled. Suppose that the numbers $n_1, \ldots, n_k$, meeting the condition $(*)$ are constructed. Since Γ is a direction without the largest element, then there exists a number m such that $\gamma_{n_k} < \gamma_m$ and $\gamma_{k+1} \leq \gamma_m$. It is easy to see that if we put $n_{k+1} = m$, then the condition $(*)$ will be fulfilled for the numbers $n_1, \ldots, n_k, n_{k+1}$.

Now we prove that the subset $\{\gamma_{n_i} | i \in \mathbb{N}\}$ is a subdirection in Γ, similar to $\mathbb{N}$.

Indeed, if $\gamma_k \in \Gamma, k \in \mathbb{N}$, then according to the construction of the sequence $\{n_i | i \in \mathbb{N}\}$, we get that $\gamma_k \leq \gamma_{n_k}$. Hence, $\{\gamma_{n_i} | i \in \mathbb{N}\}$ is a subdirection in Γ. The construction shows that the mapping $f : \mathbb{N} \to \{\gamma_{n_i} | i \in \mathbb{N}\}$, for which $f(k) = \gamma_{n_k}$, is a similarity.

3.1.9. DEFINITION. Let X be a set, Γ be a direction, $f : \Gamma \to X$ a mapping. We denote the element $f(\gamma)$ over x_γ. The set $\{x_\gamma | \gamma \in \Gamma\}$ is called a sequence of the elements of X by the direction Γ.

If Γ' is a subdirection in Γ, then the set $\{x_\gamma | \gamma \in \Gamma'\}$ is called a subsequence of the sequence $\{x_\gamma | \gamma \in \Gamma\}$.

3.1.10. REMARK. Let X be a set, Γ be a direction, $\{x_\gamma | \gamma \in \Gamma\}$ be a sequence of elements of X by this direction. The family $\mathcal{F} = \{F_\gamma | \gamma \in \Gamma\}$ of subsets (they are allowed to coincide) where $F_\gamma = \{x_\delta | \delta \geq \gamma\}$, satisfies the conditions of the definition A.12 and, hence, is a basis of a filter of the set X. We call it a filter associated with the sequence $\{x_\gamma | \gamma \in \Gamma\}$.

3.1.11. DEFINITION. Let X be a topological space, $\{x_\gamma | \gamma \in \Gamma\}$ a sequence of the elements X by a certain direction Γ. An element $a \in X$ is called a limit of the sequence $\{x_\gamma | \gamma \in \Gamma\}$ by the direction Γ (notation: is $a = \lim_{\gamma \in \Gamma} x_\gamma$), if for any neighborhood V of the point x in X there exists an element $\gamma_0 \in \Gamma$ such that $\{x_\gamma | \gamma \geq \gamma_0\} \subseteq V$ (see C.6). The sequence that has a limit is called convergent to the limit.

3.1.12. PROPOSITION. *Let X be a topological space, $a \in X, \{x_\gamma | \gamma \in \Gamma\}$ be a sequence of elements of X by a certain direction Γ and $a = \lim_{\gamma \in \Gamma} x_\gamma$. If $\{x_\gamma | \gamma \in \Gamma'\}$ is a sub-sequence of the sequence $\{x_\gamma | \gamma \in \Gamma\}$, then $a = \lim_{\gamma \in \Gamma'} x_\gamma$.*

PROOF. Let V be a neighborhood of the point a in X. Since $a = \lim_{\gamma \in \Gamma} x_\gamma$, then there exists an element $\gamma_0 \in \Gamma$ such that $\{x_\gamma | \gamma \in \Gamma, \gamma_0 \leq \gamma\} \subseteq V$. Since Γ' is a subdirection in Γ, then there exists an element $\gamma_0' \in \Gamma'$ such that $\gamma_0 \leq \gamma_0'$. Then $\{x_\gamma | \gamma \in \Gamma', \gamma_0' \leq \gamma\} \subseteq \{x_\gamma | \gamma \in \Gamma, \gamma_0 \leq \gamma\} \subseteq V$, which completes the proof.

3.1.13. THEOREM. *Let X be a set, $\{x_\gamma | \gamma \in \Gamma\}$ a sequence of elements of X by a certain direction Γ, $\{\tau_\delta | \delta \in \Delta\}$ a family of topologies on X. If there exists an element $a \in X$ such that $a = \lim_{\gamma \in \Gamma} x_\gamma$ in the topological space (X, τ_δ) for any $\delta \in \Delta$, then $a = \lim_{\gamma \in \Gamma} x_\gamma$ in the topological space $(X, \sup\{\tau_\delta | \delta \in \Delta\})$.*

PROOF. Let V be a neighborhood of the point a in the space $(X, \sup\{\tau_\delta | \delta \in \Delta\})$. Because of the definition C.9, there exist elements $\delta_1, \ldots, \delta_n \in \Delta$ and such neighborhoods V_i of the point a in the topologies τ_{δ_i} correspondingly, for $1 \leq i \leq n$, that the neighborhood V contains the subset $\bigcap_{i=1}^n V_i$.

For any $i = 1, \ldots, n$ we can find an element $\gamma_i \in \Gamma$ such that $\{x_\gamma | \gamma \in \Gamma, \gamma \geq \gamma_i\} \subseteq V_i$. Let γ_0 be such element in Γ, that $\gamma_0 \geq \gamma_i$ for all $i = 1, \ldots, n$. Then $\{x_\gamma | \gamma \in \Gamma, \gamma \geq \gamma_0\} \subseteq \bigcap_{i=1}^n V_i \subseteq V$. The proof is completed.

3.1.14. DEFINITION. Let X be a topological space, $\mathcal{F}$ a filter of the set X. An element $d \in X$ is called a limit of the filter $\mathcal{F}$ (notation is: $a = \lim \mathcal{F}$) if for any neighborhood V of the point a in X there exists a set $F \in \mathcal{F}$ such that $F \subseteq V$.

3.1.15. PROPOSITION. *Let X be a topological space, $\{x_\gamma | \gamma \in \Gamma\}$ a sequence of elements of X by a certain direction Γ, $\mathcal{F}$ the filter associated with the sequence $\{x_\gamma | \gamma \in \Gamma\}$ (see Remark 3.1.10). A point $a \in X$ is a limit of the sequence $\{x_\gamma | \gamma \in \Gamma\}$ if and only if a is a limit of the filter $\mathcal{F}$.*

PROOF. Let $a = \lim_{\gamma \in \Gamma} x_\gamma$ and V be a certain neighborhood of the point a in X. According to the definition of the limit, there exists an element $\gamma_0 \in \Gamma$ such that $\{x_\gamma | \gamma \in \Gamma, \gamma \geq \gamma_0\} \subseteq V$. But $\{x_\gamma | \gamma \in \Gamma, \gamma \geq \gamma_0\} \in \mathcal{F}$. Thus, $a = \lim \mathcal{F}$.

Vice versa, let $a = \lim \mathcal{F}$. Let V be a neighborhood of the point a in X, then there exists a set $F \in \mathcal{F}$ such that $F \subseteq V$. Since the family of sets of the type $\{x_\gamma | \gamma \in \Gamma, \gamma \geq \gamma_0\}$, $\gamma_0 \in \Gamma$, is a basis of the filter $\mathcal{F}$, then there exists an element $\gamma_1 \in \Gamma$ such that $\{x_\gamma | \gamma \in \Gamma, \gamma \geq \gamma_1\} \subseteq F$. Hence, $\{x_\gamma | \gamma \in \Gamma, \gamma \geq \gamma_1\} \subseteq F \subseteq V$. Since the neighborhood V

was chosen arbitrarily, then $a = \lim_{\gamma \in \Gamma} x_\gamma$.

3.1.16. PROPOSITION. *Let X be a topological space, $\mathcal{F} = \{F_\gamma | \gamma \in \Gamma\}$ a certain filter of the set X. Consider the direction Γ defined by this filter $\mathcal{F}$ (see Example 3.1.3). Let us construct a sequence in X by this direction the following way:*

for every $\gamma \in \Gamma$ we choose an arbitrary element x_γ from the set F_γ.

If an element $a \in X$ is a limit of the filter F, then it is a limit of the sequence $\{x_\gamma | \gamma \in \Gamma\}$ by the direction Γ.

PROOF. Let V be a neighborhood of the element a in X. Since $a = \lim \mathcal{F}$, then there exists an element $\gamma_0 \in \Gamma$ such that $F_{\gamma_0} \subseteq V$. According to the definition of the partial order on the set Γ, $F_\gamma \subseteq F_{\gamma_0}$ for $\gamma \geq \gamma_0$. Hence, $F_\gamma \subseteq V$ for $\gamma \geq \gamma_0$. Therefore, $x_\gamma \in F_\gamma \subseteq V$ for any $\gamma \geq \gamma_0$, i.e. $\{x_\gamma | \gamma \in \Gamma, \gamma \geq \gamma_0\} \subseteq V$. Hence, $a = \lim_{\gamma \in \Gamma} x_\gamma$.

3.1.17. DEFINITION. Let X be a topological Abelian group, $\{x_\gamma | \gamma \in \Gamma\}$ a sequence of elements of X by a certain direction Γ. We call $\{x_\gamma | \gamma \in \Gamma\}$ a Cauchy sequence in X if for any neighborhood V of zero in X there exists an element $\gamma_0 \in \Gamma$ such that $x_\gamma - x_{\gamma_0} \in V$ for any $\gamma \in \Gamma, \gamma \geq \gamma_0$.

3.1.18. REMARK. *Let X be a topological Abelian group. A sequence $\{x_\gamma | \gamma \in \Gamma\}$ by a certain direction Γ is a Cauchy sequence if and only if for any neighborhood U of zero in X there exists an element $\gamma_0 \in \Gamma$ such that $x_{\gamma_1} - x_{\gamma_2} \in V$ for all $\gamma_1, \gamma_2 \in \Gamma, \gamma_1 \geq \gamma_0, \gamma_2 \geq \gamma_0$.*

PROOF. Let $\{x_\gamma | \gamma \in \Gamma\}$ be a Cauchy sequence in X. For any neighborhood V of zero in X there exists a neighborhood U of zero in X such that $U - U \subseteq V$. Since $\{x_\gamma | \gamma \in \Gamma\}$ is a Cauchy sequence in X, there exists an element $\gamma_0 \in \Gamma$ such that $x_\gamma - x_{\gamma_0} \in U$ for all $\gamma \in \Gamma, \gamma \geq \gamma_0$. Then $x_{\gamma_1} - x_{\gamma_2} = (x_{\gamma_1} - x_{\gamma_0}) + (x_{\gamma_0} - x_{\gamma_2}) \in U - U \subseteq V$ for all $\gamma_1, \gamma_2 \in \Gamma, \gamma_1 \geq \gamma_0, \gamma_2 \geq \gamma_0$.

Vice versa, for the sequence $\{x_\gamma | \gamma \in \Gamma\}$ let the following condition hold:

for any neighborhood V of zero in X there exists an element $\gamma_0 \in \Gamma$ such that $x_{\gamma_1} - x_{\gamma_2} \in V$ for all $\gamma_1, \gamma_2 \in \Gamma, \gamma_1 \geq \gamma_0, \gamma_2 \geq \gamma_0$.

Then $x_{\gamma_1} - x_{\gamma_0} \in V$ for any $\gamma_1 \geq \gamma_0$.

Hence, $\{x_\gamma | \gamma \in \Gamma\}$ is a Cauchy sequence in X.

3.1.19. PROPOSITION. *Let X be a topological Abelian group, $\{x_\gamma | \gamma \in \Gamma\}$ a sequence*

of elements of X by a certain direction Γ. If $a \in X$ and $a = \lim_{\gamma \in \Gamma} x_\gamma$, then $\{x_\gamma | \gamma \in \Gamma\}$
is a Cauchy sequence in X.

PROOF. Let $\mathcal{B}_0(X)$ be the system of all neighborhoods of zero in X and $V \in \mathcal{B}_0(X)$.
There exists a subset $U \in \mathcal{B}_0(X)$ such that $U - U \subseteq V$. Since $a = \lim_{\gamma \in \Gamma} x_\gamma$, then there
exists an element $\gamma_0 \in \Gamma$ such that $x_\gamma \in a + U$ for any $\gamma \geq \gamma_0$. Then $x_\gamma \in a + U$ and
$x_{\gamma_0} \in a + U$, hence, $x_\gamma - x_{\gamma_0} \in U - U \subseteq V$ for $\gamma \geq \gamma_0$. The proof is completed.

3.1.20. DEFINITION. A filter $\mathcal{F} = \{F_\gamma | \gamma \in \Gamma\}$ of a topological Abelian group X
is called a Cauchy filter if for any neighborhood V of zero in X there exists a set $F_\gamma \in \mathcal{F}$
such that $F_\gamma - F_\gamma \subseteq V$.

3.1.21. REMARK. The system $\mathcal{B}_x(X)$ of all neighborhoods of a point x of a topological Abelian group X forms a filter which is Cauchy filter.

3.1.22. REMARK. *Let X be a topological Abelian group, A a subgroup of the topolog-*
ical group X. Let $\mathcal{F} = \{F_\gamma | \gamma \in \Gamma\}$ be a Cauchy filter of the group A and $\mathcal{T} = \{T | T \subseteq X,$
there exists $F_\gamma \in \mathcal{F}, F_\gamma \subseteq T\}$. Then the family $\mathcal{T}$ is a Cauchy filter of the group X.

PROOF. Indeed, since the family $\mathcal{F}$ is a filter of A, it satisfies, in particular, the conditions on a basis of a filter (see A.12). Therefore, taking $\mathcal{F}$ as a basis, we get that $\mathcal{T}$ is
exactly the family of sets such that any of them contains a certain set from the basis as a
subset. Hence, $\mathcal{T}$ is the filter of the group X.

Let us prove that $\mathcal{T}$ is a Cauchy filter. Let V be a neighborhood of zero in X. There
exists a set $F_\gamma \in \mathcal{F}$ such that $F_\gamma - F_\gamma \subseteq V \cap A$, because $V \cap A$ is a neighborhood of zero
in A. Therefore, $F_\gamma \in \mathcal{T}$ and $F_\gamma - F_\gamma \subseteq V$. Hence, $\mathcal{T}$ is a Cauchy filter of the group X.

3.1.23. PROPOSITION. *Let X be a topological group and A a subgroup of the*
topological group X. Let also $\mathcal{F} = \{F_\gamma | \gamma \in \Gamma\}$ be a Cauchy filter of the group X and
$F_\gamma \cap A \neq \emptyset$ for any $\gamma \in \Gamma$. Then the family $\mathcal{F}' = \{F_\gamma \cap A | \gamma \in \Gamma\}$ (may be coinciding) is a
Cauchy filter of the group A.

PROOF. It is easy to see that the family $\mathcal{F}'$ is a filter of the group A.

Now we prove that the family $\mathcal{F}'$ is a Cauchy filter of the group A. Let U be a neighborhood of zero of the group A. Then we can find a neighborhood V of zero in X such that

$U = A \cap V$. Since $\mathcal{F}$ is a Cauchy filter of the group X, then there exists an element $\gamma_0 \in \Gamma$ such that $F_{\gamma_0} - F_{\gamma_0} \subseteq V$. Then $(F_{\gamma_0} \cap A) - (F_{\gamma_0} \cap A) \subseteq (F_{\gamma_0} - F_{\gamma_0}) \cap A \subseteq V \cap A = U$.

Hence, the family $\mathcal{F}'$ is a Cauchy filter of the group A.

3.1.24. COROLLARY. *Let X be a topological group, A a dense subgroup of the topological group X and $x \in X$. Let $\mathcal{F} = \{A \cap U \big| U$ is a neighborhood of the point x in the group $X\}$. Then the family $\mathcal{F}$ is a Cauchy filter of the group A.*

PROOF. In Remark 3.1.21 it is proved that the system of all neighborhoods of the point x forms a Cauchy filter of the group A.

Since the subgroup A is dense in X, then $A \cap U \neq \emptyset$ for any neighborhood U of the point x in X. From this fact and Proposition 3.1.23 follows that the family $\mathcal{F}$ is a Cauchy filter of the group A.

3.1.25. PROPOSITION. *Let A and B be topological Abelian groups, $f : A \to B$ be a continuous homomorphism. Let $\mathcal{F} = \{F_\gamma | \gamma \in \Gamma\}$ be a Cauchy filter of the group A. Then the family $\mathcal{F}' = \{f(F_\gamma) | \gamma \in \Gamma\}$ forms a basis of a certain Cauchy filter $\mathcal{T}$ of the group B. If $a \in A$ and $a = \lim \mathcal{F}$, then $f(a) = \lim \mathcal{T}$.*

PROOF. It is easy to see that the family $\mathcal{F}'$ satisfies the conditions on a basis of filter of the group B. We denote this filter over $\mathcal{T}$.

Let us check that $\mathcal{T}$ is a Cauchy filter of the group B. Let U be a neighborhood of zero of the group B. Since the homomorphism f is continuous, then there exists a neighborhood V of zero of the group A such that $f(V) \subseteq U$. Since $\mathcal{F}$ is a Cauchy filter in A, then there exists a set $F_\gamma \in \mathcal{F}$ such that $F_\gamma - F_\gamma \subseteq V$. Then $f(F_\gamma) - f(F_\gamma) = f(F_\gamma - F_\gamma) \subseteq f(V) \subseteq U$. Hence, $\mathcal{T}$ is a Cauchy filter of the group B.

Let $a \in A$ and $a = \lim \mathcal{F}$. For any neighborhood U' of the point $f(a)$ in B there exists a neighborhood V' of the point a in A such that $f(V') \subseteq U'$. There exists a set $F_\gamma \in \mathcal{F}$ such that $F_\gamma \subseteq V'$. Then $f(F_\gamma) \subseteq U'$. Hence, $f(a) = \lim \mathcal{T}$.

3.1.26. PROPOSITION. *Let $\mathcal{F} = \{F_\gamma | \gamma \in \Gamma\}$ be a Cauchy filter of a topological Abelian group X and $a \in X$. Then $a = \lim \mathcal{F}$ if and only if $a \in \bigcap_{\gamma \in \Gamma} [F_\gamma]_X$.*

PROOF. Let $a \in \bigcap_{\gamma \in \Gamma} [F_\gamma]_X$ and U be a neighborhood of the point a in X. Then $U - a$ is a neighborhood of zero in X. There exists a set $V_0 \in \mathcal{B}_0(X)$ such that $V_0 - V_0 \in U - a$.

Since $\mathcal{F}$ is a Cauchy filter, then there exists a set $F_\gamma \in \mathcal{F}$ such that $F_\gamma - F_\gamma \subseteq V_0$. According to Proposition 1.2.25, $[F_\gamma]_X = \bigcap_{V \in \mathcal{B}_0(X)} (F_\gamma + V)$, hence, $a \in F_\gamma + V_0$. Then $F_\gamma - a \subseteq F_\gamma - (F_\gamma + V_0) \subseteq V_0 - V_0 \subseteq U - a$, hence, $F_\gamma \subseteq U$. It proves that $a = \lim \mathcal{F}$.

Vice versa, let $a = \lim \mathcal{F}$. Assume that $a \notin \bigcap_{\gamma \in \Gamma} [F_\gamma]_X$. Then there exists a set $F_{\gamma_0} \in \mathcal{F}$ such that $a \notin [F_{\gamma_0}]_X$. According to Proposition 1.2.25, there exists a neighborhood V of zero in X such that $a \notin F_{\gamma_0} + V$, i.e. $(a - V) \cap F_{\gamma_0} = \emptyset$.

Since $a - V$ is a neighborhood of the point a in X and $a = \lim \mathcal{F}$, then there exists a set F_{γ_1} in $\mathcal{F}$ such that $F_{\gamma_1} \subseteq a - V$. Hence, $F_{\gamma_0} \cap F_{\gamma_1} = \emptyset$. This contradicts the assumption. Hence, $a \in [F_\gamma]_X$ for any $\gamma \in \Gamma$, i.e. $a \in \bigcap_{\gamma \in \Gamma} [F_\gamma]_X$.

3.1.27. PROPOSITION. *Let X be a topological Abelian group, $\{x_\gamma | \gamma \in \Gamma\}$ be a sequence in X by a certain direction Γ. This sequence is a Cauchy one if and only if the filter $\mathcal{F}$, associated with the sequence $\{x_\gamma | \gamma \in \Gamma\}$ (see Remark 3.1.10), is a Cauchy filter.*

PROOF. Let $\{x_\gamma | \gamma \in \Gamma\}$ be a Cauchy sequence in X and U a neighborhood of zero in X. According to Remark 3.1.18, there exists an element $\gamma_0 \in \Gamma$ such that $x_{\gamma_1} - x_{\gamma_2} \in U$ for $\gamma_1, \gamma_2 \in \Gamma, \gamma_1 \geq \gamma_0, \gamma_2 \geq \gamma_0$. Then $F_{\gamma_0} = \{x_\gamma | \gamma \in \Gamma, \gamma \geq \gamma_0\} \in \mathcal{F}$ and $F_{\gamma_0} - F_{\gamma_0} = \{x_{\gamma_1} - x_{\gamma_2} | \gamma_1 \geq \gamma_0, \gamma_2 \geq \gamma_0\} \subseteq U$. Hence, $\mathcal{F}$ is a Cauchy filter.

Let $\mathcal{F}$ be a Cauchy filter and U a neighborhood of zero in X. Then there exists an element $\gamma_0 \in \Gamma$ such that $F_{\gamma_0} - F_{\gamma_0} \subseteq U$. If $\gamma \geq \gamma_0$, then $x_\gamma - x_{\gamma_0} \in F_{\gamma_0} - F_{\gamma_0} \subseteq U$. The arbitrariness of the choice of the neighborhood U proves that $\{x_\gamma | \gamma \in \Gamma\}$ is a Cauchy sequence in X.

3.1.28. PROPOSITION. *Let X be a topological Abelian group, $\mathcal{F} = \{F_\gamma | \gamma \in \Gamma\}$ a Cauchy filter of the group X. As in the Proposition 3.1.16, let us consider a sequence $\{x_\gamma | \gamma \in \Gamma\}$ by the direction Γ determined by this filter. In this case $\{x_\gamma | \gamma \in \Gamma\}$ is a Cauchy sequence in X.*

PROOF. Let $U \in \mathcal{B}_0(X)$. Since $\mathcal{F}$ is a Cauchy filter, then there exists a set $F_{\gamma_0} \in \mathcal{F}$ such that $F_{\gamma_0} - F_{\gamma_0} \subseteq U$. According to the definition of the partial order on Γ, we get that $F_\gamma \subseteq F_{\gamma_0}$ for any $\gamma \in \Gamma, \gamma \geq \gamma_0$. Then $x_\gamma - x_{\gamma_0} \in F_\gamma - F_{\gamma_0} \subseteq F_{\gamma_0} - F_{\gamma_0} \subseteq U$ for $\gamma \geq \gamma_0$. It proves the proposition.

3.1.29. PROPOSITION. *Let X be a topological Abelian group, $\{x_\gamma | \gamma \in \Gamma\}$ a Cauchy* sequence in X by the direction Γ and $\{x_\gamma | \gamma \in \Gamma'\}$ a sub-sequence of the sequence $\{x_\gamma | \gamma \in \Gamma\}$. Then $\{x_\gamma | \gamma \in \Gamma'\}$ is a Cauchy sequence in X by the direction Γ'.

PROOF. Let $U \in \mathcal{B}_0(X)$. Since $\{x_\gamma | \gamma \in \Gamma\}$ is a Cauchy sequence, then, according to Remark 3.1.18, there exists an element $\gamma_0 \in \Gamma$ such that $x_{\gamma_1} - x_{\gamma_2} \in U$ for any $\gamma_1, \gamma_2 \in \Gamma, \gamma_1 \geq \gamma_0$, $\gamma_2 \geq \gamma_0$. Since Γ' is a subdirection of the direction Γ, then there exists an element $\gamma_0' \in \Gamma'$ such that $\gamma_0' \geq \gamma_0$. Then for any $\gamma \in \Gamma'$ such that $\gamma \geq \gamma_0'$, we have that $\gamma \geq \gamma_0$, hence, $x_\gamma - x_{\gamma_0'} \in U$. The proof is completed.

3.1.30. THEOREM. *Let X be an Abelian group, $\{x_\gamma | \gamma \in \Gamma\}$ a sequence in X by the direction Γ. Let $\{\tau_\delta | \delta \in \Delta\}$ be a certain family of group topologies on X such that $\{x_\gamma | \gamma \in \Gamma\}$ is a Cauchy sequence in (X, τ_δ) for any $\delta \in \Delta$. Then $\{x_\gamma | \gamma \in \Gamma\}$ is a Cauchy sequence in $\left(X, \sup\{\tau_\delta | \delta \in \Delta\}\right)$.*

PROOF. Let U be a neighborhood of zero in the group $\left(X, \sup\{\tau_\delta | \delta \in \Delta\}\right)$. According to C.9, there exist such elements $\delta_1, \ldots, \delta_n \in \Delta$ and such sets $U_1, \ldots, U_n$ that U_i is a neighborhood of zero in (X, τ_{δ_i}) correspondingly and $\bigcap_{i=1}^n U_i \subseteq U$. According to Remark 3.1.18, for any $1 \leq i \leq n$ there exists an element $\gamma_i \in \Gamma$, such that $x_\gamma - x_{\gamma'} \in U_i$ for any $\gamma, \gamma' \in \Gamma, \gamma \geq \gamma_i, \gamma' \geq \gamma_i$. We can choose an element $\gamma_0 \in \Gamma$ such that $\gamma_0 \geq \gamma_i$ for all $i = 1, \ldots, n$. Then $x_\gamma - x_{\gamma_0} \in \bigcap_{i=1}^n U_i \subseteq U$ for any $\gamma \geq \gamma_0$. From the arbitrariness of the choice of neighborhood U it follows that $\{x_\gamma | \gamma \in \Gamma\}$ is a Cauchy sequence in $(X, \sup\{\tau_\delta | \delta \in \Delta\})$.

3.1.31. PROPOSITION. *Let X be a Hausdorff Abelian group, $\{x_\gamma | \gamma \in \Gamma\}$ a sequence of elements of X by a direction Γ. If $a, b \in X$ are two limits of this sequence, then $a = b$.*

PROOF. Let us assume the contrary, i.e. that $a \neq b$. Since the group X is Hausdorff, there exists a neighborhood U of zero in X such that $(a + U) \cap (b + U) = \emptyset$.

Since $a = \lim_{\gamma \in \Gamma} x_\gamma$, then there exists an element $\gamma_0 \in \Gamma$ such that $x_\gamma \in a + U$ for all $\gamma \geq \gamma_0$. By the same way we can find an element $\gamma_1 \in \Gamma$ such that $x_\gamma \in b + U$ for all $\gamma \geq \gamma_1$. We can also choose element $\gamma_2 \in \Gamma$ such that $\gamma_2 \geq \gamma_0$ and $\gamma_2 \geq \gamma_1$. Then $x_{\gamma_2} \in (a + U) \cap (b + U) = \emptyset$. The obtained contradiction completes the proof.

3.1.32. PROPOSITION. *Let X be a topological Abelian group, $a \in X$ and $\{x_\gamma | \gamma \in \Gamma\}$ be a Cauchy sequence by a direction Γ. Let $\{x_\gamma | \gamma \in \Gamma'\}$ be its sub-sequence (see*

Definition 3.1.9). Then $a = \lim_{\gamma \in \Gamma'} x_\gamma$ if and only if $a = \lim_{\gamma \in \Gamma} x_\gamma$.

PROOF. Let $a = \lim_{\gamma \in \Gamma} x_\gamma$. Then, according to Proposition 3.1.12, $a = \lim_{\gamma \in \Gamma'} x_\gamma$.

Let now $a = \lim_{\gamma \in \Gamma'} x_\gamma$ and U be a neighborhood of zero in X. There exists a neighborhood V of zero in X such that $V + V \subseteq U$. Since $a = \lim_{\gamma \in \Gamma'} x_\gamma$, then there exists an element $\gamma_0' \in \Gamma'$ such that $x_{\gamma'} \in a + V$ for all $\gamma' \in \Gamma', \gamma' \geq \gamma_0'$. Since $\{x_\gamma | \gamma \in \Gamma\}$ is a Cauchy sequence, then, according to Remark 3.1.18, there exists an element $\gamma_0 \in \Gamma$ such that $x_{\gamma_1} - x_{\gamma_2} \in V$ for $\gamma_1, \gamma_2 \in \Gamma, \gamma_1 \geq \gamma_0, \gamma_2 \geq \gamma_0$. Also there exists an element $\gamma_1' \in \Gamma'$ such that $\gamma_1' \geq \gamma_0'$ and $\gamma_1' \geq \gamma_0$. Then $x_\gamma = (x_\gamma - x_{\gamma_1'}) + x_{\gamma_1'} \in V + a + V \subseteq a + U$ for all $\gamma \in \Gamma, \gamma \geq \gamma_1'$. Since the choice of neighborhood U was arbitrary, it follows that $a = \lim_{\gamma \in \Gamma} x_\gamma$.

3.1.33. PROPOSITION. *Let A and B be topological Abelian groups, $f : A \to B$ a continuous homomorphism. Then:*

1) if $\{x_\gamma \,|\, \gamma \in \Gamma\}$ is a Cauchy sequence in A by a direction Γ, then $\{f(x_\gamma) | \gamma \in \Gamma\}$ is a Cauchy sequence in B;

2) if $a \in A$ and $a = \lim_{\gamma \in \Gamma} x_\gamma$, then $f(a) = \lim_{\gamma \in \Gamma} f(x_\gamma)$.

PROOF. Let $\mathcal{B}_0(A)$ and $\mathcal{B}_0(B)$ be the systems of all neighborhoods of zero in A and B correspondingly and $U \in \mathcal{B}_0(B)$. Then $f^{-1}(U) \in \mathcal{B}_0(A)$. That is why there exists an element $\gamma_0 \in \Gamma$ such that $x_\gamma - x_{\gamma_0} \in f^{-1}(U)$ for any $\gamma \in \Gamma, \gamma \geq \gamma_0$. Then $f(x_\gamma) - f(x_{\gamma_0}) = f(x_\gamma - x_{\gamma_0}) \in U$ for any $\gamma \in \Gamma, \gamma \geq \gamma_0$. It proves that $\{f(x_\gamma) | \gamma \in \Gamma\}$ is a Cauchy sequence in B.

The item 2) is proved the similar way.

3.1.34. COROLLARY. *Let X be an Abelian group, τ_1, τ_2 two group topologies on X and $\tau_1 < \tau_2$. Let $a \in X$ and $\{x_\gamma | \gamma \in \Gamma\}$ be a sequence of elements of X by a direction Γ, then:*

1) if $\{x_\gamma | \gamma \in \Gamma\}$ is a Cauchy sequence in (X, τ_2), then $\{x_\gamma | \gamma \in \Gamma\}$ is a Cauchy sequence in (X, τ_1);

2) if $a = \lim_{\gamma \in \Gamma} x_\gamma$ in (X, τ_2), then $a = \lim_{\gamma \in \Gamma} x_\gamma$ in the (X, τ_1).

PROOF. Since the identical mapping $f : (X, \tau_2) \to (X, \tau_1)$ is a continuous homomorphism, the proof follows from Proposition 3.1.33.

3.1.35. COROLLARY. *Let A be an Abelian group, (B, τ) a topological Abelian group, $f : A \to B$ a group homomorphism, $\{x_\gamma | \gamma \in \Gamma\}$ a sequence of elements of A by a direction Γ and $a \in A$. We denote over τ_1 the prototype of the topology τ with respect to the homomorphism f (see C.15 and Corollary 1.5.34). Then:*

1) the sequence $\{f(x_\gamma) | \gamma \in \Gamma\}$ is a Cauchy sequence in (B, τ) if and only if $\{x_\gamma | \gamma \in \Gamma\}$ is a Cauchy sequence in (A, τ_1);

2) $a = \lim_{\gamma \in \Gamma} x_\gamma$ in (A, τ_1) if and only if $f(a) = \lim_{\gamma \in \Gamma} (x_\gamma)$ in (B, τ).

PROOF. Let $\{x_\gamma | \gamma \in \Gamma\}$ be a Cauchy sequence in (A, τ_1). Since f is a continuous homomorphism, then, according to Proposition 3.1.33, $\{f(x_\gamma) | \gamma \in \Gamma\}$ is a Cauchy sequence in (B, τ).

Let $\{f(x_\gamma) | \gamma \in \Gamma\}$ be a Cauchy sequence in (B, τ) and U a neighborhood of zero in (A, τ_1). There exists a neighborhood V of zero in B such that $f^{-1}(V) \subseteq U$. We can find an element $\gamma_0 \in \Gamma$ such that $f(x_\gamma) - f(x_{\gamma_0}) \in V$ for any $\gamma \in \Gamma, \gamma \geq \gamma_0$. Then $x_\gamma - x_{\gamma_0} \in f^{-1}(V) \subseteq U$ for all $\gamma \geq \gamma_0$. Since the choice of U was arbitrary, then $\{x_\gamma | \gamma \in \Gamma\}$ is a Cauchy sequence in (A, τ_1).

The proof of the statement 2) is done the similar way.

3.1.36. THEOREM. *Let A and B be topological Abelian groups. A homomorphism $f : A \to B$ of Abelian groups is continuous if and only if for any direction Γ and any sequence $\{x_\gamma | \gamma \in \Gamma\}$ convergent to zero in A, the sequence $\{f(x_\gamma) | \gamma \in \Gamma\}$ converges to zero in B (i.e. $\lim_{\gamma \in \Gamma} x_\gamma = 0$).*

PROOF. If f is a continuous homomorphism, the proof follows from Proposition 3.1.33.

Now let the homomorphism f be not continuous. We shall prove that there exist a direction Γ and a sequence $\{x_\gamma | \gamma \in \Gamma\}$ converging to zero in A such that $\{f(x_\gamma) | \gamma \in \Gamma\}$ does not converge to zero in B.

Let Γ be a direction, defined by the system $\{U_\gamma | \gamma \in \Gamma\}$ of all neighborhoods of zero in A (see Example 3.1.3). Since the homomorphism f is not continuous, then, according to Proposition 1.5.5, there exists a neighborhood V of zero in B such that $f^{-1}(V)$ is not a neighborhood of zero in A, hence, $U_\gamma \nsubseteq f^{-1}(V)$ for any neighborhood U_γ of zero in A.

We'll construct the sequence $\{x_\gamma | \gamma \in \Gamma\}$ of elements of A by the direction Γ another way:

for any neighborhood U_γ of zero in A we choose an arbitrarily element $x_\gamma \in U_\gamma \backslash f^{-1}(V)$.

Now let us prove that $\lim_{\gamma \in \Gamma} x_\gamma = 0$ in A. Indeed, let U_{γ_0} be an arbitrary neighborhood of zero in A. Then $x_\gamma \in U_\gamma \subseteq U_{\gamma_0}$ for any $\gamma \in \Gamma, \gamma \geq \gamma_0$, i.e. $0 = \lim_{\gamma \in \Gamma} x_\gamma$.

Now we'll show that the sequence $\{f(x_\gamma) | \gamma \in \Gamma\}$ does not converge to zero in B. Indeed, if to assume that $0 = \lim_{\gamma \in \Gamma} f(x_\gamma)$, then there exists an element $\gamma_1 \in \Gamma$ such that $f(x_\gamma) \in V$ for any $\gamma \in \Gamma, \gamma \geq \gamma_1$, i.e. $x_\gamma \in f^{-1}(V)$. This contradicts the choice of the elements x_γ.

Hence, the sequence $\{f(x_\gamma) | \gamma \in \Gamma\}$ does not converge to zero in B. The proof is completed.

3.1.37. DEFINITION. Let X be an Abelian group, Γ a certain direction, $\hat{x} = \{x_\gamma | \gamma \in \Gamma\}$ and $\hat{y} = \{y_\gamma | \gamma \in \Gamma\}$ sequences in X by this direction.

The sequence $\{x_\gamma + y_\gamma | \gamma \in \Gamma\}$ we call the sum of the sequences $\hat{x}$ and $\hat{y}$ and denote over $\hat{x} + \hat{y}$. The sequence $\{-x_\gamma | \gamma \in \Gamma\}$ we call a sequence opposite to the sequence $\hat{x}$ and denote over $-\hat{x}$.

It is easy to check that the set $\widehat{X}$ of all sequences $\{x_\gamma | \gamma \in \Gamma\}$ of the elements of X with respect to the introduced operations is an Abelian group.

3.1.38. PROPOSITION. *Let X be a topological Abelian group, Γ a certain direction, and $\hat{x} = \{x_\gamma | \gamma \in \Gamma\}$ and $\hat{y} = \{y_\gamma | \gamma \in \Gamma\}$ Cauchy sequences in X by a direction Γ. Then the sum $\hat{x} + \hat{y}$ of these sequences and the opposite sequence $-\hat{x}$ is a Cauchy sequence in X.*

PROOF. Let $U \in \mathcal{B}_0(X)$. There exists a set $V \in \mathcal{B}_0(X)$ such that $V + V \subseteq U$. According to Remark 3.1.18, there exist elements $\gamma_0, \gamma_0' \in \Gamma$ such that $x_{\gamma_1} - x_{\gamma_2} \in V$ and $y_{\gamma_1'} - y_{\gamma_2'} \in V$ for any $\gamma_1, \gamma_2, \gamma_1', \gamma_2' \in \Gamma, \gamma_1 \geq \gamma_0, \gamma_2 \geq \gamma_0, \gamma_1' \geq \gamma_0', \gamma_2' \geq \gamma_0'$. There exists an element $\gamma_3 \in \Gamma$ such that $\gamma_3 \geq \gamma_0$ and $\gamma_3 \geq \gamma_0'$. Then $(x_\gamma + y_\gamma) - (x_{\gamma_3} + y_{\gamma_3}) = (x_\gamma - x_{\gamma_3}) + (y_\gamma - y_{\gamma_3}) \in V + V \subseteq U$ for any $\gamma \in \Gamma, \gamma \geq \gamma_3$.

It proves that $\hat{x} + \hat{y}$ is a Cauchy sequence.

Now consider the sequence $\{-x_\gamma | \gamma \in \Gamma\}$. Let $U \in \mathcal{B}_0(X)$ and V be a neighborhood of zero in X such that $-V \subseteq U$. Since $\{x_\gamma | \gamma \in \Gamma\}$ is a Cauchy sequence, then there exists an element $\gamma_0 \in \Gamma$ such that $x_\gamma - x_{\gamma_0} \in V$ for any $\gamma \in \Gamma, \gamma \geq \gamma_0$. Then $-x_\gamma - (-x_{\gamma_0}) \in -V \subseteq U$ for any $\gamma \in \Gamma, \gamma \geq \gamma_0$. It proves that $\{-x_\gamma | \gamma \in \Gamma\}$ is a Cauchy sequence, too.

3.1.39. COROLLARY. Let X be a topological Abelian group, Γ a direction, $\tilde{X}$ the set of all Cauchy sequences in X by this direction. Then $\tilde{X}$ is an Abelian group with respect to the operations defined in Definition 3.1.37.

3.1.40. PROPOSITION. *Let X be a topological Abelian group, $a, b \in X$, Γ a direction, $\{x_\gamma | \gamma \in \Gamma\}$ and $\{y_\gamma | \gamma \in \Gamma\}$ Cauchy sequences in X by the direction Γ. Let $a = \lim_{\gamma \in \Gamma} x_\gamma$ and $b = \lim_{\gamma \in \Gamma} y_\gamma$. Then:*

1) $a + b = \lim_{\gamma \in \Gamma}(x_\gamma + y_\gamma)$;

2) $-a = \lim_{\gamma \in \Gamma}(-x_\gamma)$.

PROOF. Let U be a neighborhood of the element $a+b$ in X. There exist neighborhoods V and W of the elements a and b in X correspondingly such that $V + W \subseteq U$. There exists an element $\gamma_0 \in \Gamma$ such that $x_\gamma \in V$ for all $\gamma \in \Gamma, \gamma \geq \gamma_0$ and an element $\gamma_1 \in \Gamma$ such that $y_\gamma \in W$ for all $\gamma \in \Gamma, \gamma \geq \gamma_1$. Also there exists an element $\gamma_2 \in \Gamma$ such that $\gamma_2 \geq \gamma_0$ and $\gamma_2 \geq \gamma_1$.

In that case for any $\gamma \in \Gamma, \gamma \geq \gamma_2$ we have $x_\gamma \in V$ and $y_\gamma \in W$, hence, $x_\gamma + y_\gamma \in V + W \subseteq U$. Since the choice of the neighborhood U was arbitrary, then $a + b = \lim_{\gamma \in \Gamma}(x_\gamma + y_\gamma)$.

Now let U be a neighborhood of the element $-a$ in X. Then $-U$ is a neighborhood of the element a in X. We can find element $\gamma_0 \in \Gamma$ such that $x_\gamma \in -U$ for all $\gamma \in \Gamma, \gamma \geq \gamma_0$. Hence, $-x_\gamma \in U$ for all $\gamma \in \Gamma, \gamma \geq \gamma_0$. It proves that $-a = \lim_{\gamma \in \Gamma}(-x_\gamma)$.

3.1.41. DEFINITION. Let R be a ring, M a right R-module, Γ a direction, and $\hat{r} = \{r_\gamma | \gamma \in \Gamma\}$ and $m = \{m_\gamma | \gamma \in \Gamma\}$ sequences by this direction of elements of R and M correspondingly. We call the sequence $\{m_\gamma \cdot r_\gamma | \gamma \in \Gamma\}$ of the elements of M a product of the sequences $\hat{m}$ and $\hat{r}$ and denote it over $\hat{m} \cdot \hat{r}$.

3.1.42. REMARK. Considering a ring R as a module over itself, we obtain a definition of the product of two sequences by certain direction in the ring R.

It is easy to see that for any direction Γ the set $\hat{R}$ of all sequences of elements of R by this direction, with respect to the introduced operations of addition, multiplication and of taking the opposite sequence is a ring.

3.1.43. PROPOSITION. *Let R be a topological ring, M a right topological R-module, Γ a direction, $\hat{r} = \{r_\gamma | \gamma \in \Gamma\}$ and $\hat{m} = \{m_\gamma | \gamma \in \Gamma\}$ Cauchy sequences in R and M*

correspondingly. Then the sequence $\hat{m} \cdot \hat{r}$ is a Cauchy sequence in M.

PROOF. Let U_0 be a neighborhood of zero in M. We can find neighborhoods U_1 and U_2 of zero in M and a neighborhood W_1 of zero in R such that $U_1 + U_1 + U_1 + U_1 + U_1 \subseteq U_0$ and $U_2 \cdot W_1 \subseteq U_1$. Since $\hat{r}$ and $\hat{m}$ are Cauchy sequences, then there exists an element $\gamma_1 \in \Gamma$ such that $r_\gamma - r_{\gamma_1} \in W_1$ and $m_\gamma - m_{\gamma_1} \in U_2$ for all $\gamma \in \Gamma, \gamma \geq \gamma_1$. Since M is a topological R-module, then there exist neighborhoods of zero W_2 in R and U_3 in M such that $W_2 \subseteq W_1$, $U_3 \subseteq U_2$, $m_{\gamma_1} \cdot W_2 \subseteq U_1$, $U_3 \cdot r_{\gamma_1} \subseteq U_1$. There exists an element $\gamma_2 \in \Gamma$ such that $r_\gamma - r_{\gamma_2} \in W_2$ and $m_\gamma - m_{\gamma_2} \in U_3$ for all $\gamma \in \Gamma, \gamma \geq \gamma_2$. We can suppose that $\gamma_2 \geq \gamma_1$. Then for any $\gamma \geq \gamma_2$ we have:

$$m_\gamma \cdot r_\gamma - m_{\gamma_2} \cdot r_{\gamma_2} \in (m_{\gamma_2} + U_3) \cdot (r_{\gamma_2} + W_2) - m_{\gamma_2} \cdot r_{\gamma_2}$$

$$\subseteq m_{\gamma_2} \cdot W_2 + U_3 \cdot r_{\gamma_2} + U_3 \cdot W_2$$

$$\subseteq (m_{\gamma_1} + U_2) \cdot W_2 + U_3 \cdot (r_{\gamma_1} + W_1) + U_3 \cdot W_2$$

$$\subseteq m_{\gamma_1} \cdot W_2 + U_2 \cdot W_2 + U_3 \cdot r_{\gamma_1} + U_3 \cdot W_1 + U_3 \cdot W_2$$

$$\subseteq U_1 + U_2 \cdot W_1 + U_1 + U_2 \cdot W_1 + U_2 \cdot W_1 \subseteq U_1 + U_1 + U_1 + U_1 + U_1 \subseteq U_0.$$

Hence, $m_\gamma \cdot r_\gamma \in m_{\gamma_2} \cdot r_{\gamma_2} + U_0$ for all $\gamma \in \Gamma, \gamma \geq \gamma_2$.

We've proved that $\hat{m} \cdot \hat{r}$ is a Cauchy sequence in M.

3.1.44. COROLLARY. *Let R be a topological ring, Γ a direction, $\hat{r} = \{r_\gamma | \gamma \in \Gamma\}$ and $\hat{s} = \{s_\gamma | \gamma \in \Gamma\}$ Cauchy sequences in R by this direction. Then the sequence $\hat{r} \cdot \hat{s}$ is a Cauchy sequence in R.*

PROOF. Consider R as a topological module over the ring R. Then, according to Proposition 3.1.43, we get that $\hat{r} \cdot \hat{s}$ is a Cauchy sequence in R.

3.1.45. PROPOSITION. *Let R be a topological ring, M a right topological R-module, Γ a direction, $\hat{r} = \{r_\gamma | \gamma \in \Gamma\}$ and $\hat{m} = \{m_\gamma | \gamma \in \Gamma\}$ sequences by this direction in R and M correspondingly. If $r_0 = \lim_{\gamma \in \Gamma} r_\gamma$ and $m_0 = \lim_{\gamma \in \Gamma} m_\gamma$, then $m_0 \cdot r_0 = \lim_{\gamma \in \Gamma} m_\gamma \cdot r_\gamma$.*

PROOF. Let U_0 be a neighborhood of zero in M. We can find a neighborhood U_1 of zero in M such that $U_1 + U_1 + U_1 \subseteq U_0$. Since M is a topological R-module, then there exist neighborhoods U_2 of zero in M and W_1 of zero in R such that $U_2 \cdot W_1 \subseteq U_1$. Then there exists a neighborhood U_3 of zero in M such that $U_3 \subseteq U_2$, and $U_3 \cdot r_0 \subseteq U_1$ and a neighborhood W_2 of zero in R such that $W_2 \subseteq W_1$ and $m_0 \cdot W_2 \subseteq U_1$. Since $r_0 = \lim_{\gamma \in \Gamma} r_\gamma$ and $m_0 = \lim_{\gamma \in \Gamma} m_\gamma$, then there exist elements $\gamma_1, \gamma_2 \in \Gamma$ such that $r_\gamma \in r_0 + W_2$ for all $\gamma \in \Gamma, \gamma \geq \gamma_1$ and $m_\gamma \in m_0 + U_3$ for all $\gamma \in \Gamma, \gamma \geq \gamma_2$. Now we choose an element $\gamma_3 \in \Gamma$ such that $\gamma_3 \geq \gamma_2$ and $\gamma_3 \geq \gamma_1$. Then for $\gamma \geq \gamma_3$ we have:

$$m_\gamma \cdot r_\gamma \in (m_0 + U_3) \cdot (r_0 + W_2) \subseteq m_0 \cdot r_0 + U_3 \cdot r_0 + m_0 \cdot W_2 + U_3 \cdot W_2$$
$$\subseteq m_0 \cdot r_0 + U_1 + U_1 + U_2 \cdot W_1 \subseteq m_0 \cdot r_0 + U_1 + U_1 + U_1 \subseteq m_0 \cdot r_0 + U_0.$$

Hence, $m_0 \cdot r_0 = \lim_{\gamma \in \Gamma} m_\gamma \cdot r_\gamma$.

3.1.46. COROLLARY. *Let R be a topological ring, Γ a direction, $\hat{r} = \{r_\gamma | \gamma \in \Gamma\}$ and $\hat{s} = \{s_\gamma | \gamma \in \Gamma\}$ sequences in R by this direction. If $r_0 = \lim_{\gamma \in \Gamma} r_\gamma$ and $s_0 = \lim_{\gamma \in \Gamma} s_\gamma$, then $r_0 \cdot s_0 = \lim_{\gamma \in \Gamma} r_\gamma \cdot s_\gamma$.*

PROOF. Consider R as a topological module over the ring R. Then, according to Proposition 3.1.45, we get $r_0 \cdot s_0 = \lim_{\gamma \in \Gamma} r_\gamma \cdot s_\gamma$.

§ 3.2. Complete Topological Groups and Their Properties

3.2.1. DEFINITION. A topological Abelian group X is called complete if any Cauchy filter $\mathcal{F}$ of the group X has its limit in X.

3.2.2. DEFINITION. A topological ring R (a topological module M) is called complete if the topological additive group of the ring R (of the module M) is complete.

3.2.3. PROPOSITION. *A locally compact Abelian group X is complete.*

PROOF. Assume the contrary, that the group X is not complete. Then there exists a Cauchy filter $\mathcal{F} = \{F_\gamma \mid \gamma \in \Gamma\}$ of the group X which does not have a limit in X. According to Proposition 3.1.26, $\bigcap_{\gamma \in \Gamma}[F_\gamma]_X = \emptyset$.

Let U be a neighborhood of zero in X such that $[U]_X$ is a compact set. Then there exists a set $F_{\gamma_0} \in \mathcal{F}$ such that $F_{\gamma_0} - F_{\gamma_0} \subseteq U$, hence, $F_{\gamma_0} - a \subseteq U$ for some element a from F_{γ_0}. Then $F_{\gamma_0} \subseteq U + a$, therefore, $[F_{\gamma_0}]_X \subseteq [U + a]_X = [U]_X + a$.

Since $[U]_X$ is compact, then $[U]_X + a$ is a compact set. Hence, $[F_{\gamma_0}]_X$ is compact. Since

$$\bigcap_{\gamma \in \Gamma}([F_\gamma]_X \cap [F_{\gamma_0}]_X) = \left(\bigcap_{\gamma \in \Gamma}[F_\gamma]_X\right) \cap [F_{\gamma_0}]_X = \bigcap_{\gamma \in \Gamma}[F_\gamma]_X = \emptyset,$$

then, because of the compactness of the set $[F_{\gamma_0}]_X$ (see C.12), there exist such elements $\gamma_1, \ldots, \gamma_n \in \Gamma$ that $\bigcap_{i=1}^n([F_{\gamma_i}]_X \cap [F_{\gamma_0}]_X) = \emptyset$. It contradicts the statement that $\mathcal{F}$ is a filter of the group X. This contradiction proves that X is a complete group.

3.2.4. COROLLARY. A compact Abelian group X is complete.

3.2.5. COROLLARY. A group endowed with the discrete topology is complete.

3.2.6. COROLLARY. *Let a topological Abelian group X have the smallest neighborhood of zero, then X is a complete group.*

PROOF. Let U be the smallest neighborhood of zero in X. Then U is a compact set, which is why X is a locally compact, hence, complete group.

3.2.7. DEFINITION. Let R be a topological ring, M_R a topological R-module which has a basis $\mathcal{B}_0(M)$ of neighborhoods of zero consisting of submodules. The module M_R is called linearly compact if any basis of a filter of the module M consisting of cosets $m_\alpha + N_\alpha$ by closed submodules N_α has a non-empty intersection, i.e. $\bigcap_{\alpha \in \Lambda}(m_\alpha + N_\alpha) \neq \emptyset$.

A topological ring R which has a basis of neighborhoods of zero consisting of right (left) ideals is called linearly compact from right (left) if the right (left) topological R-module R is linearly compact.

3.2.8. REMARK. *Let R be a topological ring, M a topological R-module which has a basis of neighborhoods of zero consisting of submodules. Then:*

1) if M is a compact module, then M is a linearly compact module;

2) if M satisfies the descending chains condition on closed submodules, then M is a linearly compact module.

PROOF. If M is a compact module, then any basis of a filter consisting of closed subsets, e.g. cosets by closed submodules, has a non–empty intersection. Hence, M is a linearly compact module. The condition 1) is proved.

Let the module M satisfy the descending chains condition on closed submodules, and assume the contrary, i.e. that M not be a linearly compact module. Then there exists a basis $\{m_\alpha + N_\alpha \mid \alpha \in \Lambda\} = \mathcal{L}$ of a filter consisting of cosets by closed submodules N_α of the module M such that $\bigcap_{\alpha \in \Lambda}(m_\alpha + N_\alpha) = \emptyset$. We construct now a chain of submodules $N_{\alpha_1} \supset N_{\alpha_2} \supset \cdots$ by induction.

Let α_1 be an arbitrary element from Λ, and suppose that the $N_{\alpha_1} \supseteq \ldots \supseteq N_{\alpha_n}$ are defined.

Since $\bigcap_{\alpha \in \Lambda}(m_\alpha + N_\alpha) = \emptyset$, then there exists an element $\alpha'_n \in \Lambda$ such that $m_{\alpha_n} + N_{\alpha_n} \not\subseteq m_{\alpha'_n} + N_{\alpha'_n}$. As $\mathcal{L}$ is a basis of a filter, then there exists an element $\alpha_{n+1} \in \Lambda$ such that $m_{\alpha_{n+1}} + N_{\alpha_{n+1}} \subseteq (m_{\alpha_n} + N_{\alpha_n}) \bigcap (m_{\alpha'_n} + N_{\alpha'_n})$, hence, $m_{\alpha_{n+1}} \in m_{\alpha_n} + N_{\alpha_n}$.

Since N_{α_n} is a submodule, then $m_{\alpha_{n+1}} + N_{\alpha_n} = m_{\alpha_n} + N_{\alpha_n}$. In that case $N_{\alpha_{n+1}} \subseteq N_{\alpha_n}$.

Also, since $m_{\alpha_{n+1}} + N_{\alpha_{n+1}} \subseteq m_{\alpha'_n} + N_{\alpha'_n}$, and $m_{\alpha_{n+1}} + N_{\alpha_n} = m_{\alpha_n} + N_{\alpha_n} \not\subseteq m_{\alpha'_n} + N_{\alpha'_n}$, then $N_{\alpha_n} \neq N_{\alpha_{n+1}}$, i.e. the constructed chain of the submodules is descending. This contradicts the condition that M satisfies the descending chains condition on closed modules. Thus, M is a linearly compact module.

3.2.9. REMARK. *Let M be a discrete vector space over a discrete skew field R. Then M is a linearly compact R-module if and only if M is a finite-dimensional space.*

PROOF. If M is a finite-dimensional vector space, then the R-module M satisfies the descending chains condition on closed submodules (subspaces). Hence, according to Remark 3.2.8, M is a linearly compact R-module.

Let M be an infinite-dimensional vector space, and assume the contrary, i.e. that M is a linearly compact module. Let $\{e_n \mid n \in \mathbb{N}\}$ be a countable linearly independent system of elements of the module M. Consider the system $\mathfrak{A} = \{(e_1 + \ldots + e_n) + \sum_{i=n+1}^{\infty} R \cdot e_i \mid n \in \mathbb{N}\}$ of the cosets by closed submodules $\sum_{i=n+1}^{\infty} R \cdot e_i$. Since

$$(e_1 + \ldots + e_{n+m}) + \sum_{i=n+m+1}^{\infty} R \cdot e_i$$
$$\subseteq \left((e_1 + \ldots + e_n) + \sum_{i=n+1}^{\infty} R \cdot e_i \right) \cap \left((e_1 + \ldots + e_m) + \sum_{i=m+1}^{\infty} R \cdot e_i \right)$$

for any $n, m \in \mathbb{N}$, then $\mathfrak{A}$ is a basis of a certain filter of the module M. Since the module M is linearly compact, then there exists an element $a \in \bigcap_{n=1}^{\infty}((e_1 + \ldots + e_n) + \sum_{i=n+1}^{\infty} R \cdot e_i)$. Then $a \in \sum_{n=1}^{\infty} R \cdot e_i$, hence, there exist natural numbers $n_1, \ldots, n_k$ such that $a = \sum_{i=1}^{k} r_i \cdot e_{n_i}$, where $0 \neq r_i \in R$ for $i = 1, \ldots, k$. Let $m \in \mathbb{N}$ be a natural number such that $m > \max\{n_1, \ldots, n_k\}$. Since $a \in (e_1 + \ldots + e_m) + \sum_{i=m+1}^{\infty} R \cdot e_i$, then $\sum_{i=1}^{k} r_i \cdot e_{n_i} \in (e_1 + \cdots + e_m) + \sum_{i=m+1}^{\infty} R \cdot e_i$. Hence, $e_m \in \sum_{i \neq m} R \cdot e_i$.

This contradicts the fact that $\{e_i \mid i \in \mathbb{N}\}$ is a linearly independent system. The proof is completed.

3.2.10. PROPOSITION. *Let R be a topological ring and M a topological R-module. If M is a linearly compact module, then M is complete.*

PROOF. Let $\{U_\gamma \mid \gamma \in \Gamma\}$ be a basis of neighborhoods of zero in M consisting of submodules and $\mathcal{F} = \{F_\alpha \mid \alpha \in \Lambda\}$ a Cauchy filter of the module M. Now we construct a family $\mathcal{N} = \{H_\gamma \mid \gamma \in \Gamma\}$ of cosets by the submodules U_γ of the module M the following way.

Since $\mathcal{F}$ is a Cauchy filter, then for any $\gamma \in \Gamma$ there exists a set $F_{\alpha(\gamma)} \in \mathcal{F}$ such that $F_{\alpha(\gamma)} - F_{\alpha(\gamma)} \subseteq U_\gamma$. We choose an arbitrary element $m_\gamma \in F_{\alpha(\gamma)}$. Then $F_{\alpha(\gamma)} - m_\gamma \subseteq U_\gamma$, hence, $F_{\alpha(\gamma)} \subseteq m_\gamma + U_\gamma$. Put $H_\gamma = m_\gamma + U_\gamma$.

Now we prove that the family $\mathcal{N} = \{H_\gamma \mid \gamma \in \Gamma\}$ is a basis of a filter of the module M. Indeed, let $\gamma, \beta \in \Gamma$, then there exists an element $\delta \in \Gamma$ such that $U_\delta \subseteq U_\gamma \cap U_\beta$. Since

$(m_\gamma + U_\gamma) \bigcap (m_\beta + U_\beta) \bigcap (m_\delta + U_\delta) \supseteq F_{\alpha(\gamma)} \bigcap F_{\alpha(\beta)} \bigcap F_{\alpha(\delta)} \neq \emptyset$, then there exists an element $m \in (m_\gamma + U_\gamma) \bigcap (m_\beta + U_\beta) \bigcap (m_\delta + U_\delta)$. Then $m + U_\gamma = m_\gamma + U_\gamma$, $m + U_\beta = m_\beta + U_\beta$ and $m + U_\delta = m_\delta + U_\delta$, hence, $m_\delta + U_\delta = m + U_\delta \subseteq (m + U_\gamma) \bigcap (m + U_\beta) = (m_\gamma + U_\gamma) \bigcap (m_\beta + U_\beta)$. Therefore, $\mathcal{N}$ is a basis of a filter of M.

Since M is a linearly compact module, then there exists an element $m_0 \in \bigcap_{\gamma \in \Gamma} (m_\gamma + U_\gamma)$. If $\gamma \in \Gamma$, then $m_0 \in m_\gamma + U_\gamma$, hence, $m_0 + U_\gamma = m_\gamma + U_\gamma$. Then $F_{\alpha(\gamma)} \subseteq m_\gamma + U_\gamma = m_0 + U_\gamma$ for any $\gamma \in \Gamma$. It means that $m_0 = \lim \mathcal{F}$.

From Definition 3.2.1 follows that M is a complete topological module.

3.2.11. COROLLARY. *Let R be a ring linearly compact from right (left). Then R is complete.*

PROOF. Consider R as the right (left) module over itself. Then the module R is linearly compact and, according to Proposition 3.2.10, complete. Hence, R is a complete ring.

3.2.12. THEOREM. *Let X be a Hausdorff Abelian group, A a subgroup of the topological group X. If A is a complete topological group, then A is a closed subgroup in X.*

PROOF. Assume the contrary, i.e. that $[A]_X \neq A$. Then there exists an element $a \in [A]_X \setminus A$. Let $\mathcal{F} = \{F_\gamma \mid \gamma \in \Gamma\}$ be the filter of all neighborhoods of the point a in X. Then $a = \lim \mathcal{F}$ in X and, according to Propositions 3.1.16 and 3.1.31, a is the only limit of the filter $\mathcal{F}$ in X.

Now consider the family $\mathcal{B} = \{F_\gamma \bigcap A \mid \gamma \in \Gamma\}$ of subsets of the group A. According to Proposition 3.1.23, $\mathcal{B}$ is a Cauchy filter of the group A, because $F_\gamma \bigcap A \neq \emptyset$ for all $\gamma \in \Gamma$. Since A is a complete topological group, then there exists an element $b \in A$ such that $b = \lim \mathcal{B}$ in A. According to Proposition 3.1.26, $b \in \bigcap_{\gamma \in \Gamma} [F_\gamma \bigcap A]_A \subseteq \bigcap_{\gamma \in \Gamma} [F_\gamma]_X = \{a\}$. Hence, $a = b$. This contradicts the choice of the element $a \in [A]_X \setminus A$. Thus, $[A]_A = A$.

3.2.13. COROLLARY. *Let X be a Hausdorff Abelian group and A a locally compact subgroup of the topological group X. Then the subgroup A is closed in X.*

PROOF. It follows from Proposition 3.2.3 that A is a complete group. Then, according to Theorem 3.2.12, A is a closed subgroup in X.

3.2.14. EXAMPLE. The group $\mathbb{R}$ of real numbers with the natural (interval) topology is locally compact, hence (see Proposition 3.2.3), complete.

3.2.15. EXAMPLE. The subgroup $\mathbb{Q}$ of rational numbers is dense in the group $\mathbb{R}$ endowed with the natural topology, i.e. $[\mathbb{Q}]_{\mathbb{R}} = \mathbb{R}$. Since $\mathbb{Q} \neq \mathbb{R}$ and the group $\mathbb{R}$ is Hausdorff, then the group $\mathbb{Q}$ in the interval topology is not complete.

3.2.16. THEOREM. *Let X be a complete Abelian group, A a subgroup of the topological group X. If A is a closed subgroup in X, then A is a complete group.*

PROOF. Let $\mathcal{F} = \{F_\gamma \mid \gamma \in \Gamma\}$ be a Cauchy filter of A and $\mathcal{B} = \{B_\delta \mid \delta \in \Delta, B_\delta \subseteq X,$ for any $\delta \in \Delta$ there exists $F_\gamma \in \mathcal{F}, F_\gamma \subseteq B_\delta\}$. According to Remark 3.1.22, $\mathcal{B}$ is a Cauchy filter of the group X. Since X is a complete group, then there exists a limit of the filter $\mathcal{B}$ in X. Then, according to Proposition 3.1.26, $\bigcap_{\delta \in \Delta}[B_\delta]_X \neq \emptyset$. Let $a \in \bigcap_{\delta \in \Delta}[B_\delta]_X$. Since $\mathcal{F} \subseteq \mathcal{B}$, then $a \in [F_\gamma]_X$ for any $\gamma \in \Gamma$. Since $F_\gamma \subseteq A$ and A is a closed subgroup in X, then $a \in [F_\gamma]_X \subseteq [A]_X = A$. In that case $a \in [F_\gamma]_X \cap A = [F_\gamma]_A$ for any $\gamma \in \Gamma$, hence, $a \in \bigcap_{\gamma \in \Gamma}[F_\gamma]_A$, i.e. $\bigcap_{\gamma \in \Gamma}[F_\gamma]_A \neq \emptyset$.

Thus, A is complete.

3.2.17. THEOREM. (Completeness criterion). *Let X be a topological Abelian group. Then the following conditions are equivalent:*

1) X is a complete Abelian group;

2) for any direction S, any Cauchy sequence $\{x_s \mid s \in S\}$ by this direction has its limit in X;

3) let T be a direction defined by a certain basis $\mathcal{B}_0(X)$ of neighborhoods of zero in X. Then any Cauchy sequence $\{x_t \mid t \in T\}$ by the direction T has its limit in X;

4) let Γ be a direction defined by the system $\mathcal{B}(X) = \{U_\gamma \mid \gamma \in \Gamma\}$ of all neighborhoods of zero of the group X. Then any Cauchy sequence $\{x_\gamma \mid \gamma \in \Gamma\}$ by the direction Γ has its limit in X.

PROOF. 1) $\Rightarrow$ 2). Let S be a direction and $\{x_s \mid s \in S\}$ a Cauchy sequence in X. Consider the Filter $\mathcal{F}$ associated with it. According to Proposition 3.1.27, $\mathcal{F}$ is a Cauchy filter in X. Then, by the condition 1), there exists an element $a \in X$ such that $a = \lim \mathcal{F}$. From Proposition 3.1.15 we get that $a = \lim_{s \in S} x_s$. Therefore, we've proved the implication 1) $\Rightarrow$ 2).

2) $\Rightarrow$ 1). Let $\mathcal{F} = \{F_\delta \mid \delta \in \Delta\}$ be a Cauchy filter of X. As in Proposition 3.1.16,

consider a sequence $\{x_\delta \mid \delta \in \Delta\}$ by the direction Δ, defined by the filter $\mathcal{F}$. According to the Proposition 3.1.28, $\{x_\delta \mid \delta \in \Delta\}$ is a Cauchy sequence. Then from the condition 2) follows that there exists an element $a \in X$ such that $a = \lim_{\delta \in \Delta} x_\delta$.

Let's show that $a \in [F_{\delta_0}]_X$ for any $\delta_0 \in \Delta$. Indeed, let U be a neighborhood of the element a in X. There exists an element $\delta_1 \in \Delta$ such that $x_\delta \in U$ for all $\delta \in \Delta, \delta > \delta_1$. Also there exists an element $\delta_2 \in \Delta$ such that $F_{\delta_2} \subseteq F_{\delta_0} \cap F_{\delta_1}$. Then, according to the Example 3.1.3, $\delta_2 \geq \delta_0$ and $\delta_2 \geq \delta_1$, hence, $x_{\delta_2} \in U$. Then $x_{\delta_2} \in F_{\delta_2} \cap U \subseteq F_{\delta_0} \cap U$, i.e. $F_{\delta_0} \cap U \neq \emptyset$. Hence, $a \in [F_{\delta_0}]_X$.

Thus, $a \in \bigcap_{\delta \in \Delta} [F_\delta]_X$. According to Proposition 3.1.26, $a = \lim \mathcal{F}$.

Therefore, implication 2) $\Rightarrow$ 1) is proved.

The implication 2) $\Rightarrow$ 3) is obvious.

3) $\Rightarrow$ 4). Let $\{x_\gamma \mid \gamma \in \Gamma\}$ be a Cauchy sequence by the direction Γ defined by the system of all neighborhoods of zero in X. Consider in Γ a subdirection T defined by a basis $\mathcal{B}_0(X)$ of neighborhoods of zero in X. Due to Proposition 3.1.29, the sub-sequence $\{x_t \mid t \in T\}$ is a Cauchy sequence in X. According to the condition 3), it has a limit in X, which is also a limit of the sequence $\{x_\gamma \mid \gamma \in \Gamma\}$ (see Proposition 3.1.32). Hence, the implication 3) $\Rightarrow$ 4) is proved.

4) $\Rightarrow$ 2). Let S be a direction, $\{x_s \mid s \in S\}$ a Cauchy sequence in X. We construct a sequence $\{y_\gamma \in X \mid \gamma \in \Gamma\}$ by the direction determined by the system $\mathcal{B}(X) = \{U_\gamma | \gamma \in \Gamma\}$ of all neighborhoods of zero in X in the following way:

for any set $U_\gamma \in \mathcal{B}(X)$ we find an element $s(\gamma) \in S$ such that $x_{s_1} - x_{s_2} \in U_\gamma$ for all $s_1, s_2 \in S, s_1 \geq s(\gamma), s_2 \geq s(\gamma)$. The element $x_{s(\gamma)}$ we denote over y_γ.

Now we prove that the sequence $\{y_\gamma \mid \gamma \in \Gamma\}$ is a Cauchy sequence in X. Indeed, let $U_{\gamma_0} \in \mathcal{B}(X)$ and U_{γ_1} be a neighborhood of zero in X such that $U_{\gamma_1} - U_{\gamma_1} \subseteq U_{\gamma_0}$.

Let's show that $y_{\gamma_2} - y_{\gamma_1} \in U_{\gamma_0}$ for any $\gamma_2 \in \Gamma$, $\gamma_2 \geq \gamma_1$. Indeed, $y_{\gamma_1} = x_{s(\gamma_1)}$ and $y_{\gamma_2} = x_{s(\gamma_2)}$. There exists an element $s_2 \in S$ such that $s_2 \geq s(\gamma_1)$ and $s_2 \geq s(\gamma_2)$. Since $\gamma_2 \geq \gamma_1$, then we get that $U_{\gamma_2} \subseteq U_{\gamma_1}$ (see Example 3.1.3.). Taking into account the definition of the elements $s(\gamma)$, we get:

$$y_{\gamma_2} - y_{\gamma_1} = x_{s(\gamma_2)} - x_{s(\gamma_1)} = \left(x_{s(\gamma_2)} - x_{s_2}\right) + \left(x_{s_2} - x_{s(\gamma_1)}\right)$$
$$\in -U_{\gamma_2} + U_{\gamma_1} \subseteq U_{\gamma_1} + U_{\gamma_1} \subseteq U_{\gamma_0}.$$

Since the choice of the neighborhood U_{γ_0} was arbitrary, then $\{y_\gamma \mid \gamma \in \Gamma\}$ is a Cauchy sequence in X.

According to the condition 4), there exists an element $a \in X$ such that $a = \lim_{\gamma \in \Gamma} y_\gamma$.

We'll prove that $a = \lim_{s \in S} x_s$. Indeed, let $U_\gamma \in \mathcal{B}(X)$ and U_{γ_0} be a neighborhood of zero in X such that $U_{\gamma_0} + U_{\gamma_0} \subseteq U_\gamma$. Then there exists an element $\gamma_1 \in \Gamma$ such that $\gamma_1 \geq \gamma_0$ and $\{y_\gamma \mid \gamma \in \Gamma, \gamma \geq \gamma_1\} \subseteq a + U_\gamma$. Since $y_{\gamma_1} = x_{s(\gamma_1)}$ and $U_{\gamma_1} \subseteq U_{\gamma_0}$, then for any $s \in S$ such that $s \geq s(\gamma_1)$ we get:

$$x_s = x_s - x_{s(\gamma_1)} + x_{s(\gamma_1)} \in U_{\gamma_1} + (a + U_{\gamma_0}) \subseteq a + U_{\gamma_0} + U_{\gamma_0} \subseteq a + U_\gamma.$$

Thus, $a = \lim_{s \in S} x_s$. The implication 4) $\Rightarrow$ 2) is proved, and the theorem is proved, too.

3.2.18. PROPOSITION. *Let X be a topological Abelian group, which satisfies the first axiom of countability. The group X is complete if and only if any Cauchy sequence $\{x_n \mid n \in \mathbb{N}\}$ has limit in X.*

PROOF. Let X be complete and $\{x_n \mid n \in \mathbb{N}\}$ be a Cauchy sequence in X. Then, according to the condition 2) of Theorem 3.2.17, the sequence $\{x_n \mid n \in \mathbb{N}\}$ has limit in X.

Let now any Cauchy sequence have a limit in X. If group X has the smallest neighborhood of zero, then, according to Corollary 3.2.6, the group X is complete. That is why we can consider that the group X does not have the smallest neighborhood of zero.

Let $\mathcal{B}_0(X)$ be a certain, not more than countable, basis of neighborhoods of zero of the group X. Let Γ be a direction determined by the basis $\mathcal{B}_0(X)$. Then, since $\mathcal{B}(X)$ does not contain the smallest element, the direction Γ does not contain the largest element. According to Proposition 3.1.8, there exists a subdirection Γ' in Γ, similar to $\mathbb{N}$. Let $\{x_\gamma \mid \gamma \in \Gamma\}$ be a Cauchy sequence by the direction Γ. Then, in agreement with the condition, the sub-sequence $\{x_\gamma \mid \gamma \in \Gamma'\}$ has a limit in X. Let $a = \lim_{\gamma \in \Gamma'} x_\gamma$. Then, according to Proposition 3.1.32, $a = \lim_{\gamma \in \Gamma} x_\gamma$.

Hence, X is a complete group (see Theorem 3.2.17).

3.2.19. PROPOSITION. *Let X be a complete topological group, M its smallest closed subgroup (see Corollary 1.4.11). Then the factor-group X/M is complete.*

PROOF. Let $\mathcal{B}(X) = \{U_\delta \mid \delta \in \Delta\}$ be the system of all neighborhoods of zero in X, then $M = \bigcap_{\delta \in \Delta} U_\delta$.

According to Proposition 1.5.6, the system $\bar{\mathcal{B}}(X) = \{\bar{U}_\delta \mid \bar{U}_\delta = \pi(U_\delta), \delta \in \Delta\}$, where $\pi : X \to X/M$ is the natural homomorphism, is a basis of neighborhoods of zero in X/M.

Let Γ be a direction and $\{\bar{x}_\gamma \mid \gamma \in \Gamma\}$ a Cauchy sequence in X/M. For any $\gamma \in \Gamma$ we choose an arbitrary element $y_\gamma \in X$ such that $\pi(y_\gamma) = \bar{x}_\gamma$. We've determined the sequence $\{y_\gamma \mid \gamma \in \Gamma\}$ in X. Now we'll prove that it is a Cauchy sequence. Indeed, let $U_0 \in \mathcal{B}(X)$, and U_1 be a neighborhood of zero such that $U_1 + U_1 \subseteq U_0$ and $\bar{U}_1 = \pi(U_1)$. Then $U_0 \supseteq U_1 + U_1 \supseteq U_1 + M = \pi^{-1}(\bar{U}_1)$. If $\gamma_1 \in \Gamma$ is such that $\bar{x}_\gamma - \bar{x}_{\gamma_1} \in \bar{U}_1$ for any $\gamma \in \Gamma$, $\gamma > \gamma_1$, then $y_\gamma - y_{\gamma_1} \in \pi^{-1}(\bar{x}_\gamma - \bar{x}_{\gamma_1}) \subseteq \pi^{-1}(\bar{U}_1) \subseteq U_0$. Hence, $\{y_\gamma \mid \gamma \in \Gamma\}$ is a Cauchy sequence in X.

Since X is a complete group, then there exists an element $a \in X$ such that $a = \lim_{\gamma \in \Gamma} y_\gamma$. Then, due to Proposition 3.1.33, $\pi(a) = \lim_{\gamma \in \Gamma} \pi(y_\gamma) = \lim_{\gamma \in \Gamma} \bar{x}_\gamma$.

According to Theorem 3.2.17, X/M is a complete group.

3.2.20. PROPOSITION. *Let X be a complete Abelian group, which satisfies the first axiom of countability, and A is a subgroup in X. Then the factor-group X/A is complete.*

PROOF. Since the group X satisfies the first axiom of countability, we can find a basis $\{U_n \mid n \in \mathbb{N}\}$ of symmetric neighborhoods of zero of the group X. By induction we can construct a sub-sequence $\{U_{n_i} \mid i \in \mathbb{N}\}$ such that $U_{n_{i+1}} + U_{n_{i+1}} \subseteq U_{n_i}$ for $i = 1, 2, \ldots$. Denote U_{n_i} over V_i. Then $V_{i+1} + V_{i+1} \subseteq V_i$ for $i = 1, 2, \ldots$, we can also consider that $V_1 = X$. By induction by l it is easy to prove that $\sum_{i=k+1}^{k+l} V_i + V_{k+l} \subseteq V_k$ for all $k, l \in \mathbb{N}$. Hence, $\sum_{i=k+1}^{k+l} V_i \subseteq V_k$ for all $k, l \in \mathbb{N}$.

Let $\pi : X \to X/A$ be the natural homomorphism. Then $\{\pi(U_i) \mid i \in \mathbb{N}\}$ is a basis of neighborhoods of zero (see Proposition 1.5.6) in X/A. Hence, the group X/A satisfies the first axiom of the countability. According to Proposition 3.2.18, to prove the proposition it is enough to check that any Cauchy sequence $\{\bar{x}_n \mid n \in \mathbb{N}\}$ has a limit in X/A.

Let $\{\bar{x}_n \mid n \in \mathbb{N}\}$ be a Cauchy sequence in the group X/A. By induction for each $i \in \mathbb{N}$ we can construct natural numbers m_i and elements $x_{m_i} \in X$ such that:

$$m_i \geq i, \; m_i \leq m_{i+1}, \; \bar{x}_j - \bar{x}_{m_i} \in \pi(V_i) \text{ for all } j \geq m_i, \; i \in \mathbb{N} \text{ and}$$

$$x_{m_{i+1}} - x_{m_i} \in V_i, \; \pi(x_{m_i}) = \bar{x}_{m_i}$$

$(*)$

for $i \in \mathbb{N}$.

Indeed, put $m_1 = 1$ and $m_2 = 2$, while for x_{m_1} and x_{m_2} choose arbitrary elements from the sets $\pi^{-1}(\bar{x}_{m_1})$ and $\pi^{-1}(\bar{x}_{m_2})$ correspondingly.

Now check that the conditions $(*)$ are fulfilled for $i = 1$. Indeed, $m_1 = 1 \geq 1$, $m_2 = 2 \geq 2$. Since $V_1 = X$, then $\bar{x}_j - \bar{x}_{m_1} \in X/A = \pi(V_1)$ for all $j \geq 1$ and $x_{m_2} - x_{m_1} \in X = V_1$. Due to the choice of the elements x_{m_1} and x_{m_2}, we have $\pi(x_{m_1}) = \bar{x}_{m_1}$ and $\pi(x_{m_2}) = \bar{x}_{m_2}$.

Suppose that the numbers $m_1, \ldots m_k \in \mathbb{N}$ and the elements $x_{m_1}, \ldots x_{m_k} \in X$ are determined, and the conditions $(*)$ are fulfilled for $i < k$.

Now we choose a number $m_{k+1} \in \mathbb{N}$ and construct an element $x_{m_{k+1}} \in X$. Since $\{\bar{x}_n \mid n \in \mathbb{N}\}$ is a Cauchy sequence in X/A, then there exists a number $m_{k+1} \in \mathbb{N}$ such that $m_{k+1} \geq k+1$, $m_{k+1} > m_k$ and $\bar{x}_j - \bar{x}_{m_{k+1}} \in \pi(V_{k+1})$ for all $j \geq m_{k+1}$. Let y be an arbitrary element from $\pi^{-1}(\bar{x}_{m_{k+1}})$. Since $\bar{x}_{m_{k+1}} - \bar{x}_{m_k} \in \pi(V_k)$, then $y - x_{m_k} \in V_k + A$, i.e. there exists an element $a \in A$ such that $y - x_{m_k} - a \in V_k$. Let $x_{m_{k+1}} = y - a$.

It is evident that the numbers $m_1, \ldots, m_{k+1}$ and the elements $x_{m_1}, \ldots, x_{m_{k+1}}$ satisfy the conditions $(*)$ for $i < k+1$. Hence, the sequences of the numbers m_i and elements x_{m_i} are constructed.

Now we prove that the sequence $\{x_{m_i} \mid i \in \mathbb{N}\}$ is a Cauchy sequence in X. Indeed, let U be a neighborhood of zero in X. We can find a number k such that $V_k \subseteq U$. Then

$$x_{m_j} - x_{m_{k+1}} = \sum_{i=k+1}^{j-1} (x_{m_{i+1}} - x_{m_i}) \in \sum_{i=k+1}^{j-1} V_i \subseteq V_k \subseteq U$$

for any $j \geq k$, i.e. $\{x_{m_i} \mid i \in \mathbb{N}\}$ is a Cauchy sequence in X.

Since X is a complete group, then there exists an element x_0 such that $x_0 = \lim_{i \in \mathbb{N}} x_{m_i}$. Then, according to the Proposition 3.1.33, $\pi(x_0) = \lim_{i \in \mathbb{N}} \pi(x_{m_i}) = \lim_{i \in \mathbb{N}} \bar{x}_{m_i}$. Since $\{\bar{x}_{m_i} \mid i \in \mathbb{N}\}$ is a sub-sequence of the Cauchy sequence $\{\bar{x}_n \mid n \in \mathbb{N}\}$, then, due to Proposition 3.1.32, $\pi(x_0) = \lim_{n \in \mathbb{N}} \bar{x}_n$.

Since the choice of the sequence $\{\bar{x}_n \mid n \in \mathbb{N}\}$ was arbitrary, then X/A is a complete group.

3.2.21. REMARK. In Theorem 4.1.48 will be shown that in general a factor– group of a complete group could be not complete.

3.2.22. PROPOSITION. *Let X be a topological Abelian group and A a subgroup of the topological group X. If A and the factor-group X/A are complete, then the group X is complete, too.*

PROOF. Let $\mathcal{F} = \{F_\gamma \mid \gamma \in \Gamma\}$ be a Cauchy filter in X and $\pi : X \to X/A$ be the natural homomorphism. Then, according to Proposition 3.1.25, the system $\bar{\mathcal{F}} = \{\pi(F_\gamma) \mid \gamma \in \Gamma\}$ forms a basis of a Cauchy filter $\bar{\Phi}$ in the group X/A. Since X/A is a complete group, then there exists an element $\bar{d} \in X/A$ such that $\bar{d} = \lim \bar{\Phi}$. Let d be an arbitrary element from $\pi^{-1}(\bar{d})$. It is easy to check that the system $\Phi = \{F_\gamma - d \mid \gamma \in \Gamma\}$ is a Cauchy filter in X. Let $\mathcal{B}(X) = \{U_\delta \mid \delta \in \Delta\}$ be a system of all neighborhoods of zero in the group X, which, according to Remark 3.1.21, is a Cauchy filter.

Now we prove that the system $\Phi + \mathcal{B}(X) = \{F_\gamma - d + U_\delta \mid \gamma \in \Gamma, \delta \in \Delta\}$ forms a basis of a certain Cauchy filter in X. Indeed, $\emptyset \notin \Phi + \mathcal{B}(X)$. If $F_{\gamma_1} - d + U_{\delta_1}$, $F_{\gamma_2} - d + U_{\delta_2} \in \Phi + \mathcal{B}(X)$, then there exist subsets $F_{\gamma_3} \in \mathcal{F}$ and $U_{\delta_3} \in \mathcal{B}(X)$ such that $F_{\gamma_3} \subseteq F_{\gamma_1} \cap F_{\gamma_2}$ and $U_{\delta_3} \subseteq U_{\delta_1} \cap U_{\delta_2}$. Then $F_{\gamma_3} - d + U_{\delta_3} \subseteq F_{\gamma_1} - d + U_{\delta_1}$ and $F_{\gamma_3} - d + U_{\delta_3} \subseteq F_{\gamma_2} - d + U_{\delta_2}$, i.e. $\Phi + \mathcal{B}(X)$ is a basis of a certain filter.

Finally, let U be a neighborhood of zero in X. There exists a neighborhood V of zero in X such that $V + V \subseteq U$. There exist subsets $F_{\gamma_0} \in \mathcal{F}$ and $U_{\delta_0} \in \mathcal{B}(X)$ such that $F_{\gamma_0} - F_{\gamma_0} \subseteq V$ and $U_{\delta_0} - U_{\delta_0} \subseteq V$. Then

$$(F_{\gamma_0} - d + U_{\delta_0}) - (F_{\gamma_0} - d + U_{\delta_0}) = F_{\gamma_0} - F_{\gamma_0} + U_{\delta_0} - U_{\delta_0} \subseteq V + V \subseteq U.$$

Therefore, we proved that the system $\Phi + \mathcal{B}(X)$ forms a basis of a Cauchy filter in X.

Now we consider the family of sets $\mathcal{T} = \{(F_\gamma - d + U_\delta) \cap A \mid \gamma \in \Gamma, \delta \in \Delta\}$ and prove that $\mathcal{T}$ is a basis of a certain Cauchy filter of the group A. To do it, it is enough to prove that $(F_{\gamma_0} - d + U_{\delta_0}) \cap A \neq \emptyset$ for all $\gamma_0 \in \Gamma$, $\delta_0 \in \Delta$ (see Proposition 3.1.23).

Indeed, there exists a neighborhood U_{δ_1} of zero in X such that $-U_{\delta_1} \subseteq U_{\delta_0}$. Since $\pi(d) = \bar{d} = \lim \bar{\Phi}$ and $\pi(U_{\delta_0})$ is a neighborhood of zero in X/A, then there exists an element $\gamma_1 \in \Gamma$ such that $\pi(F_{\gamma_1}) \subseteq \bar{d} + \pi(U_{\delta_1})$. Then $F_{\gamma_1} \subseteq d + U_{\delta_1} + A$. If $b \in F_{\gamma_0} \bigcap F_{\gamma_1}$, then there exist elements $u \in U_{\delta_1}$ and $a \in A$ such that $b = d + u + a$. Then $a = b - d - u \in b - d + U_{\delta_1} \subseteq b - d - U_{\delta_0}$. Hence, $a \in (b - d - U_{\delta_0}) \bigcap A$, i.e. $(F_{\gamma_0} - d + U_{\delta_0}) \bigcap A \neq \emptyset$.

Thus, we've proved that $\mathcal{T}$ is a basis of a Cauchy filter of the group A.

Since A is a complete group, then, according to Propositions 3.1.26 and 1.2.25,

$$\emptyset \neq \bigcap_{\substack{\gamma \in \Gamma \\ \delta \in \Delta}} \left[(F_\gamma - d + U_\delta) \cap A \right]_A \subseteq \bigcap_{\substack{\gamma \in \Gamma \\ \delta \in \Delta}} \left[F_\gamma - d + U_\delta \right]_X$$

$$= \left(\bigcap_{\gamma \in \Gamma} \bigcap_{\delta \in \Delta} (F_\gamma + U_\delta) \right) - d = \bigcap_{\gamma \in \Gamma} [F_\gamma]_X - d.$$

Hence, $\bigcap_{\gamma \in \Gamma} [F_\gamma]_X \neq \emptyset$. We've proved that X is a complete group.

3.2.23. COROLLARY. *Let X be a topological group, A its open subgroup. If A is complete, then X is complete too.*

PROOF. The factor–group X/A is a discrete group, and, according to Corollary 3.2.5, X/A is complete. Then, according to Proposition 3.2.22, X is a complete group.

3.2.24. PROPOSITION. *Let X be a topological group, A a dense subgroup in X, Γ the direction defined by the system $\mathcal{B}(X) = \{U_\gamma | \gamma \in \Gamma\}$ of all neighborhoods of zero in X. The group X is complete if and only if any Cauchy sequence $\{a_\gamma | \gamma \in \Gamma\}$ of elements of the subgroup A by the direction Γ has limit in X.*

PROOF. If group X is complete, then, according to Theorem 3.2.17, any Cauchy sequence $\{x_\gamma | \gamma \in \Gamma\}$ of elements of the group X and, in particular, any Cauchy sequence $\{a_\gamma | \gamma \in \Gamma\}$ of elements of the subgroup A has a limit in X.

Let now $\{x_\gamma | \gamma \in \Gamma\}$ be a Cauchy sequence in X by the direction Γ. According to Theorem 3.2.17, to prove the completeness of the group X it is enough to prove that this sequence has a limit in X.

Because of the density of the subgroup A in X, for any $\gamma \in \Gamma$ in the neighborhood $x_\gamma + U_\gamma$ of the element x_γ we can find an element $a_\gamma \in A$.

Let's prove that $\{a_\gamma | \gamma \in \Gamma\}$ is a Cauchy sequence in X. Indeed, let $U_{\gamma_0} \in \mathcal{B}(X)$. Then there exists a set $U_{\gamma_1} \in \mathcal{B}(X)$ such that $U_{\gamma_1} + U_{\gamma_1} - U_{\gamma_1} \subseteq U_{\gamma_0}$. Since $\{x_\gamma | \gamma \in \Gamma\}$ is a Cauchy sequence in X, then there exists an element $\gamma_2 \in \Gamma$, $\gamma_2 \geq \gamma_1$ such that $x_\gamma - x_{\gamma_2} \in U_{\gamma_1}$ for any $\gamma \in \Gamma$, $\gamma \geq \gamma_2$. Then, according to Example 3.1.3, $U_\gamma \subseteq U_{\gamma_2} \subseteq U_{\gamma_1}$. Hence:

$$a_\gamma - a_{\gamma_2} \in x_\gamma + U_\gamma - x_{\gamma_2} - U_{\gamma_2} \subseteq x_\gamma - x_{\gamma_2} + U_\gamma - U_{\gamma_2} \subseteq U_{\gamma_1} + U_{\gamma_1} - U_{\gamma_1} \subseteq U_{\gamma_0}$$

for any $\gamma \in \Gamma$, $\gamma \geq \gamma_2$. Therefore, $\{a_\gamma | \gamma \in \Gamma\}$ is a Cauchy sequence in X.

According to the condition, there exists an element $b \in X$ such that $b = \lim_{\gamma \in \Gamma} a_\gamma$. We'll prove that $b = \lim_{\gamma \in \Gamma} x_\gamma$. Indeed, for any $U_{\gamma_0} \in \mathcal{B}(X)$ there exists a set $U_{\gamma_1} \in \mathcal{B}(X)$ such that $U_{\gamma_1} - U_{\gamma_1} \in U_{\gamma_0}$. Since $b = \lim_{\gamma \in \Gamma} a_\gamma$, then there exists an element $\gamma_2 \in \Gamma$ such that $a_\gamma \in b + U_{\gamma_1}$ for any $\gamma \in \Gamma$, $\gamma > \gamma_2$. Let γ_3 be an element from Γ such that $\gamma_3 \geq \gamma_2$ and $\gamma_3 \geq \gamma_1$. Then, according to Example 3.1.3, $U_\gamma \subseteq U_{\gamma_1}$ when $\gamma \geq \gamma_3$. Hence, $x_\gamma \in a_\gamma - U_\gamma \subseteq b + U_{\gamma_1} - U_\gamma \subseteq b + U_{\gamma_1} - U_{\gamma_1} \subseteq b + U_{\gamma_0}$ for any $\gamma \geq \gamma_3$. Thus, we've proved that $b = \lim_{\gamma \in \Gamma} x_\gamma$. This completes the proof of the proposition.

3.2.25. THEOREM. *Let R be a topological ring, A and B be topological right R-modules. Let B be a complete Hausdorff module, C be a dense submodule in A and $f : C \to B$ be a continuous homomorphism of the R-modules. Then there exists the only continuous homomorphism of the R-modules $\widehat{f} : A \to B$ such that $\widehat{f}(a) = f(a)$ for any $a \in C$.*

PROOF. For every element $a \in A$ consider the system $\Phi_a = \{(a + U) \bigcap C \mid U$ is a neighborhood of zero in $A\}$. According to Corollary 3.1.24, Φ_a is a Cauchy filter A for any $a \in A$. Since f is a continuous homomorphism, then, according to Proposition 3.1.25, $\bar{\Phi}_a = \{f((a + U) \bigcap C) \mid U$ is a neighborhood of zero in $A\}$ is a basis of a certain Cauchy filter $\mathcal{F}_a$ of B. Since B is a complete module, then there exists the only (due to the fact that the group B is Hausdorff) element $b \in B$ such that $b = \lim \mathcal{F}_a$. Put $\widehat{f}(a) = b$.

Let's check that $\widehat{f}(a) = f(a)$ for all $a \in C$. Indeed, assume the contrary, i.e. that there exists an element $a \in C$ such that $\widehat{f}(a) \neq f(a)$. Since the module B is Hausdorff, then there exists a neighborhood U of zero in the module B such that $(\widehat{f}(a) + U) \bigcap (f(a) + U) = \emptyset$. Since f is a continuous homomorphism, then there exists a neighborhood V_0 of zero in A such that $f((a+V_0) \bigcap C) \subseteq f(a) + U$. Since $\widehat{f}(a) = \lim \mathcal{F}_a$, then there exists a neighborhood V_1 of zero in A such that $f((a + V_1) \bigcap C) \subseteq \widehat{f}(a) + U$. There also exists a neighborhood V_2 of zero in A such that $V_2 \subseteq V_0 \bigcap V_1$. Now we choose an arbitrary element $c \in (a + V_2) \bigcap C$. Then $f(c) \in f((a + V_1) \bigcap C) \subseteq \widehat{f}(a) + U$ and $f(c) \in f((a + V_0) \bigcap C) \subseteq f(a) + U$, i.e.

$(f(a) + U) \cap (\widehat{f}(a) + U) \neq \emptyset$. This contradicts the choice of the neighborhood U. Hence, $\widehat{f}(a) = f(a)$ for any $a \in C$.

Now we prove that $\widehat{f}$ is a group homomorphism from A to B. Assume the contrary. Then there exist elements $a_1, a_2 \in A$ such that $\widehat{f}(a_1 + a_2) \neq \widehat{f}(a_1) + \widehat{f}(a_2)$. Since the module B is Hausdorff, then there exists a neighborhood U of zero in B such that $(\widehat{f}(a_1 + a_2) + U) \cap (\widehat{f}(a_1) + \widehat{f}(a_2) + U) = \emptyset$.

There also exists a neighborhood U_1 of zero in B such that $U_1 + U_1 \subseteq U$. Then $(\widehat{f}(a_1 + a_2) + U) \cap (\widehat{f}(a_1) + U_1 + \widehat{f}(a_2) + U_1) = \emptyset$. Since $\widehat{f}(a_1 + a_2) = \lim \mathcal{F}_{a_1 + a_2}$ and $\widehat{f}(a_1) = \lim \mathcal{F}_{a_1}$, $\widehat{f}(a_2) = \lim \mathcal{F}_{a_2}$, then there exist neighborhoods V, V_1 and V_2 of zero in A such that

$$f\left((a_1 + a_2 + V) \cap C\right) \subseteq \widehat{f}(a_1 + a_2) + U,$$
$$f\left((a_1 + V_1) \cap C\right) \subseteq \widehat{f}(a_1) + U_1,$$
$$f\left((a_2 + V_2) \cap C\right) \subseteq \widehat{f}(a_2) + U_1.$$

Let V_3 be a neighborhood of zero in A such that $V_3 \subseteq V_1 \cap V_2$ and $V_3 + V_3 \subseteq V$. Since C is a dense submodule in A, then $(a_1 + V_3) \cap C \neq \emptyset$ and $(a_2 + V_3) \cap C \neq \emptyset$. Let b_1 and b_2 be elements from V_3 such that $a_1 + b_1 \in C$ and $a_2 + b_2 \in C$. Then $a_1 + b_1 + a_2 + b_2 \in (a_1 + a_2 + V_3 + V_3) \cap C \subseteq (a_1 + a_2 + V) \cap C$. Hence, $\widehat{f}(a_1 + b_1 + a_2 + b_2) = f(a_1 + b_1 + a_2 + b_2) \in f((a_1 + a_2 + V) \cap C) = f(a_1 + a_2 + V) \cap C \subseteq \widehat{f}(a_1 + a_2) + U$. Since f is a homomorphism, then $\widehat{f}(a_1 + b_1 + a_2 + b_2) = f(a_1 + b_1 + a_2 + b_2) = f(a_1 + b_1) + f(a_2 + b_2) \in f((a_1 + V_1) \cap C) + f((a_2 + V_2) \cap C) \subseteq (\widehat{f}(a_1) + U_1) + (\widehat{f}(a_2) + U_1)$. Therefore, $(\widehat{f}(a_1 + a_2) + U) \cap (\widehat{f}(a_1) + U_1 + \widehat{f}(a_2) + U_1) \neq \emptyset$. This contradicts the choice of the neighborhoods U and U_1. Therefore, $\widehat{f}$ is the group homomorphism from A to B.

Now we prove that $\widehat{f}$ is a homomorphism of R-modules. Assume the contrary. In that case there exist elements $a \in A$ and $r \in R$ such that $\widehat{f}(a \cdot r) \neq \widehat{f}(a) \cdot r$. Since the module B is Hausdorff, then there exists a neighborhood U of zero in B such that $(\widehat{f}(a \cdot r) + U) \cap (\widehat{f}(a) \cdot r + U) = \emptyset$. Since B is a topological R-module, then there exists a neighborhood W of zero in B such that $(\widehat{f}(a) + W) \cdot r \subseteq \widehat{f}(a) \cdot r + U$.

As $\widehat{f}(a \cdot r) = \lim \mathcal{F}_{a \cdot r}$ and $\widehat{f}(a) = \lim \mathcal{F}_a$, then there exist neighborhoods of zero V_1 and

V_2 in A such that:

$$f\big((a \cdot r + V_1) \cap C\big) \subseteq \widehat{f}(a \cdot r) + U,$$

and

$$f\big((a + V_2) \cap C\big) \subseteq \widehat{f}(a) + W.$$

Since A is topological module, then there exists a neighborhood V_3 of zero in A such that $(a + V_3) \cdot r \subseteq a \cdot r + V_1$. Let V be a neighborhood of zero in A such that $V \subseteq V_2 \cap V_3$. As C is a dense submodule in A, then $(a + V) \cap C \neq \emptyset$. We choose an element $b \in V$ such that $a + b \in C$. Then:

$$\widehat{f}\big((a + b) \cdot r\big) = f\big((a + b) \cdot r\big) \in f\big((a + V) \cdot r \cap C\big) \subseteq \big(f(a \cdot r + V_1) \cap C\big) \subseteq \widehat{f}(a \cdot r) + U,$$

and

$$\widehat{f}\big((a + b) \cdot r\big) = f\big((a + b) \cdot r\big) = f(a + b) \cdot r \in f\big((a + V_2) \cap C\big) \cdot r \subseteq \widehat{f}(a) \cdot r + U.$$

Hence, $\big(\widehat{f}(a \cdot r) + U\big) \cap \big(\widehat{f}(a) \cdot r + U\big) \neq \emptyset$.

It contradicts the choice of the neighborhood U. Therefore, $\widehat{f}$ is a homomorphism of R-modules.

Now we'll prove that $\widehat{f}$ is a continuous homomorphism.

Let U be a neighborhood of zero in B. Then there exists a neighborhood U_1 of zero in B such that $U_1 - U_1 \subseteq U$. Since f is a continuous homomorphism, then there exists a neighborhood V_0 of zero in A such that $f(V_0 \cap C) \subseteq U_1$. Let V_1 be such a neighborhood of zero in A that $V_1 + V_1 \subseteq V_0$.

We'll show that $\widehat{f}(V_1) \subseteq U$. Assume the contrary, i.e. that there exists an element $a \in V_1$ such that $\widehat{f}(a) \notin U$. Since $\widehat{f}(a) = \lim \mathcal{F}_a$, then there exists a neighborhood V_2 of zero in A such that $f\big((a + V_2) \cap C\big) \subseteq \widehat{f}(a) + U_1$. As C is a dense submodule in A, then $(a + (V_1 \cap V_2)) \cap C \neq \emptyset$. Choose an element $b \in V_1 \cap V_2$ such that $a + b \in C$. Then $f(a + b) \in f\big((a + V_2) \cap C\big) \subseteq \widehat{f}(a) + U_1$, hence, $\widehat{f}(a) \in f(a + b) - U_1$. Since $f(a + b) \in f\big((V_1 + V_1) \cap C\big) \subseteq f(V_0 \cap C) \subseteq U_1$, then $\widehat{f}(a) \in U_1 - U_1 \subseteq U$. It contradicts the choice of the element a. This proves that $\widehat{f}(V_1) \subseteq U$.

Thus, we've proved that $\widehat{f}$ is a continuous homomorphism.

Now we prove that $\widehat{f}$ is the only continuous homomorphism extending the homomorphism f.

Assume the contrary, i.e. that there exists a continuous homomorphism $\tilde{f} : A \to B$ of R-modules such that $\tilde{f}(a) = f(a)$ for any $a \in C$ and $\tilde{f} \neq \widehat{f}$. The last condition means that there exists an element $a \in A$ such that $\tilde{f}(a) \neq \widehat{f}(a)$. Since the module B is Hausdorff, then there exists a neighborhood U of zero in B such that $(\tilde{f}(a) + U) \bigcap (\widehat{f}(a) + U) = \emptyset$. As $\tilde{f}$ and $\widehat{f}$ are continuous homomorphisms, then there exist such neighborhoods V_1 and V_2 of zero in A that $\tilde{f}(a + V_1) \subseteq \tilde{f}(a) + U$ and $\widehat{f}(a + V_2) \subseteq \widehat{f}(a) + U$. Since C is a dense submodule in A, then $(a + (V_1 \bigcap V_2)) \bigcap C \neq \emptyset$. Let $c \in (a + (V_1 \bigcap V_2)) \bigcap C$. Then $\tilde{f}(a) + U \ni \tilde{f}(c) = f(c) = \widehat{f}(c) \in \widehat{f}(a) + U$. Hence, $(\tilde{f}(a) + U) \bigcap (\widehat{f}(a) + U) \neq \emptyset$. It contradicts the choice of the neighborhood U. It proves that $\widehat{f}$ is the only continuous homomorphism extending the homomorphism f. This completes the proof of the theorem.

3.2.26. COROLLARY. *Let A and B be topological groups, besides, B be a complete Hausdorff group. Let C be a dense subgroup of A and $f : C \to B$ be a continuous homomorphism. Then there exists the only continuous homomorphism $\widehat{f} : A \to B$ such that $\widehat{f}(a) = f(a)$ for any $a \in C$.*

PROOF. The topological groups A, B and C are topological right $\mathbb{Z}$-modules over the ring of integers $\mathbb{Z}$ endowed with the discrete topology. The group homomorphisms are homomorphisms of the corresponding $\mathbb{Z}$-modules. The corollary results from Theorem 3.2.25 applied to the groups A, B and C, considered as $\mathbb{Z}$-modules.

3.2.27. THEOREM. *Let R be a topological ring and S a complete Hausdorff ring. Let T be a dense subring in R and $f : T \to S$ a continuous ring homomorphism. Then there exists the only continuous homomorphism $\widehat{f} : R \to S$ such that $\widehat{f}(r) = f(r)$ for any $r \in T$.*

PROOF. Consider, as in Theorem 3.2.25, for any element $r \in R$ the system $\bar{\Phi}_r = \{f((r+U) \bigcap T) \mid U$ is a neighborhood of zero in $R\}$ and define $\widehat{f}(r)$ as a limit of the Cauchy filter $\mathcal{F}_r$ in S, for which $\bar{\Phi}_r$ is a basis (see Theorem 3.2.25). According to Corollary 3.2.26, $\widehat{f} : R \to S$ is a continuous group homomorphism and $\widehat{f}(r) = f(r)$ for all $r \in T$.

Now we prove that $\widehat{f} : R \to S$ is a ring homomorphism. Assume the contrary. Then there exist elements $r_1, r_2 \in R$ such that $\widehat{f}(r_1 \cdot r_2) \neq \widehat{f}(r_1) \cdot \widehat{f}(r_2)$. Since the ring S is Hausdorff, then there exists a neighborhood U of zero in S such that $(\widehat{f}(r_1 \cdot r_2) + U) \bigcap (\widehat{f}(r_1) \cdot \widehat{f}(r_2) + U) = \emptyset$. As S is a topological ring, then there exists a neighborhood U_1 of zero in S such that $(\widehat{f}(r_1) + U_1) \cdot (\widehat{f}(r_2) + U_1) \subseteq \widehat{f}(r_1) \cdot \widehat{f}(r_2) + U$. Since $\widehat{f}(r_1 \cdot r_2) = \lim \mathcal{F}_{r_1 \cdot r_2}$ and $\widehat{f}(r_1) = \lim \mathcal{F}_{r_1}$, $\widehat{f}(r_2) = \lim \mathcal{F}_{r_2}$, then there exist neighborhoods V, V_1 and V_2 of zero in R such that:

$$f\left((r_1 \cdot r_2 + V) \cap T\right) \subseteq \widehat{f}(r_1 \cdot r_2) + U;$$
$$f\left((r_1 + V_1) \cap T\right) \subseteq \widehat{f}(r_1) + U_1;$$
$$f\left((r_2 + V_2) \cap T\right) \subseteq \widehat{f}(r_2) + U_1.$$

Let V_3 be a neighborhood of zero in R such that $V_3 \subseteq V_1 \bigcap V_2$ and $(r_1 + V_3) \cdot (r_2 + V_3) \subseteq r_1 \cdot r_2 + V$. Since T is dense subring of R, then $(r_1 + V_3) \bigcap T \neq \emptyset$ and $(r_2 + V_3) \bigcap T \neq \emptyset$. Let b_1 and b_2 be elements of V_3 such that $r_1 + b_1 \in T$ and $r_2 + b_2 \in T$. Then $(r_1 + b_1) \cdot (r_2 + b_2) \in (r_1 \cdot r_2 + V) \bigcap T$. Hence, $\widehat{f}((r_1 + b_1) \cdot (r_2 + b_2)) = f((r_1 + b_1) \cdot (r_2 + b_2)) \in \widehat{f}(r_1 \cdot r_2) + U$. On the other hand, since f is a ring homomorphism, then

$$\widehat{f}((r_1 + b_1) \cdot (r_2 + b_2)) = f((r_1 + b_1) \cdot (r_2 + b_2)) = f(r_1 + b_1) \cdot f(r_2 + b_2)$$
$$\in f\left((r_1 + V_3) \cap T\right) \cdot f\left((r_2 + V_3) \cap T\right) \subseteq f\left((r_1 + V_1) \cap T\right) \cdot f\left((r_2 + V_2) \cap T\right)$$
$$\subseteq (\widehat{f}(r_1 + U_1)) \cdot (\widehat{f}(r_2) + U_1) \subseteq \widehat{f}(r_2) \cdot \widehat{f}(r_2) + U.$$

Therefore, $(\widehat{f}(r_1 \cdot r_2) + U) \bigcap (\widehat{f}(r_1) \cdot \widehat{f}(r_2) + U) \neq \emptyset$.

It contradicts the choice of the neighborhood U. Hence, $\widehat{f}$ is a ring homomorphism. Since $\widehat{f}$ is, in particular, a homomorphism of the groups $R(+)$ and $S(+)$, then the uniqueness of a continuous extension of the homomorphism $\widehat{f}$ onto the ring R follows from Corollary 3.2.26.

3.2.28. PROPOSITION. *Let C be a Hausdorff Abelian group, A and B be such complete Hausdorff groups that C is a dense subgroup in A and B. Then there exists a topological group isomorphism $\xi : A \to B$ such that $\xi(a) = a$ for all $a \in C$.*

PROOF. Let $\eta : C \to C$ be the identical mapping. According to Corollary 3.2.26, there exist continuous group homomorphisms $\xi : A \to B$ and $\xi_1 : B \to A$ such that $\xi(a) = a$

and $\xi_1(a) = a$ for all $a \in C$. Then $\xi_1\xi : A \to A$ and $\xi\xi_1 : B \to B$ are also continuous group homomorphisms, while $(\xi_1\xi)(a) = a$ and $(\xi\xi_1)(a) = a$ for all $a \in C$. Since $\epsilon(a) = a$ and $\epsilon_1(a) = a$ for the identical mapping $\epsilon : A \to A$ and $\epsilon_1 : B \to B$, then, according to Corollary 3.2.26, $\xi_1\xi = \epsilon$ and $\xi\xi_1 = \epsilon_1$. It means that $\xi = \xi_1^{-1}$, hence, $\xi : A \to B$ is a topological group isomorphism.

3.2.29. PROPOSITION. *Let T be a Hausdorff topological ring, R and S be complete Hausdorff rings such that T is a dense subring in R and S. Then there exists a topological ring isomorphism $\xi : R \to S$ such that $\xi(t) = t$ for all $t \in T$.*

PROOF. According to Theorem 3.2.27, there exists a topological ring homomorphism $\xi : R \to S$ such that $\xi(t) = t$ for all $t \in T$. Because of Proposition 3.2.28, there exists a topological group isomorphism $\eta : R(+) \to S(+)$ such that $\eta(t) = t$ for all $t \in T$. According to Corollary 3.2.26, $\eta = \xi$. Hence, $\xi : R \to S$ is a topological ring isomorphism of R and S.

3.2.30. PROPOSITION. *Let R be a topological ring, C a Hausdorff topological right R-module, and A and B complete Hausdorff R-modules such that C is a dense submodule in A and B. In that case there exists a topological isomorphism $\xi : A \to B$ of R-modules such that $\xi(a) = a$ for all $a \in C$.*

PROOF. According to Theorem 3.2.25, there exists a topological homomorphism of R-modules $\eta : A \to B$ such that $\eta(a) = a$ for all $a \in C$. Because of Proposition 3.2.28, there exists a topological group isomorphism $\xi : A \to B$ such that $\xi(a) = a$ for all $a \in C$. According to Corollary 3.2.26, $\xi = \eta$. Hence, $\xi : A \to B$ is a topological isomorphism of R-modules A and B.

3.2.31. THEOREM. *Let R be a topological ring and M a complete Hausdorff topological R-module. Let $\bar{R}$ be a topological ring which contains R as a dense subring. Then the multiplication operation of the elements of R by the elements of M can be extended to the ring $\bar{R}$ such way that M transforms into a topological $\bar{R}$-module.*

PROOF. Consider the topological R-module $\bar{R}$ containing R as a dense submodule. For any $m \in M$ we define the R-module homomorphism $\mu_m : R \to M$, where $\mu_m(r) = m \cdot r$ for all $r \in R$. According to Theorem 3.2.25, there exists the only R-module homomorphism

$\bar{\mu}_m : \bar{R} \to M$ extending μ_m. We define the multiplication operation of the elements of M by the elements of $\bar{R}$ as:

$m \cdot \bar{r} = \bar{\mu}_m(\bar{r})$ for all $m \in M$, and $\bar{r} \in \bar{R}$.

It is clear that $\bar{\mu}_m(r) = \mu_m(r) = m \cdot r$ for all $m \in M$ and $r \in R$.

Now we prove that M is an $\bar{R}$-module with respect to this multiplication operation. Indeed, let $\mathcal{B}(\bar{R}) = \{V_\gamma | \gamma \in \Gamma\}$ be the system of all neighborhoods of zero in $\bar{R}$ and Γ the direction defined by the filter $\mathcal{B}(\bar{R})$. Let $\bar{r}, \bar{s} \in \bar{R}$. Then, since R is a dense subring in $\bar{R}$, there exist such sequences $\{r_\gamma | \gamma \in \Gamma\}$ and $\{s_\gamma | \gamma \in \Gamma\}$ of elements of R that $\bar{r} = \lim_{\gamma \in \Gamma} r_\gamma$ and $\bar{s} = \lim_{\gamma \in \Gamma} s_\gamma$. According to Proposition 3.1.40, $\bar{r} + \bar{s} = \lim_{\gamma \in \Gamma}(r_\gamma + s_\gamma)$, and, due to Corollary 3.1.46, $\bar{r} \cdot \bar{s} = \lim_{\gamma \in \Gamma}(r_\gamma \cdot s_\gamma)$.

Let $m \in M$. Since the homomorphism $\bar{\mu}_m$ is continuous, then, according to Proposition 3.1.33, $m \cdot \bar{r} = \lim_{\gamma \in \Gamma}(\bar{\mu}_m(r_\gamma))$, $m \cdot \bar{s} = \lim_{\gamma \in \Gamma}(\bar{\mu}_m(s_\gamma))$ and $m \cdot (\bar{r} + \bar{s}) = \lim_{\gamma \in \Gamma}(\bar{\mu}_m(r_\gamma + s_\gamma)) = \lim_{\gamma \in \Gamma} m(r_\gamma + s_\gamma)$. Due to Proposition 3.1.40, $m \cdot \bar{r} + m \cdot \bar{s} = \lim_{\gamma \in \Gamma}(m \cdot r_\gamma + m \cdot s_\gamma)$.

Since M is an R-module, then $m \cdot (r_\gamma + s_\gamma) = m \cdot r_\gamma + m \cdot s_\gamma$ for all $\gamma \in \Gamma$. Hence, $m \cdot (\bar{r} + \bar{s}) = m \cdot \bar{r} + m \cdot \bar{s}$.

Let $m \in M$. Since the homomorphism $\bar{\mu}_m$ is continuous, then, according to Proposition 3.1.33,

$$m \cdot \bar{r} = \lim_{\gamma \in \Gamma}(\bar{\mu}_m \cdot (r_\gamma)) = \lim_{\gamma \in \Gamma} m \cdot r_\gamma, \quad (m \cdot \bar{r}) \cdot \bar{s} = \lim_{\gamma \in \Gamma}(m \cdot \bar{r}) \cdot s_\gamma$$
$$\text{and } m \cdot (\bar{r} \cdot \bar{s}) = \lim_{\gamma \in \Gamma} m \cdot (r_\gamma \cdot s_\gamma).$$

Let U_0 be a neighborhood of zero in M. As M is a topological R-module, then there exist such neighborhoods U_1 and V_1 of zero in M and R correspondingly that

$$U_1 + U_1 + U_1 + U_1 \cdot V_1 \subseteq U_0 \quad \text{and} \quad (m \cdot \bar{r}) \cdot V_1 \subseteq U_1.$$

Since $(m\bar{r}) \cdot \bar{s} = \lim_{\gamma \in \Gamma}(m\bar{r}) \cdot s_\gamma$, and $\{s_\gamma | \gamma \in \Gamma\}$ is a Cauchy sequence, then we can find an element $\gamma_0 \in \Gamma$ such that $(m\bar{r}) \cdot s_\gamma \in U_1 + (m\bar{r}) \cdot \bar{s}$ and $s_\gamma - s_{\gamma_0} \subseteq V_1$ for any $\gamma \geq \gamma_0$. As M is a topological R-module, then there exists a neighborhood U_2 of zero in M and an element $\gamma_1 \in \Gamma$ such that $U_2 \cdot s_{\gamma_0} \subseteq U_1$ and $m \cdot r_\gamma - m \cdot \bar{r} \in U_2$ for all $\gamma \geq \gamma_1$.

Since M is an R-module, then $m \cdot (r_\gamma \cdot s_\gamma) = (m \cdot r_\gamma) \cdot s_\gamma$ for any $\gamma \in \Gamma$. Hence, for $\gamma > \gamma_1$ we have:

$$m \cdot (r_\gamma \cdot s_\gamma) = (m \cdot r_\gamma) \cdot s_\gamma \in (mr_\gamma) \cdot (s_{\gamma_0} + V_1)$$
$$\subseteq (m \cdot \bar{r} + U_2) \cdot (s_{\gamma_0} + V_1)$$
$$\subseteq (m \cdot \bar{r}) \cdot s_{\gamma_0} + U_2 \cdot s_{\gamma_0} + (m \cdot \bar{r}) \cdot V_1 + U_2 \cdot V_1$$
$$\subseteq (m \cdot \bar{r}) \cdot \bar{s} + U_1 + U_1 + U_1 + U_1 \cdot V_1 \subseteq (m \cdot \bar{r}) \cdot \bar{s} + U_0.$$

Therefore, $(m \cdot \bar{r}) \cdot \bar{s} = \lim_{\gamma \in \Gamma} m \cdot (r_\gamma \cdot s_\gamma) = m \cdot (\bar{r} \cdot \bar{s})$.

Let $m_1, m_2 \in M$. Since the homomorphisms $\bar{\mu}_{m_1}$, $\bar{\mu}_{m_2}$ and $\bar{\mu}_{m_1+m_2}$ are continuous, then, according to Proposition 3.1.33, $m_1 \cdot \bar{r} = \lim_{\gamma \in \Gamma} (\bar{\mu}_{m_1}(r_\gamma)) = \lim_{\gamma \in \Gamma} m_1 \cdot r_\gamma, m_2 \cdot \bar{r} = \lim_{\gamma \in \Gamma} m_2 \cdot r_\gamma$ and $(m_1 + m_2) \cdot \bar{r} = \lim_{\gamma \in \Gamma}(m_1 + m_2) \cdot r_\gamma$. Due to Proposition 3.1.40, $m_1 \cdot \bar{r} + m_2 \cdot \bar{r} = \lim_{\gamma \in \Gamma}(m_1 \cdot r_\gamma + m_2 \cdot r_\gamma)$. Since M is an R-module, then $(m_1 + m_2) \cdot r_\gamma = m_1 \cdot r_\gamma + m_2 \cdot r_\gamma$ for all $\gamma \in \Gamma$. Hence, $(m_1 + m_2) \cdot \bar{r} = m_1 \cdot \bar{r} + m_2 \cdot \bar{r}$.

Thus, we've proved that M is an $\bar{R}$-module.

Now we prove that M is a topological $\bar{R}$-module in the initial topology. To do so it is enough to prove that the multiplication operation of the elements of M by the elements of $\bar{R}$ is continuous.

At first note that for any subset $N \subseteq M$ and for any open subset $\bar{W}$ in $\bar{R}$ it is true that $N \cdot \bar{W} \subseteq [N \cdot (\bar{W} \cap R)]_M$. Indeed, due to C.5, $\bar{W} \subseteq [\bar{W} \cap R]_{\bar{R}}$. Since the homomorphism $\mu_n : \bar{R} \to M$ is continuous, then $n \cdot [\bar{W} \cap R]_{\bar{R}} \subseteq [n \cdot (\bar{W} \cap R)]_M$ for any $n \in N$. Hence, $N \cdot \bar{W} \subseteq N \cdot [\bar{W} \cap R]_{\bar{R}} \subseteq [N \cdot (\bar{W} \cap R)]_M$.

To prove the continuity of multiplication of the elements of M by the elements of $\bar{R}$ it is enough to verify its continuity for every pair $m_0 \in M$, $\bar{r} \in \bar{R}$. Let U_0 be a neighborhood of zero in M. Then there exists a neighborhood U_1 of zero in M such that $U_1 + [U_1]_M + [U_1]_M + [U_1]_M \subseteq U_0$. Since M is a topological R-module, then there exists a neighborhood U_2 of zero in M and an open neighborhood W_1 of zero in R such that $U_2 \cdot W_1 \subseteq U_1$ and $m_0 \cdot W_1 \subseteq U_1$. As R is a dense subring in $\bar{R}$, then there exists an open neighborhood $\bar{W}_1$ of zero in $\bar{R}$ such that $\bar{W}_1 \cap R = W_1$ and an element $r_0 \in R$ such that $\bar{r} \in r_0 + \bar{W}_1$. Since M is a topological R-module, then there exists a neighborhood U_3 of zero in M such that $U_3 \subseteq U_2$ and $U_3 \cdot r_0 \subseteq U_1$. Then:

$$(m_0 + U_3) \cdot (\bar{r} + \bar{W}_1) \subseteq m_0 \cdot \bar{r} + U_3 \cdot \bar{r} + m_0 \cdot \bar{W}_1 + U_3 \cdot \bar{W}_1$$

$$\subseteq m_0 \cdot \bar{r} + U_3 \cdot (r_0 + \bar{W}_1) + \left[m_0 \cdot (\bar{W}_1 \cap R) \right]_M + U_3 \cdot \bar{W}_1$$

$$\subseteq m_0 \cdot \bar{r} + U_3 \cdot r_0 + U_2 \cdot \bar{W}_1 + \left[m_0 \cdot \bar{W}_1 \right]_M + U_2 \cdot \bar{W}_1$$

$$\subseteq m_0 \cdot \bar{r} + U_1 + \left[U_2 \cdot (\bar{W}_1 \cap R) \right]_M + \left[U_1 \right]_M + \left[U_2 \cdot (\bar{W}_1 \cap R) \right]_M$$

$$\subseteq m_0 \cdot \bar{r} + U_1 + \left[U_1 \right]_M + \left[U_1 \right]_M + \left[U_1 \right]_M \subseteq m_0 \cdot r + U_0.$$

Since $m_0 + U_3$ is a neighborhood of the element m_0 in M, and $r + \bar{W}_1$ is a neighborhood of the element $\bar{r}$ in $\bar{R}$, we get the necessary result. The theorem is proved.

3.2.32. PROPOSITION. *Let A and B be Abelian groups, $f : A \to B$ a surjective group homomorphism, τ a group topology on B such that (B, τ) is a complete group. Let τ_1 be the prototype of the topology τ with respect to the homomorphism f, then (A, τ_1) is a complete group.*

PROOF. Let Γ be a direction defined by the system of all neighborhoods of zero in (A, τ_1). According to Theorem 3.2.17, it is enough to verify that any Cauchy sequence $\{a_\gamma | \gamma \in \Gamma\}$ in the group A has its limit in A. Indeed, according to Corollary 3.1.35, $\{f(a_\gamma) | \gamma \in \Gamma\}$ is a Cauchy sequence in B. Since the group B is complete, then there exists an element $b \in B$ such that $b = \lim_{\gamma \in \Gamma} f(a_\gamma)$. If a is an element of the group A such that $b = f(a)$, then, according to Corollary 3.1.35, $a = \lim_{\gamma \in \Gamma} a_\gamma$. Hence, (A, τ_1) is a complete group.

3.2.33. THEOREM. *Let A be an Abelian group, $\{\tau_\delta \mid \delta \in \Delta\}$ a certain family of Hausdorff group topologies A that satisfy the following conditions:*

1) the groups (A, τ_δ) are complete for all $\delta \in \Delta$;

2) there exists a topology $\tau_{\delta_0}, \delta_0 \in \Delta$ such that $\tau_{\delta_0} \leq \tau_\delta$ for all $\delta \in \Delta$.

Then the group $(A, \sup\{\tau_\delta \mid \delta \in \Delta\})$ is complete.

PROOF. Let Γ be an arbitrary direction, and $\{x_\gamma \mid \gamma \in \Gamma\}$ a Cauchy sequence in the group $(A, \sup\{\tau_\delta \mid \delta \in \Delta\})$ by the direction Γ. According to the Corollary 3.1.34, $\{x_\gamma \mid \gamma \in \Gamma\}$ is a Cauchy sequence in the group (A, τ_δ) for any $\delta \in \Delta$, in particular in the group (A, τ_{δ_0}). Since the group (A, τ_{δ_0}) is complete, then there exists an element $a \in A$ such that $a = \lim_{\gamma \in \Gamma} x_\gamma$ in (A, τ_{δ_0}).

Now we prove that $a = \lim_{\gamma \in \Gamma} x_\gamma$ in (A, τ_δ) for any $\delta \in \Delta$. Indeed, since (A, τ_δ) is a complete group, then there exists an element $b \in A$ such that $b = \lim_{\gamma \in \Gamma} x_\gamma$ in (A, τ_δ). Since $\tau_{\delta_0} \le \tau_\delta$, then, due to Corollary 3.1.34, $b = \lim_{\gamma \in \Gamma} x_\gamma$ in (A, τ_{δ_0}) and $b = a$. Hence, $a = \lim_{\gamma \in \Gamma} x_\gamma$ in (A, τ_δ) for any $\delta \in \Delta$.

Then, according to Theorem 3.1.13, $a = \lim_{\gamma \in \Gamma} x_\gamma$ in the topological group $(A, \sup\{\tau_\delta \mid \delta \in \Delta\})$. Due to Theorem 3.2.17, $(A, \sup\{\tau_\delta \mid \delta \in \Delta\})$ is a complete group.

3.2.34. COROLLARY. Let A be an Abelian group, $\tau_0 \le \tau_1 \le \ldots$ an ascending countable chain of Hausdorff group topologies on A such that (A, τ_n) is a complete group for any $n \in \mathbb{N}$. Then the group $(A, \sup\{\tau_n \mid n \in \mathbb{N}\})$ is complete.

The proof follows from Theorem 3.2.33.

3.2.35. REMARK. In Example 4.1.55 we'll give an example of an Abelian group A and two Hausdorff incomparable group topologies τ_1 and τ_2 on A such that A is complete in each of them, but the group $(A, \sup\{\tau_1, \tau_2\})$ is not complete.

3.2.36. THEOREM. *Let (R, η) be a pseudonormed ring, (M, ξ) a pseudonormed module over (R, η), τ_ξ a Hausdorff topology defined by a pseudonorm ξ (see Remark 2.3.5). Let (M, τ_ξ) be a dense submodule of Hausdorff topological R-module $(\widehat{M}, \widehat{\tau})$, then there exists the only pseudonorm $\widehat{\xi}$ on $\widehat{M}$ such that $\widehat{\xi}(m) = \xi(m)$ for all $m \in M$ and $\widehat{\tau} = \tau_{\widehat{\xi}}$, where $\tau_{\widehat{\xi}}$ is a topology on $\widehat{M}$ defined by the pseudonorm $\widehat{\xi}$.*

PROOF. Let $m \in \widehat{M}$ and $\Phi_m = \{(m + \widehat{U}) \cap M \mid \widehat{U}$ a neighborhood of zero in $\widehat{M}\}$. Since M is a dense submodule in $\widehat{M}$, then, according to Corollary 3.1.24, Φ_m is a Cauchy filter of the module M.

Now we prove that for any $s \in \widehat{M}$ the family $\tilde{\Phi}_s = \{\xi((s + \widehat{U}) \cap M) \mid \widehat{U}$ a neighborhood of zero in $\widehat{M}\}$ is a basis of a certain Cauchy filter $\mathcal{F}_s$ of the ring $\mathbb{R}$ of real numbers.

Indeed, $\emptyset \notin \tilde{\Phi}_s$. Let $\xi((s + \widehat{U}_1) \cap M) \in \tilde{\Phi}_s$ and $\xi((s + \widehat{U}_2) \cap M) \in \tilde{\Phi}_s$, then there exists a neighborhood $\widehat{U}_3$ of zero in $(\widehat{M}, \widehat{\tau})$ such that $\widehat{U}_3 \subseteq \widehat{U}_1 \cap \widehat{U}_2$. Then $\xi((s + \widehat{U}_3) \cap M) \in \tilde{\Phi}_s$ and

$$\xi((s + \widehat{U}_3) \cap M) \subseteq \xi((s + \widehat{U}_1) \cap M) \cap \xi((s + \widehat{U}_2) \cap M).$$

Let $\epsilon_0 > 0$ and $V_0 = \{d \in \mathbb{R} \mid -\epsilon_0 < d < \epsilon_0\}$ be a neighborhood of zero in $\mathbb{R}$. Since the mapping $\xi : M \to \mathbb{R}$ is continuous at zero, then there exists a neighborhood U_0 of zero in M such that $\xi(u) < \epsilon_0$ for any $u \in U_0$. Let $\widehat{U}_0$ and $\widehat{U}_1$ be neighborhoods of zero in $(\widehat{M}, \widehat{\tau})$ such that $M \cap \widehat{U}_0 = U_0$ and $\widehat{U}_1 - \widehat{U}_1 \subseteq \widehat{U}_0$.

If $b_1, b_2 \in (s + \widehat{U}_1) \cap M$, then $b_1 = s + u_1$ and $b_2 = s + u_2$ for any $u_1, u_2 \in \widehat{U}_1$. In that case $b_1 - b_2 = u_1 - u_2 \in \widehat{U}_1 - \widehat{U}_2 \subseteq \widehat{U}_0$, and, as $b_1 - b_2 \in M$, then $b_1 - b_2 \in \widehat{U}_0 \cap M = U_0$. Hence, $\xi(b_1 - b_2) < \epsilon_0$. Therefore, $\left| \xi(b_1) - \xi(b_2) \right| \leq \xi(b_1 - b_2) < \epsilon_0$. This means that $\xi((s + \widehat{U}_1) \cap M) - \xi((s + \widehat{U}_1) \cap M) \subseteq V_0$.

Thus, we've proved that for any $s \in \widehat{M}$ the family $\tilde{\Phi}_s$ forms a basis of a certain Cauchy filter $\mathcal{F}_s$ in $\mathbb{R}$.

Since $\mathbb{R}$ is a locally compact group, then it is a complete group (see Proposition 3.2.3) and, hence, there exists an element $d \in \mathbb{R}$ for which $d = \lim \mathcal{F}_s$. Put $\widehat{\xi}(s) = d$.

Now we verify that $\widehat{\xi}(m) = \xi(m)$ for any $m \in M$. Assume the contrary, i.e. that $m \in M$ and $\widehat{\xi}(m) \neq \xi(m)$. Since the group $\mathbb{R}$ is Hausdorff, then there exists a neighborhood V_0 of zero in $\mathbb{R}$ such that $(\widehat{\xi}(m) + V_0) \cap (\xi(m) + V_0) = \emptyset$. As ξ is a continuous mapping from M to $\mathbb{R}$, then there exists a neighborhood U_0 of zero in M such that $\xi(m + U_0) \subseteq \xi(m) + V_0$.

Since $\widehat{\xi}(m) = \lim \mathcal{F}_m$, then there exists a set $F = \xi((m + \widehat{U}_1) \cap M)$ such that $F \subseteq \widehat{\xi}(m) + V_0$, i.e. $\xi((m + \widehat{U}_1) \cap M) \subseteq \widehat{\xi}(m) + V_0$. Since $m \in M$, then $(m + \widehat{U}_1) \cap M = m + (\widehat{U}_1 \cap M)$. Let U_2 be a neighborhood of zero in M such that $U_2 \subseteq U_0 \cap (\widehat{U}_1 \cap M)$ and $c \in m + U_2$, then $c \in m + U_0$ and $c \in (m + \widehat{U}_1) \cap M$. Hence, $\xi(c) \in \xi(m + U_0) \subseteq \xi(m) + V_0$ and $\xi(c) \in \xi((m + \widehat{U}_1) \cap M) \subseteq \widehat{\xi}(m) + V_0$, i.e.

$$\xi(c) \in \left(\xi(m) + V_0 \right) \cap \left(\widehat{\xi}(m) + V_0 \right).$$

This contradicts the choice of the neighborhood V_0. It proves that $\widehat{\xi}(m) = \xi(m)$ for all $m \in M$.

Now we prove that $\widehat{\xi}$ is a continuous mapping from $(\widehat{M}, \widehat{\tau})$ to $\mathbb{R}$.

Let $s_0 \in \widehat{M}$ and $\epsilon > 0$. Since ξ is a continuous mapping from M to $\mathbb{R}$, then there exists a neighborhood U_0 of zero in M such that $\xi(m) < \frac{\epsilon}{3}$ for all $m \in U_0$. We can choose neighborhoods of zero $\widehat{U}_0$ and $\widehat{U}_1$ in $(\widehat{M}, \widehat{\tau})$ so that $U_0 = M \cap \widehat{U}_0$ and $\widehat{U}_1 + \widehat{U}_1 + \widehat{U}_1 \subseteq \widehat{U}_0$. Let us prove that $\left| \widehat{\xi}(s) - \widehat{\xi}(s_0) \right| < \epsilon$ for all $s \in s_0 + \widehat{U}_1$. Assume the contrary, i.e.

that there exists an element $s_1 \in s_0 + \widehat{U}_1$ such that $|\widehat{\xi}(s_1) - \widehat{\xi}(s_0)| > \epsilon$. Let $V_0 = \{d \in \mathbb{R} \mid |d - \widehat{\xi}(s_0)| < \frac{\epsilon}{3}\}$ and $V_1 = \{d \in \mathbb{R} \mid |d - \widehat{\xi}(s_1)| < \frac{\epsilon}{3}\}$, then V_0 and V_1 are neighborhoods of the elements $\widehat{\xi}(s_0)$ and $\widehat{\xi}(s_1)$ in $\mathbb{R}$ correspondingly. Since $\widehat{\xi}(s_i) = \lim \mathcal{F}_{s_i}$ for $i = 0, 1$, then there exists a symmetric neighborhood $\widehat{U}_2$ of zero in $(\widehat{M}, \widehat{\tau})$ such that $\widehat{U}_2 \subseteq \widehat{U}_1$ and $\xi((s_i + \widehat{U}_2) \cap M) \subseteq V_i$ for $i = 0, 1$. Choose elements $m_0 \in (s_0 + \widehat{U}_2) \cap M$ and $m_1 \in (s_1 + \widehat{U}_2) \cap M$. Then $m_0 = s_0 + u_0$, and $m_1 = s_1 + u_1$ for some $u_0, u_1 \in \widehat{U}_2$, while

$$m_1 - m_0 = s_1 + u_1 - s_0 + u_0 = (s_1 - s_0) + u_1 - u_0$$
$$\in \widehat{U}_1 + \widehat{U}_2 - \widehat{U}_2 \subseteq \widehat{U}_1 + \widehat{U}_1 + \widehat{U}_1 \subseteq \widehat{U}_0.$$

Since $m_1 - m_0 \in M$, then $m_1 - m_0 \in U_0$. From the definition of the neighborhood U_0 in M follows that $\xi(m_1 - m_0) < \frac{\epsilon}{3}$. In that case:

$$|\widehat{\xi}(s_1) - \widehat{\xi}(s_0)| = |\widehat{\xi}(s_1) - \xi(s_1 + u_1) + \xi(s_1 + u_1) - \xi(s_0 + u_0) + \xi(s_0 + u_0) - \widehat{\xi}(s_0)|$$
$$\leq |\widehat{\xi}(s_1) - \xi(s_1 + u_1)| + |\xi(s_1 + u_1) - \xi(s_0 + u_0)| + |\xi(s_0 + u_0) - \widehat{\xi}(s_0)|.$$

Since $\xi(s_i + u_i) \in \xi((s_i + \widehat{U}_2) \cap M) \subseteq V_i$ for $i = 0, 1$, then from the definition of the neighborhoods V_i follows that $|\widehat{\xi}(s_i) - \xi(s_i + u_i)| < \frac{\epsilon}{3}$ for $i = 0, 1$. Since

$$|\xi(s_1 + u_1) - \xi(s_0 + u_0)| \leq \xi(s_1 + u_1 - s_0 - u_0) = \xi(m_1 - m_0) < \frac{\epsilon}{3},$$

then

$$|\widehat{\xi}(s_1) - \widehat{\xi}(s_0)| < \frac{\epsilon}{3} + \frac{\epsilon}{3} + \frac{\epsilon}{3} = \epsilon.$$

This contradicts the choice of the element s_1. Hence, $|\xi(s) - \xi(s_0)| < \epsilon$ for all $s \in s_0 + \widehat{U}_1$.

Since the choice of $\epsilon > 0$ was arbitrary, then $\widehat{\xi}$ is a continuous mapping from the module $\widehat{M}$ to $\mathbb{R}$.

Now we verify that $\widehat{\xi}$ is a pseudonorm on the R-module $\widehat{M}$. To do it we must verify the non-negativity of the function $\widehat{\xi}$ and the fulfillment of the conditions (NM1) – (NM3) of Definition 2.3.1.

At first prove that $\widehat{\xi} \geq 0$. Assume the contrary, i.e. that $\widehat{\xi}(s) = -\epsilon < 0$ for a certain $s \in \widehat{M}$. Then $V_0 = \{d \in \mathbb{R} \mid |d - \widehat{\xi}(s)| < \frac{\epsilon}{2}\}$ is a neighborhood of the element $\widehat{\xi}(s)$ in $\mathbb{R}$, and $a < 0$ for any $a \in V_0$.

Since $\widehat{\xi} : \widehat{M} \to \mathbb{R}$ is continuous mapping, then there exists a neighborhood $\widehat{U}_0$ of zero in $(\widehat{M}, \widehat{\tau})$ such that $\widehat{\xi}(s + \widehat{U}_0) \subseteq V_0$. There exists an element $m \in (s + \widehat{U}_0) \cap M$. Since $\xi(m) = \widehat{\xi}(s + \widehat{U}_0) \subseteq V_0$, then $\xi(m) < 0$. It contradicts the fact that ξ is a pseudonorm on M. The contradiction proves that $\widehat{\xi}(s) \geq 0$ for any $s \in \widehat{M}$.

Now we verify the fulfillment of the condition (NM1). Indeed, $\widehat{\xi}(0) = \xi(0) = 0$ since $0 \in M$.

Assume that the second part of the condition (NM1) is not fulfilled, i.e. then there exists an element $0 \neq s \in \widehat{M}$ such that $\widehat{\xi}(s) = 0$. Since $\widehat{M}$ is a Hausdorff group, then there exists a neighborhood $\widehat{U}$ of zero in $(\widehat{M}, \widehat{\tau})$ such that $(s + \widehat{U}) \cap \widehat{U} = \emptyset$. Then $\widehat{U} \cap M$ is a neighborhood of zero in M. Since τ_ξ is a topology, determined by the pseudonorm ξ, then there exists a number $\epsilon > 0$ such that $U_\epsilon = \{m \in M \mid \xi(m) < \epsilon\} \subseteq \widehat{U} \cap M$. Since $\widehat{\xi}(s) = 0$, then, due to the extension of the mapping, there exists a neighborhood $\widehat{U}_1$ of zero in $(\widehat{M}, \widehat{\tau})$ such that $\widehat{\xi}(s + \widehat{U}_1) \subseteq V_1 = \{d \in \mathbb{R} \mid -\epsilon < d < \epsilon\}$. There exists a neighborhood $\widehat{U}_2$ of zero in $\widehat{M}$ such that $\widehat{U}_2 \subseteq \widehat{U}_1 \cap \widehat{U}$. Since $(s + \widehat{U}_2) \cap M \neq \emptyset$, then there exists an element $m_0 \in (s + \widehat{U}_2) \cap M$. In that case $\xi(m_0) = \widehat{\xi}(m_0) \in \widehat{\xi}(s + \widehat{U}_1) \subseteq V_1$, i.e. $\xi(m_0) < \epsilon$, hence, $m_0 \in \{m \in M \mid \xi(m) < \epsilon\} \subseteq \widehat{U} \cap M \subseteq \widehat{U}$. Therefore, $m_0 \in \widehat{U}$ and $m_0 \in s + \widehat{U}_2 \subseteq s + \widehat{U}$, i.e. $m_0 \in \widehat{U} \cap (s + \widehat{U})$. This contradicts the choice of the neighborhood $\widehat{U}$. This contradiction proves that if $s \neq 0$, then $\xi(s) \neq 0$.

Now we'll verify the fulfillment of the condition (NM2). Assume the contrary, i.e. that there exist elements $s_1, s_2 \in \widehat{M}$ such that $\widehat{\xi}(s_1 - s_2) > \widehat{\xi}(s_1) + \widehat{\xi}(s_2)$. Since the group $\mathbb{R}$ is Hausdorff, then there exists a neighborhood V_0 of zero in $\mathbb{R}$ such that $\widehat{\xi}(s_1 - s_2) + v_0 > \widehat{\xi}(s_1) + v_1 + \widehat{\xi}(s_2) + v_2$ for any $v_0, v_1, v_2 \in V_0$. Since $\widehat{\xi}$ is continuous mapping, then there exists a neighborhood $\widehat{U}_0$ of zero in $(\widehat{M}, \widehat{\tau})$ such that $\widehat{\xi}(s_1 - s_2 + \widehat{U}_0) \subseteq \widehat{\xi}(s_1 - s_2) + V_0$, $\widehat{\xi}(s_1 + \widehat{U}_0) \subseteq \widehat{\xi}(s_1) + V_0$ and $\widehat{\xi}(s_2 + \widehat{U}_0) \subseteq \widehat{\xi}(s_2) + V_0$. Let $\widehat{U}_1$ be a neighborhood of zero in $(\widehat{M}, \widehat{\tau})$ such that $\widehat{U}_1 - \widehat{U}_1 \subseteq \widehat{U}_0$. Let $m_1 \in (s_1 + \widehat{U}_1) \cap M$ and $m_2 \in (s_2 + \widehat{U}_2) \cap M$, then $\widehat{\xi}(m_1) \in \widehat{\xi}(s_1) + V_0$ and $\widehat{\xi}(m_2) \in \widehat{\xi}(s_2) + V_0$. Since $m_1 - m_2 \in s_1 - s_2 + \widehat{U}_1 - \widehat{U}_2 \subseteq s_1 - s_2 + \widehat{U}_0$, then $\widehat{\xi}(m_1 - m_2) \in \widehat{\xi}(s_1 - s_2) + V_0$. From the definition of the neighborhood V_0 follows that

$$\widehat{\xi}(m_1 - m_2) > \widehat{\xi}(m_1) + \widehat{\xi}(m_2).$$

Since $m_1, m_2, m_1 - m_2 \in M$, then $\widehat{\xi}(m_1) = \xi(m_1)$, $\widehat{\xi}(m_2) = \xi(m_2)$ and $\widehat{\xi}(m_1 - m_2) =$

$\xi(m_1 - m_2)$. Hence, $\xi(m_1 - m_2) > \xi(m_1) + \xi(m_2)$. This contradicts the fact that ξ is a pseudonorm on M. The contradiction shows that the condition (NM2) for the mapping $\widehat{\xi}$ is fulfilled.

Now we'll verify the fulfillment of the condition (NM3). Assume the contrary, i.e. that there exist elements $r \in R$ and $s \in \widehat{M}$ such that $\widehat{\xi}(s \cdot r) > \widehat{\xi}(s) \cdot \eta(r)$. Since the group $\mathbb{R}$ is Hausdorff, then there exists a neighborhood V_0 of zero in $\mathbb{R}$ such that $d_1 > d_2$ for any $d_1 \in \widehat{\xi}(s \cdot r) + V_0$ and $d_2 \in \widehat{\xi}(s) \cdot \eta(r) + V_0$. Since $\mathbb{R}$ is a topological ring, then there exists a neighborhood V_1 of zero in $\mathbb{R}$ such that $(\widehat{\xi}(s) + V_1) \cdot \eta(r) \subseteq \widehat{\xi}(s) \cdot \eta(r) + V_0$. Since $\widehat{\xi}$ is a continuous mapping, then there exists a neighborhood $\widehat{U}_0$ of zero in $(\widehat{M}, \widehat{\tau})$ such that $\widehat{\xi}(s \cdot r + \widehat{U}_0) \subseteq \widehat{\xi}(s \cdot r) + V_0$, and $\widehat{\xi}(s + \widehat{U}_0) \subseteq \widehat{\xi}(s) + V_1$. Since $(\widehat{M}, \widehat{\tau})$ is a topological module over the topological ring (R, τ_η), then there exists a neighborhood $\bar{U}_1$ of zero in $(\widehat{M}, \widehat{\tau})$ such that $\widehat{U}_1 \subseteq \widehat{U}_0$ and $(s + U_1) \cdot r \subseteq s \cdot r + \widehat{U}_0$. Let $m \in (s + \widehat{U}_1) \bigcap M$, then $m \cdot r \in (s + \widehat{U}_1) \cdot r \subseteq s \cdot r + \widehat{U}_0$. Hence, $\widehat{\xi}(m \cdot r) \in \widehat{\xi}(s \cdot r) + V_0$ and $\widehat{\xi}(m) \in \widehat{\xi}(s) + V_1$, therefore, $\widehat{\xi}(m) \cdot \eta(r) \in (\widehat{\xi}(s) + V_1) \cdot \eta(r) \subseteq \widehat{\xi}(s) \cdot \eta(r) + V_0$. The choice of the neighborhood V_0 results that $\widehat{\xi}(m \cdot r) > \widehat{\xi}(m) \cdot \eta(r)$. Since $m \in M$ and $r \in R$, then $\xi(m \cdot r) = \widehat{\xi}(m \cdot r) > \widehat{\xi}(m) \cdot \eta(r) = \xi(m) \cdot \eta(r)$. This contradicts the fact that (M, ξ) is a pseudonormed (R, η)-module.

The contradiction proves that the mapping $\widehat{\xi}$ satisfies the condition (NM3).

Thus, we've proved that $\widehat{\xi}$ is a pseudonorm on the module $\widehat{M}$ over the pseudonormed ring (R, η).

Let $\widehat{\tau}_{\widehat{\xi}}$ be the topology on $\widehat{M}$ defined by the pseudonorm $\widehat{\xi}$ (see Remark 2.3.5).

Now we'll prove that $\widehat{\tau} = \tau_{\widehat{\xi}}$. Indeed, since $\widehat{\xi}$ is a continuous mapping, then for any number $\epsilon > 0$ the set $\{m \in \widehat{M} \mid \widehat{\xi}(m) < \epsilon\}$ is a neighborhood of zero in $(\widehat{M}, \widehat{\tau})$. Since $\tau_{\widehat{\xi}}$ and $\widehat{\tau}$ are group topologies on $\widehat{M}$, then $\tau_{\widehat{\xi}} \leq \widehat{\tau}$.

Now we'll prove that $\widehat{\tau} \leq \tau_{\widehat{\xi}}$. Let $\widehat{U}$ be a neighborhood of zero in $(\widehat{M}, \widehat{\tau})$, then there exists a neighborhood $\widehat{U}_1$ of zero in $(\widehat{M}, \widehat{\tau})$ such that $\widehat{U}_1 - \widehat{U}_1 \subseteq \widehat{U}$. Since $\widehat{U}_1 \bigcap M$ is a neighborhood of zero in (M, τ), then there exists a number $\epsilon > 0$ such that $\{m \in M \mid \xi(m) < \epsilon\} \subseteq \widehat{U}_1 \bigcap M$. The set $W = \{m \in \widehat{M} \mid \widehat{\xi}(m) < \epsilon\}$ is a neighborhood of zero in $(\widehat{M}, \widehat{\tau}_{\widehat{\xi}})$. Now we'll prove that $\widehat{W} \subseteq \widehat{U}$. Indeed, let $a \in \widehat{W}$. Then $\widehat{\xi}(a) < \epsilon$. Since $\widehat{\xi}(a) = \lim \mathcal{F}_a'$, then there exists a neighborhood $\widehat{U}_2$ of zero in $(\widehat{M}, \widehat{\tau})$ such that $\widehat{U}_2 \subseteq \widehat{U}_1$ and $\xi((a + \widehat{U}_2) \bigcap M) \subseteq (-\epsilon; \epsilon)$, hence, $(a + \widehat{U}_2) \bigcap M \subseteq \widehat{U}_1 \bigcap M$. Let $m \in (a + \widehat{U}_2) \bigcap M$, then $m = a + u$ for a certain element $u \in \widehat{U}_2$ and $m \in \widehat{U}_1 \bigcap M$. Then $a = m - u \in \widehat{U}_1 - \widehat{U}_2 \subseteq \widehat{U}_1 - \widehat{U}_1 \subseteq \widehat{U}$. Since the choice of the element $a \in \widehat{W}$ was arbitrary, then $\widehat{W} \subseteq \widehat{U}$. Since $\tau_{\widehat{\xi}}$ and $\widehat{\tau}$ are group topologies on $\widehat{M}$, then $\widehat{\tau} \leq \tau_{\widehat{\xi}}$. Hence, $\widehat{\tau} = \tau_{\widehat{\xi}}$.

Thus, we only have to prove that if $\tilde{\xi}$ is such pseudonorm on $\widehat{M}$ that $\tau_{\tilde{\xi}} = \widehat{\tau}$ and $\tilde{\xi}(m) = \xi(m)$ for all $m \in M$, then $\tilde{\xi} = \widehat{\xi}$.

Assume the contrary, i.e. that $\tilde{\xi} \neq \widehat{\xi}$, and let s be an element of $\widehat{M}$ such that $\tilde{\xi}(s) \neq \widehat{\xi}(s)$. To be more definite, suppose that $\tilde{\xi}(s) < \widehat{\xi}(s)$. In that case there exists a positive number ϵ such that $\tilde{\xi}(s) + \epsilon < \widehat{\xi}(s) - \epsilon$. Since $\tau_{\tilde{\xi}} = \widehat{\tau} = \tau_{\widehat{\xi}}$, then the set $W = \{d \in \widehat{M} \mid \tilde{\xi}(d) < \tilde{\xi}(s) + \epsilon\} \bigcap \{d \in \widehat{M} \mid \widehat{\xi}(d) > \widehat{\xi}(s) - \epsilon\}$ is a neighborhood of the element s in $(\widehat{M}, \widehat{\tau})$, hence, $W \bigcap M \neq \emptyset$. If $m \in W \bigcap M$, then

$$\xi(m) = \tilde{\xi}(m) < \tilde{\xi}(s) + \epsilon < \widehat{\xi}(s) - \epsilon < \widehat{\xi}(m) = \xi(m).$$

The contradiction obtained completes the proof.

3.2.37. THEOREM. *Let* (R, ξ) *be a pseudonormed ring,* τ_ξ *a topology on R defined by the pseudonorm ξ. Let* $(\widehat{R}, \widehat{\tau})$ *be a Hausdorff topological ring which contains* (R, τ_ξ) *as a dense subring. Then there exists only one pseudonorm $\widehat{\xi}$ on $\widehat{R}$ such that $\widehat{\tau} = \tau_{\widehat{\xi}}$ and $\xi(r) = \widehat{\xi}(r)$ for all $r \in R$. Moreover, if ξ is a norm, then $\widehat{\xi}$ is a norm, too.*

PROOF. It is clear that $(R(+), \xi)$ is a pseudonormed R-module over the ring (R, ξ). We can consider $(\widehat{R}(+), \widehat{\tau})$ as a topological $\widehat{R}$-module and we can qualify it as a topological module over the topological ring (R, τ_ξ) containing $(R(+), \tau_\xi)$ as a dense R-submodule. Due to Theorem 3.2.36, there exists only one pseudonorm $\widehat{\xi}$ on the R-module $\widehat{R}(+)$ such that $\tau_{\widehat{\xi}} = \widehat{\tau}$ and $\widehat{\xi}(r) = \xi(r)$ for all $r \in R$.

Now we prove that $\widehat{\xi}$ is a pseudonorm on the ring $\widehat{R}$. The fulfillment for $\widehat{\xi}$ of the conditions (NR1) and (NR2) of Definition 1.1.12 follows from the fact that $(\widehat{R}, \widehat{\xi})$ is a pseudonormed module.

We verify the fulfillment of the condition (NR3) for $\widehat{\xi}$. Assume the contrary, i.e. that there exist elements $s_1, s_2 \in \widehat{R}$ such that $\widehat{\xi}(s_1 \cdot s_2) > \widehat{\xi}(s_1) \cdot \widehat{\xi}(s_2)$. Since the additive group $\mathbb{R}(+)$ is Hausdorff, then there exists a neighborhood V_0 of zero in $\mathbb{R}$ such that $d_1 > d_2$ for any $d_1 \in \widehat{\xi}(s_1 \cdot s_2) + V_0$ and $d_2 \in \widehat{\xi}(s_1) \cdot \widehat{\xi}(s_2) + V_0$. Since $\mathbb{R}$ is a topological ring, then there exists a neighborhood V_1 of zero in $\mathbb{R}$ such that $(\widehat{\xi}(s_1) + V_1) \cdot (\widehat{\xi}(s_2) + V_1) \subseteq \widehat{\xi}(s_1) \cdot \widehat{\xi}(s_2) + V_0$.

Since $\widehat{\xi}$ is a continuous mapping, then there exists a neighborhood $\widehat{U}_0$ of zero in $\widehat{R}$ such that:

$$\widehat{\xi}(s_1 \cdot s_2 + \widehat{U}_0) \subseteq \widehat{\xi}(s_1 \cdot s_2) + V_0;$$

$$\widehat{\xi}(s_1 + \widehat{U}_0) \subseteq \widehat{\xi}(s_1) + V_1;$$

$$\widehat{\xi}(s_2 + \widehat{U}_0) \subseteq \widehat{\xi}(s_2) + V_1.$$

Since $\widehat{R}$ is a topological ring, then there exists a neighborhood $\widehat{U}_1 \subseteq \widehat{U}_0$ of zero in $\widehat{R}$ such that $(s_1 + \widehat{U}_1) \cdot (s_2 + \widehat{U}_1) \subseteq s_1 \cdot s_2 + \widehat{U}_0$.

Let $r_1 \in (s_1 + \widehat{U}_1) \bigcap R$ and $r_2 \in (s_2 + \widehat{U}_1) \bigcap R$, then $r_1 \cdot r_2 \in s_1 \cdot s_2 + \widehat{U}_0$. Further, $\widehat{\xi}(r_1 \cdot r_2) \in \widehat{\xi}(s_1 \cdot s_2) + V_0$ and $\widehat{\xi}(r_i) \in \widehat{\xi}(s_i) + V_1$ for $i = 1,2$, hence, $\widehat{\xi}(r_1) \cdot \widehat{\xi}(r_2) \in (\widehat{\xi}(s_1) + V_1) \cdot (\widehat{\xi}(s_2) + V_1) \subseteq \widehat{\xi}(s_1) \cdot \widehat{\xi}(s_2) + V_0$. From the definition of the neighborhood V_0 it follows that $\widehat{\xi}(r_1 \cdot r_2) > \widehat{\xi}(r_1) \cdot \widehat{\xi}(r_2)$. Since $r_1, r_2, r_1 \cdot r_2 \in R$, then $\xi(r_1 \cdot r_2) = \widehat{\xi}(r_1 \cdot r_2) > \widehat{\xi}(r_1) \cdot \widehat{\xi}(r_2) = \xi(r_1) \cdot \xi(r_2)$. This contradicts the fact that ξ is a pseudonorm on R. This contradiction proves that the mapping $\widehat{\xi}$ satisfies the condition (NR3). Thus, we've proved that $\widehat{\xi}$ is a pseudonorm on the ring $\widehat{R}$.

Now we'll show that if ξ is a norm on R, then $\widehat{\xi}$ is a norm on $\widehat{R}$. Assume the contrary, i.e. that $\widehat{\xi}(a \cdot b) < \widehat{\xi}(a) \cdot \widehat{\xi}(b)$ for some $a, b \in \widehat{R}$. There exists a positive number ϵ such that $\widehat{\xi}(a \cdot b) + \epsilon < \widehat{\xi}(a) \cdot \widehat{\xi}(b) - \epsilon$. It is clear that the sets $\{r \in R \mid r < \widehat{\xi}(a \cdot b) + \epsilon\}$ and $\{r \in R \mid r > \widehat{\xi}(a) \cdot \widehat{\xi}(b) - \epsilon\}$ are open in $\mathbb{R}$, hence, they are neighborhoods of the numbers $\widehat{\xi}(a \cdot b)$ and $\widehat{\xi}(a) \cdot \widehat{\xi}(b)$ correspondingly in $\mathbb{R}$.

Since $\mathbb{R}$ is a topological ring, then there exist neighborhoods V_0 and V_1 of the numbers $\widehat{\xi}(a)$ and $\widehat{\xi}(b)$ correspondingly in $\mathbb{R}$ such that $V_0 \cdot V_1 \subseteq \{r \in R \mid \widehat{\xi}(a) \cdot \widehat{\xi}(b) - \epsilon\}$.

Since $\widehat{\xi}$ is a continuous mapping of the ring $\widehat{R}$ to $\mathbb{R}$, then there exist neighborhoods U, U_0, U_1 of the elements $a \cdot b, a, b$ correspondingly in $\widehat{R}$ such that $\widehat{\xi}(U) \subseteq \{r \in \mathbb{R} \mid r < \widehat{\xi}(a \cdot b) + \epsilon\}$ and $\widehat{\xi}(U_0) \subseteq V_0$, $\widehat{\xi}(U_1) \subseteq V_1$. As $\widehat{R}$ is a topological ring, then there exist neighborhoods U_0' and U_1' of the elements a and b correspondingly in $\widehat{R}$ such that $U_0' \cdot U_1' \subseteq U$. Since R is a dense subset in $\widehat{R}$, then $(U_0 \bigcap U_0') \bigcap R \neq \emptyset$ and $(U_1 \bigcap U_1') \bigcap R \neq \emptyset$. Let $c \in (U_0 \bigcap U_0') \bigcap R$ and $d \in (U_1 \bigcap U_1') \bigcap R$. Then $\xi(c \cdot d) = \widehat{\xi}(c \cdot d) \in \widehat{\xi}(U_0' \cdot U_1') \subseteq \widehat{\xi}(U)$, hence, $\xi(c \cdot d) < \widehat{\xi}(a \cdot b) + \epsilon$. On the other hand, since ξ is a norm on R, then $\xi(c \cdot d) = \xi(c) \cdot \xi(d) = \widehat{\xi}(c) \cdot \widehat{\xi}(d) \in \widehat{\xi}(U_0) \cdot \widehat{\xi}(U_1) \subseteq V_0 \cdot V_1 \subseteq \{r \in \mathbb{R} \mid r > \widehat{\xi}(a) \cdot \widehat{\xi}(b) - \epsilon\}$, hence,

$\xi(c \cdot d) \geq \widehat{\xi}(a) \cdot \widehat{\xi}(b) - \epsilon$. Comparing the two inequalities, we get $\widehat{\xi}(a) \cdot \widehat{\xi}(b) - \epsilon < \widehat{\xi}(a \cdot b) + \epsilon$. This contradicts the choice of the number ϵ. Hence, $\widehat{\xi}$ is a norm.

3.2.38. THEOREM. *Let (R, ξ) be a pseudonormed ring, (M, η) a pseudonormed (R, ξ)-module, τ_ξ and τ_η the topologies on R and M defined by the pseudonorms ξ and η correspondingly. Let $(\widehat{M}, \widehat{\tau}_0)$ be a complete topological module over the topological ring (R, τ_ξ) containing (M, τ_η) as a dense submodule, $\widehat{\eta}$ a pseudonorm on the R-module $\widehat{M}$, constructed in Theorem 3.2.36, $(\widehat{R}, \widehat{\tau})$ a topological ring such that (R, τ_ξ) is its dense subring, $\widehat{\xi}$ a pseudonorm on the ring $\widehat{R}$, constructed in Theorem 3.2.37. Consider, according to Theorem 3.2.31, $\widehat{M}$ as a module over the ring $\widehat{R}$. Then:*

1) $(\widehat{M}, \widehat{\eta})$ is a pseudonormed $(\widehat{R}, \widehat{\xi})$-module;

2) if ξ and η are norms, then $\widehat{\xi}$ and $\widehat{\eta}$ are norms, too.

PROOF. Since $\widehat{\eta}$ is a pseudonorm on the R-module $\widehat{M}$, then the conditions (NM1) and (NM2) of Definition 2.3.1 are fulfilled for it.

Now we'll verify the fulfillment of the condition (NM3) of this definition for the $\widehat{R}$-module $\widehat{M}$. Assume the contrary, i.e. that there exist elements $m \in \widehat{M}$ and $\widehat{r} \in \widehat{R}$ such that $\widehat{\eta}(m \cdot \widehat{r}) > \widehat{\eta}(m) \cdot \widehat{\xi}(\widehat{r})$. Since the additive group $\mathbb{R}(+)$ is Hausdorff, then there exists a neighborhood V_0 of zero in $\mathbb{R}$ such that $d_1 > d_2$ for any $d_1 \in \widehat{\eta}(m \cdot \widehat{r}) + V_0$ and $d_2 \in \widehat{\eta}(m) \cdot \widehat{\xi}(\widehat{r}) + V_0$. As $\mathbb{R}$ is a topological ring, then there exists a neighborhood V_1 of zero in $\mathbb{R}$ such that $\widehat{\eta}(m) \cdot (\widehat{\xi}(\widehat{r}) + V_1) \subseteq \widehat{\eta}(m) \cdot \widehat{\xi}(\widehat{r}) + V_0$. Since $\widehat{\eta}$ and $\widehat{\xi}$ are continuous mappings, then there exist a neighborhood $\widehat{U}_0$ of zero in $(\widehat{M}, \widehat{\tau}_0)$ and a neighborhood $\widehat{W}_0$ of zero in $(\widehat{R}, \widehat{\tau})$ such that $\widehat{\eta}(m \cdot \widehat{r} + U_0) \subseteq \widehat{\eta}(m \cdot \widehat{r}) + V_0$, and $\widehat{\xi}(\widehat{r} + \widehat{W}_0) \subseteq \widehat{\xi}(\widehat{r}) + V_1$.

Since, due to Theorem 3.2.31, $(\widehat{M}, \widehat{\tau}_0)$ is a topological module over the topological ring $(\widehat{R}, \widehat{\tau})$, then there exists a neighborhood $\widehat{W}_1$ of zero in $\widehat{R}$ such that $\widehat{W}_1 \subseteq W_0$ and $m \cdot (\widehat{r} + \widehat{W}_1) \subseteq m \cdot \widehat{r} + \widehat{U}_0$. Let $r \in (\widehat{r} + W_1) \bigcap R$, then $m \cdot r \in m \cdot (\widehat{r} + \widehat{W}_1) \subseteq m \cdot \widehat{r} + \widehat{U}_0$. Then $\widehat{\eta}(m \cdot r) \in \widehat{\eta}(m \cdot \widehat{r}) + V_0$ and $\widehat{\xi}(r) \in \widehat{\xi}(\widehat{r}) + V_1$, hence, $\widehat{\eta}(m) \cdot \widehat{\xi}(r) \in \widehat{\eta}(m) \cdot (\widehat{\xi}(\widehat{r}) + V_1) \subseteq \widehat{\eta}(m) \cdot \widehat{\xi}(r) + V_0$. From the definition of the neighborhood V_0 it follows that $\widehat{\eta}(m \cdot r) > \widehat{\eta}(m) \cdot \widehat{\xi}(r)$. Since $r \in R$, then $\widehat{\xi}(r) = \xi(r)$, hence, $\widehat{\eta}(m \cdot r) > \widehat{\eta}(m) \cdot \xi(r)$. That contradicts the condition that $(\widehat{M}, \eta)$ is a pseudonormed module over (R, ξ). The contradiction proves that the mapping $\widehat{\eta}$ on the $\widehat{R}$-module $\widehat{M}$ satisfies the condition (NM3). Thus, we've proved that $\widehat{\eta}$ is a pseudonorm on the $\widehat{R}$-module $\widehat{M}$, i.e. 1) is fulfilled.

Let η and ξ be norms. Then, according to Theorem 3.2.37, $\widehat{\xi}$ is a norm on $\widehat{R}$. Now we'll prove that the pseudonorm $\widehat{\eta}$ is a norm on the $\widehat{R}$-module $\widehat{M}$. So it is enough to check that $\widehat{\eta}(m \cdot \widehat{r}) = \widehat{\eta}(m) \cdot \widehat{\xi}(\widehat{r})$ for all elements $m \in \widehat{M}, \widehat{r} \in \widehat{R}$. Assume the contrary, i.e. that $\widehat{\eta}(m \cdot \widehat{r}) < \widehat{\eta}(m) \cdot \widehat{\xi}(r)$ for some $m \in \widehat{M}, r \in \widehat{R}$. There exists a positive number ϵ that $\widehat{\eta}(m \cdot \widehat{r}) + \epsilon < \widehat{\eta}(m) \cdot \widehat{\xi}(\widehat{r}) - \epsilon$. It is clear that the sets $\{a \in \mathbb{R} \mid a < \widehat{\eta}(m \cdot \widehat{r}) + \epsilon\}$ and $\{a \in \mathbb{R} \mid a > \widehat{\eta}(m) \cdot \widehat{\xi}(\widehat{r}) - \epsilon\}$ are open in $\mathbb{R}$ and, hence, are neighborhoods of the numbers $\widehat{\eta}(m \cdot \widehat{r})$ and $\widehat{\eta}(m) \cdot \widehat{\xi}(\widehat{r})$ correspondingly in $\mathbb{R}$. Since $\mathbb{R}$ is a topological ring, then there exist neighborhoods V_0 and V_1 of the numbers $\widehat{\eta}(m)$ and $\widehat{\xi}(\widehat{r})$ correspondingly in $\mathbb{R}$ such that $V_0 \cdot V_1 \subseteq \{a \in \mathbb{R} \mid a > \widehat{\eta}(m) \cdot \widehat{\xi}(\widehat{r}) - \epsilon\}$. Since $\widehat{\eta}$ and $\widehat{\xi}$ are continuous mappings, then there exist neighborhoods U and U_0 of the elements $m \cdot \widehat{r}$ and m correspondingly in $\widehat{M}$ and a neighborhood W_0 of the element $\widehat{r}$ in $\widehat{R}$ such that $\widehat{\eta}(U_0) \subseteq V_0$, $\widehat{\eta}(U) \subseteq \{a \in R \mid |a < \widehat{\eta}(m \cdot \widehat{r}) + \epsilon\}$ and $\widehat{\xi}(W_0) \subseteq V_1$. Since $\widehat{M}$ is a topological $\widehat{R}$-module, then there exist a neighborhood U_1 of the element m in $\widehat{M}$ and a neighborhood W_1 of the element $\widehat{r}$ in $\widehat{R}$ such that $U_1 \cdot W_1 \subseteq U$. Since M is a dense submodule in $\widehat{M}$ and R is a dense subring in $\widehat{R}$, then $(U_0 \cap U_1) \cap M \neq \emptyset$ and $(W_0 \cap W_1) \cap R \neq \emptyset$. Let $m_0 \in (U_0 \cap U_1) \cap M$ and $r_0 \in (W_0 \cap W_1) \cap R$. Then $\eta(m_0 \cdot r_0) = \widehat{\eta}(m_0 \cdot r_0) \in \widehat{\eta}(U_1 \cdot W_1) \subseteq \widehat{\eta}(U)$, hence, $\eta(m_0 \cdot r_0) < \widehat{\eta}(m \cdot \widehat{r}) + \epsilon$. On the other hand, since (M, η) is a normed module over the ring (R, ξ), then $\eta(m_0 \cdot r_0) = \eta(m_0) \cdot \xi(r_0) = \widehat{\eta}(m_0) \cdot \widehat{\xi}(r_0) \in \widehat{\eta}(U_0) \cdot \widehat{\xi}(W_0) \subseteq V_0 \cdot V_1$, hence, $\eta(m_0 \cdot r_0) > \widehat{\eta}(m) \cdot \widehat{\xi}(\widehat{r}) - \epsilon$. Comparing the two inequalities, we see that $\widehat{\eta}(m) \cdot \widehat{\xi}(r) - \epsilon < \widehat{\eta}(m \cdot \widehat{r}) + \epsilon$. This contradicts the choice of the number ϵ. Hence, $\widehat{\eta}$ is a norm on the $\widehat{R}$-module $\widehat{M}$.

§ 3.3. The Completion of Topological Abelian Groups, Rings and Modules

3.3.1. PROPOSITION. *For any topological Abelian group* (X, τ) *there exists a complete topological group* $(\tilde{X}, \tilde{\tau})$ *such that* (X, τ) *is a dense subgroup of* $(\tilde{X}, \tilde{\tau})$.

PROOF. We'll construct such topological Abelian group $(\tilde{X}, \tilde{\tau})$. Let $\mathcal{B}(X) = \{U_\gamma | \gamma \in \Gamma\}$ be the system of all neighborhoods of zero of the group (X, τ), Γ a direction defined by the filter $\mathcal{B}(X)$. Consider the Abelian group $\tilde{X}$ of all Cauchy sequences $\{x_\gamma | \gamma \in \Gamma\}$ by the direction Γ (see Corollary 3.1.39). For any neighborhood U_{γ_0} of zero in X consider in $\tilde{X}$ the subset $P_{\gamma_0} = \left\{ \bar{x} \, \middle| \, \bar{x} = \{x_\gamma \, | \, \gamma \in \Gamma\} \in \tilde{X}, \text{ there exists an element } \gamma_1 \in \Gamma \text{ such that } x_\gamma \in U_{\gamma_0} \text{ for all } \gamma \geq \gamma_1 \right\}$.

Note that if $\gamma_1 \neq \gamma_2$, then $P_{\gamma_1} \neq P_{\gamma_2}$. Indeed, $U_{\gamma_1} \neq U_{\gamma_2}$. Let, for example, $U_{\gamma_1} \not\subseteq U_{\gamma_2}$ and $x \in U_{\gamma_1} \backslash U_{\gamma_2}$. If $\bar{x} = \{x_\gamma \, | \, \gamma \in \Gamma\}$, where $x_\gamma = x$ for all $\gamma \in \Gamma$, then $\bar{x} \in P_{\gamma_1}$ and $\bar{x} \notin P_{\gamma_2}$. Hence, $P_{\gamma_1} \neq P_{\gamma_2}$.

Now we prove that the conditions (BN1) – (BN4) of the Proposition 1.2.1 are fulfilled for the system $\mathcal{P}(\tilde{X}) = \{P_\gamma \, | \, \gamma \in \Gamma\}$.

The sequence $\bar{0} = \{x_\gamma \, | \, \gamma \in \Gamma\}$, where $x_\gamma = 0$ for all $\gamma \in \Gamma$, is a neutral element of the group $\tilde{X}$. Since $0 \in U_\gamma$ for any $U_\gamma \in \mathcal{B}(X)$, then $\bar{0} \in P_\gamma$ for any $P_\gamma \in \mathcal{P}(\tilde{X})$. Hence, we've proved that the condition (BN1) is fulfilled.

Let $P_{\gamma_1}, P_{\gamma_2} \in \mathcal{P}(\tilde{X})$. There exists $U_{\gamma_3} \in \mathcal{B}(X)$ such that $U_{\gamma_3} = U_{\gamma_1} \cap U_{\gamma_2}$. Then from the definition of the sets P_γ it follows that $P_{\gamma_3} \subseteq P_{\gamma_2} \cap P_{\gamma_1}$, i.e. the condition (BN2) is also fulfilled.

Let $P_{\gamma_0} \in \mathcal{P}(\tilde{X})$. There exists $U_{\gamma_1} \in \mathcal{B}(X)$ such that $U_{\gamma_1} + U_{\gamma_1} \subseteq U_{\gamma_0}$. Now we verify that $P_{\gamma_1} + P_{\gamma_1} \subseteq P_{\gamma_0}$. Indeed, let $\{x_\gamma \, | \, \gamma \in \Gamma\} \in P_{\gamma_1}$ and $\{y_\gamma \, | \, \gamma \in \Gamma\} \in P_{\gamma_1}$. Then there exist elements γ_2 and γ_3 from Γ such that $x_\gamma \in U_{\gamma_1}$ for all $\gamma \geq \gamma_2$ and $y_\gamma \in U_{\gamma_1}$ for all $\gamma \geq \gamma_3$. Also there exists an element $\gamma_4 \in \Gamma$ such that $\gamma_4 \geq \gamma_2$ and $\gamma_4 \geq \gamma_3$. Then $x_\gamma, y_\gamma \in U_{\gamma_1}$ for all $\gamma \in \Gamma$, $\gamma \geq \gamma_4$. Hence, $x_\gamma + y_\gamma \in U_{\gamma_1} + U_{\gamma_1} \subseteq U_{\gamma_0}$ for all $\gamma \geq \gamma_4$, i.e. $\{x_\gamma \, | \, \gamma \in \Gamma\} + \{y_\gamma \, | \, \gamma \in \Gamma\} \in P_{\gamma_0}$. Thus, $P_{\gamma_1} + P_{\gamma_1} \subseteq P_{\gamma_0}$, i.e. the condition (BN3) is fulfilled, too.

Let $P_{\gamma_0} \in \mathcal{P}(\tilde{X})$ and U_{γ_1} be a neighborhood of zero in X such that $-U_{\gamma_1} \subseteq U_{\gamma_0}$, then, according to the definition of the sets $\mathcal{P}_\gamma$, $-\mathcal{P}_{\gamma_1} \subseteq \mathcal{P}_{\gamma_0}$, i.e. the condition (BN4) is fulfilled.

Hence, the family $\{P_\gamma \mid \gamma \in \Gamma\}$ defines a certain group topology $\tilde{\tau}$ on the group $\tilde{X}$, where it is a basis of neighborhoods of zero.

Now we define the mapping $\psi : X \to \tilde{X}$ in another way: $\psi(x) = \{x_\gamma \mid \gamma \in \Gamma\}$, where $x_\gamma = x$ for all $\gamma \in \Gamma$. It is easy to see that ψ is a homomorphism of the Abelian groups.

Let's prove that ψ is a monomorphism. Assume the contrary, i.e. that there exists an element $0 \neq a \in X$ such that $\psi(a) = 0$. Since $\psi(a) = \{x_\gamma \mid x_\gamma = a, \gamma \in \Gamma\}$ and the neutral element of the group $\tilde{X}$ is the sequence $\{y_\gamma \mid y_\gamma = 0, \gamma \in \Gamma\}$, then $a = 0$. The contradiction obtained proves that ψ is a monomorphism.

It is clear that $\psi(U_\gamma) = \mathcal{P}_\gamma \cap \psi(X)$ for any $\gamma \in \Gamma$. Since the system $\{\mathcal{P}_\gamma \cap \psi(X) \mid \gamma \in \Gamma\}$ forms a basis of neighborhoods of zero in $\psi(X)$, then, according to Proposition 1.5.5, ψ is a topological homomorphism of the groups (X, τ) and $\left(\psi(X), \tilde{\tau}\big|_{\psi(X)}\right)$. It means that the group (X, τ) is topologically isomorphic to the subgroup $\psi(X)$ of the topological group $(\tilde{X}, \tilde{\tau})$.

Let's prove that the subgroup $\psi(X)$ is a dense subgroup of $(\tilde{X}, \tilde{\tau})$. Indeed, let $\bar{z} = \{z_\gamma \mid \gamma \in \Gamma\} \in \tilde{X}$ and $\mathcal{P}_{\gamma_0}$ a neighborhood of zero from the basis of neighborhoods $\mathcal{P}(\tilde{X})$ of zero of the group $\tilde{X}$. Since $\bar{z}$ is a Cauchy sequence in X, then there exists an element $\gamma_1 \in \Gamma$ such that $z_\gamma - z_{\gamma_1} \in U_{\gamma_0}$ for all $\gamma \geq \gamma_1$. Then $\bar{x} = \{x_\gamma \mid x_\gamma = z_{\gamma_1}, \gamma \in \Gamma\} \in \psi(X)$, while $z_\gamma - x_\gamma = z_\gamma - z_{\gamma_1} \in U_{\gamma_0}$ for all $\gamma \geq \gamma_1$. Then, according to the definition of the neighborhood $\mathcal{P}_{\gamma_0}$ of zero in $\tilde{X}$, $\bar{z} - \bar{x} \in \mathcal{P}_{\gamma_0}$, i.e. $\bar{x} \in \bar{z} + \mathcal{P}_{\gamma_0}$. Since the choice of the element $\bar{z}$ and of the neighborhood $\mathcal{P}_{\gamma_0}$ was arbitrary, then $\psi(X)$ is a dense subgroup of the group $\tilde{X}$.

Now we'll verify that the topological group $(\tilde{X}, \tilde{\tau})$ is complete.

At first we show that the basis of the neighborhoods $\mathcal{P}(\tilde{X}) = \{\mathcal{P}_\gamma \mid \gamma \in \Gamma\}$ of zero of the group $(\tilde{X}, \tilde{\tau})$ defines the direction, coinciding with the direction Γ. Indeed, the index set in $\mathcal{B}(X)$ and in $\mathcal{P}(\tilde{X})$ is the same, and, as we saw it above, $\mathcal{P}_{\gamma_1} \neq \mathcal{P}_{\gamma_2}$ for $\gamma_1 \neq \gamma_2$. It is clear, that from $U_{\gamma_1} \subseteq U_{\gamma_2}$ follows that $\mathcal{P}_{\gamma_1} \subseteq \mathcal{P}_{\gamma_2}$. If $\mathcal{P}_{\gamma_1} \subseteq \mathcal{P}_{\gamma_2}$, then for any $x \in U_{\gamma_1}$ we get that $\bar{x} = \{x_\gamma \mid \gamma \in \Gamma\} \in \mathcal{P}_{\gamma_1}$, where $x_\gamma = x$ for all $\gamma \in \Gamma$, hence, $\bar{x} \in \mathcal{P}_{\gamma_2}$. According to the definition of $\mathcal{P}_\gamma$, we get that $x \in U_{\gamma_2}$. Therefore, from $\mathcal{P}_{\gamma_1} \subseteq \mathcal{P}_{\gamma_2}$ follows that $U_{\gamma_1} \subseteq U_{\gamma_2}$, hence, the orders on Γ, defined by $\mathcal{B}(X)$ and $\mathcal{P}(\tilde{X})$, coincide.

According to Proposition 3.2.24, it is enough to check that any Cauchy sequence $\{\bar{x}_\gamma \,|\, \gamma \in \Gamma\}$ of the elements from $\psi(X)$ has the limit by the direction Γ in $(\tilde{X}, \tilde{\tau})$. Let $\{x_\gamma \,|\, \gamma \in \Gamma\}$ be a sequence of elements of X such that $\bar{x}_\gamma = \psi(x_\gamma)$ for $\gamma \in \Gamma$, then it is a Cauchy sequence by the direction Γ in (X, τ), i.e. $\bar{x} = \{x_\gamma \,|\, \gamma \in \Gamma\} \in \tilde{X}$.

Let us check that $\bar{x} = \lim_{\gamma \in \Gamma} \bar{x}_\gamma$ in $(\tilde{X}, \tilde{\tau})$. Indeed, let $\mathcal{P}$ be a neighborhood of zero in $(\tilde{X}, \tilde{\tau})$ and γ_0 be an element of Γ such that $\mathcal{P}_{\gamma_0} \subseteq \mathcal{P}$. Since $\{x_\gamma \,|\, \gamma \in \Gamma\}$ is a Cauchy sequence in (X, τ), then there exists $\gamma_1 \in \Gamma$ such that $x_\gamma - x_\delta \in U_{\gamma_0}$ for $\gamma, \delta \in \Gamma$, $\gamma \geq \gamma_1$, $\delta \geq \gamma_1$ (see Remark 3.1.18). Since $\bar{x} - \bar{x}_\sigma = \bar{x} - \psi(x_\sigma) = \{x_\gamma - x_\sigma \,|\, \gamma \in \Gamma\}$ for any $\sigma \in \Gamma$, then $\bar{x} - \psi(x_\delta) \in \mathcal{P}_{\gamma_0} \in \mathcal{P}$ for any $\delta \in \Gamma$, $\delta > \gamma_1$. Hence, $\bar{x} = \lim_{\gamma \in \Gamma} \bar{x}_\gamma$, i.e. $(\tilde{X}, \tilde{\tau})$ is a complete group.

Since we identified the group X with the topologically isomorphic subgroup $\psi(X)$ of $\tilde{X}$, then we get the necessary result. The proof is completed.

3.3.2. THEOREM. *Let X be a Hausdorff topological Abelian group. Then:*

1) there exists a Hausdorff complete topological group $\widehat{X}$ such that X is a dense subgroup in $\widehat{X}$;

2) if X' is a Hausdorff complete topological group containing X as a dense subgroup, then there exists a topological isomorphism $\xi : \widehat{X} \to X'$ such that $\xi(x) = x$ for all $x \in X$.

PROOF. Let $\tilde{X}$ be a complete topological group, constructed in Proposition 3.3.1, which contains the group X as a dense subgroup. Let H be the smallest closed subgroup of the group $\tilde{X}$, then, since the group X is Hausdorff, $X \cap H = \{0\}$. Then, according to Proposition 1.5.22, the group X is topologically isomorphic to the subgroup $(X + H)/H$ of the topological group $\tilde{X}/H = \widehat{X}$. Since X is a dense subgroup in $\tilde{X}$, then $(X + H)/H$ is a dense subgroup in $\widehat{X}$. According to Proposition 3.2.19, the topological group $\widehat{X}$ is complete. Let us identify the groups $(X + H)/H$ and X. Then 1) is proved.

Due to Proposition 3.2.28, any Hausdorff complete topological Abelian group X' that contains a Hausdorff topological Abelian group X as a dense subgroup is topologically isomorphic to the constructed Hausdorff complete topological Abelian group $\widehat{X} = \tilde{X}/H$, and this isomorphism is identical on the group X.

This completes the proof of the theorem.

3.3.3. DEFINITION. Let X be a Hausdorff topological Abelian group. We'll call a completion and denote over $\hat{X}$ the Hausdorff complete topological Abelian group, constructed in Theorem 3.3.2 (unique up to topological isomorphism), which contains X as a dense subgroup.

3.3.4. PROPOSITION. *Let R be a topological ring. Then there exists a complete topological ring $\tilde{R}$ such that it contains R as a dense subring.*

PROOF. Let $\mathcal{B}(R) = \{U_\gamma \,|\, \gamma \in \Gamma\}$ be the system of all neighborhoods of zero in R and Γ the direction defined by the filter $\mathcal{B}(R)$. Let $\tilde{R}$ be the system of all Cauchy sequences of elements of R, which is a ring with respect to the operations defined in Proposition 3.1.38 and Remark 3.1.42.

According to Proposition 3.3.1, the topology defined on $\tilde{R}(+)$ by the system $\mathcal{P}(\tilde{R}) = \{\mathcal{P}_\gamma \,|\, \gamma \in \Gamma\}$, as a basis of neighborhoods of zero, is a group topology and $\tilde{R}(+)$ is a complete group in this topology.

Now we'll show that $\tilde{R}$ is a topological ring in this topology. For this it is enough to verify the continuity of the operation of multiplication on $\tilde{R}$.

Let $\bar{r} = \{r_\gamma \,|\, \gamma \in \Gamma\}$ and $\bar{s} = \{s_\gamma \,|\, \gamma \in \Gamma\} \in \tilde{R}$. Let $\mathcal{P}$ be a neighborhood of the element $\bar{r} \cdot \bar{s}$, then there exist indexes $\gamma_1, \gamma_2 \in \Gamma$ such that $\bar{r} \cdot \bar{s} + \mathcal{P}_{\gamma_1} \subseteq \mathcal{P}$ and $U_{\gamma_2} + U_{\gamma_2} + U_{\gamma_2} \cdot U_{\gamma_2} + U_{\gamma_2} \cdot U_{\gamma_2} + U_{\gamma_2} \cdot U_{\gamma_2} \subseteq U_{\gamma_1}$. Since $\bar{r}$ and $\bar{s}$ are Cauchy sequences in R, then there exists an index $\gamma_3 \in \Gamma$ such that $r_\gamma - r_{\gamma_3} \in U_{\gamma_2}$ and $s_\gamma - s_{\gamma_3} \in U_{\gamma_2}$ for any $\gamma \geq \gamma_3$. Since R is a topological ring, then there exists an element $\gamma_4 \in \Gamma$ such that $U_{\gamma_4} + U_{\gamma_4} \subseteq U_{\gamma_2}$, $r_{\gamma_3} \cdot U_{\gamma_4} \subseteq U_{\gamma_2}$ and $U_{\gamma_4} \cdot s_{\gamma_3} \in U_{\gamma_2}$. We can consider $\gamma_4 \geq \gamma_2$, hence $U_{\gamma_4} \subseteq U_{\gamma_2}$. We'll prove that $(\bar{r} + \mathcal{P}_{\gamma_4}) \cdot (\bar{s} + \mathcal{P}_{\gamma_4}) \subseteq \mathcal{P}$.

Indeed, let $\bar{r}' = \{r'_\gamma \,|\, \gamma \in \Gamma\}$, $\bar{s}' = \{s'_\gamma \,|\, \gamma \in \Gamma\} \in \mathcal{P}_{\gamma_4}$. Then there exists an element γ_5 of Γ such that $\gamma_5 \geq \gamma_3$, and $r'_\gamma, s'_\gamma \in U_{\gamma_4}$ for $\gamma \geq \gamma_5$. Then for any $\gamma \geq \gamma_5$:

$$(r_\gamma + r'_\gamma) \cdot (s_\gamma + s'_\gamma) = r_\gamma \cdot s_\gamma + r'_\gamma \cdot s_\gamma + r_\gamma \cdot s'_\gamma + r'_\gamma \cdot s'_\gamma$$

$$\in r_\gamma \cdot s_\gamma + (r_{\gamma_3} + U_{\gamma_2}) \cdot U_{\gamma_4} + U_{\gamma_4} \cdot (s_{\gamma_3} + U_{\gamma_2}) + U_{\gamma_4} \cdot U_{\gamma_4}$$

$$\subseteq r_\gamma \cdot s_\gamma + r_{\gamma_3} \cdot U_{\gamma_4} + U_{\gamma_2} \cdot U_{\gamma_4} + U_{\gamma_4} \cdot s_{\gamma_3} + U_{\gamma_4} \cdot U_{\gamma_2} + U_{\gamma_4} \cdot U_{\gamma_4}$$

$$\subseteq r_\gamma \cdot s_\gamma + U_{\gamma_2} + U_{\gamma_2} \cdot U_{\gamma_2} + U_{\gamma_2} + U_{\gamma_2} \cdot U_{\gamma_2} + U_{\gamma_2} \cdot U_{\gamma_2} \subseteq r_\gamma \cdot s_\gamma + U_{\gamma_1}.$$

Hence, $(\bar{r} + \bar{r}') \cdot (\bar{s} + \bar{s}') \in \bar{r} \cdot \bar{s} + \mathcal{P}_{\gamma_1} \subseteq \mathcal{P}$. Therefore, $(\bar{r} + \mathcal{P}_{\gamma_4}) \cdot (\bar{s} + \mathcal{P}_{\gamma_4}) \subseteq \mathcal{P}$.

Thus, the operation of the multiplication on $\tilde{R}$ is continuous, hence $\tilde{R}$ is a topological ring.

As it was shown in the proof of Proposition 3.3.1, the mapping $\psi : R \rightarrow \tilde{R}$, which corresponds an element $x \in R$ to the element $\{x_\gamma \mid x_\gamma = x, \gamma \in \Gamma\}$, is a topological isomorphism of the group $R(+)$ on $\psi(R)$, where $\psi(R)$ is a dense subset in $\tilde{R}$.

It is clear that ψ is also a ring homomorphism. The proof of the proposition is completed.

3.3.5. THEOREM. *Let R be a Hausdorff topological ring. Then:*

1) there exists a complete Hausdorff topological ring $\widehat{R}$ that contains R as a dense subring;

2) if R' is a complete Hausdorff topological ring containing R as a dense subring, then there exists a topological isomorphism $\xi : \widehat{R} \rightarrow R'$ such that $\xi(r) = r$ for all $r \in R$.

PROOF. Let $\tilde{R}$ be a complete topological ring constructed in Proposition 3.3.4. Let D be the smallest closed ideal of the ring $\tilde{R}$, then D is the smallest closed subgroup in the additive group of the ring $\tilde{R}$, and, since R is Hausdorff, then $D \cap R = \{0\}$. Then, according to Proposition 1.5.22, the additive group $R(+)$ is topologically isomorphic to the subgroup $(R + D)/D$ of the topological group $\widehat{R}(+) = \tilde{R}/D$, and this isomorphism is an isomorphism of rings R and $(R+D)/D$, since D is ideal in $\tilde{R}$. As R is a dense subring in $\tilde{R}$, then the ring $(R+D)/D$ is a dense subring in $\widehat{R} = \tilde{R}/D$. According to Proposition 3.2.19, the topological ring $\widehat{R}$ is complete. Let's identify the rings $(R + D)/D$ and R. Then 1) is proved.

Due to Proposition 3.2.29, any Hausdorff complete topological ring R' that contains a Hausdorff topological ring R as a dense subring is topologically isomorphic to the constructed Hausdorff topological ring $\widehat{R}$, and this isomorphism is identical on the ring R. This completes the proof of the theorem.

3.3.6. DEFINITION. Let R be a Hausdorff topological ring. We call a completion of the ring R and denote over $\widehat{R}$ the Hausdorff complete topological ring constructed in Theorem 3.3.5 (unique up to topological isomorphism), which contains R as a dense subring.

3.3.7. PROPOSITION. *Let R be a topological ring, M_R a topological right R-module and $\tilde{M}$ complete topological Abelian group constructed in Proposition 3.3.1 from the topological Abelian group M. Then $\tilde{M}$ can be converged into the topological right R-module, and R-module M will be its dense submodule.*

PROOF. Let $\mathcal{B}(M) = \{U_\gamma \,|\, \gamma \in \Gamma\}$ be the system of all neighborhoods of zero in M, Γ the direction defined by the filter $\mathcal{B}(M)$. Then, due to Proposition 3.3.1, Cauchy sequences $\bar{m} = \{m_\gamma \,|\, \gamma \in \Gamma\}$ of the elements of M are elements of $\tilde{M}$. We define the multiplication of the elements of $\tilde{M}$ by the elements of R as:

$$\{m_\gamma \,|\, \gamma \in \Gamma\} \cdot r = \{m_\gamma r \,|\, \gamma \in \Gamma\}.$$

Now we prove that the sequence $\{m_\gamma \cdot r \,|\, \gamma \in \Gamma\}$ belongs to $\tilde{M}$, i.e. that it is a Cauchy sequence in M. Indeed, since M_R is a topological R-module, then the group homomorphism $\mu_r : M_R \to M_R$, acting as: $\mu_r(m) = m \cdot r$, is continuous. Then, due to Proposition 3.1.33, it transforms any Cauchy sequence $\{m_\gamma \,|\, \gamma \in \Gamma\}$ to the Cauchy sequence $\{m_\gamma \cdot r \,|\, \gamma \in \Gamma\}$.

It is easy to see that with respect to this operation of multiplication the group $\tilde{M}$ is a right R-module, and the subgroup of the stationary sequences identified with M is an R-submodule in $\tilde{M}_R$.

Now we prove that $\tilde{M}$ is a topological module over the topological ring R.

Really, due to Proposition 3.3.1, $\tilde{M}$ is a topological group. Hence, it is enough to verify that the operation of multiplication by elements of R is continuous in $\tilde{M}$.

Let $\bar{m} = \{m_\gamma \,|\, \gamma \in \Gamma\} \in \tilde{M}$, $r \in R$ and $\mathcal{P}$ be a neighborhood of the element $\bar{m} \cdot r = \{m_\gamma \cdot r \,|\, \gamma \in \Gamma\}$ in $\tilde{M}$. Since the system $\{\mathcal{P}_\gamma \,|\, \gamma \in \Gamma\}$ of the sets $\mathcal{P}_\gamma$, constructed due to Proposition 3.3.1, is a basis of neighborhoods of zero in $\tilde{M}$ and M is a topological R-module, then there exist elements $\gamma_1, \gamma_2 \in \Gamma$ and a neighborhood V of zero in R such that $\bar{m} \cdot r + \mathcal{P}_{\gamma_1} \subseteq \mathcal{P}$ and $U_{\gamma_2} + U_{\gamma_2} \cdot V \subseteq U_{\gamma_1}$. Since $\bar{m}$ is a Cauchy sequence, then there exists an element $\gamma_3 \in \Gamma$ such that $m_\gamma - m_{\gamma_3} \in U_{\gamma_2}$ for $\gamma \geq \gamma_3$. Since M is a topological module over the topological ring R, then there exists a neighborhood U_{γ_4} of zero in M and a neighborhood V_1 of zero in R such that $V_1 \subseteq V$ and $m_{\gamma_3} \cdot V_1 + U_{\gamma_4} \cdot V_1 + U_{\gamma_4} \cdot r \subseteq U_{\gamma_2}$.

Now we verify that $(\bar{m} + \mathcal{P}_{\gamma_4}) \cdot (r + V_1) \subseteq \mathcal{P}$.

Let $\bar{m}' = \{m'_\gamma \,|\, \gamma \in \Gamma\} \in \mathcal{P}_{\gamma_4}$ and $r' \in V_1$. There exists an element $\gamma_5 \in \Gamma$ such that

$\gamma_5 \geq \gamma_3$ and $m'_\gamma \in U_{\gamma_4}$ for all $\gamma \geq \gamma_5$. Thus, for any $\gamma \geq \gamma_5$:

$$(m_\gamma + m'_\gamma) \cdot (r + r') = m_\gamma \cdot r + m_\gamma \cdot r' + m'_\gamma \cdot r + m'_\gamma \cdot r'$$

$$\in m_\gamma \cdot r + (m_{\gamma_3} + U_{\gamma_2}) \cdot V_1 + U_{\gamma_4} \cdot r + U_{\gamma_4} \cdot V_1$$

$$\subseteq m_\gamma \cdot r + m_{\gamma_3} \cdot V_1 + U_{\gamma_2} \cdot V_1 + U_{\gamma_4} \cdot r + U_{\gamma_4} \cdot V_1$$

$$\subseteq m_\gamma \cdot r + U_{\gamma_2} + U_{\gamma_2} \cdot V \subseteq m_\gamma \cdot r + U_{\gamma_1}.$$

Hence, $(\bar{m} + \bar{m}') \cdot (r + r') \in \bar{m} \cdot r + \mathcal{P}_{\gamma_1} \subseteq \mathcal{P}$. Therefore, $(\bar{m} + \mathcal{P}_{\gamma_4}) \cdot (r + V_1) \subseteq \mathcal{P}$.

Thus, we've proved that $\tilde{M}$ is a topological module over the topological ring R.

3.3.8. THEOREM. *Let R be a topological ring and M a Hausdorff topological right R-module. Then:*

1) there exists Hausdorff complete topological right R-module $\widehat{M}$ such that M is a dense submodule in $\widehat{M}$;

2) if M' is a Hausdorff complete topological right R-module that contains M as a dense submodule, then there exists a topological isomorphism $\xi : \widehat{M} \to M'$ such that $\xi(m) = m$ for all $m \in M$.

PROOF. Let $\tilde{M}$ be a complete topological module constructed in Proposition 3.3.7 from the module M. Let N be the smallest closed submodule of the module $\tilde{M}$, then N is the smallest closed subgroup in the group $\tilde{M}$, and, since the module M is Hausdorff, then $N \cap M = \{0\}$. Thus, according to Proposition 1.5.22, the Abelian group M is topologically isomorphic to the subgroup $(M + N)/N$ of the topological group $\tilde{M}/N$, and this isomorphism is an isomorphism of modules, since N is a submodule in $\tilde{M}$. As M is a dense submodule in $\tilde{M}$, then the module $(M + N)/N$ is a dense submodule in $\tilde{M}/N$. Finally, due to Proposition 3.2.19, the topological module $\tilde{M}/N$ is complete. Now we identify the modules M and $(M + N)/N$. Then 1) is proved.

According to Proposition 3.2.30, any Hausdorff complete topological R-module M' that contains a Hausdorff topological R-module M as a dense submodule is topologically isomorphic to the constructed Hausdorff topological R-module $\widehat{M} = \tilde{M}/N$, and this isomorphism is the identical mapping on the module M. This completes the proof of the theorem.

3.3.9. DEFINITION. Let R be a topological ring, M a Hausdorff topological R-module. We call a completion of the module M and denote over $\widehat{M}$ the Hausdorff topological R-module constructed in Theorem 3.3.8 (unique up to topological isomorphism) that contains M as a dense submodule.

3.3.10. PROPOSITION. *Let R be a topological ring, M a Hausdorff topological R-module and $\widehat{M}$ its completion. Then:*

1) if M is a bounded module, then $\widehat{M}$ is a bounded module too;

2) if M is a connected module, then $\widehat{M}$ is a connected module too;

3) let $\mathfrak{m}$ be a cardinal number and M have a basis of neighborhoods of zero, the cardinality of which does not exceed $\mathfrak{m}$. Then $\widehat{M}$ has a basis of neighborhoods of zero with the cardinality not more than $\mathfrak{m}$;

4) if the module M satisfies the ascending chains condition on open submodules, then $\widehat{M}$ satisfies the ascending chains condition on open submodules.

PROOF. The validity of the item 1) is a consequence of Propositions 3.2.30, 1.1.42 and 1.4.24.

Let M be a connected module, then $M \subseteq C(\widehat{M})$, where $C(\widehat{M})$ is the connected component of zero of the module $\widehat{M}$. Since $C(\widehat{M}) = [C(\widehat{M})]_{\widehat{M}}$, then $\widehat{M} = [M]_{\widehat{M}} \subseteq [C(M)]_{\widehat{M}} = C(\widehat{M})$. Thus, $\widehat{M} = C(\widehat{M})$, i.e. $\widehat{M}$ is a connected module. The item 2) is proved.

Let M have a basis $\mathcal{B}$ of neighborhoods of zero, the cardinality of which does not exceed $\mathfrak{m}$. Consider the system $\widehat{\mathcal{B}} = \{[U]_{\widehat{M}} \mid U \in \mathcal{B}\}$. According to Proposition 1.4.24, the system $\widehat{\mathcal{B}}$ consists of neighborhoods of zero in $\widehat{M}$ and $|\widehat{\mathcal{B}}| \leq \mathfrak{m}$. Now we prove that $\widehat{\mathcal{B}}$ is a basis of neighborhoods of zero in $\widehat{M}$. Indeed, let V be a neighborhood of zero in $\widehat{M}$ and V_1 a neighborhood of zero in $\widehat{M}$ such that $[V_1]_{\widehat{M}} \subseteq V$. Since $V_1 \cap M$ is a neighborhood of zero in M, then there exists a neighborhood $U \in \mathcal{B}$ such that $U \subseteq V_1 \cap M$. Therefore, $[U]_{\widehat{M}} \subseteq [V_1]_{\widehat{M}} \subseteq V$. The item 3) is proved.

Let M satisfy the ascending chains condition on open submodules. Assume that $\widehat{M}$ does not satisfy this condition. This means that there exists a sequence $\{N_k \mid k \in \mathbb{N}\}$ of open submodules in $\widehat{M}$ such that $N_k \subseteq N_{k+1}$ and $N_k \neq N_{k+1}$. Therefore, $\{N_k \cap M \mid k \in \mathbb{N}\}$ is an ascending chain of open submodules in M. In this case there exists a number $k_0 \in \mathbb{N}$ such that $N_{k_0} \cap M = N_k \cap M$ for all $k \geq k_0$. Since M is a dense submodule in $\widehat{M}$ and N_k

are open submodules in M, then, due to C.5:

$$N_{k_0} = [N_{k_0}]_{\widehat{M}} = [N_{k_0} \cap M]_{\widehat{M}} = [N_k \cap M]_{\widehat{M}} = [N_k]_{\widehat{M}} = N_k$$

for all $k \geq k_0$, i.e. $N_k = N_{k_0}$ for all $k \geq k_0$.

It contradicts the choice of the chain $\{N_k \mid k \in \mathbb{N}\}$, therefore the module $\widehat{M}$ satisfies the ascending chains condition on open submodules.

3.3.11. COROLLARY. *Let R be a Hausdorff topological ring, $\widehat{R}$ its completion. Then:*

1) if R is a bounded (from right, from left) ring, then $\widehat{R}$ is a bounded ring (from right, from left), too;

2) if R is an connected ring, $\widehat{R}$ is a connected ring, too;

3) if $\mathfrak{m}$ is a cardinal number and R has a basis of the neighborhoods of zero, whose cardinality does not exceed $\mathfrak{m}$, then $\widehat{R}$ has the basis of neighborhoods of zero, whose cardinality is not more than $\mathfrak{m}$.

PROOF. Consuder R as a module over itself. Then $\widehat{R}$ is a completion of the R-module R and, according to Proposition 3.3.10 and 3.2.30, we get that the items 1), 2), 3) are valid.

3.3.12. PROPOSITION. *Let (R, τ) be a Hausdorff topological ring, $(\widehat{R}, \widehat{\tau})$ its completion. Then:*

1) if the ring R satisfies the ascending chains condition on open left (right, two-sided) ideals, then the ring $\widehat{R}$ satisfies this condition too;

2) if the topology τ on the ring R is non-weakable in the class of Hausdorff ring topologies, then the topology $\widehat{\tau}$ on the ring $\widehat{R}$ is non-weakable in the class of Hausdorff ring topologies, too.

PROOF. Let the ring R satisfy the ascending chains condition on open left (right, two-sided) ideals. Assume that ring $\widehat{R}$ does not satisfy this condition. This means that there exists a sequence $\{I_k \mid k \in \mathbb{N}\}$ of open left (right, two-sided) ideals in $\widehat{R}$ such that $I_k \subseteq I_{k+1}$ and $I_k \neq I_{k+1}$ for all $k \in \mathbb{N}$. Then $\{I_k \cap R \mid k \in \mathbb{N}\}$ is the ascending chain of open left (right,two-sided) ideals in R. Due to the condition, there exists a number k_0 such that

$I_k \cap R = I_{k_0} \cap R$ for all $k \geq k_0$. Then as in the proof of the item 4) of Proposition 3.3.10, we verify that $I_k = I_{k_0}$ for all $k \geq k_0$. The contradiction with the choice of the chain $\{I_k \cap R \,|\, k \in \mathbb{N}\}$ proves that the ring $\widehat{R}$ satisfies the ascending chains condition on open left (right, two-sided) ideals. The item 1) is proved.

Let the topology τ on the ring R be non-weakable. Assume that the topology $\widehat{\tau}$ on the ring $\widehat{R}$ is weakable. It means that there exists a topology τ_1 on the ring $\widehat{R}$ such that $(\widehat{R}, \tau_1)$ is a Hausdorff topological ring and $\tau_1 < \widehat{\tau}$. Then there exists a neighborhood U of zero in $(\widehat{R}, \widehat{\tau})$ such that it does not contain a neighborhood of zero of the ring $(\widehat{R}, \tau_1)$.

There exists a neighborhood of zero U_1 in the $(\widehat{R}, \widehat{\tau})$ such that $[U_1]_{(\widehat{R}, \widehat{\tau})} \subseteq U$. Therefore, $U_1 \cap R$ is a neighborhood of zero in (R, τ). Let τ_2 be a restriction of the topology τ_1 on the ring R. Then (R, τ_2) is a Hausdorff topological ring and $\tau_2 \leq \tau$, because $\tau_1 < \widehat{\tau}$. Since the topology τ of the ring R is non-weakable, then $\tau_2 = \tau$. It means that there exists an open neighborhood V of zero in $(\widehat{R}, \tau_1)$ such that $V \cap R \subseteq U_1 \cap R$. Then $[V \cap R]_{(\widehat{R}, \widehat{\tau})} \subseteq [U_1]_{(\widehat{R}, \widehat{\tau})} \subseteq U$. Since $\widehat{\tau} > \tau_1$, then V is an open neighborhood of zero in $(\widehat{R}, \widehat{\tau})$. Since R is a dense subring in $(\widehat{R}, \widehat{\tau})$, then, due to C.5, $[V]_{(\widehat{R}, \widehat{\tau})} = [V \cap R]_{(\widehat{R}, \widehat{\tau})}$. Therefore, $V \subseteq [V \cap R]_{(\widehat{R}, \widehat{\tau})} \subseteq U$.

This contradicts the choice of the neighborhood U of zero in $(\widehat{R}, \widehat{\tau})$. Thus, the topology $\widehat{\tau}$ on $\widehat{R}$ is non-weakable.

3.3.13. PROPOSITION. *Let R be a topological ring, (M, τ) be a topological R-module, $(\widehat{M}, \widehat{\tau})$ its completion. If the topology τ on the module M is non-weakable in the class of Hausdorff module topologies, then the topology $\widehat{\tau}$ on $\widehat{M}$ is non-weakable in the class of Hausdorff module topologies, too.*

PROOF. Assume the contrary, i.e. that the topology $\widehat{\tau}$ on the module $\widehat{M}$ is weakable. It means that there exists a topology τ_1 on the module $\widehat{M}$ such that $(\widehat{M}, \tau_1)$ is a Hausdorff topological R-module and $\tau_1 < \widehat{\tau}$. Then there exists a neighborhood U of zero in $(\widehat{M}, \widehat{\tau})$ such that it does not contain any neighborhood of zero of the module $(\widehat{M}, \tau_1)$.

There exists a neighborhood U_1 of zero in $(\widehat{M}, \widehat{\tau})$ such that $[U_1]_{(\widehat{M}, \widehat{\tau})} \subseteq U$. Therefore, $U_1 \cap M$ is a neighborhood of zero in M. Let τ_2 be a restriction of the topology τ_1 on the module M. Then (M, τ_2) is a Hausdorff topological R-module and $\tau_2 \leq \tau$, because $\tau_1 < \widehat{\tau}$. Since the topology τ on the module M is non-weakable, then $\tau_2 = \tau$. It means that there exists an open neighborhood V of zero in $(\widehat{M}, \widehat{\tau})$ such that $V \cap M \subseteq U_1 \cap M$. Thus, $[V \cap M]_{(\widehat{M}, \widehat{\tau})} \subseteq [U_1]_{(\widehat{M}, \widehat{\tau})} \subseteq U$. Since $\widehat{\tau} > \tau_1$, then V is an open neighborhood of zero in $(\widehat{M}, \widehat{\tau})$. Since M is a dense submodule in $(\widehat{M}, \widehat{\tau})$, then $[V]_{(\widehat{M}, \widehat{\tau})} = [V \cap M]_{(\widehat{M}, \widehat{\tau})}$. Thus, $V \subseteq [V]_{(\widehat{M}, \widehat{\tau})} \subseteq U$

This contradicts the choice of the neighborhood U of zero in $(\widehat{M}, \widehat{\tau})$, therefore the topology $\widehat{\tau}$ on the R-module $\widehat{M}$ is non-weakable.

3.3.14. PROPOSITION. *Let R be a Hausdorff topological ring, $\widehat{R}$ its completion, M a Hausdorff topological R-module, $\widehat{M}$ its completion. Consider $\widehat{M}$ as a topological $\widehat{R}$-module, according to Theorem 3.2.31. If M is a bounded R-module, then $\widehat{M}$ is a bounded $\widehat{R}$-module, too.*

PROOF. Let U be a neighborhood of zero in $\widehat{M}$ and U_1 a neighborhood of zero in $\widehat{M}$ such that $[U_1]_{\widehat{M}} \subseteq U$. Since M is a bounded R-module, then there exists a neighborhood V of zero in R such that $M \cdot V \subseteq U_1 \cap M$. Due to Proposition 1.4.24, $[V]_{\widehat{R}}$ is a neighborhood of zero in $\widehat{R}$. Since $\widehat{M}$ is a topological $\widehat{R}$-module, then, due to Proposition 1.1.42, $\widehat{M} \cdot [V]_{\widehat{R}} = [M]_{\widehat{M}} \cdot [V]_{\widehat{R}} \subseteq [M \cdot V]_{\widehat{M}} \subseteq [U_1 \cap M]_{\widehat{M}} \subseteq [V_1]_{\widehat{M}} \subseteq U$. Since the choice of the neighborhood U of zero in $\widehat{M}$ was arbitrary, then the $\widehat{R}$-module $\widehat{M}$ is bounded.

§ 3.4 Precompactness in the Completion of Topological
Rings and Modules

3.4.1. DEFINITION. A subset A of a topological Abelian group X is called precompact in X if for any neighborhood V of zero in X there exists a finite set of elements $a_1, \ldots, a_n \in A$ such that $A \subseteq \bigcup_{i=1}^{n}(a_i + V)$.

3.4.2. DEFINITION. A topological Abelian group X is called precompact if the set X is precompact in the group X.

3.4.3. REMARK. *Let a set A of an Abelian group X be compact in the topology, induced from the group X. Then A is precompact set.*

PROOF. Let U be a neighborhood of zero in X, and V be an open subset in X such that $0 \in V \subseteq U$. Then for any $a \in A$ the subset $(a + V) \bigcap A$ is open in A. It is clear that when the element a runs through all the set A, then the sets $(a + V) \bigcap A$ make an open cover of the compact set A. Thus, there exists a finite subcover, i.e. a finite set of elements $a_1, \ldots, a_n \in A$ such that $A \subseteq \bigcup_{i=1}^{n}((a_i + V) \bigcap A)$, then $A \subseteq \bigcup_{i=1}^{n}(a_i + U)$.

3.4.4. PROPOSITION. *Let X be a topological group, A be a subset in X. Then the following conditions are equivalent:*

1) A is precompact in X;

2) for any neighborhood U of zero in X there exists a finite set $x_1, \ldots, x_n \in X$ such that $A \subseteq \bigcup_{i=1}^{n}(x_i + U)$.

PROOF. It is clear that if A is a precompact set in X, then the condition 2) is fulfilled.

Let U be a neighborhood of zero in X. There exists a neighborhood V of zero in X such that $V - V \subseteq U$. Due to 2), there exist elements $x_1, \ldots, x_n \in X$ such that $A \subseteq \bigcup_{i=1}^{n}(x_i + V)$. We can consider that $(x_i + V) \bigcap A \neq \emptyset$ for any $i = 1, \ldots, n$ (if the contrary, it is possible to remove the element x_i from the set $\{x_1, \ldots, x_n\}$ but to conserve the inclusion $A \subseteq \bigcup_{j \neq i}(x_j + V)$ and consider the reduced set). We choose for any $i = 1, \ldots, n$ an element $a_i \in (x_i + V) \bigcap A$. Then $x_i \in a_i - V$, hence $x_i + V \subseteq a_i + V - V \subseteq a_i + U$ for any $i = 1, \ldots, n$. Thus, $A \subseteq \bigcup_{i=1}^{n}(x_i + V) \subseteq \bigcup_{i=1}^{n}(a_i + U)$, i.e. A is a precompact subset in X.

3.4.5. PROPOSITION. *Let X be a topological Abelian group, A and B precompact subsets in X. Then:*

1) if $C \subseteq A$, then C is a precompact subset in X;

2) $x + A$ is a precompact subset in X for any $x \in X$;

3) $A + B$ is a precompact subset in X.

PROOF. Let U be a neighborhood of zero in X, then there exists a finite set of elements $x_1, \ldots, x_n \in X$ such that $A \subseteq \bigcup_{i=1}^{n}(x_i + U)$.

If $C \subseteq A$, then $C \subseteq \bigcup_{i=1}^{n}(x_i + U)$, and, due to Proposition 3.4.4, the item 1) is proved. Since $A \subseteq \bigcup_{i=1}^{n}(x_i + U)$, then $x + A \subseteq \bigcup_{i=1}^{n}(x + x_i + U)$. Denote $x + x_i$ over y_i for $i = 1, \ldots, n$. Then $y_i \in X$ for $i = 1, \ldots, n$ and $x + A \subseteq \bigcup_{i=1}^{n}(y_i + U)$, that means, due to Proposition 3.4.4, that the set $x + A$ is precompact in the group X, i.e. the item 2) is proved.

Let U be a neighborhood of zero in X. Then there exists a neighborhood V of zero in X such that $V + V \subseteq U$. Let $\{a_1, \ldots, a_n\}$ and $\{b_1, \ldots, b_k\}$ be finite subsets in A and B correspondingly such that $A \subseteq \bigcup_{i=1}^{n}(a_i + V)$ and $B \subseteq \bigcup_{i=1}^{k}(b_i + V)$. Then $\{a_i + b_j \mid 1 \le i \le n, 1 \le j \le k\}$ is a finite subset in $A + B$, and

$$A + B \subseteq \left(\bigcup_{i=1}^{n}(a_i + V)\right) + \left(\bigcup_{j=1}^{k}(b_j + V)\right) = \bigcup_{i=1}^{n}\bigcup_{j=1}^{k}(a_i + b_j + V + V)$$

$$\subseteq \bigcup_{i=1}^{n}\bigcup_{j=1}^{k}(a_i + b_j + U).$$

Since the choice of the neighborhood U was arbitrary, the set $A + B$ is precompact.

3.4.6. PROPOSITION. *Let X be a topological group, A a precompact subset in the X. Then the set $[A]_X$ is precompact, too.*

PROOF. Let U be a neighborhood of zero in X. Then there exists a neighborhood V of zero in X such that $[V]_X \subseteq U$. Since the set A is precompact, then there exists a finite set of elements $a_1, \ldots, a_n \in A$ such that $A \subseteq \bigcup_{i=1}^{n}(a_i + V)$. Then

$$[A]_X \subseteq \left[\bigcup_{i=1}^{n}(a_i + V)\right]_X = \bigcup_{i=1}^{n}[a_i + V]_X = \bigcup_{i=1}^{n}(a_i + [V]_X) \subseteq \bigcup_{i=1}^{n}(a_i + U),$$

which proves that $[A]_X$ is precompact in X.

3.4.7. PROPOSITION. *Let X and Y be topological Abelian groups and $f : X \to Y$ be a continuous homomorphism. If A is a precompact subset in X, then $f(A)$ is a precompact subset in Y.*

PROOF. Let V be a neighborhood of zero in Y, then $U = f^{-1}(V)$ is a neighborhood of zero in X. Since the set A is precompact, then there exists a finite set of elements $a_1, \dots, a_n \in A$ such that $A \subseteq \bigcup_{i=1}^{n}(a_i + U)$. Then $f(A) \subseteq \bigcup_{i=1}^{n} f(a_i + U) = \bigcup_{i=1}^{n}(f(a_i) + V)$. Thus, $f(A)$ is precompact in Y.

3.4.8. COROLLARY. *Let X be an Abelian group, A a subset in X, and τ_1 and τ_2 group topologies on X such that $\tau_1 \geq \tau_2$. If A is a precompact subset in (X, τ_1), then A is precompact in (X, τ_2).*

PROOF. Since the identical mapping is a continuous homomorphism from (X, τ_1) to (X, τ_2), then the corollary is a consequence of Proposition 3.4.7.

3.4.9. PROPOSITION. *Let X and Y be Abelian groups, $f : X \to Y$ a group homomorphism, τ a group topology on Y and τ_1 the prototype of the topology τ with respect to the homomorphism f. If C is a subset in X such that $f(C)$ is precompact in (Y, τ), then C is precompact in (X, τ_1).*

PROOF. Let U be a neighborhood of zero in (X, τ_1), then there exists a neighborhood V of zero in (Y, τ) such that $f^{-1}(V) \subseteq U$. Since $f(C)$ is a precompact subset in (Y, τ), then there exists a finite set $\{b_1, \dots, b_n\}$ of elements of $f(C)$ such that $f(C) \subseteq \bigcup_{i=1}^{n}(b_i + V)$. Let $c_1, \dots, c_n$ be elements of C such that $b_i = f(c_i)$. Then

$$C \subseteq f^{-1}\big(f(C)\big) = f^{-1}\left(\bigcup_{i=1}^{n}(b_i + V)\right) = \bigcup_{i=1}^{n}(c_i + f^{-1}(V)) \subseteq \bigcup_{i=1}^{n}(c_i + U),$$

i.e. C is a precompact subset in (X, τ_1).

3.4.10. PROPOSITION. *Let X be an Abelian group, A a subset in X and $\tau_1, \dots, \tau_n$ such group topologies on X that A is a precompact in (X, τ_i) for any $i = 1, \dots, n$. Let $\tau = \sup\{\tau_1, \dots, \tau_n\}$, then A is precompact in (X, τ).*

PROOF. It is clear that it is enough to prove the statement for $n = 2$. Let $\tau = \sup\{\tau_1, \tau_2\}$ and W be a neighborhood of zero in (X, τ). Then there exist neighborhoods U_0 of zero in (X, τ_1) and V_0 in (X, τ_2) such that $U_0 \bigcap V_0 \subset W$. There exists a neighborhood

U_1 of zero in (X, τ_1) such that $U_1 - U_1 \subseteq U_0$. Since the set A is precompact in (X, τ_1), then there exist elements $x_1, \ldots, x_n \in X$ such that $A \subseteq \bigcup_{i=1}^{n}(x_i + U_1)$.

Due to Proposition 3.4.5, the set $(A - x_i) \bigcap U_1$ is precompact in (X, τ_2) for any $i = 1, \ldots, n$. Hence, there exist elements $y_{i,1}, \ldots, y_{i,k_i} \in (A - x_i) \bigcap U_1$ such that $(A - x_i) \bigcap U_1 \subseteq \bigcup_{j=1}^{k_i}(y_{i,j} + V_0)$ for $i = 1, \ldots, n$. Let $\{y_1, \ldots, y_k\} = \bigcup_{i=1}^{n}\{y_{i,1}, \ldots, y_{i,k_i}\}$.

Now we show that $A \subseteq S + W$, where $S = \{x_i + y_i \mid i = 1, \ldots, n, j = 1, \ldots, k\}$. Indeed, for any element $a \in A$ there exists a number i_0 such that $1 \leq i_0 \leq n$ and $a \in x_{i_0} + U_1$, i.e. $a - x_{i_0} = u \in U_1$. Since $u \in (A - x_{i_0}) \bigcap U_1$, then there exists a number j_0 such that $1 \leq j_0 \leq k_{i_0}$ and $u \in y_{j_0} + V_0$, i.e. $u = y_{j_0} + v$, where $v \in V_0$. Therefore, $v = u - y_{j_0} \in U_1 - U_1 \subseteq U_0$, hence $v \in U_0 \bigcap V_0 = W$. Thus, $a = x_{i_0} + y_{i_0} + v \in x_{i_0} + y_{i_0} + W \subseteq S + W$. Due to Proposition 3.4.4, A is a precompact subset in (X, τ).

The proposition is proved.

3.4.11. PROPOSITION. *Let X be an Abelian group, A a subset in X, $\{\tau_i \mid i \in I\}$ a family of group topologies on X such that A is a precompact subset in (X, τ_i) for any $i \in I$. Let $\tau = \sup\{\tau_i \mid i \in I\}$, then A is a precompact subset in (X, τ).*

PROOF. Let U be a neighborhood of zero in (X, τ), then there exist a finite set $i_1, \ldots, i_n \in I$ and neighborhoods V_k of zero in (X, τ_{i_k}), $k = 1, \ldots, n$, such that $U \supseteq \bigcap_{k=1}^{n} V_k$. Due to Proposition 3.4.10, A is a precompact subset in $(X, \sup\{\tau_{i_1}, \ldots, \tau_{i_n}\})$, hence there exists a finite set $a_1, \ldots, a_m \in A$ such that $A \subseteq \bigcup_{p=1}^{m}(a_p + \bigcap_{k=1}^{n} V_k)$. Therefore, $A \subseteq \bigcup_{p=1}^{m}(a_p + U)$, which proves that the set A is precompact in the group (X, τ).

3.4.12. REMARK. We should note that the strengthening of the topology does not necessarily conserve the precompactness of the sets. For example, a segment [-1;1] is compact, hence, a precompact subset in the additive group of real numbers $\mathbb{R}$ with the interval topology. But if we consider on $\mathbb{R}$ the discrete topology, then the segment [-1;1] will not be precompact.

3.4.13. PROPOSITION. *Let X be a topological Abelian group, A a subgroup of the topological group X and B a subset in A. The set B is precompact in A if and only if B is precompact in X.*

PROOF. Let B be precompact in A and W be a neighborhood of zero in X. Since

$V = W \bigcap A$ is a neighborhood of zero in A, then there exists a finite set of elements $b_1, \ldots, b_n \in B$ such that $B \subseteq \bigcup_{i=1}^{n}(b_i + V)$. Therefore, $B \subseteq \bigcup_{i=1}^{n}(b_i + W)$. It proves that the set B is precompact in the group X.

Let B be a precompact set in X and V be a neighborhood of zero in A. Then there exists a neighborhood W of zero in X such that $W \bigcap A = V$. Since the set B is precompact in X, then there exists a finite set $b_1, \ldots, b_n \in B$ such that $B \subseteq \bigcup_{i=1}^{n}(b_i + W)$. Since $b_i \in B \subseteq A$, then $(b_i + W) \bigcap A = b_i + (W \bigcap A) = b_i + V$, hence, $B \subseteq \bigcup_{i=1}^{n}((b_i + W) \bigcap A) = \bigcup_{i=1}^{n}(b_i + V)$. It proves that the set B is precompact in the group A.

3.4.14. COROLLARY. A subgroup A of a topological Abelian group X is a precompact group if and only if the set A is precompact in the group X.

3.4.15. PROPOSITION. *Let X be a topological Abelian group, A a precompact subset in X. Let Φ be an ultrafilter (see A.13) of the set A, then the family $\mathcal{B} = \{C \mid C \subseteq X,$ there exists a subset $F \in \Phi$, $F \subseteq C\}$ is a Cauchy filter of the group X.*

PROOF. It is easily verified that $\mathcal{B}$ is a filter of the set X and Φ is its basis. Now we prove that $\mathcal{B}$ is a Cauchy filter. Let U be a neighborhood of zero in X and V be a neighborhood of zero in X such that $V - V \subseteq U$. Since the set A is precompact, then there exists a finite subset $\{a_1, \ldots, a_n\}$ of elements of A such that $A \subseteq \bigcup_{i=1}^{n}(a_i + V)$, hence, $A = \bigcup_{i=1}^{n}((a_i + V) \bigcap A)$. Since Φ is an ultrafilter of the set A, then there exists a number $1 \leq i_0 \leq n$ such that $(a_{i_0} + V) \bigcap A \in \Phi$ (see A.13). Then $a_{i_0} + V \in \mathcal{B}$, and $(a_{i_0} + V) - (a_{i_0} + V) = V - V \subseteq U$. Since the choice of the neighborhood U of zero in X was arbitrary, then $\mathcal{B}$ is a Cauchy filter of the group X.

3.4.16. THEOREM. *Let X be a complete topological Abelian group, A a subset in X. The set A is precompact in X if and only if $[A]_X$ is a compact subset in X.*

PROOF. If $[A]_X$ is a compact subset in X, then, due to Remark 3.4.3, $[A]_X$ is a precompact subset in X, and, according to Proposition 3.4.5, A is a precompact subset in X.

Let A be a precompact set, then, due to Proposition 3.4.6, $[A]_X$ is a precompact subset in X. Assume that the set $[A]_X$ is not compact in X. It means that there exists such open cover $\{U_\omega \mid \omega \in \Omega\}$ of the set $[A]_X$, that we cannot choose a finite subcover. Let

$\mathcal{N} = \{\Gamma \mid \Gamma \subseteq \Omega, \mid \Gamma \mid < \infty\}$, then the system of all sets of the type $[A]_X \backslash (\bigcup_{\omega \in \Gamma} U_\omega)$, where $\Gamma \in \mathcal{N}$, is a basis of a certain filter $\mathcal{F}$ of the set $[A]_X$. There exists an ultrafilter Φ of the set $[A]_X$, majoring the filter $\mathcal{F}$. According to Proposition 3.4.15, the family $\mathcal{B} = \{C \mid C \subseteq X, \text{ there exists a set } F \in \Phi, F \subseteq C\}$ is a Cauchy filter of the group X. Since X is a complete group, then there exists $x \in X$ such that $x = \lim \mathcal{B}$, hence $x \in \bigcap_{C \in \mathcal{B}} [C]_X \subseteq [A]_X$ (see Proposition 3.1.26).

Since $([A]_X \backslash U_\omega) \in \mathcal{F} \subseteq \Phi \subseteq \mathcal{B}$ and $[[A]_X \backslash U_\omega]_X = [A]_X \backslash U_\omega$ for any $\omega \in \Omega$, then $x \notin U_\omega$ for any $\omega \in \Omega$. This contradicts the fact that the family $\{U_w \mid \omega \in \Omega\}$ is a cover of the set $[A]_X$. Then $[A]_X$ is compact in X.

The theorem is proved.

3.4.17. COROLLARY. Let X be a complete topological Abelian group and A a dense subgroup of the topological group X. A subset B is precompact in the group A if and only if $[B]_X$ is a compact set.

We get the proof using Proposition 3.4.13 and Theorem 3.4.16.

3.4.18. COROLLARY. The completion of a Hausdorf precompact Abelian group is a Hausdorf compact group.

3.4.19. DEFINITION. A topological Abelian group X is called locally precompact if there exists a neighborhood U of zero in X such that U is a precompact subset in X.

3.4.20. REMARK. A locally compact Abelian group is locally precompact.

3.4.21. PROPOSITION. *The completion $\widehat{X}$ of a Hausdorff locally precompact Abelian group X is a Hausdorf locally compact Abelian group.*

PROOF. Let U be a precompact neighborhood of zero in X, then, according to Proposition 1.4.24, the set $[U]_{\widehat{X}}$ is a neighborhood of zero in $\widehat{X}$. According to Corollary 3.4.17, $[U]_{\widehat{X}}$ is a compact set.

3.4.22. PROPOSITION. *Let X be an Abelian group and $\{\tau_1, \ldots, \tau_n\}$ such family of group topologies on X that (X, τ_i) is a locally precompact group for any $1 \le i \le n$. Let $\tau = \sup\{\tau_1, \ldots, \tau_n\}$, then (X, τ) is a locally precompact group.*

PROOF. Let U_i be a precompact neighborhood of zero in (X, τ_i) for $i = 1, \dots, n$. Then $U = \bigcap_{i=1}^{n} U_i$ is a neighborhood of zero in (X, τ) and, according to Proposition 3.4.5, the set U is precompact in (X, τ_i) for any $i = 1, \dots, n$. According to Proposition 3.4.10, the set U is precompact in (X, τ).

3.4.23. REMARK. In Example 4.1.67 will be presented an example of an Abelian group G and an infinite family $\{\tau_i \mid i \in I\}$ of group topologies on G such that the group (G, τ_i) is locally precompact for any $i \in I$, but the group $(G, \sup\{\tau_i \mid i \in I\})$ is not locally precompact.

3.4.24. PROPOSITION. *Let R be a topological ring, M be a topological right R-module. If A is a precompact subset in M, then A is a bounded set.*

PROOF. Let U be a neighborhood of zero in M, then there exist neighborhoods U_1 and U_2 of zero in M and a neighborhood V_0 of zero in R such that $U_1 + U_1 \subseteq U$ and $U_2 \cdot V_0 \subseteq U_1$. Since A is a precompact set, then there exists a finite set of $a_1, \dots, a_n \in A$ such that $A \subseteq \bigcup_{i=1}^{n}(a_i + U_2)$. For any $i = 1, \dots, n$ there exists a neighborhood V_i of zero in R such that $a_i \cdot V_i \subseteq U_1$. Then $V = \bigcap_{i=0}^{n} V_i$ is a neighborhood of zero in R and $A \cdot V \subseteq \bigcup_{i=1}^{n}(a_i \cdot V_i + U_2 \cdot V_0) \subseteq U_1 + U_1 \subseteq U$. It proves that the set A is bounded in M.

3.4.25. COROLLARY. *Let R be a topological ring. If A is a precompact set in R, then A is a bounded set.*

PROOF. Consider R as a right R-module. Then, due to Proposition 3.4.24, A is a set bounded from right in the topological ring R. Considering the same way R as a left R-module, we get that A is a set bounded from left. Therefore, A is a bounded set in the ring R.

§ 3.5. Invertibility and Quasi-invertibility
in the Completion

3.5.1. THEOREM. *Let R be a Hausdorff topological associative ring and $\widehat{R}$ its completion. For ring $\widehat{R}$ to have the unitary element and for any non-zero element from R to be invertable in $\widehat{R}$ it is necessary and sufficient that the ring R would contain neither left nor right non-zero topological divisors of zero (see Definition 1.8.1) and any non-zero one-sided ideal in R would be totally dense.*

PROOF. At first we'll show the necessity of the conditions. Let $0 \neq a \in R$. Since the element a is invertible in $\widehat{R}$, then, due to Proposition 1.8.9, the element a is neither a left nor right topological divisor of zero in R.

Let I be a non-zero one-sided ideal in R. For example, suppose that I is a left ideal. Due to Corollary 1.4.9, $[I]_{\widehat{R}}$ is a left ideal in $\widehat{R}$. If a is a certain non-zero element from I, then, since it is invertible in $\widehat{R}$, we get $\widehat{R} = \widehat{R} \cdot a \subseteq [I]_{\widehat{R}}$, hence, $[I]_{\widehat{R}} = \widehat{R}$. Then $[I]_R = [I]_{\widehat{R}} \cap R = R$, i.e. I is a totally dense ideal of the ring R. The case when I is a right ideal is considered the same way. The necessity of the theorem conditions is proved.

Now we'll verify the sufficiency of the conditions. At first we show that any non-zero element of R is neither a right nor left topological divisor of zero in $\widehat{R}$. Assume the contrary, i.e. that a certain non-zero element $a \in R$ is, for example, a left topological divisor of zero in $\widehat{R}$. Then there exists a subset M in $\widehat{R}$ such that $0 \notin [M]_{\widehat{R}}$ and $0 \in [a \cdot M]_{\widehat{R}}$. There exist neighborhoods $\widehat{V}_1$ and $\widehat{V}_2$ of zero in $\widehat{R}$ such that $\widehat{V}_1 \cap M = \emptyset$ and $\widehat{V}_2 - \widehat{V}_2 \subseteq \widehat{V}_1$. It is clear that $\widehat{V}_2 \cap (M + \widehat{V}_2) = \emptyset$, hence, $0 \notin [M + \widehat{V}_2]_{\widehat{R}}$. Let $M_1 = (M + \widehat{V}_2) \cap R$, then $[M_1]_R \subseteq [M + \widehat{V}_2]_{\widehat{R}}$. Thus, $0 \notin [M_1]_R$.

Since the element a is not a left topological divisor of zero in R, then $0 \notin [a \cdot M_1]_R$, hence, $(a \cdot M_1) \cap V = \emptyset$ for a certain neighborhood V of zero in R. There exist neighborhoods $\widehat{U}_1, \widehat{U}_2$ of zero in $\widehat{R}$ such that $\widehat{U}_1 \cap R \subseteq V_1, \widehat{U}_2 \subseteq \widehat{V}_2$ and $a \cdot \widehat{U}_2 + \widehat{U}_2 \subseteq \widehat{U}_1$. Since $0 \in [a \cdot M]_{\widehat{R}}$, then there exists an element $m \in M$ such that $a \cdot m \in \widehat{U}_2$. Since the ring R is totally dense in $\widehat{R}$, then there exists an element $r \in (m + \widehat{U}_2) \cap R$. Therefore, $r \in (M + \widehat{V}_2) \cap R = M_1$, where

$$a \cdot r \in \big(a \cdot (m + \widehat{U}_2)\big) \cap R \subseteq (a \cdot m + a \cdot \widehat{U}_2) \cap R \subseteq (\widehat{U}_2 + a \cdot \widehat{U}_2) \cap R \subseteq \widehat{U}_1 \cap R \subseteq V.$$

This contradicts the choice of the neighborhood V. Hence, the element a is not a left topological divisor of zero in $\widehat{R}$.

The way to verify that the element a is not a right topological divisor of zero in $\widehat{R}$ is the same.

Now we fix a certain non-zero element $a_0 \in R$. According to Proposition 1.8.7, the subgroup $a_0 \cdot \widehat{R}(+)$ is topologically isomorphic to the additive group $\widehat{R}(+)$, hence, $a_0 \cdot \widehat{R}(+)$ is a complete group. Due to Theorem 3.2.12, $a_0 \cdot \widehat{R}(+)$ is a closed subgroup in $\widehat{R}(+)$. Since $a_0 \cdot R \neq 0$, then the right ideal $a_0 \cdot R$ is totally dense in R, thus, $a_0 \cdot R$ is a totally dense subset in $\widehat{R}$. Therefore, $\widehat{R} = [a_0 \cdot R]_{\widehat{R}} \subseteq [a_0 \cdot \widehat{R}]_{\widehat{R}} = a_0 \cdot \widehat{R}$, i.e. $\widehat{R} = a_0 \cdot \widehat{R}$. Thus, $a_0 = a_0 \cdot e$ for a certain element $e \in \widehat{R}$.

Now we show that the element e is the unitary element of the ring $\widehat{R}$. Let $b \in \widehat{R}$. Then $a_0 \cdot (e \cdot b - b) = a_0 \cdot e \cdot b - a_0 \cdot b = a_0 \cdot b - a_0 \cdot b = 0$. Since the element a_0 is not a left divisor of zero in $\widehat{R}$, then $e \cdot b - b = 0$, i.e. $e \cdot b = b$. Since b is an arbitrary element of $\widehat{R}$, hence, $e \cdot a_0 = a_0$. Then $(b \cdot e - b) \cdot a_0 = b \cdot e \cdot a_0 - b \cdot a_0 = b \cdot a_0 - b \cdot a_0 = 0$. Since the element a_0 is not a right divisor of zero in the $\widehat{R}$, then $b \cdot e - b = 0$, i.e. $b \cdot e = b$. Therefore, we've proved that the element e is the unitary element of the ring $\widehat{R}$.

Now we show that any non-zero element $a \in R$ is invertible in the ring $\widehat{R}$. Analogously to the proof of the equality $a_0 \cdot \widehat{R} = \widehat{R}$ for the element a_0 it is verified that $a \cdot \widehat{R} = \widehat{R}$. Then there exists an element $r \in \widehat{R}$ such that $a \cdot r = e$.

Analogously it is verified that $\widehat{R} \cdot a = \widehat{R}$, hence, $e = r_1 \cdot a$ for a certain element $r_1 \in R$. Then $r_1 = r_1 \cdot e = r_1 \cdot a \cdot r = e \cdot r = r$, i.e the element a is invertible in $\widehat{R}$.

The theorem is proved.

3.5.2. THEOREM. *Let R be a Hausdorff topological ring with the unitary element 1, possessing a topologically nilpotent (see Definition 1.8.16) neighborhood of zero, and $\widehat{R}$ be its completion. If in the ring R every non-zero one-sided ideal is totally dense, then every non-zero element of R is invertible in $\widehat{R}$.*

PROOF. Due to Theorem 3.5.1, it is enough to verify that any non-zero element of R is neither a right nor left topological divisor of zero. Assume the contrary, i.e. that an element $0 \neq a \in R$ is, for example, a right topological divisor of zero, and let M be a subset in R such that $0 \notin [M]_R$, but $0 \in [M \cdot a]_R$. Then there exists a neighborhood V_0

of zero in R such that $V_0 \bigcap M = \emptyset$. We can consider that V_0 is a topologically nilpotent neighborhood of zero. Then there exist neighborhoods V_1 and V_2 of zero in R and a natural number n_0 such that $V_1 \cdot V_1 + V_1 \subseteq V_0$, $V_2 + V_2 \subseteq V_1$ and $V_0^{(k)} \subseteq V_2$ (see B.12) for any $k \geq n_0$.

Since any non-zero right ideal is totally dense in the ring R, then there exists an element $x \in R$ such that $a \cdot x + 1 \in V_0$.

Let

$$y = \sum_{i=1}^{n_0} C_{n_0}^i \cdot (a \cdot x)^{i-1} \cdot x,$$

where

$$C_n^i = \frac{n \cdot (n-1) \cdot \ldots \cdot (n-i+1)}{1 \cdot 2 \cdot \ldots \cdot i}.$$

Therefore,

$$a \cdot y + 1 = \sum_{i=1}^{n_0} C_{n_0}^i \cdot (a \cdot x)^i + 1 = (a \cdot x + 1)^{n_0} \in V_0^{(n_0)} \subseteq V_2 \subseteq V_1.$$

Since $(a \cdot y + 1) + (a \cdot y + 1) \cdot V_0 \subseteq V_2 + V_0^{(n_0+1)} \subseteq V_2 + V_2 \subseteq V_1 \subseteq V_0$, then realizing induction by the number t, it is easy to verify that $\sum_{k=1}^{t}(a \cdot y + 1)^k \in V_1$ for any $t \in \mathbb{N}$. Then

$$y \cdot \left(\sum_{k=0}^{t}(a \cdot y + 1)^k \right) = y + y \cdot \sum_{k=1}^{t}(a \cdot y + 1)^k \in y + y \cdot V_1 \quad \text{for any} \quad t \in \mathbb{N}. \qquad (*)$$

Let U be a neighborhood of zero in R such that $U \cdot y \subseteq V_1$, then since $0 \in [M \cdot a]_R$, there exists an element $b \in M$ such that $b \cdot a \in U$. Let U_1 be a neighborhood of zero in R such that $b \cdot U_1 \subseteq V_1$, and m be a natural number such that $V_0^{(k)} \subseteq U_1$ for $k \geq m$. Since $(a \cdot y + 1)^{(t+1)} = 1 + a \cdot y \cdot \sum_{i=0}^{t}(a \cdot y + 1)^i$ for $t = 1$, then realizing the induction by the number t, it is easy to see that

$$(a \cdot y + 1)^{t+1} = (a \cdot y + 1) \cdot (a \cdot y + 1)^t = (a \cdot y + 1) \cdot \left(1 + a \cdot y \cdot \sum_{i=0}^{t-1}(a \cdot y + 1)^i\right)$$

$$= 1 + a \cdot y + a \cdot y \cdot \sum_{i=1}^{t}(a \cdot y + 1)^i = 1 + a \cdot y \cdot \sum_{i=0}^{t}(a \cdot y + 1)^i$$

for any $t \in \mathbb{N}$. Then, taking into account $(*)$, we get

$$b = b \cdot \left(1 + a \cdot y \cdot \sum_{i=0}^{m-1}(a \cdot y + 1)^i\right) - b \cdot a \cdot y \cdot \left(\sum_{i=0}^{m-1}(a \cdot y + 1)^i\right)$$

$$= b \cdot (a \cdot y + 1)^m - (b \cdot a) \cdot y \cdot \sum_{i=0}^{m-1}(a \cdot y + 1)^i \in b \cdot V_0^{(mno)} + (b \cdot a) \cdot (y + y \cdot V_1)$$

$$\subseteq b \cdot U_1 + U \cdot (y + y \cdot V_1) \subseteq V_1 + V_1 + V_1 \cdot V_1 \subseteq V_0.$$

It contradicts the fact that $V_0 \cap M = \emptyset$. Hence, the element a is not a right topological divisor of zero.

Analogously we prove that the element a is not a left topological divisor of zero. The theorem is proved.

3.5.3. THEOREM. *Let (R, τ_0) be a Hausdorff topological commutative ring, and the topology τ_0 is non-weakable in the class of all Hausdorff ring topologies. For the completion $(\widehat{R}, \widehat{\tau})$ of the topological ring (R, τ_0) to be a topological field it is necessary and sufficient that ring R is topologically simple (see Definition 1.4.15), i.e. that all its non-zero ideals are totally dense.*

PROOF. The necessity of the conditions. If the completion $\widehat{R}$ of the ring R is a field, then any non-zero element of R is invertible in $\widehat{R}$, and, due to Theorem 3.5.1, any non-zero ideal in R is totally dense.

The sufficiency of the conditions. At first we show that the ring $(\widehat{R}, \widehat{\tau})$ is topologically simple. Assume the contrary, i.e. that there is a non-zero ideal I in ring $(\widehat{R}, \widehat{\tau})$ such that $[I]_{\widehat{R}} \neq \widehat{R}$. It is easy to see that the family $\{(I + V) \cap R \mid V$ is a neighborhood of zero in $(\widehat{R}, \widehat{\tau})\}$ satisfies the conditions (BN1) - (BN6) (see Propositions 1.2.1 and 1.2.2). Then, according to Theorem 1.2.5, this family defines a certain ring topology τ_1 on R, forming a basis of neighborhoods of zero of the ring (R, τ_1).

Since $(I+V) \cap R \supseteq V \cap R$ for any neighborhood V of zero in $\widehat{R}$, then the set $(I+V) \cap R$ is a neighborhood of zero in (R, τ_0). It means that $\tau_1 \leq \tau_0$.

Since $[I]_{\widehat{R}} \neq \widehat{R}$ and $[R]_{\widehat{R}} = \widehat{R}$, then $[I]_{\widehat{R}} \cap R \neq R$. Since the ring R does not contain non-trivial closed ideals, then $[I]_{\widehat{R}} \cap R = \{0\}$. According to Proposition 1.2.25, $[I]_{\widehat{R}} = \bigcap_V (I+V)$, hence, $\bigcap_V ((I+V) \cap R) = (\bigcap_V (I+V)) \cap R = [I]_{\widehat{R}} \cap R = \{0\}$, i.e. the family

$\{(I+V)\bigcap R\,|\,V$ is a neighborhood of zero in $(\widehat{R},\widehat{\tau})\}$ satisfies the condition (BN1$'$) (see Corollary 1.3.6). It means that the topology τ_1 is a Hausdorff ring topology. Since the topology τ_0 is non-weakable, then $\tau_0 = \tau_1$.

Let a be a certain non-zero element of I. There exists a neighborhood V_0 of zero in $\widehat{R}$ such that $V_0\bigcap(a+V_0) = \emptyset$. Since $\tau_0 = \tau_1$, then there exists a neighborhood V_1 of zero in $\widehat{R}$ such that $(I+V_1)\bigcap R \subseteq V_0\bigcap R$. Then $(a+(V_1\bigcap V_0))\bigcap R \subseteq V_0$. Since R is a totally dense subring in $\widehat{R}$, then $(a+(V_1\bigcap V_0))\bigcap R \neq \emptyset$, hence, $V_0\bigcap(a+V_0) \supseteq V_0\bigcap(a+(V_1\bigcap V_0))\bigcap R = (a+(V_1\bigcap V_0))\bigcap R \neq \emptyset$. This contradicts the choice of the neighborhood V_0 of zero in $\widehat{R}$. Hence, we've proved that every non-zero ideal in the ring $(\widehat{R},\widehat{\tau})$ is totally dense, i.e. $(\widehat{R},\widehat{\tau})$ is a topologically simple ring.

Due to Corollary 1.7.14, the topology $\widehat{\tau}$ is non-weakable in the class of all Hausdorff ring topologies in $\widehat{R}$. According to Theorem 1.8.15, the topological ring $(\widehat{R},\widehat{\tau})$ does not contain topological divisors of zero. Then, due to Theorem 3.5.1, any non-zero element of $\widehat{R}$ is invertible in the completion $\tilde{R}$ of the topological ring $\widehat{R}$. Since, according to Proposition 3.3.5, the ring $\tilde{R}$ coincides with $\widehat{R}$. Then the ring $\widehat{R}$ is a field.

To complete the proof we must verify that $(\widehat{R},\widehat{\tau})$ is a topological field. Indeed, as we've noted above, the topology $\widehat{\tau}$ is non-weakable in the class of all Hausdorff ring topologies on the ring $\widehat{R}$. Then, due to Theorem 1.7.8, $(\widehat{R},\widehat{\tau})$ is a topological field.

3.5.4. COROLLARY. Let R be a field, τ_0 a Hausdorff ring topology on R. If τ_0 is a non-weakable topology in the class of all Hausdorff ring topologies on R, then the completion $\widehat{R}$ of the topological ring (R,τ_0) is a topological field.

3.5.5. EXAMPLE. Now we show that in general the completion of the topological field is not a topological field even having a topologically nilpotent neighborhood of zero.

Let $\mathbb{Q}$ be the field of rational numbers. For any natural number n consider the set $U_n = \{q\,|\,q = \frac{2^n \cdot 3^n \cdot t}{s}$, s,t are integers and s is not dividable by 2 and 3$\}$.

It is easy to verify that the family $\{U_n\,|\,n \in \mathbb{N}\}$ satisfies the conditions (BN1$'$), (BN2) - (BN6) (see Corollary 1.3.7, Proposition 1.2.1 and 1.2.2). Hence, there exists a certain Hausdorff ring topology τ on $\mathbb{Q}$ with the basis $\{U_n\,|\,n \in \mathbb{N}\}$ of neighborhoods of zero.

Since $U_1^n \subseteq U_n$ for any $n \in \mathbb{N}$, then U_1 is a topologically nilpotent neighborhood of zero. Then, due to Theorem 1.8.44, $(\mathbb{Q}, \tau)$ is a topological field. Denote over $\widehat{\mathbb{Q}}$ its completion. We construct by induction sequences of integers $a_1, a_2, \ldots$ and $b_1, b_2, \ldots$ such that $a_i \in 2^i \cdot \mathbb{Z}$, $b_i \in 3^i \cdot \mathbb{Z}$, and $a_{i+1} - a_i \in 2^i \cdot 3^i \cdot \mathbb{Z}$, $b_{i+1} - b_i \in 2^i \cdot 3^i \cdot \mathbb{Z}$ for any $i \in \mathbb{N}$.

Let $a_1 = 2$ and $b_1 = 3$. Suppose that numbers a_i and b_i are determined for $1 \leq i \leq n$. Since the numbers 2^{n+1} and 3^{n+1} are mutually prime, then there exist integers k and t such that $1 = 2^{n+1} \cdot k + 3^{n+1} \cdot t$. Let $a_{n+1} = 2^{n+1} \cdot k \cdot a_n$ and $b_{n+1} = 3^{n+1} \cdot t \cdot b_n$. Then $a_{n+1} \in 2^{n+1} \cdot \mathbb{Z}$ and $b_{n+1} \in 3^{n+1} \cdot \mathbb{Z}$, where $a_{n+1} - a_n = -3^{n+1} \cdot t \cdot a_n \in 3^n \cdot 2^n \cdot \mathbb{Z}$ and $b_{n+1} - b_n = -2^{n+1} \cdot k \cdot b_n \in 3^n \cdot 2^n \cdot \mathbb{Z}$. Hence, we've constructed the necessary sequences $a_1, a_2, \ldots$ and $b_1, b_2, \ldots$ of integers.

Since

$$a_n - a_k = (a_n - a_{n-1}) + (a_{n-1} - a_{n-2}) + \ldots + (a_{k+1} - a_k) \in 2^k \cdot 3^k \cdot \mathbb{Z} \subseteq U_k$$

and

$$b_n - b_k = (b_n - b_{n-1}) + \ldots + (b_{k+1} - b_k) \in 2^k \cdot 3^k \cdot \mathbb{Z} \subseteq U_k$$

for all natural numbers $n \geq k$, then the constructed sequences are Cauchy sequences in $(\mathbb{Q}, \tau)$. There exist elements $a, b \in \widehat{\mathbb{Q}}$ such that $a = \lim_{i \in \mathbb{N}} a_i$ and $b = \lim_{i \in \mathbb{N}} b_i$.

Since $a_1 = 2 \notin 3 \cdot \mathbb{Z}$ and $b_1 = 3 \notin 2 \cdot \mathbb{Z}$, then it could be verfied by induction that $a_k \notin 3 \cdot \mathbb{Z}$ and $b_k \notin 2 \cdot \mathbb{Z}$ for any $k \in \mathbb{N}$. Then $a_k, b_k \notin U_1$ for any $k \in \mathbb{N}$, hence, $a \neq 0$ and $b \neq 0$.

Since $a_k \cdot b_k \in 2^k \cdot 3^k \cdot \mathbb{Z} \subseteq U_k$ for any $k \in \mathbb{N}$, then

$$a \cdot b = \lim_{k \in \mathbb{N}} a_k \cdot \lim_{k \in \mathbb{N}} b_k = \lim_{k \in \mathbb{N}} (a_k \cdot b_k) = 0.$$

Therefore, we've proved that the ring $\widehat{\mathbb{Q}}$ has a non-zero divisor of zero. Thus, $\widehat{\mathbb{Q}}$ is not a field.

3.5.6. DEFINITION. Let $\mathbb{Q}$ be the field of rational numbers, p a prime natural number, τ_p be the p-adic topology on $\mathbb{Q}$ (see Example 1.1.19). Due to Corollary 1.7.12, the topology τ_p is non-weakable in the class of all Hausdorff ring topologies on ring $\mathbb{Q}$. Due to Corollary 3.5.4, the completion $\widehat{\mathbb{Q}}_p$ of the topological ring $(\mathbb{Q}, \tau_p)$ is a topological field. This field we call the field of p-adic numbers.

3.5.7. THEOREM. *The field $\widehat{\mathbb{Q}}_p$ of p-adic numbers is a locally compact field and its topology is determined by a certain norm.*

PROOF. Due to the definition of topology τ_p on $\mathbb{Q}$, the family $\{V_n \mid n \in \mathbb{N}\}$ of sets, where $V_n = \{\frac{q}{r} \mid q \in p^n \cdot \mathbb{Z} \text{ and } r \in \mathbb{Z}, \ r \text{ is not dividable by } p\}$ is a basis of neighborhoods of zero in $(\mathbb{Q}, \tau_p)$. Since $V_0 = \{1, \ldots, p^n\} + V_n$ for any natural number n, then V_0 is a precompact subset in $(\mathbb{Q}, \tau_p)$. According to Corollary 3.4.17, $[V_0]_{\widehat{\mathbb{Q}}}$ is compact, and, due to Proposition 1.4.24, it is a neighborhood of zero in $\widehat{\mathbb{Q}}_p$. Thus, $\widehat{\mathbb{Q}}_p$ is a locally compact field.

Besides, due to Theorem 3.2.37, the norm ξ_p, that determines the topology τ_p on $\mathbb{Q}$ (see 1.1.19) is extended up to the norm $\widehat{\xi}_p$ on the field $\widehat{\mathbb{Q}}_p$, and the topology determined by the norm $\widehat{\xi}_p$ coincides with the existing topology on the field $\widehat{\mathbb{Q}}_p$.

3.5.8 DEFINITION. An element a of a topological ring R is called topologically quasi-invertible (or, in other terms, topologically quasi-regular) from right, if for any neighborhood V of zero in R there exists an element $b \in R$ such that $a \circ b = a + b - a \cdot b \in V$.

If for the element $a \in R$ there exists an element $a' \in R$ such that $a \circ a' = 0$, then the element a is called quasi-invertible (or, in other terms, quasi-regular) from right (see B.4).

Analogously is defined a quasi-invertible from left and topologically quasi-invertible from left element of R. If an element is quasi-invertible (topologically quasi-invertible) from right and from left, then we call it quasi-invertible (topologically quasi-invertible).

3.5.9. DEFINITION. If all elements of a topological ring R are quasi-invertible, topologically quasi-invertible (from left, from right), then we call ring R quasi-regular, topologically quasi-regular (from left, from right) corespondingly.

3.5.10. PROPOSITION. *Any topologically nilpotent element in a topological ring R is topologically quasi-regular.*

PROOF. Let a be a topologically nilpotent element of R and V a neighborhood of zero. There exists a natural number n such that $a^k \in V$ for any $k \geq n$. Let $b = -\sum_{i=1}^{n-1} a^i$, then $a \circ b = a + (-\sum_{i=1}^{n-1} a^i) - a \cdot (-\sum_{i=1}^{n-1} a^i) = a^n \in V$ and, analogously, $b \circ a = a^n \in V$. The proposition is proved.

3.5.11. PROPOSITION. *Let R be a bounded (see Definition 1.6.2) topological ring* *and A its totally dense subring. The ring A is topologically quasi-regular if and only if the ring R is topologically quasi-regular.*

PROOF. The necessity of the conditions. Let $r \in R$ and V_0 be a neighborhood of zero in R. Then there exists a neighborhood V_1 of zero in R such that $V_1 + V_1 + V_1 \cdot R \subseteq V_0$. Since A is a totally dense subring, then there exists an element $a \in A$ such that $r \in a + V_1$. Since the ring A is topologically quasi-regular from right, then there exists an element $b \in A$ such that $a \circ b = a + b - a \cdot b \in V_1$. Then

$$r \circ b = r + b - r \cdot b \in a + V_1 + b - (a + V_1) \cdot b$$
$$\subseteq a + b - a \cdot b + V_1 + V_1 \cdot R \subseteq V_1 + V_1 + V_1 \cdot R \subseteq V_0,$$

i.e. in R the element r is topologically quasi-invertible from right.

Analogously we prove the topological quasi-invertibility from left in R of the element r. Hence, R is a topologically quasi-regular ring.

The sufficiency of the conditions. Let $a \in A$ and U_0 be a neighborhood of zero in A. There exist neighborhoods V_0 and V_1 of zero in R such that $U_0 = V_0 \cap A$ and $V_1 + V_1 + a \cdot V_1 \subseteq V_0$. Since R is a topologically quasi-regular ring, then there exists an element $b \in R$ such that $a \circ b \in V_1$. Since the ring A is totally dense in R, then there exists an element $a_1 \in A$ such that $a_1 \in b + V_1$. Then

$$a \circ a_1 = a + a_1 - a \cdot a_1 \in a + b + V_1 - a \cdot b + a \cdot V_1$$
$$\subseteq a \circ b + V_1 + a \cdot V_1 \subseteq V_1 + V_1 + a \cdot V_1 \subseteq V_0,$$

i.e $a \circ a_1 \in V_0 \cap A = U_0$.

Hence, we've proved the topological quasi-invertibility from right in A of the element a.

The topological quasi-invertibility from left in A of the element a is proved analogously. Thus, A is a topologically quasi-regular ring.

3.5.12. THEOREM. *Let a Hausdorf topological ring R be topologically nilpotent, then its completion $\widehat{R}$ is a quasi-regular ring.*

PROOF. Let $a \in \widehat{R}$. For any natural number n consider the element $b_n = -\sum_{i=1}^n a^i$. Now we show that the sequence $\{b_1, b_2, \cdots\}$ is a Cauchy sequence in $\widehat{R}$. Indeed, let V

be a neighborhood of zero in $\widehat{R}$. Due to Proposition 1.8.33, the ring $\widehat{R}$ is topologically nilpotent. Hence, there exists a natural number n such that $\widehat{R}^{(n+1)} \subseteq V$ (see B.12). Then

$$b_n - b_k = \sum_{i=n+1}^{k} a^i = a^n \cdot \sum_{i=1}^{k-n} a^i \in \widehat{R}^{(n)} \cdot \widehat{R} = \widehat{R}^{(n+1)} \subseteq V$$

for any $k \geq n$. Hence, the sequence $\{b_1, b_2, \dots\}$ is a Cauchy sequence in $\widehat{R}$.

Since $\widehat{R}$ is a complete ring, then there exists an element $b' \in \widehat{R}$ such that $b' = \lim_{n \to \infty} b_n$.

Now we show that $a \circ b' = 0$. Assume the contrary, i.e. that $a \circ b' \neq 0$. Then there exist symmetric neighborhoods V_0 and V_1 of zero in $\widehat{R}$ such that $V_0 \cap ((a \circ b') + V_0) = \emptyset$ and $V_1 - a \cdot V_1 \subseteq V_0$. Since $b' = \lim_{n \to \infty} b_n$, then there exists a natural number m_0 such that $b' - b_n \in V_1$ for all $n \geq m_0$. Since $\widehat{R}$ is a topologically nilpotent ring, then there exists a natural number m_1 such that $\widehat{R}^{(k)} \subseteq V_0$ for $k \geq m_1$. Let $n > \max\{m_0, m_1\}$, then

$$a \circ b_n = a + b_n - a \cdot b_n \in a + (b' + V_1) - a \cdot (b' + V_1)$$
$$= a + b' - a \cdot b' + V_1 - a \cdot V_1 \subseteq a \circ b' + V_0$$

and

$$a \circ b_n = a + \left(-\sum_{i=1}^{n} a^i \right) - a \cdot \left(-\sum_{i=1}^{n} a^i \right) = a^{n+1} \in \widehat{R}^{(n+1)} \subseteq V_0.$$

Hence, $V_0 \cap (a \circ b' + V_0) \neq \emptyset$. This contradicts the choice of the neighborhood V_0. Thus $a \circ b' = 0$. Analogously we prove that $b' \circ a = 0$. Hence, $\widehat{R}$ is a quasi-regular ring.

3.5.13. THEOREM. *Let R be a Hausdorff topological ring bounded from left (right), and $\widehat{R}$ be its completion. If R is a topologically quasi-regular ring, then any element of R is quasi-invertible in $\widehat{R}$.*

PROOF. According to Theorem 1.6.31, the topological ring R has a basis $\mathcal{B}$ of neighborhoods of zero consisting of right ideals of the multiplicative semigroup $R(\cdot)$. Let $a \in R$. For any neighborhood $V \in \mathcal{B}$ put:

$$F_V = \{b \in \widehat{R} \mid a \circ b \in V\}.$$

Now we show that the family $\{F_V \mid V \in \mathcal{B}\}$ is a basis of a certain Cauchy filter of $\widehat{R}$. Since R is a topologically quasi-regular ring, then $\emptyset \neq \{b \in R \mid a \circ b \in V\} \subseteq F_V$. Therefore,

$F_V \neq \emptyset$ for any $V \in \mathcal{B}$. Besides, $F_{V_3} \subseteq F_{V_2} \cap F_{V_1}$ for such elements $V_1, V_2, V_3 \in \mathcal{B}$, that $V_3 \subseteq V_1 \cap V_2$. Hence, the family $\{F_V | V \in \mathcal{B}\}$ is a basis of a certain filter $\mathcal{A}$ of $\widehat{R}$.

Now we verify that $\mathcal{A}$ is a Cauchy filter. Indeed, let U be a neighborhood of zero in $\widehat{R}$. There exist $V_0, V_1 \in \mathcal{B}$ such that $V_0 - V_0 \subseteq U$ and $V_1 - V_1 - V_1 \subseteq V_0$. Since the ring R is topologically quasi-regular, there exists an element $b \in R$ such that $b \circ a \in V_1$.

Let V_2 be a neighborhood of zero from $\mathcal{B}$ such that $V_2 + b \cdot V_2 \subseteq V_1$, and $c \in F_{V_2}$. Then

$$(b \circ a) \circ c = (b \circ a) + c - (b \circ a) \cdot c \in V_1 + c - V_1 \cdot c \subseteq c + V_1 + V_1$$

and

$$b \circ (a \circ c) = b + (a \circ c) - b \cdot (a \circ c) \in b + V_2 - b \cdot V_2 \subseteq b + V_1.$$

Since $(b \circ a) \circ c = b \circ (a \circ c)$, then $c \in b + V_1 - V_1 - V_1 \subseteq b + V_0$. Thus, $F_{V_2} \subseteq b + V_0$. Then $F_{V_2} - F_{V_2} \subseteq V_0 - V_0 \subseteq U$. Hence, we've proved that $\mathcal{A}$ is a Cauchy filter of $\widehat{R}$.

Since the ring $\widehat{R}$ is complete, then there exists an element $a' \in \widehat{R}$ such that $a' = \lim \mathcal{A}$.

Now we show that $a \circ a' = 0$. Assume the contrary, i.e. that $a \circ a' \neq 0$. Then there exist neighborhoods U, U_1 of zero in $\widehat{R}$ such that $a \circ a' \notin U$ and $U_1 - U_1 + a \cdot U_1 \subseteq U$.

Since $a' = \lim \mathcal{A}$, then, according to Definition 3.1.14, there exists a neighborhood V_1 from $\mathcal{B}$ such that $V_1 \subseteq U_1$ and $F_{V_1} \subseteq a' + U_1$. Let $b \in F_{V_1}$. Then $b \in a' + U_1$, hence, $a' \in b - U_1$.

Since $b \in F_{V_1}$, then

$$a \circ a' = a + a' - a \cdot a' \in a + b - U_1 - a \cdot b + a \cdot U_1$$
$$= a \circ b - U_1 + a \cdot U_1 \subseteq V_1 - U_1 + a \cdot U_1 \subseteq U.$$

It contradicts the choice of the neighborhood U. Hence, $a \circ a' = 0$.

Now we show that $a' \circ a = 0$. Assume the contrary, i.e. that $a' \circ a \neq 0$. Let U be a neighborhood of zero in $\widehat{R}$ such that $a' \circ a \notin U$. We can consider that U is a closed neighborhood of zero in $\widehat{R}$. There exist neighborhoods V_0 and V_1 of zero from $\mathcal{B}$ such that $V_0 + V_0 + V_0 \subseteq U$ and $V_1 + V_1 \subseteq V_0$. Then, due to Proposition 1.1.43, $[V_0]_{\widehat{R}}$ and $[V_1]_{\widehat{R}}$ are right ideals of the multiplicative semigroup of the ring $\widehat{R}$.

Since R is a topologically quasi-regular ring, then there exists an element $c \in R$ such that $c \circ a \in V_1$. Then

$$c = c \circ 0 = c \circ a \circ a' = (c \circ a) + a' - (c \circ a) \cdot a' \in a' + V_1 - V_1 \cdot a'$$
$$\subseteq a' + V_1 + [V_1]_{\widehat{R}} \subseteq a' + [V_0]_{\widehat{R}},$$

hence, $a' \in c - [V_0]_{\widehat{R}}$.

Due to the choice of the neighborhood V_0 and of the element c,

$$a' \circ a = a' + a - a' \cdot a \in c - [V_0]_{\widehat{R}} + a - c \cdot a + [V_0]_{\widehat{R}} \cdot a$$
$$= c \circ a - [V_0]_{\widehat{R}} + [V_0]_{\widehat{R}} \subseteq [U]_{\widehat{R}} = U.$$

It contradicts the choice of the neighborhood U of zero in $\widehat{R}$. Hence, $a' \circ a = 0$. Therefore, we've shown that any element of R is quasi-invertible in $\widehat{R}$.

Analogously is considered the case when R is the ring bounded from right.

The theorem is proved.

3.5.14. COROLLARY. The completion $\widehat{R}$ of a bounded Hausdorff topologically quasi-regular ring R is a quasi-regular ring.

PROOF. Indeed, according to Corollary 3.3.11, $\widehat{R}$ is a bounded ring. Then, due to Proposition 3.5.11, $\widehat{R}$ is a topologically quasi-regular ring. Due to Theorem 3.5.13, any element of $\widehat{R}$ is quasi-invertible in the completion of the ring $\widehat{R}$. Since the completion of the ring $\widehat{R}$ coincides with $\widehat{R}$, then $\widehat{R}$ is a quasi-regular ring.

4. Products of Topological Rings and Modules

The constructions of the products of topological rings and modules are as important for the topological algebra as the construction of completion considered in the previous chapter.

We consider in this chapter such constructions of topological rings and modules as Tychonoff direct product, a direct product with the brick topology, local product, semidirect products and the limit of an inverse spectrum. We investigate the connections between these products. Using the semidirect product and completion constructions, we get the criterion of the embedding of a precompact ring into a compact ring with the unitary element.

Mostly results included in § 4.1 are well known to the specialists in topological algebra. Original results included in this section could be found in [148, 151, 156, 157] as well as the original results of § 4.2 could be found in [242] and of § 4.3 in [82, 254, 432].

§ 4.1. Direct Products

4.1.1. DEFINITION. Let $\{X_i \mid i \in I\}$ be a family of sets. By the direct product $X = \prod_{i \in I} X_i$ of the sets X_i we mean the set of all mappings $f : I \to \bigcup_{i \in I} X_i$ from the set I to $\bigcup_{i \in I} X_i$ such that $f(i) \in X_i$ for all $i \in I$.

Let $I = \{1, \dots, n\}$ be a finite set, $J = \{1, 2, \dots\}$ a countable set. Then we denote $\prod_{i \in I} X_i$ over $\prod_{i=1}^{n} X_i$ or $X_1 \times X_2 \times \dots \times X_n$, and $\prod_{i \in J} X_i$ over $\prod_{i=1}^{\infty} X_i$. The elements of $\prod_{i=1}^{n} X_i$ we denote as the sets $(f(1), \dots, f(n))$, and the elements of the $\prod_{i=1}^{\infty} X_i$ as the sequences $(f(1), f(2), \dots)$.

We call X_i the factor of the direct product $X = \prod_{i \in I} X_i$, and the mappings $\pi_i : X \to X_i$, acting according to the rule: $\pi_i(f) = f(i)$ for any $f \in X$, we call the canonical projections from X to the factors X_i.

If X_i is an Abelian group for any $i \in I$, then we define on X the operation of addition and taking of the opposite element as:

$$(f_1 + f_2)(i) = f_1(i) + f_2(i) \text{ and } (-f)(i) = -f(i) \text{ for } f, f_1, f_2 \in X \text{ and } i \in I.$$

It is easy to verify that the set X is an Abelian group with respect to these operations, and the neutral element is such mapping $f_0 : I \to \prod_{i \in I} X_i$ that $f_0(i) = 0_i$, where 0_i is the neutral element of the group X_i. It is evident that for any $i \in I$ the projection π_i is a homomorphism of Abelian groups. We call the set X, with the operations determined above, the direct product of the family $\{X_i \mid i \in I\}$ of Abelian group.

4.1.2. PROPOSITION. *Let* $\mathfrak{m}$ *be an infinite cardinal number,* $\{X_i \mid i \in I\}$ *a family of topological Abelian groups and* $\mathcal{B}(X_i)$ *the system of all neighborhoods of zero in the group* X_i, $i \in I$. *Let* $\mathcal{A}$ *be the system of all subsets in* X *of the type* $\prod_{i \in I} U_i$, *where* $U_i \in \mathcal{B}(X_i)$ *and* $\left| \{i \in I \mid U_i \neq X_i\} \right| < \mathfrak{m}$ *(see A.7), then there exists the only group topology* τ *on the group* X *such that* $\mathcal{A}$ *is a basis of neighborhoods of zero in* (X, τ). *All canonical projections* $\pi_i : X \to X_i$ *are continuous and open in this topology.*

PROOF. Due to Theorem 1.2.4, it is enough to verify the fulfillment of the conditions (BN1) - (BN4).

Now we verify the fulfillment of the condition (BN1). The neutral element of the group X is the element f_0 such that $f_0(i) = 0_i$ for all $i \in I$, where 0_i is the neutral element of the Abelian group X_i. Then $f_0 \in \prod_{i \in I} U_i$ for any family of the sets $U_i \in \mathcal{B}(X_i)$, since $f_0(i) = 0_i \in U_i$ for all $i \in I$.

Now we verify the fulfillment of the condition (BN2). Let $\prod_{i \in I} U_i$, $\prod_{i \in I} V_i \in \mathcal{A}$. Let also $W_i = U_i \cap V_i$ for $i \in I$, then $W_i \in \mathcal{B}(X_i)$ and

$$\left| \{i \mid W_i \neq X_i\} \right| = \left| \{i \mid U_i \neq X_i\} \cup \{i \mid V_i \neq X_i\} \right| < \mathfrak{m} + \mathfrak{m} = \mathfrak{m},$$

since $\mathfrak{m}$ is an infinite cardinal. Thus, $\prod_{i \in I} W_i \in \mathcal{A}$ and

$$\prod_{i \in I} W_i \subseteq \left(\prod_{i \in I} U_i \right) \cap \left(\prod_{i \in I} V_i \right).$$

Now we verify the fulfillment of the condition (BN3). Let $\prod_{i \in I} U_i \in \mathcal{A}$. For any $i \in I$ we determine $V_i \in \mathcal{B}(X_i)$ such that :

if $U_i \neq X_i$, then we take a neighborhood V_i of zero in X_i such that $V_i + V_i \subseteq U_i$;

if $U_i = X_i$, then we take $V_i = X_i$.

Then $\prod_{i \in I} V_i \in \mathcal{A}$, (since $\{i \mid V_i \neq X_i\} = \{i \mid U_i \neq X_i\}$), and $\prod_{i \in I} V_i + \prod_{i \in I} V_i = \prod_{i \in I} (V_i + V_i) \subseteq \prod_{i \in I} U_i$.

Now we verify the fulfillment of the condition (BN4). Let $\prod_{i \in I} U_i \in \mathcal{A}$. For any $i \in I$ put $V_i = -U_i$. Then $\prod_{i \in I} V_i \in \mathcal{A}$, (since $\{i \mid V_i \neq X_i\} = \{i \mid U_i \neq X_i\}$), and $-\prod_{i \in I} V_i = \prod_{i \in I}(-V_i) \subseteq \prod_{i \in I} U_i$.

Hence, we've proved that the system $\mathcal{A}$ is a basis of neighborhoods of zero of a certain group topology τ on the group X.

Now we prove that all canonical projections $\pi_i : X \to X_i$ are continuous and open for all $i \in I$. Indeed, let $i_0 \in I$ and V_{i_0} be a neighborhood of zero in X_{i_0}. Then the subset $U = \prod_{i \in I} W_i$ in X, where $W_{i_0} = V_{i_0}$ and $W_i = X_i$ for $i \in I$, and $i \neq i_0$, is a neighborhood of zero in X and $\pi_{i_0}(U) = V_{i_0}$. It proves that the mapping π_{i_0} is continuous for any $i_0 \in I$. If $V = \prod_{i \in I} V_i$ is a neighborhood of zero in X, then $\pi_i(V) = V_i$ for any $i \in I$ according to the determination of the topology on X. It proves that the mapping π_i is open for $i \in I$.

4.1.3. DEFINITION. Let $\{X_i \mid i \in I\}$ be a family of topological Abelian groups, $\mathfrak{m}$ a certain infinite cardinal number. The direct product $X = \prod_{i \in I} X_i$ of the groups X_i, endowed with the topology from Proposition 4.1.2, is called the $\mathfrak{m}$-product of the groups X_i, $i \in I$. The $\aleph_0$-product is called a Tychonoff product, or, in other terms, the product with the Tychonoff topology τ_T.

4.1.4. REMARK. *Let $\{X_i \mid i \in I\}$ be a family of topological Abelian groups, (X, τ_T) their Tychonoff product. Denote over τ_i the topology on X that is the prototype of the group topology on X_i with respect to the projection π_i. Then $\tau_T = \sup\{\tau_i \mid i \in I\}$.*

PROOF. Indeed, let $i_0 \in I$ and V be a neighborhood of zero in X_{i_0}, then $\pi_{i_0}^{-1}(V) = \prod_{i \in I} V_i$, where $V_i = V$ if $i = i_0$ and $V_i = X_i$ in all other cases, while $\{\pi_{i_0}^{-1}(V) \mid V$ is a neighborhood of zero in $X_{i_0}\}$ is a basis of neighborhoods of zero in (X, τ_{i_0}). Then the system of all sets of the type $\bigcap_{k=1}^{n} \pi_{i_k}^{-1}(V_k) = \prod_{i \in I} U_i$, where V_k is a neighborhood of zero in X_{i_k} and $U_i = V_k$ where $i = i_k$ for $k = 1, \dots, n$, and $U_i = X_i$ in all other cases, forms a basis of neighborhoods of zero in X with the topology $\sup\{\tau_i \mid i \in I\}$. According to the definition of the Tychonoff product, such sets also form a basis of the neighborhoods of zero of the Tychonoff topology on the group X.

4.1.5. PROPOSITION. *Let $\mathfrak{m}$ be a cardinal number, $\{X_i \mid i \in I\}$ be a family of topological Abelian groups, $X = \prod_{i \in I} X_i$ their $\mathfrak{m}$-product. For $i_0 \in I$ let $X'_{i_0} = \{f \in$*

$X \mid f(i) = 0_i$ for $i \neq i_0\}$, where 0_i is the neutral element of the group X_i for $i \in I$. Then X'_{i_0} is a subgroup of the topological group X, topologically isomorphic to the group X_{i_0}. If for any $i \in I$ the group X_i is Hausdorff, then X'_{i_0} is a closed subgroup in X.

PROOF. Let π'_{i_0} be the restriction on X'_{i_0} of the canonical projection π_{i_0}. It is clear that π'_{i_0} is an isomorphism of the groups X'_{i_0} and X_{i_0} and, due to Proposition 4.1.2, it is a continuous mapping.

Now we prove that π'_{i_0} is an open isomorphism. Indeed, let V be a neighborhood of zero in X'_{i_0}. Then there exists a neighborhood W of zero in X such that $W \cap X'_{i_0} \subseteq V$. Without loss of generality we can consider that $W = \prod_{i \in I} U_i$, where U_i is a neighborhood of zero in X_i. Then $\pi'_{i_0}(V) \supseteq U_{i_0}$, hence, $\pi'_{i_0}(V)$ is a neighborhood of zero in X_{i_0}. Hence, π'_{i_0} is an open isomorphism, therefore π'_{i_0} is a topological isomorphism.

Now we prove that if the group X_i is Hausdorff for every $i \in I$, then X'_{i_0} is a closed subgroup in X. Assume the contrary, i.e. that $X'_{i_0} \neq [X'_{i_0}]_X$. It means that there exists an element $f \in [X'_{i_0}]_X \backslash X'_{i_0}$. Then there exists an element $i_1 \in I$, $i_1 \neq i_0$, such that $f(i_1) \neq 0_{i_1}$. Since X_{i_1} is a Hausdorff group, then there exists a neighborhood V of zero in X_{i_1} such that $0_{i_1} \notin (f(i_1) + V)$. Let $W = \prod_{i \in I} U_i$, where $U_{i_1} = V$ and $U_i = X_i$ for $i \in I$ and $i \neq i_1$. Then W is a neighborhood of zero in X, hence $(f + W) \cap X'_{i_0} \neq \emptyset$. If $g \in (f + W) \cap X'_{i_0}$, then $g(i_1) \in f(i_1) + V$, therefore $g(i_1) \neq 0_{i_1}$. It contradicts the fact that $g \in X'_{i_0}$. This contradiction proves that X'_{i_0} is a closed subgroup in X.

4.1.6. PROPOSITION. *Let* $\{X_i \mid i \in I\}$ *be a family of topological Abelian groups,* $\mathfrak{m}$ *a certain cardinal number. For any* $i \in I$ *we fix up a certain basis* $\mathcal{B}_0(X_i)$ *of neighborhoods of zero in* X_i*, where* $X_i \in \mathcal{B}_0(X_i)$*. Then the family* $\mathcal{L}$ *of all subsets in* $X = \prod_{i \in I} X_i$ *of the type* $\prod_{i \in I} V_i$*, where* $V_i \in \mathcal{B}_0(X_i)$ *and* $\left| \{ i \mid V_i \neq X_i \} \right| < \mathfrak{m}$*, forms a basis of neighborhoods of zero in the* $\mathfrak{m}$*-product* $\prod_{i \in I} X_i$*.*

PROOF. It is clear that any of the subsets of the family $\mathcal{L}$ is a neighborhood of zero in the $\mathfrak{m}$-product $\prod_{i \in I} X_i$. If $W = \prod_{i \in I} U_i$ is a subset from the family $\mathcal{A}$ in X (see Proposition 4.1.2), then for any $i \in I$ such that $U_i \neq X_i$ there exists a set $V_i \in \mathcal{B}_0(X_i)$ such that $V_i \subseteq U_i$. For all other $i \in I$ put $V_i = X_i$. Then $V = \prod_{i \in I} V_i \subseteq W$ and $V \in \mathcal{L}$. The proposition is proved.

4.1.7. PROPOSITION. *Let* $\{X_i \mid i \in I\}$ *be a family of topological Abelian groups, where any of the groups* X_i *has a basis* $\mathcal{B}_0(X_i)$ *of neighborhoods of zero consisting of subgroups, and* $\mathfrak{m}$ *be a certain cardinal number. Then the* $\mathfrak{m}$*-product* $\prod_{i \in I} X_i$ *also has a basis of neighborhoods of zero consisting of subgroups.*

PROOF. In any of the groups $X_i, i \in I$, fix up a basis $\mathcal{B}_0(X_i)$ of neighborhoods of zero consisting of subgroups such that $X_i \in \mathcal{B}_0(X_i)$. Then, due to Proposition 4.1.6, the family $\mathcal{L}$ of subsets of the type $\prod_{i \in I} V_i$ in $\prod_{i \in I} X_i$, where $V_i \in \mathcal{B}_0(X_i)$ and $\left| \{i \mid V_i \neq X_i\} \right| < \mathfrak{m}$, forms a basis of neighborhoods of zero in $\prod_{i \in I} X_i$. It is easy to see that any of the sets $\prod_{i \in I} V_i$ is a subgroup in $\prod_{i \in I} X_i$, i.e. $\mathcal{L}$ is a basis of neighborhoods of zero consisting of subgroups.

4.1.8 DEFINITION. Let $\{R_i \mid i \in I\}$ be a family of rings. Consider the direct product $R = \prod_{i \in I} R_i$ of the Abelian groups $R_i(+)$ and define the multiplication operation on it as: $(f_1 \cdot f_2)(i) = f_1(i) \cdot f_2(i)$ for $f_1, f_2 \in R$ and $i \in I$.

It is easy to see that with respect to this operation the group $R(+)$ is a ring. We call this ring the direct product of the family $\{R_i \mid i \in I\}$ of rings.

4.1.9. PROPOSITION. *Let* $\{R_i \mid i \in I\}$ *be a family of topological rings and* $\mathfrak{m}$ *be a cardinal number. Consider the topology of the* $\mathfrak{m}$*-product of Abelian groups* $R_i(+)$, $i \in I$, *on the direct product* $R = \prod_{i \in I} R_i$ *of the rings* R_i. *Then* R *is a topological ring.*

PROOF. According to the Theorem 1.2.5, it is enough to show that the family $\mathcal{A}$ of subsets of the type $\prod_{i \in I} U_i$, where U_i is a neighborhood of zero in R_i and $\left| \{i \mid U_i \neq R_i\} \right| < \mathfrak{m}$, satisfies the conditions (BN1) - (BN6).

We showed in Proposition 4.1.2 that this family satisfies the conditions (BN1) - (BN4).

Now we verify the fulfillment of the condition (BN5). Let $\prod_{i \in I} U_i \in \mathcal{A}$. For any $i \in I$ we define the neighborhood V_i of zero in R_i as the following:

if $U_i \neq R_i$, then we choose V_i such that $V_i \cdot V_i \subseteq U_i$;

if $U_i = R_i$, then we take $V_i = R_i$.

Then $\prod_{i \in I} V_i \in \mathcal{A}$, since $\{i \mid V_i \neq R_i\} \subseteq \{i \mid U_i \neq R_i\}$, and

$$\left(\prod_{i \in I} V_i\right) \cdot \left(\prod_{i \in I} V_i\right) \subseteq \prod_{i \in I} (V_i \cdot V_i) \subseteq \prod_{i \in I} U_i.$$

Now we verify the fulfillment of the condition (BN6). Let $f \in R$ and $\prod_{i \in I} U_i \in \mathcal{A}$. For any $i \in I$ we define the neighborhood V_i of zero in R_i as:

if $U_i \neq R_i$, then we choose V_i such that $f(i) \cdot V_i \subseteq U_i$ and $V_i \cdot f(i) \subseteq U_i$;

if $U_i = R_i$, then we take $V_i = R_i$.

Since $\{i \mid V_i \neq R_i\} \subseteq \{i \mid U_i \neq R_i\}$, then $\prod_{i \in I} V_i \in \mathcal{A}$, and $f \cdot \prod_{i \in I} V_i \subseteq \prod_{i \in I} (f(i) \cdot V_i) \subseteq \prod_{i \in I} U_i$ and $(\prod_{i \in I} V_i) \cdot f \subseteq \prod_{i \in I} (V_i \cdot f(i)) \subseteq \prod_{i \in I} U_i$.

The proposition is proved.

4.1.10. DEFINITION. Let $\{R_i \mid i \in I\}$ be a family of topological rings, $\mathfrak{m}$ a certain cardinal number. We call the topological ring $R = \prod_{i \in I} R_i$, endowed with the topology from Proposition 4.1.9, the $\mathfrak{m}$-product of the family $\{R_i \mid i \in I\}$ of topological rings.

4.1.11. PROPOSITION. Let $\{R_i \mid i \in I\}$ be a family of topological rings, where each of them has a basis of neighborhoods of zero consisting of right (left, two-sided) ideals. Then the $\mathfrak{m}$-product $R = \prod_{i \in I} R_i$ has a basis of the neighborhoods of zero consisting of right (left, two-sided) ideals, too.

The proof is analogous to the proof of Proposition 4.1.7.

4.1.12. DEFINITION. Let (X, τ) be a topological space, $\mathfrak{m}$ be a cardinal number. We call the topology τ a $\mathfrak{m}$-topology if for any set of indexes I, such that $|I| < \mathfrak{m}$ (see A.7), and any family $\{U_i \mid i \in I\}$ of open sets U_i in (X, τ), the set $\bigcap_{i \in I} U_i$ is open in (X, τ), too.

4.1.13. REMARK. A topology τ is a $\aleph_0$-topology for any topological space (X, τ). The discrete topology is a $\mathfrak{m}$-topology for any cardinal number $\mathfrak{m}$.

4.1.14. DEFINITION. Let R be a ring and $\{M_i \mid i \in I\}$ a family of right R-modules. Consider the direct product $M = \prod_{i \in I} M_i$ of the Abelian groups M_i and define on it the multiplication operation by the elements of the ring R as the following:

$(f \cdot r)(i) = f(i) \cdot r$ for $f \in M, r \in R$ and $i \in I$.

It is easy to verify that with respect to this operation the group M is a right R-module. We call this module a direct product of the family $\{M_i \mid i \in I\}$ of modules.

4.1.15. PROPOSITION. *Let $\mathfrak{m}$ be a certain cardinal number, R a topological ring whose topology is a $\mathfrak{m}$-topology and $\{M_i \mid i \in I\}$ be a family of topological right R-modules.*

Consider the direct product $M = \prod_{i \in I} M_i$ of the modules M_i endowed with the topology of the $\mathfrak{m}$-product of the topological Abelian groups M_i. Then M is a topological right R-module.

PROOF. According to Proposition 4.1.2, the family $\mathcal{A}$ of the subsets $\prod_{i \in I} U_i$, where U_i is a neighborhood of zero in M_i and $\left| \{i \,|\, U_i \neq M_i\} \right| < \mathfrak{m}$, satisfies the conditions (BN1) - (BN4). According to Theorem 1.2.6, to complete the proof it is enough to verify the fulfillment of the conditions (BN5$'$), (BN6$'$) and (BN6$''$) for the family $\mathcal{A}$.

Now we verify the fulfillment of the condition (BN5$'$). Let $\prod_{i \in I} U_i \in \mathcal{A}$. For every $i \in I$ we define the neighborhood V_i of zero in M_i and the neighborhood W_i of zero in R as the following:

if $U_i \neq M_i$, then we choose V_i and W_i such that $V_i \cdot W_i \subseteq U_i$;

if $U_i = M_i$, then we take $V_i = M_i$ and $W_i = R$.

Then $\prod_{i \in I} V_i \in \mathcal{A}$, since $\{i \,|\, V_i \neq M_i\} \subseteq \{i \,|\, U_i \neq M_i\}$. Let $W = \bigcap_{i \in I} W_i$. Since the topology on R is a $\mathfrak{m}$-topology and $W = \bigcap_{i \in I} W_i = \bigcap_{i \in J} W_i$, where $J = \{i \,|\, U_i \neq M_i\}$, then W is a neighborhood of zero in R, and

$$\left(\prod_{i \in I} V_i \right) \cdot W \subseteq \prod_{i \in I} (V_i \cdot W) \subseteq \prod_{i \in I} (V_i \cdot W_i) \subseteq \prod_{i \in I} U_i.$$

Now we verify the fulfillment of the condition (BN6$'$). Let $\prod_{i \in I} U_i \in \mathcal{A}$ and $r \in R$. For every $i \in I$ we define the neighborhood V_i of zero in M_i as the following:

if $U_i \neq M$, then we take a neighborhood V_i of zero in M_i such that $V_i \cdot r \subseteq U_i$;

if $U_i = M$, then we take $V_i = M_i$.

Since $\{i \,|\, V_i \neq M_i\} \subseteq \{i \,|\, U_i \neq M_i\}$, then $\prod_{i \in I} V_i \in \mathcal{A}$ and $(\prod_{i \in I} V_i) \cdot r = \prod_{i \in I} (V_i \cdot r) \subseteq \prod_{i \in I} U_i$.

Now we verify the fulfillment of the condition (BN6$''$). Let $\prod_{i \in I} U_i \in \mathcal{A}$ and $f \in M$. For every $i \in I$ we define the neighborhood W_i of zero in R as the following:

if $U_i \neq M_i$, then we take a neighborhood W_i of zero in R such that $f(i) \cdot W_i \subseteq U_i$;

if $U_i = M_i$, then we take $W_i = R$.

Let $W = \bigcap_{i \in I} W_i$. Since the topology on R is a $\mathfrak{m}$-topology, and $W = \bigcap_{i \in I} W_i = \bigcap_{i \in J} W_i$, where $J = \{i \,|\, W_i \neq R\} \subseteq \{i \,|\, U_i \neq M_i\}$, then W is a neighborhood of zero in R, and $f \cdot W \subseteq \prod_{i \in I} U_i$, since $f(i) \cdot W \subseteq f(i) \cdot W_i \subseteq U_i$ for any $i \in I$.

The proposition is proved.

4.1.16. DEFINITION. Let $\mathfrak{m}$ be a cardinal number, R be a topological ring, the topology of which is a $\mathfrak{m}$-topology and $\{M_i \,|\, i \in I\}$ be a family of topological R-modules. The topological R-module $M = \prod_{i \in I} M_i$ endowed with the topology from Proposition 4.1.15 we call a $\mathfrak{m}$-product of the family $\{M_i \,|\, i \in I\}$ of topological modules.

4.1.17. REMARK. As we noted in Remark 4.1.13, a topology of any topological space is a $\aleph_0$-topology, then the Tychonoff product of an arbitrary family of topological modules over a topological ring is a topological module.

4.1.18 EXAMPLE. Let $\mathbb{R}$ be the field of real numbers endowed with the interval topology, and $\mathfrak{m}$ be a cardinal number, where $\mathfrak{m} > \aleph_0$. It is easy to verify that the interval topology on $\mathbb{R}$ is not a $\mathfrak{m}$-topology.

Let $\{M_i \,|\, i \in \mathbb{N}\}$ be a countable family of the copies of the field $\mathbb{R}$ considered as a topological module over itself and $M = \prod_{i \in \mathbb{N}} M_i$ be its $\mathfrak{m}$-product.

Now we prove that the $\mathbb{R}$-module M is not a topological $\mathbb{R}$-module (and show that the condition for the topology on R to be a $\mathfrak{m}$-topology, used in Proposition 4.1.15, is important).

Indeed, let $U = (-1; 1)$ be the open interval in $\mathbb{R}$. Due to the definition of $\mathfrak{m}$-product, the subset $\prod_{i \in \mathbb{N}} U_i$, where $U_i = U$ for all $i \in \mathbb{N}$, is open in M. Let f be an element of M such that $f(i) = 2^i$ for $i \in \mathbb{N}$. Assume that $\mathbb{R}$-module M is topological. Then there exists a neighborhood V of zero in $\mathbb{R}$ such that $f \cdot V \subseteq \prod_{i \in \mathbb{N}} U_i$. Let $0 < r \in V$, then there exists a natural number n such that $2^n \cdot r > 1$, i.e. $f(n) \cdot r \notin U$. It means that $f \cdot V \nsubseteq \prod_{i \in \mathbb{N}} U_i$. The contradiction received proves that the module M is not topological.

4.1.19. PROPOSITION. Let $\mathfrak{m}$ be a cardinal number, R a topological ring, the topology of which is a $\mathfrak{m}$-topology, and $\{M_i \,|\, i \in I\}$ be a family of topological R-modules, each of them having a basis of neighborhoods of zero consisting of submodules. Then their $\mathfrak{m}$-product $\prod_{i \in I} M_i$ has a basis of neighborhoods of zero consisting of submodules, too.

The proof is analogous to the proof of Proposition 4.1.7.

4.1.20. DEFINITION. Let $\{X_i \,|\, i \in I\}$ be a family of Abelian groups (rings, modules over a certain ring). Consider the subset of their direct product $X = \prod_{i \in I} X_i$, consisting of such mappings $f : I \to \bigcup_{i \in I} X_i$ that $\left| \{i \mid f(i) \neq 0_i\} \right| < \aleph_0$ (see A.7), where 0_i is the neutral element of the additive group X_i.

It is easy to verify that this subset is a subgroup (two-sided ideal, submodule) in X. We call it the direct sum of groups (rings, modules) X_i and denote it over $\sum_{i \in I} X_i$. If there is no misunderstanding, we denote the restriction of the canonic projections π_j on $\sum_{i \in I} X_i$ over π_j, too. If $I = \{1, \dots, n\}$ is finite, and $J = \{1, 2, \dots\}$ is a countable set, then we denote $\sum_{i \in I} X_i$ over $\sum_{i=1}^{n} X_i$ or $X_1 \oplus X_2 \oplus \dots \oplus X_n$, and $\sum_{j \in J} X_j$ over $\sum_{j=1}^{\infty} X_j$.

4.1.21. REMARK. Let $\mathfrak{m}$ be a cardinal number, $\{X_i \mid i \in I\}$ a family of topological Abelian groups (rings) and $X = \prod_{i \in I} X_i$ their $\mathfrak{m}$-product. Then the direct sum $\sum_{i \in I} X_i$ endowed with the topology, induced from X, is a topological group (ring). We call it a direct sum of groups (rings) X_i with the $\mathfrak{m}$-product topology.

If $U = \prod_{i \in I} U_i$ is a neighborhood of zero in $X = \prod_{i \in I} X_i$, then a neighborhood of the type $Y \cap U$ of zero in $Y = \sum_{i \in I} X_i$ we sometimes denote over $\sum_{i \in I} U_i$.

Note that if $\{X_i' \mid i \in I\}$ is a family of subgroups (ideals) in $X = \prod_{i \in I} X_i$, defined in Proposition 4.1.5, then the group (ring) $\sum_{i \in I} X_i$ coincides with the algebraic sum $\sum_{i \in I} X_i'$ of the family of its subgroups (ideals) X_i'.

4.1.22. PROPOSITION. *Let $\mathfrak{m}$ be a cardinal number, R be a locally precompact topological ring, $\{M_i \mid i \in I\}$ a family of topological R-modules. Consider the direct sum $N = \sum_{i \in I} M_i$ of the modules M_i endowed with the topology of the $\mathfrak{m}$-product of the Abelian groups M_i. Then N is a topological R-module. (We also call the module N the direct sum of the modules M_i with the $\mathfrak{m}$-product topology).*

PROOF. According to Remark 4.1.21, N is a topological Abelian group, i.e. the family $\mathcal{L}$ of the subsets $(\prod_{i \in I} U_i) \cap N$, where U_i is a neighborhood of zero in M_i and $\left| \{i \mid U_i \neq M_i\} \right| < \mathfrak{m}$, satisfies the conditions (BN1) - (BN4) of Proposition 1.2.1. According to Theorem 1.2.6, to complete the proof it is enough to verify the fulfillment for the family $\mathcal{L}$ of the conditions (BN5′), (BN6′) and (BN6″).

Now we verify the fulfillment of the condition (BN5′). Let $(\prod_{i \in I} U_i) \cap N \in \mathcal{L}$ and W be a precompact neighborhood of zero in R. For every $i \in I$ there exists a neighborhood $\widehat{U}_i$ of zero in M_i such that $\widehat{U}_i + \widehat{U}_i \subseteq U_i$ and there exist a neighborhood V_i of zero in M_i

and a neighborhood W_i of zero in R such that $V_i \cdot W_i \subseteq \widehat{U}_i$. Since W is a precompact set, then there exist finite subsets S_i in R such that $W \subseteq S_i + W_i$.

If $U_i \neq M_i$, then we choose a neighborhood $\widehat{V}_i$ of zero in M_i such that $\widehat{V}_i \subseteq V_i$ and $\widehat{V}_i \cdot S_i \subseteq \widehat{U}_i$. Then for this $i \in I$ we get

$$\widehat{V}_i \cdot W \subseteq \widehat{V}_i \cdot (S_i + W_i) \subseteq \widehat{V}_i \cdot S_i + V_i \cdot W_i \subseteq \widehat{U}_i + \widehat{U}_i \subseteq U_i.$$

If $U_i = M_i$, then we take $\widehat{V}_i = M_i$. Then $\widehat{V}_i \cdot W \subseteq U_i$ for any $i \in I$.

Since $\{i \,|\, \widehat{V}_i \neq M_i\} \subseteq \{i \,|\, U_i \neq M_i\}$, then $(\prod_{i \in I} \widehat{V}_i) \bigcap N \in \mathcal{L}$ and

$$\left((\prod_{i \in I} \widehat{V}_i) \cap N \right) \cdot W \subseteq \left(\prod_{i \in I} \widehat{V}_i \cdot W \right) \cap N \subseteq \left(\prod_{i \in I} U_i \right) \cap N.$$

Now we verify the fulfillment of the condition (BN6′). Let $(\prod_{i \in I} U_i) \bigcap N \in \mathcal{L}$ and $r \in R$. For every $i \in I$ we define the neighborhood V_i of zero in M_i as the following:

if $U_i \neq M_i$, then we take a neighborhood V_i of zero in M_i such that $V_i \cdot r \subseteq U_i$;

if $U_i = M_i$, then we take $V_i = M_i$.

Since $\{i \,|\, V_i \neq M_i\} \subseteq \{i \,|\, U_i \neq M_i\}$, then $(\prod_{i \in I} V_i) \bigcap N \in \mathcal{L}$, and

$$\left((\prod_{i \in I} V_i) \cap N \right) \cdot r \subseteq \left(\prod_{i \in I} (V_i \cdot r) \right) \cap N \subseteq \left(\prod_{i \in I} U_i \right) \cap N.$$

Now we verify the fulfillment of the condition (BN6″). Let $(\prod_{i \in I} U_i) \bigcap N \in \mathcal{L}$ and $f \in N$. There exists a finite subset J in I such that $f(i) = 0_i$, when $i \notin J$, where 0_i is the neutral element of the group M_i. For every $j \in J$ we can find a neighborhood W_j of zero in R such that $f(j) \cdot W_j \subseteq U_j$. Then $W = \bigcap_{j \in J} W_j$ is a neighborhood of zero in R.

Since $f(i) = 0_i$ for $i \in I \backslash J$, then

$$f \cdot W \subseteq \left(\prod_{i \in I} U_i \right) \cap N.$$

The proposition is proved.

4.1.23. REMARK. Let $\{X_i \,|\, i \in I\}$ be a family of topological Abelian groups, where the topological group X_i is not anti-discrete for every $i \in I$; $\mathfrak{m}_1$ and $\mathfrak{m}_2$ be such cardinal numbers that $\aleph_0 \leq \mathfrak{m}_1 < \mathfrak{m}_2$. If $\mathfrak{m}_1 \leq |I|$, then the topology τ_1 of the $\mathfrak{m}_1$-product

$\prod_{i\in I} X_i$ and the topology τ_2 of the $\mathfrak{m}_2$-product $\prod_{i\in I} X_i$ are different, and $\tau_1 < \tau_2$. If $|I| < \mathfrak{m}_1$, then the topologies τ_1 and τ_2 coincide, having as a basis of neighborhoods of zero the family of sets of the type $\prod_{i\in I} U_i$, where U_i is a neighborhood of zero in X_i (in particular, if I is a finite set, then the topologies of the $\mathfrak{m}$-products on $\prod_{i\in I} X_i$ coincide for all $\mathfrak{m}$).

4.1.24. DEFINITION. Let $\{X_i \,|\, i \in I\}$ be a family of topological Abelian groups, $\mathfrak{m}$ be a cardinal number and $\mathfrak{m} > |I|$. We call a brick topology on $\prod_{i\in I} X_i$ and denote over τ_B the $\mathfrak{m}$-topology on $\prod_{i\in I} X_i$ that, as we mentioned above, has as a basis of neighborhoods of zero the family of sets of the type $\prod_{i\in I} U_i$, where U_i is a neighborhood of zero in X_i.

4.1.25. DEFINITION. Let X be a topological Abelian group (ring, module), $\{X_i \,|\, i \in I\}$ be a certain family of its subgroups (two-sided ideals, submodules), that have the topology, induced from X, and $\mathfrak{m}$ a certain cardinal number. Consider the $\mathfrak{m}$-product $\prod_{i\in I} X_i$. If there exists a topological isomorphism of the Abelian groups (rings, modules) $\xi : X \to \prod_{i\in I} X_i$ such that for any $i \in I$:

$\pi_j(\xi(x)) = x$ if $j = i$ and $\pi_j(\xi(x)) = 0$ if $j \neq i$, for any $x \in X_i$,

then we say that the topological group (ring, module) X is decomposed into a $\mathfrak{m}$-product of the family $\{X_i \,|\, i \in I\}$ of its subgroups (two-sided ideals, submodules). Also we denote it over the equality $X = \prod_{i\in I} X_i$ if there is no misunderstanding.

Consider the direct sum $\sum_{i\in I} X_i$ endowed with the topology induced from their $\mathfrak{m}$-product, then, if there exists a topological isomorphism of the Abelian groups (rings, modules) $\xi : X \to \sum_{i\in I} X_i$ such that for any $i \in I$:

$\pi_j(\xi(x)) = x$ for $j = i$ and $\pi_j(\xi(x)) = 0$ for $j \neq i$, for any $x \in X_i$,

we say that the topological group (ring, module) X is decomposed into the direct sum of the family $\{X_i \,|\, i \in I\}$ of its subgroups (two-sided ideals, submodules). Also we denote it over the equality $X = \sum_{i\in I} X_i$, if there is no misunderstanding.

4.1.26. PROPOSITION. *Let (X, τ) be a topological Abelian group, $\{X_1, \ldots, X_n\}$ - a finite family of its subgroups such that:*

1) the group X, considered without topology, is decomposed into the direct product $X_1 \times X_2 \times \ldots \times X_n$ of the family $\{X_1, \ldots, X_n\}$ of its subgroups;

2) *if U_i is a neighborhood of zero of topological group $(X_i, \tau|_{X_i})$, for $i = 1, \ldots, n$, then $U_1 + \ldots + U_n$ is a neighborhood of zero in (X, τ).*

Then the topological Abelian group (X, τ) is decomposed into the direct product of the family $\{(X_i, \tau|_{X_i}) \mid i = 1, \ldots, n\}$ of its topological subgroups.

PROOF. The condition 1) means that there exists an isomorphism of the groups $\xi :$ $X \to \prod_{i \in I} X_i$ such that for any $i \in I$ and $x \in X_i$ we get that $\pi_j(\xi(x)) = x$ if $j = i$ and $\pi_j(\xi(x)) = 0$ if $j \neq i$. Then the isomorphism $\xi^{-1} : \prod_{i \in I} X_i \to X$ corresponds the element $(x_1, \ldots, x_n) \in \prod_{i \in I} X_i$ to the element $x_1 + x_2 + \ldots + x_n \in X$. Since the sum operation in (X, τ) is continuous, then the isomorphism ξ^{-1} is continuous. The isomorphism ξ is continuous due to the condition 2) and the definition of the topology on $\prod_{i \in I} X_i$.

The proposition is proved.

4.1.27. DEFINITION. Let (X, τ) be a topological group (ring, module), Y a subgroup (two-sided ideal, submodule) in X. We say that Y is *separeted as a direct summand* in (X, τ) if there exists a subgroup (two-sided ideal, submodule) Z in X such that (X, τ) is decomposed into the direct sum $(Y, \tau|_Y) \oplus (Z, \tau|_Z)$.

4.1.28. PROPOSITION. *Let (R, τ) be a topological ring, A be a two-sided ideal in R that has a unitary element as a ring. Then A can be separated as a direct summand in R.*

PROOF. Let $B = \{r \in R \mid r \cdot A = 0\}$. It is clear that B is a two-sided ideal in R. Now we prove that $A \cap B = 0$ and that R coincides with the sum of the ideals A and B. Indeed, let $r \in A \cap B$ and e be the unitary element of the ring A. Then $r = r \cdot e \in B \cdot A = \{0\}$, i.e. $r = 0$. It means that $A \cap B = \{0\}$.

Let $r \in R$, then $r = r \cdot e + (r - r \cdot e)$, where $r \cdot e \in R \cdot A \subseteq A$. Since $(r - r \cdot e) \cdot a = r \cdot a - r \cdot e \cdot a = r \cdot a - r \cdot a = 0$ for any $a \in A$, then $r - r \cdot e \in B$. Thus, R, considered without topology, is decomposed into the direct product of its ideals A and B.

To complete the proof of the proposition we have to show that the topological Abelian group $(R(+), \tau)$ is decomposed into the topological direct sum of its subgroups $(A, \tau|_A)$ and $(B, \tau|_B)$. According to Proposition 4.1.26, it is enough to show that for any neighborhood U of zero in R the set $(U \cap A) + (U \cap B)$ is a neighborhood of zero in (R, τ). Indeed, since (R, τ) is a topological ring, then there exists a neighborhood V of zero in R such that $V \cdot e \subseteq U$ and $V - V \cdot e \subseteq U$. Then for any $v \in V$ we get $v = v \cdot e + (v - v \cdot e) \in (U \bigcap A) + (U \bigcap B)$. It means that $V \subseteq (U \bigcap A) + (U \bigcap B)$, i.e. $(U \bigcap A) + (U \bigcap B)$ is a neighborhood of zero in (R, τ).

The proposition is proved.

4.1.29. PROPOSITION. *Let $\{R_i \,|\, i \in I\}$ be a family of topological rings, $\mathfrak{m}$ be a cardinal number and $R = \prod_{i \in I} R_i$ be the $\mathfrak{m}$-product of the family $\{R_i \,|\, i \in I\}$. The subset Γ in R is bounded (from left, right) if and only if for any $i \in I$ the subset $\pi_i(\Gamma)$ is bounded (from left, right) in R_i.*

PROOF. If the subset Γ is bounded from left in R, then, due to Propositions 1.6.28 and 4.1.2, the subset $\pi_i(\Gamma)$ is bounded from left in R_i, for any $i \in I$.

Now let for any $i \in I$ the subset $\pi_i(\Gamma)$ be bounded in R_i and U be a certain neighborhood of zero in R. We can consider that $U = \prod_{i \in I} U_i$, where U_i is a certain neighborhood of zero in R_i, and $\left| \{i \,|\, U_i \neq R_i\} \right| < \mathfrak{m}$. For any $i \in I$ we choose the neighborhood V_i of zero in R_i the following way:

if $U_i \neq R_i$, then we take a neighborhood V_i of zero in R_i such that $V_i \cdot \pi_i(\Gamma) \subseteq U_i$;

if $U_i = R_i$, then we take $V_i = R_i$.

Since $\{i \,|\, V_i \neq R_i\} \subseteq \{i \,|\, U_i \neq R_i\}$, then $\prod_{i \in I} V_i$ is a neighborhood of zero in R, and $\left(\prod_{i \in I} V_i \right) \cdot \Gamma \subseteq \prod_{i \in I} (V_i \cdot \pi_i(\Gamma)) \subseteq \prod_{i \in I} U_i$.

The boundedness from right is proved analogously.

4.1.30. PROPOSITION. *Let $\mathfrak{m}$ be a cardinal number, R be a topological ring, the topology on which is a $\mathfrak{m}$-topology (see Definition 4.1.12), and $\{M_i \,|\, i \in I\}$ be a family of topological R-modules. Let $M = \prod_{i \in I} M_i$ be a $\mathfrak{m}$-product, then a subset Γ in M is bounded if and only if for any $i \in I$ the subset $\pi_i(\Gamma)$ is bounded in M_i.*

PROOF. If the subset Γ is bounded in M, then, due to Propositions 1.6.27 and 4.1.2, the subset $\pi_i(\Gamma)$ is bounded in M_i, for any $i \in I$.

Now for any $i \in I$ let the subset $\pi_i(\Gamma)$ be bounded in M_i and U be a neighborhood of zero in M. As in Proposition 4.1.29, we can consider then $U = \prod_{i \in I} U_i$. If $J = \{i \,|\, U_i \neq M_i\}$, then $\left| J \right| < \mathfrak{m}$. For $i \in J$ we choose a neighborhood W_i of zero in R such that

$\pi_i(\Gamma) \cdot W_i \subseteq U_i$.

Since the topology on R is a $\mathfrak{m}$-topology, then $W = \bigcap_{i \in J} W_i$ is a neighborhood of zero in R, and

$$\Gamma \cdot W \subseteq \prod_{i \in I} (\pi_i(\Gamma) \cdot W) \subseteq \prod_{i \in I} U_i.$$

This completes the proof.

4.1.31. COROLLARY. Let R be a topological ring, $\{M_i \,|\, i \in I\}$ be a family of topological R-modules and $M = \prod_{i \in I} M_i$ be their Tychonoff product. A subset Γ is bounded in M if and only if for any $i \in I$ the subset $\pi_i(\Gamma)$ is bounded in M_i.

4.1.32. PROPOSITION. *Let $\{R_i \,|\, i \in I\}$ be a family of topological rings and $R = \prod_{i \in I} R_i$ be their Tychonoff product. A subset $S \subseteq R$ is topologically nilpotent (Σ-nilpotent) (see Definitions 1.8.16 and 1.8.20) if and only if for every $i \in I$ the subset $\pi_i(S)$ is topologically nilpotent (Σ-nilpotent) in R_i.*

PROOF. If the subset S is topologically nilpotent (Σ-nilpotent), then, due to Proposition 1.8.34, for any $i \in I$ the subset $\pi_i(S)$ is topologically nilpotent (Σ-nilpotent) in R_i.

Now let the subset $\pi_i(S)$ be topologically nilpotent in R_i for any $i \in I$, and U be a neighborhood of zero in R. As above, we can consider that $U = \prod_{i \in I} U_i$. Let $J = \{i \,|\, U_i \neq R_i\}$, then J is a finite subset in I.

For any $i \in J$ there exists a natural number n_i such that $(\pi_i(S))^{(k)} \subseteq U_i$ (see B.12) for $k \geq n_i$. Let $n = \max\{n_i \,|\, i \in J\}$, then $S^{(k)} \subseteq \prod_{i \in I} (\pi_i(S))^{(k)} \subseteq \prod_{i \in I} U_i$ for all $k \geq n$. Hence, we have shown that the subset S is nilpotent. Analogously is proved that the subset S is Σ-nilpotent. The proposition is proved.

4.1.33. PROPOSITION. *Let $\mathfrak{m}$ be a cardinal number, where $\mathfrak{m} > \aleph_0$ and $\{R_i \,|\, i \in I\}$ be a family of Haussdorff topological rings. Let $R = \prod_{i \in I} R_i$ be their $\mathfrak{m}$-product, then a subset S is topologically nilpotent (Σ-nilpotent) in R if and only if each of the sets $\pi_i(S)$ is topologically nilpotent (Σ-nilpotent) in R_i, and there exists a finite subset $J \subseteq I$ and a natural number k such that $(\pi_j(S))^{(k)} = 0$ for all $j \in I \backslash J$.*

PROOF. Suppose that each of the subsets $\pi_i(S)$ is topologically nilpotent (Σ-nilpotent) in R_i and there exist a finite subset $J \subseteq I$ and a natural number k such that $(\pi_j(S))^{(k)} = 0$ for any $j \in I \backslash J$.

Let U be a neighborhood of zero in R. Then there exist neighborhoods of zero U_i in R_i such that $\prod_{i \in I} U_i \subseteq U$. For any $i \in I$ there exists a natural number k_i such that $(\pi_i(S))^{(n)} \subseteq U_i$ $\left((\pi_i(S))^n \subseteq U_i\right)$ for any $n \geq k_i$. Let $m = \max\{k, k_i \,|\, i \in J\}$. If now $n \geq m$, then $(\pi_i(S))^{(n)} = (\pi_i(S))^n = 0 \in U_i$ for $i \in I \backslash J$ and $(\pi_i(S))^{(n)} \subseteq U_i$ $\left((\pi_i(S))^n \subseteq U_i\right)$ for $i \in J$. Therefore, $(\pi_i(S))^{(n)} \subseteq U_i\left((\pi_i(S))^n \subseteq U_i\right)$ for any $i \in I$. Thus, $S^{(n)} \subseteq \prod_{i \in I} U_i \subseteq U$ $\left(S^n \subseteq \prod_{i \in I} U_i \subseteq U\right)$, which proves that the subset S in R is topologically nilpotent $\left(\Sigma\text{-nilpotent}\right)$.

Now let the subset S be topologically nilpotent (Σ-nilpotent) in R. Then, according to Proposition 1.8.34, any of the subsets $\pi_i(S)$ is topologically nilpotent (Σ-nilpotent) in R. Assume that for any natural number $n \in \mathbb{N}$ the set $\{i \,|\, \pi_i(S)^n \neq 0\}$ is infinite. Then, by induction by the number k it is easy to construct a countable subset $\{i_k \,|\, k \in \mathbb{N}\} \subseteq I$ such that $(\pi_{i_k}(S))^k \neq 0$, therefore $(\pi_{i_k}(S))^{(k)} \neq 0$ for any $k \in \mathbb{N}$.

Since the ring R_{i_k} is Hausdorff, then for any $k \in \mathbb{N}$ there exists a neighborhood U_k of zero in R_{i_k} such that $(\pi_{i_k}(S))^{(k)} \nsubseteq U_k$. Let $V_i = U_k$ for $i = i_k$ and $V_i = R_i$ for $i \notin \{i_k \,|\, k \in \mathbb{N}\}$, then $V = \prod_{i \in I} V_i$ is a neighborhood of zero in R. Since the set S is topologically nilpotent (Σ-nilpotent), then there exists a number $n \in \mathbb{N}$ such that $S^{(n)} \subseteq V$ $(S^n \subseteq V)$. Thus, $(\pi_{i_n}(S))^{(n)} \subseteq \pi_{i_n}(V) = U_n$. This contradicts the choice of the neighborhood U_n of zero in R_{i_n}. The contradiction obtained completes the proof.

4.1.34. COROLLARY. *Let* $\mathfrak{m}$ *be a cardinal number,* $\{R_i \,|\, i \in I\}$ *be a family of Hausdorff topological rings,* $S = \sum_{i \in I} R_i$ *their direct sum with the* $\mathfrak{m}$*-product topology. An element* $r \in S$ *is topologically nilpotent (*Σ*-nilpotent) if and only if for any* $i \in I$ *the element* $\pi_i(r)$ *is topologically nilpotent (*Σ*-nilpotent) in* R_i.

PROOF. If the element r is topologically nilpotent (Σ-nilpotent) in S, then the subset $\{r\}$ is topologically nilpotent (Σ-nilpotent) in the $\mathfrak{m}$-product $R = \prod_{i \in I} R_i$. Then, according to Proposition 4.1.33, for any $i \in I$ the set $\{\pi_i(r)\}$ is topologically nilpotent (Σ-nilpotent) in R_i.

Now let $\pi_i(r)$ be a topologically nilpotent (Σ-nilpotent) element in R_i for every $i \in I$.

Let $J = \{i \in I \mid \pi_i(r) \neq 0_i\}$, then J is a finite set. Since $\pi_i(r) = 0_i$ for any $i \in I \setminus J$, then, according to Proposition 4.1.33, the one-element set $\{r\}$ is topologically nilpotent (Σ-nilpotent) in the $\mathfrak{m}$-product $\prod_{i \in I} R_i$, therefore is topologically nilpotent (Σ-nilpotent) in S.

4.1.35. PROPOSITION. *Let* $\mathfrak{m}$ *be a regular cardinal number**, $\{I_\omega \mid \omega \in \Omega\}$ *a certain system of pairwise disjoint sets,* $I = \bigcup_{\omega \in \Omega} I_\omega$ *and for any* $i \in I$ *is determined a topological Abelian group* X_i. *Let* $Y_\omega = \prod_{i \in I_\omega} X_i$ *and* $X = \prod_{i \in I} X_i$ *be the* $\mathfrak{m}$-*products of the topological groups* X_i. *Then the group* X *is topologically isomorphic to the* $\mathfrak{m}$-*product* $Y = \prod_{\omega \in \Omega} Y_\omega$.

PROOF. Consider the mapping $\xi : X \to Y$, constructed as the following:

for every $\omega \in \Omega$ let $y_\omega \in Y_\omega$ be such that $y_\omega(i) = x(i)$ for any $i \in I_\omega$. Let $y \in Y$ be such that $y(\omega) = y_\omega$ for any $\omega \in \Omega$. Put $\xi(x) = y$.

Let's show that the mapping $\xi : X \to Y$ so defined is a topological isomorphism of the topological groups X and Y. From the definition of ξ follows that ξ is an isomorphic mapping of the discrete group X to the discrete group Y.

Now we prove that ξ is an open isomorphism. Let U be a neighborhood of zero in X. As above, we can consider that $U = \prod_{i \in I} U_i$, where U_i is a neighborhood of zero in X_i, besides, if $J = \{i \mid U_i \neq X_i\}$, then $|J| < \mathfrak{m}$. Since $U_i = X_i$ for any $\omega \in \Omega$ and $i \in I_\omega \setminus J$, then $W_\omega = \prod_{i \in I_\omega} U_i$ is a neighborhood of zero in Y_ω. It is easy to see that if ω is such element from Ω that $J \cap I_\omega = \emptyset$, then $W_\omega = \prod_{i \in I_\omega} X_i = Y_\omega$. Since $|\{\omega \mid J \cap I_\omega \neq \emptyset\}| < \mathfrak{m}$, then $W = \prod_{\omega \in \Omega} W_\omega$ is a neighborhood of zero in $Y = \prod_{\omega \in \Omega} Y_\omega$. From the definition of the mapping ξ follows that $\xi(U) = W$, i.e. ξ is an open isomorphism.

Now we prove that ξ is a continuous isomorphism. Let W be a neighborhood of zero in Y. As above, we can consider the $W = \prod_{\omega \in \Omega} W_\omega$, where W_ω is a neighborhood of zero in Y_ω. Let $B = \{\omega \mid W_\omega \neq Y_\omega\}$, then $|B| < \mathfrak{m}$. For any $\omega \in B$ there exists a neighborhood V_ω of zero in Y_ω such that $V_\omega \subseteq W_\omega$ and $V_\omega = \prod_{i \in I_\omega} U_i$, where U_i is a neighborhood of zero in X_i. Let $B_\omega = \{i \in I_\omega \mid U_i \neq X_i\}$, then $|B_\omega| < \mathfrak{m}$. It is clear that if $i \notin \bigcup_{\omega \in B} B_\omega$, then

---

* We call the cardinal number $\mathfrak{m}$ regular if it cannot be represented as $\sum_{\lambda \in \Lambda} \mathfrak{m}_\lambda$, where $\mathfrak{m}_\lambda$ are cardinal numbers, $\mathfrak{m}_\lambda < \mathfrak{m}$ for any $\lambda \in \Lambda$ and $|\Lambda| < \mathfrak{m}$. It is clear that $\aleph_0$ is a regular cardinal number.

$U_i = X_i$. Since the cardinal number $\mathfrak{m}$ is regular, then $\left| \bigcup_{\omega \in B} B_\omega \right| \leq \sum_{\omega \in B} |B_\omega| < \mathfrak{m}$. Then $U = \prod_{i \in I} U_i$ is a neighborhood of zero in X. From the definition of the mapping ξ follows that $\xi(U) \subseteq W$. Therefore, ξ is a continuous isomorphism.

4.1.36. PROPOSITION. *Let $\mathfrak{m}$ be a cardinal number, I_1 and I_2 be non-intersecting sets, $I = I_1 \bigcup I_2$ and for any $i \in I$ is defined a topological group X_i. Let $Y_1 = \prod_{i \in I_1} X_i$ and $Y_2 = \prod_{i \in I_2} X_i$ be the $\mathfrak{m}$-products of the topological groups, then the product $Y_1 \times Y_2$ is topologically isomorphic to the $\mathfrak{m}$-product $X = \prod_{i \in I} X_i$.*

PROOF. Let $\Omega = \{1, 2\}$ be a two-element set. As in Proposition 4.1.35, consider the mapping $\xi : X \to Y_1 \times Y_2$. In Proposition 4.1.35 we used the regularity of $\mathfrak{m}$ only to prove that $\left| \bigcup_{\omega \in B} B_\omega \right| < \mathfrak{m}$. Since the cardinal number $\mathfrak{m}$ is infinite, then for any two subsets $B_1, B_2 \subseteq I$ such that $|B_1| < \mathfrak{m}$ and $|B_2| < \mathfrak{m}$ we get $|B_1 \bigcup B_2| < \mathfrak{m}$. Using this fact and repeating the proof of Proposition 4.1.35, we obtain that ξ is an open and continuous isomorphism.

4.1.37. PROPOSITION. *Let X be an Abelian group, $\{\tau_i \mid i \in I\}$ be a family of group topologies on X and $\tau = \sup\{\tau_i \mid i \in I\}$. Let $(Y, \widehat{\tau}) = \prod_{i \in I}(X, \tau_i)$ be the Tychonoff product of the family $\{(X, \tau_i) \mid i \in I\}$ of topological groups. Then the homomorphism $f : (X, \tau_i) \to Y$, which corresponds to the element $x \in X$ the element $y \in Y$ such that $y(i) = x$ for any $i \in I$, determines a topological isomorphism between the topological groups (X, τ) and $(f(X), \widehat{\tau}\big|_{f(X)})$.*

PROOF. It is evident that the homomorphism $f : X \to Y$ determines the algebraic isomorphism of the Abelian groups X and $f(X)$.

Now we prove that f is a continuous isomorphism. Indeed, let U be a neighborhood of zero in $(f(X), \widehat{\tau}\big|_{f(X)})$, then there exist neighborhoods U_i of zero in (X, τ_i) for $i \in I$ such that the set $J = \{i \mid U_i \neq X\}$ is finite, and $(\prod_{i \in I} U_i) \bigcap f(X) \subseteq U$. Then $V = \bigcap_{i \in J} U_i$ is a neighborhood of zero in (X, τ), and $f(V) \subseteq (\prod_{i \in I} U_i) \bigcap f(X) \subseteq U$. The continuity of the isomorphism is proved.

Now let V be a neighborhood of zero in (X, τ). Since $\tau = \sup\{\tau_i \mid i \in I\}$, then, according to C.9, there exist a finite subset $J \subseteq I$ and neighborhoods U_i of zero in (X, τ_i) for $i \in J$ such that $V = \bigcap_{i \in J} U_i$. Let $W = \prod_{i \in I} W_i$, where $W_i = U_i$ for $i \in J$ and $W_i = X_i$ in the contrary case. Then W is a neighborhood of zero in $(Y, \hat{\tau})$, therefore $W \cap f(X)$ is a neighborhood of zero in $(f(X), \hat{\tau}|_{f(X)})$. Let $y \in W \cap f(X)$, then $y(i) \in W_i$ for any $i \in I$. Since $y \in f(X)$, then $y(i) = y(j)$ for any $i, j \in I$. Thus, $y(i) \in \bigcap_{j \in I} W_j = \bigcap_{j \in J} U_j = V$ for any $i \in I$. It means that $y \in f(V)$. Since the choice of the element y was arbitrary, then $W \cap f(X) \subseteq f(V)$, i.e. $f(V)$ is a neighborhood of zero in $(f(X), \hat{\tau}|_{f(X)})$. Hence, we've proved that the isomorphism f is open.

4.1.38. PROPOSITION. *Let $\mathfrak{m}$ be a cardinal number, $\{X_i \,|\, i \in I\}$ and $\{Y_i \,|\, i \in I\}$ be families of topological Abelian groups, and for any $i \in I$ is determined a continuous homomorphism $f_i : X_i \to Y_i$. Then the natural homomorphism $f : \prod_{i \in I} X_i \to \prod_{i \in I} Y_i$ of the $\mathfrak{m}$-products, where $f(x)$ is such that $(f(x))(i) = f_i(x(i))$ for $i \in I$ and $x \in \prod_{i \in I} X_i$, is continuous.*

PROOF. It is clear that f is a homomorphism of the discrete groups. Now we prove that f is a continuous homomorphism.

Indeed, let U be a neighborhood of zero in $\prod_{i \in I} Y_i$. We can consider that $U = \prod_{i \in I} U_i$, where U_i is a neighborhood of zero in Y_i for $i \in I$, and $\left| \{i \,|\, U_i \neq Y_i\} \right| < \mathfrak{m}$. For any $i \in I$ we define the neighborhood V_i of zero in X_i as the following:

if $U_i \neq Y_i$, then we take a neighborhood V_i of zero in X_i such that $f_i(V_i) \subseteq U_i$;

if $U_i = Y_i$, we take $V_i = X_i$.

Since $\{i \,|\, V_i \neq X_i\} \subseteq \{i \,|\, U_i \neq Y_i\}$, then $\prod_{i \in I} V_i$ is a neighborhood of zero in $\prod_{i \in I} X_i$ and $f(\prod_{i \in I} V_i) \subseteq \prod_{i \in I} f_i(V_i) \subseteq \prod_{i \in I} U_i$. It proves the continuity of the homomorphism f.

4.1.39. COROLLARY. *Let X be a topological Abelian group, $\{Y_i \,|\, i \in I\}$ be a family of topological Abelian groups, where for every $i \in I$ is determined a continuous homomorphism $\varphi_i : X \to Y_i$. Let $Y = \prod_{i \in I} Y_i$ be the Tychonoff product of the groups Y_i. Then the natural homomorphism $\varphi : X \to Y$, that corresponds to the element $x \in X$ an element $y \in Y$ such that $y(i) = \varphi_i(x)$ for any $i \in I$, is continuous.*

PROOF. Let $\prod_{i \in I} X_i$ be the Tychonoff product of the topological groups X_i, where $X_i = X$ for all $i \in I$. Due to Proposition 4.1.37, the homomorphism $f : X \to \prod_{i \in I} X_i$, where $(f(x))(i) = x$ for all $x \in X$ and $i \in I$, is continuous. Due to Proposition 4.1.38, the homomorphism $h : \prod_{i \in I} X_i \to \prod_{i \in I} Y_i$, where $(h(z))(j) = \varphi_j(z(j))$ for any $z \in \prod_{i \in I} X_i$ and $j \in I$, is continuous. Since $\varphi = hf$, then the homomorphism φ is continuous, too.

4.1.40. PROPOSITION. *Let* $\mathfrak{m}$ *be a cardinal number and* $\{Y_i \,|\, i \in I\}$ *be a family of topological Abelian groups, where for every* $i \in I$ *is determined a subgroup* X_i *of the topological group* Y_i. *Then the topology* τ_1 *of the* $\mathfrak{m}$-*product* $\prod_{i \in I} X_i$ *coincides with the topology* τ_2, *induced from the* $\mathfrak{m}$-*product* $\prod_{i \in I} Y_i$.

PROOF. Let $f_i : X_i \to Y_i$ be the natural embedding, for $i \in I$. Then the natural homomorphism $f : \prod_{i \in I} X_i \to \prod_{i \in I} Y_i$, that corresponds to the element $x \in \prod_{i \in I} X_i$ an element $y \in \prod_{i \in I} Y_i$ such that $x(i) = y(i)$ for $i \in I$, is the monomorphism of groups. Due to Proposition 4.1.38, the monomorphism f is continuous, hence $\tau_1 \geq \tau_2$.

Now we prove that $\tau_1 \leq \tau_2$. Indeed, let V be a neighborhood of zero in the $\mathfrak{m}$-product $(\prod_{i \in I} X_i, \tau_1)$. As above, we can consider $V = \prod_{i \in I} V_i$, where V_i is a neighborhood of zero in X_i for $i \in I$, and $\left| \{i \,|\, V_i \neq X_i\} \right| < \mathfrak{m}$. For any $i \in I$ we define the neighborhood U_i of zero in Y_i as the following:

if $V_i \neq X_i$, then U_i is a neighborhood of zero in Y_i such that $X_i \cap U_i = V_i$;

if $V_i = X_i$, then we take $U_i = Y_i$.

Since $\{i \,|\, U_i \neq Y_i\} = \{i \,|\, V_i \neq X_i\}$, then $\prod_{i \in I} U_i$ is a neighborhood of zero in the $\mathfrak{m}$-product $\prod_{i \in I} Y_i$. Therefore $(\prod_{i \in I} U_i) \cap (\prod_{i \in I} X_i)$ is a neighborhood of zero in $(\prod_{i \in I} X_i, \tau_2)$, and

$$\left(\prod_{i \in I} U_i \right) \cap \left(\prod_{i \in I} X_i \right) = \prod_{i \in I} V_i = V.$$

It means that $\tau_1 \leq \tau_2$. Hence, $\tau_1 = \tau_2$. The proposition is proved.

4.1.41. PROPOSITION. *Let* $\mathfrak{m}$ *be a cardinal number,* $\{Y_i \,|\, i \in I\}$ *be a family of topological Abelian groups, where for any* $i \in I$ *is determined a subgroup* X_i *of the group* Y_i, *and* $Z_i = Y_i/X_i$ *be the factor-group, endowed with the factor-topology. Let* $Y = \prod_{i \in I} Y_i$ *be the* $\mathfrak{m}$-*product of the topological groups* Y_i *and* $X = \prod_{i \in I} X_i$ *be its subgroup. Then the factor-group* Y/X *is topologically isomorphic to the* $\mathfrak{m}$-*product* $\prod_{i \in I}(Y_i/X_i)$ *of topological groups* Y_i/X_i.

PROOF. Let $f_i : Y_i \to Y_i/X_i$ be natural homomorphisms $i \in I$. Then, according to Proposition 4.1.38, the natural homomorphism $f : \prod_{i \in I} Y_i \to \prod_{i \in I}(Y_i/X_i)$ is continuous, and $\ker f = \prod_{i \in I} X_i = X$. Due to Theorem 1.5.12, the natural isomorphism $\varphi : Y/X \to \prod_{i \in I}(Y_i/X_i)$ is continuous.

Now we prove that φ is an open isomorphism. Let $\widetilde{U}$ be a neighborhood of zero in Y/X. According to the definition of the factor-topology, there exists a neighborhood U of zero in Y such that $\psi(U) \subseteq \widetilde{U}$, where $\psi : Y \to Y/X$ is the natural homomorphism. We can consider that $\prod_{i \in I} U_i = U$, where U_i is a neighborhood of zero in Y_i, and $\big|\{i \mid U_i \neq Y_i\}\big| < \mathfrak{m}$. Then $\widetilde{V}_i = f_i(U_i)$ is a neighborhood of zero in Y_i/X_i, for $i \in I$. Since $\{i \mid \widetilde{V}_i \neq Y_i/X_i\} \subseteq \{i \mid U_i \neq Y_i\}$, then $\prod_{i \in I} \widetilde{V}_i$ is a neighborhood of zero in $\prod_{i \in I}(Y_i/X_i)$. Since

$$\varphi(\widetilde{U}) \supseteq \varphi(\psi(U)) = f(U) = f\left(\prod_{i \in I} U_i\right) = \prod_{i \in I} f_i(U_i) = \prod_{i \in I} \widetilde{V}_i,$$

then $\varphi(\widetilde{U})$ is a neighborhood of zero in $\prod_{i \in I}(Y_i/X_i)$, i.e. φ is an open isomorphism.

The proposition is proved.

4.1.42. PROPOSITION. *Let $\mathfrak{m}$ be a cardinal number, $\{X_i \mid i \in I\}$ be a family of the topological Abelian groups and for every $i \in I$ is defined a subset $A_i \subseteq X_i$. Let $X = \prod_{i \in I} X_i$ be the $\mathfrak{m}$-product of the family $\{X_i \mid i \in I\}$, then $[\prod_{i \in I} A_i]_X = \prod_{i \in I}[A_i]_{X_i}$.*

PROOF. Let $f \in [\prod_{i \in I} A_i]_X$ and assume that $f(i_0) \notin [A_{i_0}]_{X_{i_0}}$ for a certain $i_0 \in I$. Then $f \notin \pi_{i_0}^{-1}([A_{i_0}]_{X_{i_0}})$. Since $\prod_{i \in I} A_i \subseteq \pi_{i_0}^{-1}(A_{i_0})$, and the projection π_{i_0} is continuous, then $[\prod_{i \in I} A_i]_X \subseteq [\pi_{i_0}^{-1}([A_{i_0}]_{X_{i_0}})]_X = \pi_{i_0}^{-1}([A_{i_0}]_{X_{i_0}})$, therefore $f \notin [\prod_{i \in I} A_i]_X$. The obtained contradiction shows that $[\prod_{i \in I} A_i]_X \subseteq \prod_{i \in I}[A_i]_{X_i}$.

Now let $f(i) \in [A_i]_{X_i}$ for any $i \in I$ and let U be a neighborhood of zero in X. As above, we can consider that $U = \prod_{i \in I} U_i$. Then for any $i \in I$ there exists an element $a_i \in A_i$ such that $a_i \in (f(i) + U_i) \cap A_i$. Let g be an element of X such that $g(i) = a_i$ for any $i \in I$. Therefore, $g \in (f + \prod_{i \in I} U_i) \cap (\prod_{i \in I} A_i) = (f + U) \cap (\prod_{i \in I} A_i)$, hence $(f + U) \cap (\prod_{i \in I} A_i) \neq \emptyset$. Since the choice of the neighborhood U of zero in X was arbitrary, then $f \in [\prod_{i \in I} A_i]_X$.

Thus, $\prod_{i \in I}[A_i]_{X_i} \subseteq [\prod_{i \in I} A_i]_X$. Therefore, $\prod_{i \in I}[A_i]_{X_i} = [\prod_{i \in I} A_i]_X$.

4.1.43. COROLLARY. Let $\mathfrak{m}$ be a cardinal number, $\{X_i \mid i \in I\}$ be a family of topological Abelian groups and A_i be a subset in X_i, for $i \in I$. Let $X = \prod_{i \in I} X_i$ be the $\mathfrak{m}$-product of the family $\{X_i \mid i \in I\}$, then:

1) the set $\prod_{i\in I} A_i$ is closed in X if and only if for any $i \in I$ the set A_i is closed in X_i;

2) the set $\prod_{i\in I} A_i$ is dense in X if and only if for any $i \in I$ the set A_i is dense in X_i.

4.1.44. COROLLARY. *Let* $\mathfrak{m}$ *be a cardinal number,* $\{X_i \,|\, i \in I\}$ *be a family of topological Abelian groups and* $X = \prod_{i\in I} X_i$ *be their* $\mathfrak{m}$-product. *The group* X *is Hausdorff if and only if for any* $i \in I$ *the group* X_i *is Hausdorff.*

PROOF. Due to Theorem 1.3.2, a topological group is Hausdorff if and only if the one-element set, consisting of the neutral element, is closed. Let 0_i be the neutral element of the group X_i, then the element f_0 such that $f_0(i) = 0_i$ for any $i \in I$, is neutral in X, and $\{f_0\} = \prod_{i\in I}\{0_i\}$. Due to Corollary 4.1.43, the subset $\{f_0\}$ is closed in X if and only if for any $i \in I$ the subset $\{0_i\}$ is closed in X_i.

4.1.45. PROPOSITION. *Let* $\{X_i \,|\, i \in I\}$ *be a family of topological Abelian groups and* X *their Tychonoff product. Then the subgroup* $Y = \sum_{i\in I} X_i$ *is a dense subgroup of the group* X.

PROOF. Let $f \in X$ and U be a neighborhood of zero in X. There exists a family $\{U_i \,|\, i \in I\}$ of neighborhoods U_i of zero in X_i such that $\prod_{i\in I} U_i \subseteq U$ and the set $J = \{i \,|\, U_i \neq X_i\}$ is finite. Let g be an element of the group X such that $g(i) = -f(i)$ for $i \in I \backslash J$ and $g(i) = 0_i$ for $i \in J$, where 0_i is the neutral element of the group X_i. Then $g \in \prod_{i\in I} U_i \subseteq U$. Since $f(i) + g(i) = 0_i$ for any $i \in I \backslash J$, then $f + g \in Y$, hence $(f + U) \bigcap Y \neq \emptyset$. Since the choice of the element f and the neighborhood U of zero in X was arbitrary, then Y is a dense subgroup in X.

4.1.46. PROPOSITION. *Let* $\mathfrak{m}$ *be a non-countable cardinal number,* $\{X_i \,|\, i \in I\}$ *be a family of Hausdorff topological Abelian groups and* $X = \prod_{i\in I} X_i$ *be their* $\mathfrak{m}$-product. *Then the subgroup* $Y = \sum_{i\in I} X_i$ *is closed in the topological group* X.

PROOF. Assume the contrary, i.e. that there exists an element $f \in [Y]_X \backslash Y$. Then there exists a subset $J \subseteq I$ such that $|J| = \aleph_0$ and $f(i) \neq 0_i$ for $i \in J$, where 0_i is the neutral element of the group X_i. Since each of the groups X_j, $j \in J$, is Hausdorff, then there exists a neighborhood U_j of zero in X_j such that $0_j \notin f(j) + U_j$.

Since $\mathfrak{m}$ is a non-countable cardinal number, then, according to the definition of the topology on a $\mathfrak{m}$-product, the set $V = \prod_{i\in I} V_i$, where $V_i = U_i$ for $i \in J$ and $V_i = X_i$ for

$i \notin J$, is a neighborhood of zero. Since $f \in [Y]_X$, then $(f + V) \cap Y \neq \emptyset$, i.e. there exists an element $g \in Y$ such that $g \in f + V$, hence $g(i) \in f(i) + U_i$ for all $i \in J$. Therefore, $g(i) \neq 0_i$ for any $i \in J$. It contradicts the fact that $g \in Y = \sum_{i \in I} X_i$. The obtained contradiction proves that the subgroup Y is closed in X.

4.1.47. PROPOSITION. *Let* $\mathfrak{m}$ *be a cardinal number,* $\{X_i \,|\, i \in I\}$ *be a family of topological Abelian groups and* $X = \prod_{i \in I} X_i$ *be their* $\mathfrak{m}$-*product. The group* X *is complete if and only if each of the groups* X_i *is complete.*

PROOF. Let X be complete group and $i_0 \in I$. Let $\mathcal{F}_{i_0}$ be a Cauchy filter of X_{i_0}, then it is easy to see that the family $\mathcal{F} = \{\prod_{j \in I} U_j \,|\, U_j = \{0\}$ for $j \neq i_0$ and $U_{i_0} \in \mathcal{F}_{i_0}\}$ forms a basis of a certain Cauchy filter Φ of X. Since X is a complete group, then there exists an element $x \in X$ such that $x = \lim \Phi$. Then $\pi_{i_0}(x) = \lim\{\pi_{i_0}(F) \,|\, F \in \Phi\}$ (see Proposition 3.1.25).

Since $\mathcal{F}_{i_0} = \{\pi_{i_0}(F) \,|\, F \in \mathcal{F}\}$ and $\mathcal{F}$ is a basis of the filter Φ, then $\pi_{i_0}(x) = \lim\{\pi_{i_0}(F) \,|\, F \in \Phi\} = \lim \mathcal{F}_{i_0}$, i.e. X_{i_0} is a complete group.

Let now X_i be a complete topological group for any $i \in I$ and Φ be a Cauchy filter of X. Then for any $i \in I$ the system $\Phi_i = \{\pi_i(F) \,|\, F \in \Phi\}$ is a Cauchy filter in X_i. Let $x_i = \lim \Phi_i$ in X_i and f be the element of X such that $f(i) = x_i$ for any $i \in I$.

Now we prove that $f = \lim \Phi$. Indeed, let U be a neighborhood of zero in X. As above, we can consider that $U = \prod_{i \in I} U_i$, where U_i is a neighborhood of zero in X_i for $i \in I$. For every $i \in I$ we choose the neighborhood V_i of zero in X_i as the following:

if $U_i \neq X_i$, then we take a neighborhood of zero V_i in X_i such that $[V_i]_{X_i} \subseteq U_i$;

if $U_i = X_i$, then we take $V_i = X_i$.

Since $\{i \,|\, V_i \neq X_i\} \subseteq \{i \,|\, U_i \neq X_i\}$, then $\prod_{i \in I} V_i$ is a neighborhood of zero in X. Since Φ is a Cauchy filter of X, then there exists a set $F \in \Phi$ such that $F - F \subseteq \prod_{i \in I} V_i$. Since $f(i) = x_i = \lim \Phi_i$ in X_i, then, according to Proposition 3.1.26, $f(i) \in [\pi_i(F)]_{X_i}$ for any $i \in I$, hence $f \in \prod_{i \in I} [\pi_i(F)]_{X_i}$. Therefore, using Proposition 4.1.42, we get:

$$F - f \subseteq \prod_{i \in I} \pi_i(F) - \prod_{i \in I} [\pi_i(F)]_{X_i} \subseteq \Big[\prod_{i \in I} \pi_i(F) \Big]_X - \Big[\prod_{i \in I} \pi_i(F) \Big]_X$$
$$\subseteq \Big[\prod_{i \in I} \pi_i(F - F) \Big]_X \subseteq \Big[\prod_{i \in I} V_i \Big]_X = \prod_{i \in I} [V_i]_{X_i} \subseteq \prod_{i \in I} U_i = U,$$

hence $F \subseteq f + U$. It proves that $f = \lim \Phi$ in X, i.e. that X is a complete group.

4.1.48. THEOREM. *Let (X, τ) be a Hausdorff topological Abelian group. Then there exists a complete Hausdorff topological Abelian group $(Y, \bar{\tau})$ such that the group (X, τ) is topologically isomorphic to the factor-group of the group $(Y, \bar{\tau})$ by one of its closed subgroups.*

PROOF. Let $Z = \prod_{i=1}^{\infty} X_i$ be the direct product of the family $\{X_i \mid i \in \mathbb{N}\}$ of Abelian groups, where $X_i = X$ for any $i \in \mathbb{N}$, and $Y = \sum_{i=1}^{\infty} X_i$. For any $i, j \in \mathbb{N}$ put:

$$\left(X_i, \tau_{j,i}\right) = \begin{cases} (X, \tau), & \text{if} \quad i \neq j; \\ X \text{ with the discrete topology, if} \quad i = j. \end{cases}$$

For any number $n \in \mathbb{N}$ consider the brick topology $\hat{\tau}_n$, defined on the group Z, as on the direct product of the family $\{(X_i, \tau_{n,i}) \mid i \in \mathbb{N}\}$. According to Corollary 4.1.44, $(Z, \hat{\tau}_n)$ is a Hausdorff topological group for any $n \in \mathbb{N}$. Let $\bar{\tau}_n$ be a topology, induced on Y by the topology from $(Z, \hat{\tau}_n)$. Let $\hat{\tau} = \sup\{\hat{\tau}_n \mid n \in \mathbb{N}\}$ and $\bar{\tau} = \sup\{\bar{\tau}_n \mid n \in \mathbb{N}\}$. Then $\bar{\tau}$ coincides with the restriction at Y of the topology $\hat{\tau}$. The groups $(Z, \hat{\tau})$ and $(Y, \bar{\tau})$ also are Hausdorff topological Abelian groups.

Now we prove that $(Y, \bar{\tau})$ is a complete group. Indeed, let Γ be an arbitrary direction and $\{y_\gamma \mid \gamma \in \Gamma\}$ be a Cauchy sequence in $(Y, \bar{\tau})$ by the direction Γ. Since $\bar{\tau} \geq \bar{\tau}_n$ for any $n \in \mathbb{N}$, then $\{y_\gamma \mid \gamma \in \Gamma\}$ is a Cauchy sequence in $(Y, \bar{\tau}_n)$ for any $n \in \mathbb{N}$.

Let $\pi_k : Z \to X_k = X$ be the natural projection of the direct product Z onto X_k then, due to Proposition 4.1.2, $\pi_k : (Z, \hat{\tau}_n) \to (X_k, \tau_{n,k})$ is a continuous homomorphism for any $k, n \in \mathbb{N}$. Then, according to Proposition 3.1.33, $\{\pi_k(y_\gamma) \mid \gamma \in \Gamma\}$ is a Cauchy sequence in $(X_k, \tau_{k,k})$. Since the topology $\tau_{k,k}$ is discrete, then there exists $\gamma_k \in \Gamma$ such that $\pi_k(y_\gamma) = \pi_k(y_{\gamma_k})$ for all $\gamma \in \Gamma, \gamma \geq \gamma_k$. For any $k \in \mathbb{N}$ we fix this element γ_k.

Let f be the element of Z such that $f(k) = \pi_k(y_{\gamma_k})$ for $k \in \mathbb{N}$. Now we prove that $f = \lim_{\gamma \in \Gamma} y_\gamma$ in the group $(Z, \hat{\tau})$.

Due to Theorem 3.1.13, it is easy to verify that $f = \lim_{\gamma \in \Gamma} y_\gamma$ in $(Z, \hat{\tau}_n)$ for any $n \in \mathbb{N}$. Let $n \in \mathbb{N}$ and U be a neighborhood of zero in $(Z, \hat{\tau}_n)$. Without loss of generality we can consider that $U = \prod_{i=1}^{\infty} U_i$, where U_i is a neighborhood of zero in (X, τ) for $i \neq n$ and $U_n = \{0\}$. Since $\{y_\gamma \mid \gamma \in \Gamma\}$ is a Cauchy sequence in $(Z_n, \hat{\tau}_n)$, then there exists an element $\delta \in \Gamma, \delta > \gamma_n$ such that $y_\alpha - y_\beta \in U$ for any $\alpha, \beta \in \Gamma, \ \alpha > \delta, \ \beta > \delta$. Let $\gamma \in \Gamma$ and $\gamma > \delta$.

Then $\pi_n(y_\gamma - f) = \pi_n(y_\gamma) - \pi_n(f) = \pi_n(y_\gamma) - \pi_n(y_{\gamma_n}) = 0$. Now let $k \neq n$. Let $\gamma' \in \Gamma$ be such that $\gamma' \geq \delta$ and $\gamma' \geq \gamma_k$, then

$$\pi_k(y_\gamma - f) = \pi_k(y_\gamma - y_{\gamma'}) + \pi_k(y_{\gamma'} - f)$$
$$= \pi_k(y_\gamma - y_{\gamma'}) + \pi_k(y_{\gamma'}) - \pi_k(y_{\gamma_k}) = \pi_k(y_\gamma - y_{\gamma'}) \in \pi_k(U) = U_k,$$

since $\pi_k(y_{\gamma'}) = \pi_k(y_k)$. Therefore, $y_\gamma - f \in \prod_{i=1}^{\infty} U_i = U$ for any $\gamma \geq \delta$. Hence, $f = \lim_{\gamma \in \Gamma} y_\gamma$ in $(Z, \hat{\tau}_n)$ for any $n \in \mathbb{N}$. Since $\hat{\tau} = sup\{\hat{\tau}_n \mid n \in \mathbb{N}\}$, then $f = \lim_{\gamma \in \Gamma} y_\gamma$ in $(Z, \hat{\tau})$ (see Theorem 3.1.13).

Due to Proposition 4.1.46 and Definition 4.1.24, the subgroup Y is closed in $(Z, \hat{\tau}_n)$ for all $n \in \mathbb{N}$. Since $\hat{\tau} \geq \hat{\tau}_n$ for all $n \in \mathbb{N}$, then Y is a closed subgroup in $(Z, \hat{\tau})$. Since $\{y_\gamma \mid \gamma \in \Gamma\} \subseteq Y$, then $f \in [Y]_{(Z, \hat{\tau})} = Y$. Therefore, the Cauchy sequence $\{y_\gamma \mid \gamma \in \Gamma\}$ in (Y, τ) has the limit f. Due to Theorem 3.2.17, the group (Y, τ) is complete.

Let now $p : Y \to X$ be the mapping that corresponds an element $f \in Y$ to the element $\sum_{i=1}^{\infty} f(i) \in X$. Since $\{i \mid f(i) \neq 0\}$ is finite, then the mapping p is defined correctly. It is easy to verify that p is a surjective homomorphism of Abelian groups.

Now we prove that p is a continuous homomorphism from $(Y, \bar{\tau})$ to (X, τ). Indeed, let V be a neighborhood of zero in (X, τ). We can construct by induction the sequence $V_0, \ldots, V_n, \ldots$ of such neighborhoods of zero in (X, τ) that $V_0 = V$ and $V_{i+1} + V_{i+1} \subseteq V_i$ for any $i = 0, 1, 2, \ldots$ Realizing the induction by n we can verify that $\sum_{i=1}^{n} V_i + V_n \subseteq V_0$. Then $\sum_{i=1}^{n} V_i \subseteq V$. Let $U = \prod_{i=1}^{\infty} V_i$, then $U \cap Y$ is a neighborhood of zero in $(Y, \bar{\tau})$. If $f \in U \cap Y$, then $f(i) \in V_i$ for any $i \in \mathbb{N}$ and there exists a number $n \in \mathbb{N}$ such that $f(j) = 0$ for all $j > n$. Then $p(f) = \sum_{i=1}^{n} f(i) \in \sum_{i=1}^{n} V_i \subseteq V_0 = V$. Hence, $p(U \cap Y) \subseteq V$. Therefore, p is a continuous homomorphism.

Now we prove that p is an open homomorphism. Indeed, let U be a neighborhood of zero in $(Y, \bar{\tau})$. Since $\bar{\tau} = sup\{\hat{\tau}_i \mid i \in \mathbb{N}\}$, then $U = (\bigcap_{i=1}^{n} W_i) \cap Y$, where W_i is a neighborhood of zero in $(Z, \hat{\tau}_i)$. Then, due to Proposition 4.1.2, $\pi_{n+1}(W_i)$ is a neighborhood of zero in $(X, \tau) = (X_{n+1}, \tau_{i, n+1})$ for any $i = 1, \ldots, n$ (see the difinition of the topologies $\tau_{i,j}$ above). Let $V = \bigcap_{i=1}^{n} \pi_{n+1}(W_i)$, then V is a neighborhood of zero in (X, τ). Let $U' = \{f \in Y \mid f(n+1) \in V, f(k) = 0$ for $k \in \mathbb{N} \setminus \{n+1\}\}$. Then $U' \subseteq U$ and $V = \pi_{n+1}(U')$. If $f \in U'$, then $p(f) = \sum_{i=1}^{\infty} f(i) = f(n+1) = \pi_{n+1}(f)$. Therefore, $p(U) \supseteq p(U') = V$. Hence, we've proved that p is an open homomorphism, then p is a topological homomorphism.

Due to Theorem 1.5.12, the factor-group $(Y, \tau)/\ker p$ is topologically isomorphic to (X, τ). The theorem is proved.

4.1.49. THEOREM. *Let R be a Hausdorff topological ring. Then there exists a complete Hausdorff topological ring S such that the ring R is topologically isomorphic to the factor-ring of S by a certain closed ideal.*

PROOF. Let $S = \sum_{i=1}^{\infty} R_i$, where $R_i = R$ for any $i \in \mathbb{N}$. Let $\bar{\tau}_n$ for $n \in \mathbb{N}$ and $\bar{\tau}$ be the topologies constructed in Theorem 4.1.48 on the Abelian group $S(+)$. According to Proposition 4.1.9 and Definition 4.1.24, $(S, \bar{\tau}_n)$ is a topological ring for any $n \in \mathbb{N}$. Due to Theorem 1.2.21, $(S, \bar{\tau})$ is a topological ring, too. It is easy to see that the mapping $p : S \to R$, defined in Theorem 4.1.48, is a ring homomorphism.

Since S is a complete ring, and the homomorphism p is an open and continuous surjective mapping, then the ring R is topologically isomorphic to the factor-ring of the complete ring S by the kernel of the homomorphism p, that is a closed ideal.

4.1.50. THEOREM. *Let R be a topological ring, in which the intersection of any countable family of open subsets is open (i.e. the topology on R is a $\mathfrak{m}$-topology for a certain non-countable cardinal number $\mathfrak{m}$), M be a Hausdorff topological R-module such that M is a topological R-module in the discrete topology (see Remark 1.2.9). Then there exists a complete Hausdorff topological R-module N such that the module M is topologically isomorphic to the factor-module of the module N by one of its closed submodules.*

PROOF. Let $N = \sum_{i=1}^{\infty} M_i$, where $M_i = M$ for any $i \in \mathbb{N}$. Let $\bar{\tau}_i$ for $i \in \mathbb{N}$ and $\bar{\tau}$ be the topologies constructed in Theorem 4.1.48 on the Abelian group N. Since M in both discrete and initial topologies is a topological module over the topological ring R, then, due to Proposition 4.1.15 and Definition 4.1.24, $(N, \bar{\tau}_i)$ is a topological module for any $i \in \mathbb{N}$, and, due to Theorem 1.2.22, $(N, \bar{\tau})$ is a topological R-module, too. It is easy to see that the mapping $p : N \to M$, defined in Theorem 4.1.48, is a homomorphism of modules.

Since N is a complete module and the homomorphism p is a continuous and open surjective mapping, then the module M is topologically isomorphic to the factor-module of the complete module N by the kernel of the homomorphism p, which is a closed submodule.

4.1.51. THEOREM. *Let R be a locally precompact topological ring, M be a Hausdorff topological R-module such that M in the discrete topology is also a topological R-module (see Remark 1.2.9). Then there exists a complete Hausdorff topological R-module N such that the module M is topologically isomorphic to the factor-module of the module N by its certain closed submodule.*

PROOF. Let $N = \sum_{i=1}^{\infty} M_i$, where $M_i = M$ for any $i \in \mathbb{N}$. Let $\bar{\tau}_i$ for $i \in \mathbb{N}$ and $\bar{\tau}$ be the topologies constructed in Theorem 4.1.48 on the Abelian group N. Since M in both discrete and initial topologies is a topological module over the topological ring R, then, according to Proposition 4.1.22 and Definition 4.1.24, $(N, \bar{\tau}_i)$ is a topological R-module for any $i \in \mathbb{N}$, and, due to Theorem 1.2.22, $(N, \bar{\tau})$ is a topological R-module, too. As we noted in Theorem 4.1.50, the mapping $p : N \to M$ from Theorem 4.1.48 is a homomorphism of modules. Repeating the end of the proof of Theorem 4.1.50, we get the proof of the theorem.

4.1.52. PROPOSITION. *Let $\{X_\gamma \,|\, \gamma \in \Gamma\}$ be a family of topological Abelian groups, X be their Tychonoff product, and A_i be a subset in X_i for each $i \in I$. The subset $A = \prod_{i=1}^{\infty} A_i$ is precompact in X if and only if for any $i \in I$ the subset A_i is precompact in X_i.*

PROOF. Let A be a precompact set in X. Since for any $i \in I$ the homomorphism π_i is continuous, and $\pi_i(A) = A_i$, then, due to Proposition 3.4.7, A_i is a precompact set in X_i.

Now let the set A_i be precompact in X_i for any $i \in I$. Denote over τ_i the prototype in X of the group topology X_i with respect to the homomorphism π_i. Since $\pi_i(A) = A_i$, then, due to Proposition 3.4.9, A is a precompact set in (X, τ_i) for any $i \in I$, hence, according to Proposition 3.4.11, A is a precompact set in $(X, \sup\{\tau_i \,|\, i \in I\})$. Since, due to Remark 4.1.4, $\sup\{\tau_i \,|\, i \in I\}$ coincides with the Tychonoff topology on X, then A is a precompact set in X.

4.1.53. COROLLARY (Tychonoff theorem). *Let $\{X_i \,|\, i \in I\}$ be a family of topological Abelian groups, X be their Tychonoff product. The group X is compact if and only if each of the groups X_i is compact.*

PROOF. Using consequently Theorem 3.4.16, Proposition 4.1.47, Proposition 4.1.52 and Theorem 3.4.16, we get the necessary result.

4.1.54. REMARK. Since the topological properties of $\mathfrak{m}$-products of topological rings and modules are determined by the properties of their $\mathfrak{m}$-products as topological Abelian groups, then for $\mathfrak{m}$-products of topological rings and modules the analogs of statements 4.1.35 - 4.1.47, 4.1.52 and 4.1.53 are fulfilled.

4.1.55. EXAMPLE. We give the example of a group X and two group topologies τ_1 and τ_2 on X such that (X, τ_1) and (X, τ_2) are compact, hence complete, groups, but the group $(X, \sup\{\tau_1, \tau_2\})$ is not complete (see Remark 3.2.35).

Let $\{X_i \mid i \in \mathbb{N}\}$ be a countable family of two-element groups, i.e. $X_i = \{0, 1\}$, that have the discrete topology, (X, τ_1) be their Tychonoff product. Due to Corollaries 4.1.44 and 4.1.53, (X, τ_1) is a compact Hausdorff group.

For every $i \in \mathbb{N}$ choose such elements $x_i, y_i \in X$ that:

$$\pi_j(x_i) = \begin{cases} 1, & \text{for} \quad j = i; \\ 0, & \text{for} \quad j \neq i, \end{cases}$$

$$\pi_j(y_i) = \begin{cases} 0, & \text{for} \quad j \in 2^i \cdot \mathbb{N}; \\ 1, & \text{in other cases}. \end{cases}$$

The group X can be considered as a vector space over the two-element field $\mathbb{Z}_2$. It is easy to see that the family $\{x_i, y_i \mid i \in \mathbb{N}\}$ is linearly independent in X. Then it can be completed up to a certain basis B of the vector space X. Since the set B has the continuum cardinality, then we can consider that it is enumerated by the transfinite numbers from 1 to $\omega(c)$ (see A.7), i.e. $B = \{z_\alpha \mid 1 \leq \alpha < \omega(c)\}$, where $x_i = z_i, y_i = z_{w_0+i-1}$ (see A.7) for any $i \in \mathbb{N}$. We define the mapping $\psi : B \to B$ as the following:

$$\psi(z_\alpha) = \begin{cases} y_\alpha, & \text{for} \quad \alpha \in \mathbb{N}; \\ x_i, & \text{for} \quad \alpha = \omega_0 + i - 1; \\ z_\alpha, & \text{in other cases}. \end{cases}$$

It is clear that ψ is a one-to-one mapping of the set B to itself. Then ψ could be extended up to the isomorphic mapping $\widehat{\psi}$ of the group X to itself. Let τ_2 be the prototype of the topology τ_1 with respect to the isomorphism $\widehat{\psi}$. Then, due to Proposition 3.4.9, (X, τ_2) is a precompact group and, due to Proposition 3.2.32, (X, τ_2) is a complete group. According to Theorem 3.4.16, (X, τ_2) is a compact group. Since $\widehat{\psi}$ is a one-to-one mapping of X to itself, then (X, τ_2) is a Hausdorff group.

Now we show that $\tau_2 \not\le \tau_1$. Assume the contrary, i.e. that $\tau_2 \le \tau_1$. Since $\pi_1^{-1}(0)$ is a neighborhood of zero in (X, τ_1), then $\widehat{\psi}^{-1}(\pi_1^{-1}(0))$ is a neighborhood of zero in (X, τ_2). Since $\tau_2 \le \tau_1$, then $\widehat{\psi}^{-1}(\pi_1^{-1}(0))$ is a neighborhood of zero in (X, τ_1). It means that there exists a natural number n such that $\bigcap_{i=1}^{n} \pi_i^{-1}(0) \subseteq \widehat{\psi}^{-1}(\pi_1^{-1}(0))$, hence, $\widehat{\psi}(\bigcap_{i=1}^{n} \pi_i^{-1}(0)) \subseteq \pi_1^{-1}(0)$. Since $z_{n+1} = x_{n+1} \in \bigcap_{i=1}^{n} \pi_i^{-1}(0)$, then $y_{n+1} = \psi(z_{n+1}) \in \pi_1^{-1}(0)$, hence, $\pi_1(y_{n+1}) = 0$. It contradicts the choice of the elements y_j for $j \in \mathbb{N}$. The obtained contradiction shows that $\tau_1 \not\le \tau_2$. Thus, $\tau_2 < \sup\{\tau_1, \tau_2\}$.

According to Proposition 3.4.10, the group $(X, \sup\{\tau_1, \tau_2\})$ is precompact, and it is evident that it is a Hausdorff group. Since $\tau_2 < \sup\{\tau_1, \tau_2\}$ and the topology of a Hausdorff compact group cannot be weakened to the Hausdorff topology, then $(X, \sup\{\tau_1, \tau_2\})$ is not a complete group.

4.1.56. PROPOSITION. *Let R be a topological ring, M and N be topological R-modules that have a basis of neighborhoods of zero consisting of submodules, $f : M \to N$ be a continuous homomorphism of the modules. If M is a linearly compact module, then the submodule $\varphi(M)$ of the topological module N is linearly compact, too.*

PROOF. Let $\{n_\alpha + N_\alpha \,|\, \alpha \in \Lambda\}$ be a basis of filter of the module $\varphi(M)$, consisting of the cosets by the closed submodules N_α of the module $\varphi(M)$, and $\{m_\alpha \,|\, \alpha \in \Lambda\}$ the set of elements of the module M such that $\varphi(m_\alpha) = n_\alpha$ for any $\alpha \in \Lambda$. Then the family $\{m_\alpha + \varphi^{-1}(N_\alpha) \,|\, \alpha \in \Lambda\}$ is a basis of a certain filter of the module M, consisting of the residue classes by the closed submodules $\varphi^{-1}(N_\alpha)$ of the module M. Since M is a linearly compact module, then $\bigcap_{\alpha \in \Lambda}(m_\alpha + \varphi^{-1}(N_\alpha)) \ne \emptyset$. If $m \in \bigcap_{\alpha \in \Lambda}(m_\alpha + \varphi^{-1}(N_\alpha))$, then $\varphi(m) \in \bigcap_{\alpha \in \Lambda}(n_\alpha + N_\alpha)$. Therefore, $\varphi(M)$ is a linearly compact module.

4.1.57. PROPOSITION. *Let R be a topological ring, $\{M_i \,|\, i \in I\}$ be a family of topological R-modules that have a basis of neighborhoods of zero consisting of submodules, M be their Tychonoff product. The module M is linearly compact if and only if the module M_i is linearly compact for each $i \in I$.*

PROOF. Let M be a linearly compact module, then, according to Proposition 4.1.56, the module M_i is linearly compact for every $i \in I$.

Now let the module M_i be linearly compact for every $i \in I$, and assume that the module M is not linearly compact. Then there exists a basis $\{b_\alpha + N_\alpha \mid \alpha \in \Lambda\}$ of a filter $\mathcal{F}$ of M consisting of the cosets by the closed submodules N_α such that $\bigcap_{\alpha \in \Lambda}(b_\alpha + N_\alpha) = \emptyset$.

Consider the set $\mathcal{A}$ of all filters of the module M that have a basis consisting of the cosets by closed submodules. Let $\{\mathcal{H}_\gamma \mid \gamma \in \Gamma\}$ be a linearly ordered set of the filters from $\mathcal{A}$, and $\mathcal{B}_\gamma$ be a basis of the filter $\mathcal{H}_\gamma$ consisting of the cosets by closed submodules, for $\gamma \in \Gamma$. The system $\mathcal{B} = \bigcup_{\gamma \in \Gamma} \mathcal{B}_\gamma$ is a basis of the filter $\mathcal{H} = \bigcup_{\gamma \in \Gamma} \mathcal{H}_\gamma$ and consists of the cosets by closed submodules. Hence, $\mathcal{H} \in \mathcal{A}$. Therefore, $\mathcal{A}$ is an inductive set (see A.11).

According to Zorn lemma, in the set $\mathcal{A}$ exists a maximal element Φ, majoring $\mathcal{F}$. Let $\{m_\delta + J_\delta \mid \delta \in \Delta\}$ be a basis of Φ, consisting of the cosets by closed modules J_δ. Then

$$\bigcap_{\delta \in \Delta}(m_\delta + J_\delta) \subseteq \bigcap_{F \in \mathcal{F}} F = \bigcap_{\alpha \in \Lambda}(b_\alpha + N_\alpha) = \emptyset.$$

For any $i \in I$ the family $\{\pi_i(m_\delta) + [\pi_i(J_\delta)]_{M_i} \mid \delta \in \Delta\}$ is a basis of a certain filter of M_i consisting of the cosets by the closed submodules $[\pi_i(J_\delta)]_{M_i}$. Since M_i is a linearly compact module, then there exists $a_i \in \bigcap_{\delta \in \Delta}(\pi_i(m_\delta) + [\pi_i(J_\delta)]_{M_i})$.

Now let's consider the element $f \in M$ such that $f(i) = a_i$ for any $i \in I$. Since $\bigcap_{\delta \in \Delta}(m_\delta + J_\delta) = \emptyset$, then there exists $\delta_0 \in \Delta$ such that $f \notin m_{\delta_0} + J_{\delta_0}$. Since J_{δ_0} is a closed submodule in M, then there exists a neighborhood U of zero in M such that $(f + U) \cap (m_{\delta_0} + J_{\delta_0}) = \emptyset$. We can consider that $U = \prod_{i \in I} U_i$, where U_i is an open submodule in M_i for $i \in I$, and the set $I_1 = \{i \mid U_i \neq M_i\}$ is finite. Since $\pi_i(f) = a_i \in \pi_i(m_\delta) + [\pi_i(J_\delta)]_{M_i}$ for any $i \in I$ and $\delta \in \Delta$, then for any $i \in I$ the family $\{m_\delta + J_\delta \mid \delta \in \Delta\} \bigcup \{f + \pi_i^{-1}(U_i)\}$ is a basis of a certain filter $\Phi_i \in \mathcal{A}$, and $\Phi_i \geq \Phi$. Since the filter Φ is maximal, then $\Phi_i = \Phi$ for any $i \in I$, hence $f + \prod_{i \in I} U_i = \bigcap_{i \in I_1}(f + \pi_i^{-1}(U_i)) \in \Phi$, therefore

$$(f + U) \cap (m_{\delta_0} + J_{\delta_0}) = \left(f + \prod_{i \in I} U_i\right) \cap (m_{\delta_0} + J_{\delta_0}) \neq \emptyset.$$

It contradicts the choice of the neighborhood U of zero in M.

The obtained contradiction proves that the module M is linearly compact.

4.1.58. EXAMPLE. Now we present the example showing that in the general case the $\mathfrak{m}$-product of compact completely disconnected (therefore, linearly compact) modules is neither linearly compact, nor a compact module.

Let R be the two-element field that has the discrete topology, $\mathfrak{m}$ be an infinite non-countable cardinal number. According to the Remark 4.1.13, the topology on R is a $\mathfrak{m}$-topology. Let $\{M_i \,|\, i \in \mathbb{N}\}$ be a family of finite topological R-modules that have the discrete topology. Then M_n is a completely disconnected compact module, therefore, a linearly compact module for any $n \in \mathbb{N}$. Let $M = \prod_{i=1}^{\infty} M_i$ be their $\mathfrak{m}$-product, then M is an infinite-dimensional discrete vector space over R. According to Remark 3.2.9, the module M is not linearly compact and not even a compact module.

4.1.59. PROPOSITION. *Let $\mathfrak{m}$ be a cardinal number, $\{X_i \,|\, i \in I\}$ be a family of topological Abelian groups, $Y = \sum_{i \in I} X_i$ be their direct sum with the $\mathfrak{m}$-product topology. The group Y is connected if and only if any of X_i is connected.*

PROOF. If Y is a connected group, then $X_i = \pi_i(Y)$ is connected for any $i \in I$ as a continuous image of the connected set (see C.13).

Now let any of X_i be connected. Then, due to Proposition 4.1.5, for every $i \in I$ the subgroup $Y_i = \{f \in Y \,|\, f(j) = 0_j \text{ for } j \neq i\}$ is connected. Since the connected component $C(Y)$ of zero in Y is a subgroup in Y, then $(C(Y)) \supseteq \sum_{i \in I} Y_i = Y$.

Therefore, $C(Y) = Y$, i.e. Y is a connected group.

4.1.60. PROPOSITION. *Let $\{X_i \,|\, i \in I\}$ be a family of topological Abelian groups, $X = \prod_{i \in I} X_i$ be their Tychonoff product. The group X is connected if and only if any of X_i is connected.*

PROOF. If X is a connected group, then $X_i = \pi_i(X)$ is a connected group for any $i \in I$ as a continuous mapping of a connected set.

Now let X_i be a connected group for any $i \in I$. Then, according to Proposition 4.1.59, $\sum_{i \in I} X_i$ is a connected subgroup in X. Due to Proposition 4.1.45, $X = [\sum_{i \in I} X_i]_X$, therefore X is a connected group.

4.1.61. PROPOSITION. *Let $\mathfrak{m}$ be a cardinal number, $\{X_i \,|\, i \in I\}$ be a family of topological Abelian groups, $X = \prod_{i \in I} X_i$ be their $\mathfrak{m}$-product. The group X is completely* *disconnected if and only if each X_i is completely disconnected.*

PROOF. If the group X is completely disconnected, then for any $i \in I$ the subgroup $Y_i = \{f \in X \mid f(j) = 0 \text{ for } j \neq i\}$ of the group X is completely disconnected, too. Then, according to Proposition 4.1.5, the group X_i is completely disconnected for any $i \in I$.

Now let X_i be completely disconnected for any $i \in I$. Assume the contrary, i.e. that the connected component $C(X)$ of zero of the group X is not equal to $\{0\}$. Then there exists $f \in C(X)$ such that $f \neq 0$, hence $f(i) \neq 0$ for a certain $i \in I$. Then $\pi_i(C(X))$ is a connected set in X_i and $0 \in \pi_i(C(X)) \neq \{0\}$. This contradicts the fact that X_i is completely disconnected. Thus, $C(X) = 0$, i.e. the group X is completely disconnected.

4.1.62. PROPOSITION. *Let $\mathfrak{m}$ be a non-countable cardinal number, $\{X_i \mid i \in I\}$ be a family of connected Hausdorff topological groups, X be their $\mathfrak{m}$-product. Then the connected component $C(X)$ of zero in X is equal to $\sum_{i \in I} X_i$.*

PROOF. According to Proposition 4.1.59, $C(X) \supseteq \sum_{i \in I} X_i$. Assume that $C(X) \neq \sum_{i \in I} X_i$ and let $x \in C(X) \setminus \sum_{i \in I} X_i$. Then there exists a countable subset $\{i_k \mid k \in \mathbb{N}\} \subseteq I$ such that $\pi_{i_k}(x) \neq 0$ for any $k \in \mathbb{N}$. Since X_{i_k} is Hausdorff, then for any $k \in \mathbb{N}$ there exist neighborhoods V_k and U_k of zero in X_{i_k} such that $\pi_{i_k}(x) \notin V_k$ and $\underbrace{U_k + \ldots + U_k}_{k \text{ times}} \subseteq V_k$.

Let
$$W_i = \begin{cases} U_k, & \text{for} \quad i = i_k, \ k \in \mathbb{N}; \\ X_i, & \text{in other cases.} \end{cases}$$

Then $W = \prod_{i \in I} W_i$ is a neighborhood of zero in X. Let A be the subgroup in X generated by the set W, then A is an open-closed subgroup in X (see Proposition 1.4.18). Since $C(X)$ is a connected set, then $C(X) \subseteq A$ (see C.13), hence, $x \in A$. Then there exist $x_1, \ldots, x_n \in W$ such that $x = x_1 + x_2 + \ldots + x_n$. Therefore,

$$\pi_{i_n}(x) = \sum_{k=1}^{n} \pi_{i_n}(x_k) \in \underbrace{\pi_{i_n}(W) + \ldots + \pi_{i_n}(W)}_{n \text{ times}} = \underbrace{U_n + \ldots + U_n}_{n \text{ times}} \subseteq V_n.$$

It contradicts the choice of the neighborhood V_n of zero in X_n, which shows that $C(X) = \sum_{i \in I} X_i$.

4.1.63. PROPOSITION. *Let $\mathfrak{m}$ be a cardinal number, $\{X_i \mid i \in I\}$ be a family of Hausdorff topological Abelian groups, X be their $\mathfrak{m}$-product. Then:*

1) if $\mathfrak{m} = \aleph_0$, *then* $C(X) = \prod_{i \in I} C(X_i)$;

2) if $\mathfrak{m} > \aleph_0$, *then* $C(X) = \sum_{i \in I} C(X_i)$.

PROOF. According to Proposition 4.1.41, the factor-group $X / \prod_{i \in I} C(X_i)$, which has the factor-topology, is topologically isomorphic to the $\mathfrak{m}$-product $\prod_{i \in I} (X_i / C(X_i))$ of the topological groups $X_i / C(X_i)$. Since for any $i \in I$ the group $X_i / C(X_i)$ is totally disconnected, then, due to Proposition 4.1.61, the group $\prod_{i \in I} (X_i / C(X_i))$ and, hence, the group $X / \prod_{i \in I} C(X_i)$, is totally disconnected, too. Therefore, $C(X) \subseteq \prod_{i \in I} C(X_i)$. Then

$$C(X) = C(C(X)) \subseteq C\left(\prod_{i \in I} C(X_i) \right) \subseteq C(X),$$

i.e. $C(X) = C(\prod_{i \in I} C(X_i))$.

If $\mathfrak{m} = \aleph_0$, then, according to Proposition 4.1.60, $C(\prod_{i \in I} C(X_i)) = \prod_{i \in I} C(X_i)$, hence, $C(X) = \prod_{i \in I} C(X_i)$.

If $\mathfrak{m} > \aleph_0$, then, according to Proposition 4.1.62, $C(\prod_{i \in I} C(X_i)) = \sum_{i \in I} C(X_i)$, therefore $C(X) = \sum_{i \in I} C(X_i)$.

4.1.64. COROLLARY. *Let* $\mathfrak{m}$ *be a cardinal number,* $\{X_i \mid i \in I\}$ *be a family of Hausdorff topological Abelian groups,* $Y = \sum_{i \in I} X_i$ *be their direct sum with the* $\mathfrak{m}$-*product topology. Then* $C(Y) = \sum C(X_i)$.

PROOF. Let X be the $\mathfrak{m}$-product of the family $\{X_i \mid i \in I\}$. Since Y is a subgroup of the topological group X, then $C(Y) \subseteq C(X) \cap Y$. Due to Proposition 4.1.63, $C(X) \cap Y = \sum_{i \in I} C(X_i)$, and due to Proposition 4.1.59, $\sum_{i \in I} C(X_i) \subseteq C(Y)$. Thus, $C(Y) = \sum_{i \in I} C(X_i)$.

4.1.65. PROPOSITION. *Let* $\{X_i \mid i \in I\}$ *be a family of topological Abelian groups. The Tychonoff product* $X = \prod_{i \in I} X_i$ *is locally precompact if and only if any of the groups* X_i *is locally precompact, and there exists a finite subset* $J \subseteq I$ *such that the group* X_i *is precompact for any* $i \in I \setminus J$.

PROOF. Let X be a locally precompact group and U be a precompact neighborhood of zero in X, then $\pi_i(U)$ is a precompact neighborhood of zero in X_i (see Proposition 3.4.7) for every $i \in I$. Therefore, X_i is a locally precompact group for every $i \in I$. As above, we can consider that $U = \prod_{i \in I} U_i$, where U_i is a neighborhood of zero in X_i, and the set

$J = \{i \mid U_i \neq X_i\}$ is finite. Then, according to Proposition 3.4.7, for any $i \in I \backslash J$ the set $\pi_i(X_i) = X_i$ is precompact in X_i, i.e. X_i is a precompact group for $i \notin J$.

Now let the group X_i be locally precompact for every $i \in I$, and there exists a finite subset $J \subseteq I$ such that X_i is a precompact group for $i \notin J$. For every $i \in J$ choose a precompact neighborhood V_i of zero in X_i. Then the set $U = \prod_{i \in I} U_i$, where $U_i = V_i$ for $i \in J$ and $U_i = X_i$ for $i \notin J$, is a neighborhood of zero in X, and, according to Proposition 4.1.52, is a precompact set. Thus, X is a locally precompact group.

4.1.66. PROPOSITION. *Let* $\mathfrak{m}$ *be a cardinal number,* $\mathfrak{m} > \aleph_0$ *and* $\{X_i \mid i \in I\}$ *be a family of topological Abelian groups that have non-antidiscrete topology. Let X be their* $\mathfrak{m}$*-product, then the group X is locally precompact if and only if:*

1) $|I| < \mathfrak{m}$;

2) for every $i \in I$ *the group* X_i *is locally precompact;*

3) there exists a finite subset $J \subseteq I$ *such that for any* $i \in I \backslash J$ *the group* X_i *has the smallest neighborhood of zero.*

PROOF. Let X be a locally precompact group and U be a precompact neighborhood of zero in X. As above, we can consider that $U = \prod_{i \in I} U_i$, where U_i is a neighborhood of zero in X_i, and $\left| \{i \mid U_i \neq X_i\} \right| < \mathfrak{m}$.

Assume the contrary of condition 1), i.e. that $|I| \geq \mathfrak{m}$. Since $\mathfrak{m} > \aleph_0$ and $|I| \geq \mathfrak{m}$, then $\left| \{i \in I \mid U_i = X_i\} \right| \geq \mathfrak{m}$. Let $J_1 = \{i \in I \mid U_i \neq X_i\}$ and $J_2 = \{i \in I \mid U_i = X_i\}$. Consider the $\mathfrak{m}$-products $Y = \prod_{i \in J_1} X_i$ and $Z = \prod_{i \in J_2} X_i$. According to Proposition 4.1.36, the group X is topologically isomorphic to the direct product $Y \times Z$. We identify X and $Y \times Z$. According to Proposition 3.4.7, the natural projection $\pi : Y \times Z \to Z$ transfers the precompact set U to the precompact set. Hence, $Z = \prod_{i \in J_2} X_i$ is a precompact set. Let $J = \{i_k \mid k \in \mathbb{N}\}$ be a countable subset in J_2. For any $k \in \mathbb{N}$ let V_k be a symmetric neighborhood of zero in X_{i_k} such that $V_k + V_k \neq X_{i_k}$. Then the set $W = \prod_{i \in J_2} W_i$, where $W_i = V_k$ for $i = i_k$ and $W_i = X_i$ for $i \in J_2 \backslash J$, is a neighborhood of zero in Z. Since Z is precompact, then there exists a finite set of elements $f_1, \ldots, f_n \in Z$ such that $Z = \cup_{t=1}^{n} (f_t + W)$.

Now we prove that $f_t(i_k) + V_k \neq X_{i_k}$ for any $k \in \mathbb{N}$ and $t = 1, \ldots, n$. Indeed, assume the contrary, i.e. that $f_t(i_k) + V_k = X_{i_k}$ for a certain $k \in \mathbb{N}$ and a certain $1 \leq t \leq n$.

Then there exists $v \in V_k$ such that $f_t(i_k) + v = 0$. Since the neighborhood V_k of zero is symmetric in X_{i_k}, then $f_t(i_k) = -v \in V_k$, therefore, $X_{i_k} = f_t(i_k) + V_k \subseteq V_k + V_k$. It contradicts the choice of the numbers k and t. Hence, $f_t(i_k) + V_k \neq X_{i_k}$ for any $k \in \mathbb{N}$ and $1 \leq t \leq n$.

For any $1 \leq t \leq n$ choose $g_t \in X_{i_t}$ such that $g_t \notin f_t(i_t) + V_t$. Let g be an element of Z such that $g(i_t) = g_t$ for any $1 \leq t \leq n$. Then $g \in f_l + W$ for any $1 \leq l \leq n$ (as $Z = \bigcup_{t=1}^{n}(f_t + W)$), therefore $g_l = g(i_l) \in f_l(i_l) + V_l$. It contradicts the choice of the element g_l. The obtained contradiction shows that $\mathfrak{m} > |I|$, i.e. the condition 1) is proved.

Since $U_i = \pi_i(U)$ is a neighborhood of zero in X_i and, due to Proposition 3.4.7, is a precompact set, then X_i is a locally precompact group for any $i \in I$, i.e. the condition 2) is proved.

Assume the contrary to the condition 3), i.e. that there exists a countable subset $J = \{i_k \mid k \in \mathbb{N}\}$ in I such that for any $k \in \mathbb{N}$ the group X_{i_k} does not have the smallest neighborhood of zero. Then U_{i_k} is not the smallest neighborhood of zero in X_{i_k} for every $k \in \mathbb{N}$. Then there exists a symmetric neighborhood V_k of zero in X_{i_k} such that $U_{i_k} \supseteq V_k + V_k$.

Let $W = \prod_{i \in I} W_i$, where $W_i = X_{i_k}$ if $i \notin J$ and $W_i = V_k$ if $i = i_k$, for $k \in \mathbb{N}$. Then W is a neighborhood of zero in X. Since U is a precompact set in X, then there exists a finite set of elements $f_1, \dots, f_n \in X$ such that $U = \bigcup_{t=1}^{n}(f_t + W)$. As above (see proof that $X_{i_k} \neq f_t(i_k) + V_k$), for any $1 \leq t \leq n$ it is verified that $U_{i_k} \not\subseteq f_t(i_k) + V_k$ for any $k \in \mathbb{N}$. For any $1 \leq t \leq n$ choose an element $g_t \in U_{i_t}$ such that $g_t \notin f_t(i_t) + V_t$. There exists an element g of $U = \prod_{i \in I} U_i$ such that $g(i_t) = g_t$ for any $1 \leq t \leq n$. Then $g \in f_l + W$ for a certain $1 \leq l \leq n$ (as $U = \bigcup_{t=1}^{n}(f_t + W)$) and, hence, $g_l = g(i_l) \in f(i_l) + V_l$. It contradicts the choice of the elements g_t. The obtained contradiction shows that the group X_i has the smallest neighborhood of zero almost for all, i.e. except not more that a finite number, $i \in I$. Then the condition 3) is proved.

Now let the conditions 1)–3) be fulfilled (i.e. $\mathfrak{m} > |I|$; for any $i \in I$ the group X_i is locally precompact and there exists a finite subset $J \subseteq I$ such that for any $i \in I \backslash J$ the group X_i has the smallest neighborhood of zero). Let $U = \prod_{i \in I} U_i$, where U_i is a precompact neighborhood of zero in X_i for each $i \in J$, and U_i be the smallest neighborhood of zero for $i \in I \backslash J$. Since $\mathfrak{m} > |I|$, then U is a neighborhood of zero in X.

Now we prove that U is a precompact set. Indeed, let $\Gamma = I \backslash J$, and $Y = \prod_{i \in J} X_i$ and $Z = \prod_{i \in \Gamma} X_i$ be $\mathfrak{m}$-products of the groups X_i. According to Proposition 4.1.36, the group X is topologically isomorphic to the direct product $Y \times Z$. Now we identify the groups X and $Y \times Z$. Then $U = W_1 \times W_2$, where $W_1 = \prod_{i \in J} U_i$ and $W_2 = \prod_{i \in \Gamma} U_i$.

Since J is a finite set, then, according to Proposition 4.1.23, the $\mathfrak{m}$-product topology on Y coincides with the Tychonoff topology, therefore, according to Proposition 4.1.52, W_1 is a precompact set in Y. It is clear that W_2 is the smallest neighborhood of zero in Z, hence, W is a precompact set in Z. Then, according to Proposition 4.1.52, $U = W_1 \times W_2$ is a precompact set in X.

4.1.67. EXAMPLE. We present an example of an Abelian group G and a family $\{\tau_k \,|\, k \in \mathbb{N}\}$ of group topologies on G such that (G, τ_k) is a locally precompact group for any $k \in \mathbb{N}$, but the group $(G, \sup\{\tau_k \,|\, k \in \mathbb{N}\}\,)$ is not locally precompact.

Let $G_i = \mathbb{Z}$ with the discrete topology for $i \in \mathbb{N}$ and $G = \prod_{i \in \mathbb{N}} G_i$. Let τ_i be the prototype on G of the discrete topology with respect to the projection π_i. Since for any $i \in \mathbb{N}$ the set $\{0\}$ is a precompact in G_i, then, according to Proposition 3.4.9, $\pi_i^{-1}(0)$ is a precompact neighborhood of zero in (G, τ_i). Therefore, (G, τ_i) is a locally precompact group for any $i \in \mathbb{N}$. Let $\tau = \sup\{\tau_i \,|\, i \in \mathbb{N}\}$, then, according to Remark 4.1.4, τ is a topology of the Tychonoff product. Since $\mathbb{Z}$ in the discrete topology is not a precompact group, then, according to Proposition 4.1.65, the group (G, τ) is not locally precompact.

4.1.68. DEFINITION. Let X be a topological Abelian group, I be a set of indexes, The direct product $Y = \prod_{i \in I} X_i$ of the family $\{X_i \,|\, i \in I\}$, where $X_i = X$ for any $i \in I$, we call a direct power of the group X. Consider the family $\mathcal{L}$ of all subsets in Y of the type $\prod_{i \in I} V_i$, where $V_i = V$ for any $i \in I$ and V is a neighborhood of zero in X. It is easy to see that the family $\mathcal{L}$ satisfies the conditions (BN1) – (BN4), therefore, according to Theorem 1.2.4, there is the only one topology on Y such that Y is a topological Abelian group and $\mathcal{L}$ is a basis of neighborhoods of zero. We call this topology a cubic topology and denote it over τ_c.

4.1.69. REMARK. Let X be a topological Abelian group, I be a set of indexes, $Y = \prod_{i \in I} X_i$, where $X_i = X$ for $i \in I$. Let τ_T be the Tychonoff topology (see Definition 4.1.3), τ_C be the cubic topology and τ_B be the brick (see Definition 4.1.24) topology on Y. Then, due to the type of a basis of neighborhoods of zero in τ_T, τ_C, τ_B we get that:

1) if I is a finite set or X has the antidiscrete topology, then $\tau_T = \tau_C = \tau_B$;

2) if I is an infinite set, the group X has the smallest neighborhood of zero and X has a non-antidiscrete topology, then $\tau_T < \tau_C = \tau_B$;

3) if I is an infinite set and the group X does not have the smallest neighborhood of zero, then $\tau_T < \tau_C < \tau_B$.

4.1.70. PROPOSITION. *Let X be a topological Abelian group, I a set of indexes, $Y = \prod_{i \in I} X_i$, where $X_i = X$ for $i \in I$. Let the group Y have the cubic topology τ_C and for any $i \in I$ is determined a subset A_i in the group X_i. Then $[\prod_{i \in I} A_i]_Y = \prod_{i \in I} [A_i]_{X_i}$.*

PROOF. Let $K = (\prod_{i \in I} X_i, \tau_B)$, where τ_B is the brick topology on $\prod_{i \in I} X_i$, and $D = (\prod_{i \in I} X_i, \tau_T)$, where τ_T is the Tychonoff topology on $\prod_{i \in I} X_i$. Since, due to Remark 4.1.69, $\tau_T \leq \tau_C \leq \tau_B$, then

$$\left[\prod_{i \in I} A_i\right]_D \supseteq \left[\prod_{i \in I} A_i\right]_Y \supseteq \left[\prod_{i \in I} A_i\right]_K.$$

According to Proposition 4.1.42,

$$\left[\prod_{i \in I} A_i\right]_D = \prod_{i \in I} [A_i]_{X_i} = \left[\prod_{i \in I} A_i\right]_K.$$

Therefore,

$$\left[\prod_{i \in I} A_i\right]_Y = \prod_{i \in I} [A_i]_{X_i}.$$

4.1.71. COROLLARY. Let X be a topological Abelian group, I be a set of indexes, $Y = \prod_{i \in I} X_i$, where $X_i = X$ for $i \in I$. Let the group Y have the cubic topology and for any $i \in I$ in the group X_i is determined a subset A_i. Then:

1) the subset $\prod_{i \in I} A_i$ is closed in Y if and only if for any $i \in I$ the subset A_i is closed in X_i;

2) the subset $\prod_{i \in I} A_i$ is dense in Y if and only if for any $i \in I$ the subset A_i is dense in X_i;

3) the group Y is Hausdorff if and only if the group X is Hausdorff;

4) the group Y is complete if and only if the group X is complete.

The proof of each of the statements is analogous to the proof of Corollaries 4.1.43, 4.1.44 and Proposition 4.1.47 correspondingly.

4.1.72. EXAMPLE. We present an example showing that in a topological Abelian group $Y = \prod_{i \in I} X_i$, where $X_i = X$, that has the cubic topology, the subgroup $\sum_{i \in I} X_i$ is not closed and is not dense.

Let X be an additive group of real numbers with the interval topology, $X_i = X$ for $i \in \mathbb{N}$. Now we prove that in the group $Y = \prod_{i \in I} X_i$ endowed with the cubic topology the equality $[\ \sum_{i=1}^{\infty} X_i\ ]_Y = \{f \in Y \mid \lim_{n \to \infty} f(n) = 0 \text{ in } X\ \}$ is fulfilled.

Indeed, let $f \in [\ \sum_{i=1}^{\infty} X_i\ ]_Y$. Then for any neighborhood U of zero in X the set $W = \prod_{i \in I} U_i$, where $U_i = U$ for $i \in I$, is a neighborhood of zero in Y. Hence $(f + W) \cap \sum_{i=1}^{\infty} X_i \neq \emptyset$. Let $g \in (f + W) \cap (\sum_{i=1}^{\infty} X_i)$, then there exists a natural number n such that $g(i) = 0$ for all $i \geq n$. Then $-f(i) \in U_i = U$ for all $i \geq n$. Therefore, the sequence $f(1), f(2), \ldots$ converges to zero in X.

Now let $f \in Y$ and the sequence $f(1), f(2), \ldots$ converges to zero in X. Let W be a neighborhood of zero in Y, then, as above, we can consider that $W = \prod_{i \in I} U_i$, where $U_i = U$ for $i \in I$ and U is a neighborhood of zero in X. There exists a natural number n such that $-f(i) \in U$ for all $i \geq n$. Let g be the element of Y such that $g(i) = f(i)$ for all $i \leq n$ and $g(i) = 0$ for all $i > n$. Then $g \in \sum_{i=1}^{\infty} X_i$ and $g - f \in \prod_{i \in I} U_i = W$, i.e. $g \in f + W$. Therefore, $f \in [\ \sum_{i=1}^{\infty} X_i\ ]_Y$.

4.1.73. REMARK. Let X be a topological Abelian group, I be a set of indexes, $X_i = X$ for $i \in I$ and $\prod_{i \in I} X_i$ be the direct power of the group X with the cubic topology. Example 4.1.72 shows that it is possible that $\sum_{i \in I} X_i$ is not closed and not dense in Y. Thus, for Y neither analog of Proposition 4.1.45, nor the analog of Proposition 4.1.46 is fulfilled.

Example 4.1.58 also shows that the construction of the direct power with the cubic topology does not inherit compactness or linear compactness.

4.1.74. PROPOSITION. *Let R be a topological ring, M be a bounded topological* *right R-module (see Definition 1.6.1),* $M_i = M$ *for* $i \in I$, *and* $N = \prod_{i \in I} M_i$ *be the direct power of the group* M *with the cubic topology* τ_c. *According to Proposition 4.1.14,* N *is a right R-module. Then* (N, τ_c) *is a bounded topological R-module.*

PROOF. According to Definition 4.1.68, (N, τ_c) is a topological Abelian group. Now we prove that the multiplication operation of the elements of N by the elements of the ring R is continuous.

Let $f \in N$, $r \in R$ and U be a neighborhood of zero in (N, τ_c). Without loss of generality we can consider that $U = \prod_{i \in I} U_i$, where $U_i = V$ for any $i \in I$, and V is a neighborhood of zero in M. There exist neighborhoods V_0 and V_1 of zero in M such that $V_0 + V_0 + V_0 \subseteq V$ and $V_1 \cdot r \subseteq V_0$. Since the module M is bounded, then there exists a neighborhood W of zero in R such that $M \cdot W \subseteq V_0$. Then $V_2 = V_0 \cap V_1$ is a neighborhood of zero in M, and for any $i \in I$ we have:

$$(f(i) + V_2) \cdot (r + W) = f(i) \cdot r + V_2 \cdot r + f(i) \cdot W + V_2 \cdot W$$
$$\subseteq f(i) \cdot r + V_1 \cdot r + M \cdot W + M \cdot W \subseteq f(i) \cdot r + V_0 + V_0 + V_0 \subseteq f(i) \cdot r + V.$$

Let $U_i' = V_2$ for $i \in I$, then $P = \prod_{i \in I} U_i'$ is a neighborhood of zero in M and $(f + P) \cdot (r + W) \subseteq f \cdot r + \prod_{i \in I} U_i = f \cdot r + U$. The continuity of the multiplication operation is proved.

Now we prove that (N, τ_c) is a bounded topological R-module. Indeed, let U be a neighborhood of zero in N. As above, we can consider that $U = \prod_{i \in I} U_i$, where $U_i = V$ for $i \in I$, and V is a certain neighborhood of zero in M. Since M is a bounded topological module, then there exists a neighborhood W of zero in R such that $M \cdot W \subseteq V$. Then $N \cdot W \subseteq \prod_{i \in I} U_i = U$. Thus, (N, τ_c) is a bounded R-module.

The proposition is proved.

4.1.75. PROPOSITION. Let R be a bounded topological ring (see Definition 1.6.2), $R_i = R$ for $i \in I$, and $S = \prod_{i \in I} R_i$ be the direct power of the additive group $R(+)$, that has the cubic topology τ_c. Consider S as a ring, according to Definition 4.1.8. Then (S, τ_c) is a bounded topological ring.

PROOF. According to Definition 4.1.68, (S, τ_c) is a topological Abelian group. We prove the continuity of the multiplication operation in (S, τ_c).

Let $f, g \in S$ and U be a neighborhood of zero in (S, τ_c). Without loss of generality we can consider that $U = \prod_{i \in I} U_i$, where $U_i = V$ for any $i \in I$, and V is a neighborhood of zero in R. Let V_1 and V_2 be neighborhoods of zero in R such that $V_1 + V_1 + V_1 \subseteq V$, $V_2 \cdot R \subseteq V_1$ and $R \cdot V_2 \subseteq V_1$. Then for any $i \in I$ we have:

$$(f(i) + V_2) \cdot (g(i) + V_2) \subseteq f(i) \cdot g(i) + f(i) \cdot V_2 + V_2 \cdot g(i) + V_2 \cdot V_2$$
$$\subseteq f(i) \cdot g(i) + V_1 + V_1 + V_1 \subseteq f(i) \cdot g(i) + V.$$

Let $W_i = V_2$ for all $i \in I$. Then $W = \prod_{i \in I} W_i$ is a neighborhood of zero in (S, τ_c), and

$$(f + W) \cdot (g + W) \subseteq \prod_{i \in I}((f(i) + W_i) \cdot (g(i) + W_i)) \subseteq f \cdot g + \prod_{i \in I} U_i = f \cdot g + U.$$

Hence, we've proved the continuity of the multiplication operation in the ring (S, τ_c).

Now we prove that (S, τ_c) is a bounded ring. Indeed, let U be a neighborhood of zero in (S, τ_c). As above, we can consider that $U = \prod_{i \in I} U_i$, where $U_i = V$ for $i \in I$, and V is a certain neighborhood of zero in R. Since R is a bounded topological ring, then there exists a neighborhood V_1 of zero in R such that $V_1 \cdot R \subseteq V$ and $R \cdot V_1 \subseteq V$. Let $W = \prod_{i \in I} W_i$, where $W_i = V_1$ for any $i \in I$. Then W is a neighborhood of zero in (S, τ_c) and $S \cdot W \subseteq \prod_{i \in I}(R_i \cdot W_i) = \prod_{i \in I}(R_i \cdot V) \subseteq \prod_{i \in I} U_i \subseteq U$. Analogously it is checked that $W \cdot S \subseteq U$. Thus, (S, τ_c) is a bounded ring.

The proposition is proved.

4.1.76. REMARK. As it was shown in Example 4.1.18, if we omit the condition of the boundedness of the module in Proposition 4.1.74, then it would be false. Also, the module, considered in Example 4.1.18, is a ring in correspondence to the multiplication operation defined there, but the same example shows that if the condition of the boundedness of the ring is omitted, then Proposition 4.1.75 would be false, too. It is possible to avoid the condition of the boundedness for the cubic topology by the transfer from the consideration of the direct products of the rings and modules to their subdirect products, as we'll do in Propositions 4.3.32 and 4.3.37.

4.1.77. PROPOSITION. *Let R be a bounded topological ring, I be a set of indexes, $R_i = R$ for every $i \in I$ and $S = \prod_{i \in I} R_i$ be the direct power of the topological ring R with* the cubic topology τ_c. *A subset A of the topological ring (S, τ_c) is topologically nilpotent (Σ-nilpotent) (see Definitions 1.8.16 and 1.8.20) if and only if for any neighborhood V of zero in R there exists a number $n \in N$ such that for any $k \geq n$ we have $\left(\pi_i(A)\right)^{(k)} \subseteq V$ $\left(\left(\pi_i(A)\right)^k \subseteq V\right)$ for all $i \in I$.*

PROOF. Let the subset A satisfy the condition formulated above, and U be a neighborhood of zero in (S, τ_c). As above, we can consider that $U = \prod_{i \in I} U_i$, where $U_i = V$ for $i \in I$, and V is a certain neighborhood of zero in R. If n is the natural number, mentioned in the condition for the neighborhood V, then for any $k \in \mathbb{N}$, $k \geq n$ we have:

$$A^{(k)} \subseteq \prod_{i \in I} \left(\pi_i(A)\right)^{(k)} \subseteq \prod_{i \in I} U_i = U$$
$$\left(A^k \subseteq \prod_{i \in I} \left(\pi_i(A)\right)^k \subseteq \prod_{i \in I} U_i = U\right).$$

It means that the set A is topologically nilpotent (Σ-nilpotent).

Now let the set A be nilpotent (Σ-nilpotent) in (S, τ_c) and V be a neighborhood of zero in R. Since $U = \prod_{i \in I} U_i$, where $U_i = V$ for $i \in I$, is a neighborhood of zero in S, then there exists a number $n \in \mathbb{N}$ such that $A^{(k)} \subseteq U$ $(A^k \subseteq U)$ for $k \geq n$. Then $\left(\pi_i(A)\right)^{(k)} = \pi_i(A^{(k)}) \subseteq U_i = V$ $\left(\left(\pi_i(A)\right)^k = \pi_i(A^k) \subseteq U_i = V\right)$ for any $k \geq n$ and $i \in I$. The proof is completed.

4.1.78. PROPOSITION. *Let X be a topological Abelian group, I be a set of indexes, $X_i = X$ for $i \in I$, and $Y = \prod_{i \in I} X_i$ be the direct power of the group X with the cubic topology τ_c; $C(X)$ and $C(Y)$ be the connected components of zero of the topological groups X and (Y, τ_c) correspondingly. Let $K(Y) = \{f \in Y \mid f(i) \in C(X) \text{ for all } i \in I \text{ and the set } \{f(i) \mid i \in I\} \text{ is precompact in } X\}$, then:*

1) $K(Y) \subseteq C(Y)$;

2) if X is a locally precompact group, then $K(Y) = C(Y)$.

PROOF. For any subset $J \subseteq I$ let $\varphi_J : X \to Y$ be the mapping, defined as the following:

$$\varphi_J(x)(i) = \begin{cases} x, & \text{if } i \in J; \\ 0, & \text{if } i \in I \backslash J. \end{cases}$$

It is clear that φ_J is a homomorphism. Now we prove that φ_J is a continuous homomorphism for any $J \subseteq I$.

Indeed, let U be a neighborhood of zero in (Y, τ_c). As above, we can consider that $U = \prod_{i \in I} U_i$, where V is a neighborhood of zero in X, and $U_i = V$ for $i \in I$. Then $\varphi_J(V) \subseteq \prod_{i \in I} U_i = U$. Therefore, φ_J is a continuous homomorphism.

Thus, $\varphi_J(C(X))$ is a connected subgroup in the (Y, τ_c) for any $J \subseteq I$, hence $\left[\sum_{J \subseteq I} \varphi_J(C(X)) \right]_{(Y, \tau_c)} \subseteq C(Y)$.

Now we prove that $\left[\sum_{J \subseteq I} \varphi_J(C(X)) \right]_{(Y, \tau_c)} = K(Y)$. Indeed, let $g \in \left[\sum_{J \subseteq I} \varphi_J(C(X)) \right]_{(Y, \tau_c)}$. Let V be a neighborhood of zero in X and $U = \prod_{i \in I} U_i$, where $U_i = V$ for any $i \in I$. Then U is a neighborhood of zero in (Y, τ_c). Hence, there exists $f \in \sum_{J \subseteq I} \varphi_J(C(X))$ such that $g - f \in U$. There exist subsets $J_1, \ldots, J_n \subseteq I$ such that $f = \sum_{k=1}^{n} f_k$, where $f_k \in \varphi_{J_k}(C(X))$ for $1 \leq k \leq n$. According to the definition of the homomorphism $\varphi_J : X \to Y$, the set $\{f_k(i) \mid i \in I\}$ is a two-element subset in $C(X)$ for any $k \in \{1, \ldots, n\}$. Hence $P = \{f(i) \mid i \in I\}$ is a finite subset of the $C(X)$. Since $g - f \in U$, then $\{g(i) - f(i) \mid i \in I\} \subseteq V$, i.e.

$$\{g(i) \mid i \in I\} \subseteq \{f(i) \mid i \in I\} + V = P + V.$$

Therefore, according to Proposition 3.4.4, $\{g(i) \mid i \in I\}$ is a precompact set in X, and $\{g(i) \mid i \in I\} \subseteq C(X)$. It means that $g \in K(Y)$, hence

$$\left[\sum_{J \subseteq I} \varphi_J(C(X)) \right]_{(Y, \tau_c)} \subseteq K(Y).$$

Now let $g \in K(Y)$ and U be a neighborhood of zero in (Y, τ_c). Then, as above, we can consider that $U = \prod_{i \in I} U_i$, where V is a neighborhood of zero in X and $U_i = V$ for every $i \in I$.

Since $\{g(i) \mid i \in I\}$ is a precompact set in X and $\{g(i) \mid i \in I\} \subseteq C(X)$, then, due to Definition 3.4.1, there exist elements $i_1, \ldots, i_n \in I$ such that $\{g(i) \mid i \in I\} \subseteq \bigcup_{k=1}^{n}(g(i_k) + V)$. It means that for any $j \in I$ there exists $i_k \in \{i_1, \ldots, i_n\}$ such that $g(j) \in g(i_k) + U$. For any $j \in I$ choose one such element i_k. Hence, we get the partition of the set I to pairwise disjoint subsets $J_1, \ldots, J_n$ such that $g(j) \in g(i_k) + V$ for all $j \in J_k$, $k \in \{1, \ldots, n\}$. Let $f = \sum_{k=1}^{n} \varphi_{J_k}(g(i_k))$, then $f \in \sum_{J \subseteq I} \varphi_J(C(X))$. As for any $j_0 \in I$ there exists only one number $k_0 \in \{1, \ldots, n\}$ such that $j_0 \in J_{k_0}$, then

$$(\varphi_{J_k}(g(i_k)))(j_0) = \begin{cases} g(i_{k_0}), & \text{if} \quad k = k_0; \\ 0, & \text{if} \quad k \neq k_0. \end{cases}$$

Therefore,

$$g(j_0) - f(j_0) \in g(i_{k_0}) + V - \sum_{k=1}^{n} \left(\varphi_{J_k} \left(g(i_k) \right) \right)(j_0)$$

$$= g(i_{k_0}) + V - \left(\varphi_{J_{k_0}} \left(g(i_{k_0}) \right) \right)(j_0) = g(i_{k_0}) + V - g(i_{k_0}) = V.$$

Then $g - f \in \prod_{i \in I} U_i = U$, hence $f \in g - U$. Thus,

$$(g - U) \cap \sum_{J \subseteq I} \varphi_J(C(X)) \neq \emptyset$$

for any neighborhood U of zero in (Y, τ_c). It means that

$$g \in \left[\sum_{J \subseteq I} \varphi_J(C(X)) \right]_{(Y, \tau_C)},$$

hence

$$K(Y) \subseteq \left[\sum_{J \subseteq I} \varphi_J(C(X)) \right]_{(Y, \tau_C)}.$$

Thus, we have proved that

$$K(Y) = \left[\sum_{J \subseteq I} \varphi + J\big(C(X) \big) \right]_Y \subseteq C(Y),$$

i.e. item 1) is proved.

Now let X be a locally precompact Abelian group, $f \in C(Y)$ and $i_0 \in I$. Since $\pi_{i_0} : Y \to X_{i_0} = X$ is a continuous homomorphism, then $\pi_{i_0}(C(Y))$ is a connected subgroup in X. Hence, $\pi_{i_0}(C(Y)) \subseteq C(X)$. Then $f(i_0) = \pi_{i_0}(f) \in C(X)$ for any $i_0 \in I$, thus $\{ f(i) \mid i \in I \} \subseteq C(X)$.

Now we show that $\{ f(i) \mid i \in I \}$ is a precompact set. Let V be a precompact neighborhood of zero in X. Then $U = \prod_{i \in I} U_i$, where $U_i = V$ for any $i \in I$, is a neighborhood of zero in Y. Let A be the subgroup in Y generated by the set U, then A is an open-closed subgroup. Then $f \in A$, hence there exists a finite set of elements $u_1, \dots, u_n \in U$ such that $f = u_1 + \dots + u_n$. Then

$$\underbrace{V + \dots + V}_{n \text{ summands}} \ni f(i)$$

for any $i \in I$, i.e.

$$\{f(i) \mid i \in I\} \subseteq \underbrace{V + \ldots + V}_{n \text{ summands}}.$$

Since V is a precompact set in X, then, according to Proposition 3.4.5, the set

$$\underbrace{V + \ldots + V}_{n \text{ summands}}$$

is precompact in X, too. Then $\{f(i) \mid i \in I\}$ is precompact as a subset of a precompact set

$$\underbrace{V + \ldots + V}_{n \text{ summands}}.$$

Therefore, $f \in K(Y)$, i.e. $C(Y) \subseteq K(Y)$. Hence, $C(Y) = K(Y)$. The proof is completed.

4.1.79. THEOREM. *Let $\{(R_i, \tau_i) \mid i \in I\}$ be a family of such Hausdorff topological rings that $a_i \cdot R_i \neq 0$ for any $0 \neq a_i \in R_i$ and any $i \in I$. Let $(R, \tau_{_T}) = \prod_{i \in I} R_i$ be the Tychonoff product of the family $\{(R_i, \tau_i) \mid i \in I\}$. Then the topology $\tau_{_T}$ is non-weakable in the class of all Hausdorff ring topologies on R if and only if for any $i \in I$ the topology τ_i on the ring R_i is non-weakable in the class of all Hausdorff ring topologies on R_i.*

PROOF. **SUFFICIENCY.** Assume the contrary, i.e that the topology $\tau_{_T}$ on the ring R is weakable and let τ_1 be a certain topology on R, strongly weaker then the topology $\tau_{_T}$. Let $\{W'_\gamma \mid \gamma \in \Gamma\}$ be a basis of neighborhoods of zero of the ring (R, τ_1).

Now we show that for any $i \in I$ the canonical projection $\pi_i : (R, \tau_1) \to (R_i, \tau_i)$ is a continuous homomorphism. Indeed, let $i_0 \in I$. It is easy to see that the family $\{\pi_{i_0}(W'_\gamma) \mid \gamma \in \Gamma\}$ of subsets of the ring R_{i_0} satisfies the conditions (BN1) - (BN6) of Propositions 1.2.1 and 1.2.2. Then, according to Theorem 1.2.5, on the ring R_{i_0} exists a topology τ'_{i_0} such that (R_{i_0}, τ'_{i_0}) is a topological ring and the family $\{\pi_{i_0}(W'_\gamma) \mid \gamma \in \Gamma\}$ is a basis of neighborhoods of zero in (R_{i_0}, τ'_{i_0}).

Now we show that (R_{i_0}', τ_{i_0}') is a Hausdorff topological ring. Let $R'_{i_0} = \{f \in R \mid f(i) = 0$ for $i \neq i_0\}$, then R'_{i_0} is a subring in R. Since $a_{i_0} \cdot R_{i_0} \neq 0$ for any $0 \neq a_{i_0} \in R_{i_0}$, then $\ker \pi_{i_0} = \{f \in R \mid f(i_0) = 0\} = \{f \in R \mid f \cdot R'_{i_0} = 0\}$, hence, according to Corollary 1.4.31, $\ker \pi_{i_0}$ is a closed ideal in (R, τ_1) (as an annulator of the subring R'_{i_0}). Then

$$\bigcap_{\gamma \in \Gamma} \pi_{i_0}^{-1}\left(\pi_{i_0}(W_\gamma)\right) = \bigcap_{\gamma \in \Gamma} (\ker \pi_{i_0} + W'_\gamma) = \ker \pi_{i_0},$$

hence

$$\bigcap_{\gamma \in \Gamma} \pi_{i_0}(W'_\gamma) = \bigcap_{\gamma \in \Gamma} \pi_{i_0}\left(\pi_{i_0}^{-1}\left(\pi_{i_0}(W'_\gamma)\right)\right)$$
$$\subseteq \pi_{i_0}\left(\bigcap_{\gamma \in \Gamma} \pi_{i_0}^{-1}\left(\pi_{i_0}(W'_\gamma)\right)\right) = \pi_{i_0}\left(\ker \pi_{i_0}\right) = \{0\}.$$

Due to Theorem 1.3.2, (R_{i_0}, τ'_{i_0}) is Hausdorff ring for any $i_0 \in I$.

Since the topology on the ring (R, τ_1) is weaker then the topology on the ring (R, τ_T) and for any $i_0 \in I$ the projection $\pi_{i_0} : (R, \tau_1) \to (R_{i_0} \tau_{i_0})$ is an open mapping (since, according to Proposition 4.1.2, the projection $\pi_{i_0} : (R, \tau_T) \to (R_{i_0}, \tau_{i_0})$ is an open mapping), then the topology τ'_{i_0} on the ring R_{i_0} is weaker then the topology τ_{i_0}. Since the topology on the ring (R_{i_0}, τ_{i_0}) is non-weakable, then the topology τ_{i_0} on the ring R_{i_0} coincides with the topology τ'_{i_0}. It means that for any neighborhood V of zero of the ring (R_{i_0}, τ_{i_0}) there exists a neighborhood W'_γ of zero of the ring (R, τ_1) such that $\pi_{i_0}(W'_\gamma) \subseteq V$. Hence, we have shown that $\pi_{i_0} : (R, \tau_1) \to (R_{i_0}, \tau_{i_0})$ is a continuous homomorphism for any $i_0 \in I$.

Since the topology τ_1 on the ring R does not coincide with the topology τ_T, then there exists a neighborhood W of zero in the ring (R, τ_T) such that $W'_\gamma \not\subseteq W$ for any neighborhood W'_γ of zero in the ring (R, τ_1). According to Remark 4.1.4, there exist a finite subset $J \subseteq I$ and neighborhoods V_j of zero in the rings (R_j, τ_j), for any $j \in J$, such that $\bigcap_{j \in J} \pi_j^{-1}(V_j) \subseteq W$. Since the homomorphisms $\pi_i : (R, \tau_1) \to (R_i, \tau_i)$ are continuous for any $i \in I$, then the sets $\pi_j^{-1}(V_j)$ are neighborhoods of zero in (R, τ_1) for any $j \in J$. Since $\bigcap_{j \in J} \pi_j^{-1}(V_j) \subseteq W$, then W is a neighborhood of zero in (R, τ_1). We obtain the contradiction with the choice of the neighborhood W of zero in (R, τ_T). The contradiction shows that the topology τ_T on the ring R is non-weakable in the class of all Hausdorff ring topologies on R.

NECESSITY. Assume the contrary, i.e. that for a certain $i_0 \in I$ the topology τ_{i_0} on the ring R_{i_0} is weakable, and let $\hat{\tau}_{i_0}$ be a weaker topology on the ring R_{i_0}. Consider the family $\{(R_i, \tau'_i) \mid i \in I\}$ of topological rings, where $\tau'_i = \tau_i$ for $i \in I \backslash \{i_0\}$ and $\tau'_{i_0} = \hat{\tau}_{i_0}$. Let $(R, \hat{\tau}) = \prod_{i \in I}(R_i, \tau'_i)$ be the Tychonoff product of the family $\{(R_i, \tau'_i) \mid i \in I\}$. Then the topology $\hat{\tau}$ is a Hausdorff topology on the ring R, which is weaker than the the topology τ_T. Hence, $\tau_T = \hat{\tau}$, since the topology τ_T on the ring R is non-weakable.

Since $\tau_{i_0} \neq \widehat{\tau}_{i_0}$, then there exists a neighborhood V of zero of the ring (R_{i_0}, τ_{i_0}) such that $U \not\subseteq V$ for any neighborhood U of zero of the ring $(R_{i_0}, \widehat{\tau}_{i_0})$. Since $\widehat{\tau} = \tau_T$, then there exists a neighborhood W of zero in $(R, \widehat{\tau})$ such that $W \subseteq \pi_{i_0}^{-1}(V)$. Then $\pi_{i_0}(W) \subseteq V$. Since, according to Proposition 4.1.2, the homomorphism $\pi_{i_0} : (R, \widehat{\tau}) \to (R_{i_0}, \widehat{\tau}_{i_0})$ is open, then $\pi_{i_0}(W)$ is a neighborhood of zero in $(R_{i_0}, \widehat{\tau}_{i_0})$, and $\pi_{i_0}(W) \subseteq V$. This contradicts the choice of the neighborhood V of zero in (R_{i_0}, τ_{i_0}).

Thus, the theorem is proved.

§ 4.2. Inverse Limits

4.2.1. DEFINITION. Let Γ be a direction, $\{X_\gamma | \gamma \in \Gamma\}$ be a family of topological Abelian groups, indexed by the set Γ. Let also for any pair of elements $\alpha, \beta \in \Gamma$ such that $\alpha \le \beta$, be defined a continuous homomorphism $f_{\alpha,\beta} : X_\beta \to X_\alpha$ such that $f_{\alpha,\gamma} = f_{\alpha,\beta} \circ f_{\beta,\gamma}$ (a composition of homomorphisms) for $\alpha, \beta, \gamma \in \Gamma$, $\alpha \le \beta \le \gamma$, and $f_{\alpha,\alpha}$ be the identical homomorphism for $\alpha \in \Gamma$. Then we say that it is defined an inverse spectrum $S = \{X_\alpha, f_{\alpha,\beta} | \alpha, \beta \in \Gamma\}$. The groups X_α, $\alpha \in \Gamma$, we call elements of the inverse spectrum S and the homomorphisms $f_{\alpha,\beta}$, $\alpha, \beta \in \Gamma$, we call projections.

An element x of the direct product $\prod_{\gamma \in \Gamma} X_\gamma$ is called a thread of the spectrum S if $\pi_\lambda(x) = f_{\lambda,\beta}(\pi_\beta(x))$ for any $\lambda, \beta \in \Gamma$, $\beta \ge \lambda$, where π_λ and π_β are the canonical projections. We call the set X of all the threads of the spectrum S, that have the topology, induced from the Tychonoff product $(\prod_{\gamma \in \Gamma} X_\gamma, \tau_T)$ a limit of an inverse spectrum S (or inverse limit). It is easy to see that X is a subgroup in $\prod_{\gamma \in \Gamma} X_\gamma$. We denote the topological group $(X, \tau_T|_X)$ over $\varprojlim S$.

The restriction of the canonical projections $\pi_\alpha : \prod_{\gamma \in \Gamma} X_\gamma \to X_\alpha$ on X we denote over f_α and call the canonical projections. Due to the definition of the threads of spectrum S, $f_\alpha = f_{\alpha,\beta} \circ f_\beta$ for any $\alpha, \beta \in \Gamma$, $\alpha \le \beta$. The inverse limit of topological Abelian groups is not empty, since it always contains a zero element of the group $\prod_{\gamma \in \Gamma} X_\gamma$.

4.2.2. REMARK. Let Γ be a direction, $\{X_\gamma | \gamma \in \Gamma\}$ be a family of topological rings (modules over a topological ring R), $S = \{X_\alpha, f_{\alpha,\beta} | \alpha, \beta \in \Gamma\}$ be an inverse spectrum of the topological groups X_α, where any of the homomorphisms $f_{\alpha,\beta}$ is a ring (R-module) homomorphism. Then $(X, \tau_T|_X) = \varprojlim S$ is a subring (submodule) of the topological ring (R-module) $\prod_{\gamma \in \Gamma} X_\gamma$, with Tychonoff topology τ_T.

4.2.3. REMARK. Since the Tychonoff product of Hausdorff Abelian groups is a Hausdorff group, then the inverse limit of Hausdorff groups is Hausdorff group.

4.2.4. PROPOSITION. *Let $S = \{X_\alpha, f_{\alpha,\beta} | \alpha, \beta \in \Gamma\}$ be an inverse spectrum of topological groups X_α and $(X, \tau_T|_X) = \varprojlim S$. Then the system $\mathcal{B}_0(X)$ of subsets of the type $f_\alpha^{-1}(U_\alpha)$ in X, where U_α is a neighborhood of zero in X_α, $\alpha \in \Gamma$, is a basis of neighborhoods of zero in $(X, \tau_T|_X)$.*

PROOF. Let U be a neighborhood of zero in $(X, \tau_T|_X)$. Then, according to Remark 4.1.4, there exists a neighborhood W of zero in $(\prod_{\gamma \in \Gamma} X_\gamma, \tau_T)$ such that $W \cap X \subseteq U$ and $W = \bigcap_{k=1}^n \pi_{\alpha_k}^{-1}(U_{\alpha_k})$, where $\{\alpha_1, \ldots, \alpha_n\}$ is a certain finite subset in Γ and U_{α_k} is a neighborhood of zero in X_{α_k}, $1 \le k \le n$. Choose an element $\alpha \in \Gamma$ such that $\alpha \ge \alpha_k$, $k = 1, \ldots, n$. Then the set $U_\alpha = \bigcap_{k=1}^n f_{\alpha_k, \alpha}^{-1}(U_{\alpha_k})$ is a neighborhood of zero of X_α, and

$$U \supseteq \bigcap_{k=1}^n (X \cap \pi_{\alpha_k}^{-1}(U_{\alpha_k})) = \bigcap_{k=1}^n f_{\alpha_k}^{-1}(U_{\alpha_k})$$

$$= \bigcap_{k=1}^n f_\alpha^{-1}(f_{\alpha_k, \alpha}^{-1}(U_{\alpha_k})) \supseteq f_\alpha^{-1}\left(\bigcap_{k=1}^n f_{\alpha_k, \alpha}^{-1}(U_{\alpha_k})\right) = f_\alpha^{-1}(U_\alpha).$$

Since f_α is a continuous homomorphism, then $f_\alpha^{-1}(U_\alpha)$ is a neighborhood of zero in $(X, \tau_T|_X)$. The proposition is proved.

4.2.5. PROPOSITION. *Let elements of an inverse spectrum $S = \{X_\alpha, f_{\alpha,\beta} | \alpha, \beta \in \Gamma\}$ be Hausdorff topological Abelian groups, then the inverse limit $X = \varprojlim S$ is a closed subgroup in $(\prod_{\gamma \in \Gamma} X_\gamma, \tau_T)$.*

PROOF. Assume the contrary, i.e. that X is not closed and let $y \in [X]_{(\prod_{\gamma \in \Gamma} X_\gamma, \tau_T)} \setminus X$. Then there exists $\alpha, \beta \in \Gamma$ such that $\beta \ge \alpha$ and $y_\alpha' = f_{\alpha,\beta}(\pi_\beta(y)) \ne \pi_\alpha(y) = y_\alpha$. Choose a neighborhood V of zero in X_α such that $(y_\alpha' + V) \cap (y_\alpha + V) = \emptyset$. Since $f_{\alpha,\beta}$ is a continuous homomorphism, then there exists a neighborhood U of zero in X_β such that $f_{\alpha,\beta}(\pi_\beta(y) + U) \subseteq y_\alpha' + V$. Since $W = \pi_\alpha^{-1}(V) \cap \pi_\beta^{-1}(U)$ is a neighborhood of zero in $(\prod_{\gamma \in \Gamma} X_\gamma, \tau_T)$, then $(y + W) \cap X \ne \emptyset$. Let $z \in (y + W) \cap X$, then $\pi_\alpha(z) \in \pi_\alpha(y) + \pi_\alpha(W) \subseteq y_\alpha + V$ and $\pi_\beta(z) \in \pi_\beta(y) + \pi_\beta(W) \subseteq \pi_\beta(y) + U$. Since $z \in X$, then $\pi_\alpha(z) = f_{\alpha,\beta}(\pi_\beta(z)) \in f_{\alpha,\beta}(\pi_\beta(y)) + U \subseteq y_\alpha' + V$, hence, $(y_\alpha + V) \cap (y_\alpha' + V) \ne \emptyset$. This contradicts the choice of the neighborhood V of zero in X_α. Thus, the group X is closed in $(\prod_{\gamma \in \Gamma} X_\gamma, \tau_T)$.

4.2.6. COROLLARY. *The limit of an inverse spectrum $S = \{X_\alpha, f_{\alpha,\beta} | \alpha, \beta \in \Gamma\}$ of Hausdorff compact groups is a Hausdorff compact group.*

PROOF. Due to Corollaries 4.1.44 and 4.1.53, the group $(\prod_{\gamma \in \Gamma} X_\gamma, \tau_T)$ is a Hausdorff compact group, and due to Proposition 4.2.5, $X = \varprojlim S$ is a closed subgroup in $(\prod_{\gamma \in \Gamma} X_\gamma, \tau_T)$. Then X is a Hausdorff compact group.

4.2.7. COROLLARY. *The limit of an inverse spectrum $S = \{X_\alpha, f_{\alpha,\beta}|\alpha, \beta \in \Gamma\}$ of complete Hausdorff Abelian groups is a complete Hausdorff group.*

PROOF. Due to Corollary 4.1.44 and Proposition 4.1.47, the group $(\prod_{\gamma \in \Gamma} X_\gamma, \tau_T)$ is a complete Hausdorff group, and due, to Proposition 4.2.5, $X = \lim\limits_{\leftarrow} S$ is a closed subgroup in $(\prod_{\gamma \in \Gamma} X_\gamma, \tau_T)$. Then, according to Theorem 3.2.16, X is a complete group.

4.2.8. PROPOSITION. *Let $S = \{X_\alpha, f_{\alpha,\beta}|\alpha, \beta \in \Gamma\}$ be an inverse spectrum of Abelian topological groups, Y be a topological Abelian group, and let for any $\alpha \in \Gamma$ be defined a continuous homomorphism $\varphi_\alpha : Y \to X_\alpha$, and $\varphi_\alpha = f_{\alpha,\beta} \circ \varphi_\beta$ for any $\alpha, \beta \in \Gamma$, $\alpha \leq \beta$. Then there exists a continuous homomorphism $\varphi : Y \to X = \lim\limits_{\leftarrow} S$ such that $\varphi_\alpha = f_\alpha \circ \varphi$ for any $\alpha \in \Gamma$, where $f_\alpha : X \to X_\alpha$ is the homomorphism, defined in Definition 4.2.1.*

PROOF. We define the mapping $\varphi : Y \to (\prod_{\gamma \in \Gamma} X_\gamma, \tau_T)$ the following way:

an element $y \in Y$ corresponds to the element $\varphi(y) \in \prod_{\gamma \in \Gamma} X_\gamma$ such that $(\varphi(y))(\alpha) = \varphi_\alpha(y)$ for any $\alpha \in \Gamma$.

According to Corollary 4.1.39, φ is a continuous homomorphism. Besides, since

$$(\varphi(y))(\alpha) = \varphi_\alpha(y) = f_{\alpha,\beta}(\varphi_\beta(y)) = f_{\alpha,\beta}\Big((\varphi(y))(\beta)\Big)$$

for any $\alpha, \beta \in \Gamma$, $\alpha \leq \beta$, then $\varphi(y) \in \lim\limits_{\leftarrow} S$ for any $y \in Y$, i.e $\varphi : Y \to \lim\limits_{\leftarrow} S$. Then φ is required mapping.

4.2.9. PROPOSITION. *Let $S = \{X_\alpha, f_{\alpha,\beta}|\alpha, \beta \in \Gamma\}$ be an inverse spectrum of topological rings (modules over a topological ring R), Y be a topological ring (topological R-module) and let for any $\alpha \in \Gamma$ be defined a continuous homomorphism of rings (R-modules) $\varphi_\alpha : Y \to X_\alpha$, where $\varphi_\alpha = f_{\alpha,\beta} \circ \varphi_\beta$ for any $\alpha, \beta \in \Gamma$, $\alpha \leq \beta$. Then there exists a continuous homomorphism of rings (R-modules) $\varphi : Y \to \lim\limits_{\leftarrow} S$ such that $\varphi_\alpha = f_\alpha \circ \varphi$ for any $\alpha \in \Gamma$, where $f_\alpha : \lim\limits_{\leftarrow} S \to X_\alpha$ is the homomorphism, defined in Definition 4.2.1.*

PROOF. It is enough to observe that the continuous group homomorphism $\varphi : Y \to \lim\limits_{\leftarrow} S$, constructed in Proposition 4.2.8, in this case is a ring (R-module) homomorphism.

4.2.10. PROPOSITION. *Let $S = \{X_\alpha, f_{\alpha,\beta}|\alpha, \beta \in \Gamma\}$ and $T = \{Y_\alpha, h_{\alpha,\beta}|\alpha, \beta \in \Gamma\}$ be inverse spectra of topological groups, and for any $\alpha \in \Gamma$ is defined a continuous homomorphism $\psi_\alpha : X_\alpha \to Y_\alpha$ such that $h_{\alpha,\beta} \circ \psi_\beta = \psi_\alpha \circ f_{\alpha,\beta}$ for any $\alpha, \beta \in \Gamma$, $\alpha \leq \beta$.*

Then there exists a continuous homomorphism $\varphi : \lim\limits_{\leftarrow} S \to \lim\limits_{\leftarrow} T$ *such that* $h_\alpha \circ \varphi = \psi_\alpha \circ f_\alpha$

for any $\alpha \in \Gamma$, *where* $f_\alpha : \lim\limits_{\leftarrow} S \to X_\alpha$ *and* $h_\alpha : \lim\limits_{\leftarrow} T \to Y_\alpha$ *are the canonical projections*

for each $\alpha \in \Gamma$.

PROOF. Let $\varphi_\lambda = \psi_\lambda \circ f_\lambda : \lim\limits_{\leftarrow} S \to Y_\lambda$. Then φ_λ is a continuous homomorphism

for any $\lambda \in \Gamma$, and $\varphi_\lambda = \psi_\lambda \circ f_\lambda = \psi_\lambda \circ f_{\lambda,\beta} \circ f_\beta = h_{\lambda,\beta} \circ \psi_\beta \circ f_\beta = h_{\lambda,\beta} \circ \varphi_\beta$ for any

$\lambda, \beta \in \Gamma$, $\lambda \leq \beta$. According to Proposition 4.2.8, there exists a continuous homomorphism

$\varphi : \lim\limits_{\leftarrow} S \to \lim\limits_{\leftarrow} T$ such that $\varphi_\lambda = h_\lambda \circ \varphi$ for any $\lambda \in \Gamma$. Hence, $\psi_\lambda \circ f_\lambda = h_\lambda \circ \varphi$ for any

$\lambda \in \Gamma$.

The proposition is proved.

4.2.11. PROPOSITION. *Let* $S = \{X_\lambda, f_{\lambda,\beta} | \lambda, \beta \in \Gamma\}$ *and* $T = \{Y_\lambda h_{\lambda,\beta} | \lambda, \beta$

$\in \Gamma\}$ *be inverse spectra of topological rings (modules over topological ring R), and for*

any $\alpha \in \Gamma$ *is defined a continuous homomorphism* $\psi_\lambda : X_\lambda \to Y_\lambda$ *of rings (R-modules)*

such that $h_{\lambda,\beta} \circ \psi_\beta = \psi_\lambda \circ f_{\lambda,\beta}$ *for any* $\lambda, \beta \in \Gamma$, $\lambda \leq \beta$. *Then there exists a continuous*

homomorphism $\varphi : \lim\limits_{\leftarrow} S \to \lim\limits_{\leftarrow} T$ *of rings (R-modules) such that* $\psi_\lambda \circ f_\lambda = h_\lambda \circ \varphi$ *where*

$f_\lambda : \lim\limits_{\leftarrow} S \to X_\lambda$ *and* $h_\lambda : \lim\limits_{\leftarrow} T \to Y_\lambda$ *are the canonical projections for each* $\lambda \in \Gamma$.

PROOF. It is enough to observe that the continuous group homomorphism $\varphi : \lim\limits_{\leftarrow} S \to$

$\lim\limits_{\leftarrow} T$, constructed in Proposition 4.2.10, in this case is a homomorphism of rings (R-

modules).

4.2.12. PROPOSITION. *Let* $S = \{X_\alpha, f_{\alpha,\beta} | \alpha, \beta \in \Gamma\}$ *be an inverse spectrum of*

topological Abelian groups, Γ_1 *be a subdirection of the direction* Γ, $T_1 = \{X_\alpha, f_{\alpha,\beta} | \alpha, \beta \in$

$\Gamma_1\}$ *be an inverse spectrum of topological Abelian groups. Then the topological group* $\lim\limits_{\leftarrow} S$

is topologically isomorphic to the topological group $\lim\limits_{\leftarrow} T_1$.

PROOF. Since for any $\alpha \in \Gamma_1$ the projection $f_\alpha : \lim\limits_{\leftarrow} S \to X_\alpha$ is continuous and

$f_\alpha = f_{\alpha,\beta} \circ f_\beta$ for any $\beta \in \Gamma_1$, $\beta \geq \alpha$, then, according to Proposition 4.2.8, there exists a

continuous homomorphism $\varphi : \lim\limits_{\leftarrow} S \to \lim\limits_{\leftarrow} T$ such that $f'_\alpha \circ \varphi = f_\alpha$ for any $\alpha \in \Gamma_1$, where

f'_α is the projection of $\lim\limits_{\leftarrow} T$ on to X_α.

For any $\alpha \in \Gamma$ we define the mapping $\psi_\alpha : \lim\limits_{\leftarrow} T \to X_\alpha$ as the following:

since Γ_1 is a subdirection in Γ, then there exists an element $\beta \in \Gamma_1$ such that $\beta \geq \alpha$.

Put $\psi_\alpha = f_{\alpha,\beta} \circ f'_\beta$.

Now we prove that this definition is correct. Indeed, let $\beta, \gamma \in \Gamma_1$ $\beta \geq \alpha$ and $\gamma \geq \alpha$. Then there exists an element $\delta \in \Gamma_1$ such that $\delta \geq \gamma$ and $\delta \geq \beta$. Then $f_{\alpha,\beta} \circ f'_\beta = f_{\alpha,\beta} \circ f_{\beta,\delta} \circ f'_\delta = f_{\alpha,\delta} \circ f'_\delta$.

Analogously $f_{\alpha,\gamma} \circ f'_\gamma = f_{\alpha,\delta} \circ f'_\delta$. Then $f_{\alpha,\gamma} \circ f'_\gamma = f_{\alpha,\beta} \circ f'_\beta$. Hence, we've proved that the mapping ψ_α is defined correctly.

Since $f_{\alpha,\beta}$ and f'_β are continuous homomorphisms for any α, $\beta \in \Gamma$, then ψ_α is a continuous homomorphism for any $\alpha \in \Gamma$.

Let $\alpha, \beta \in \Gamma$ and $\beta \geq \alpha$. Since Γ_1 is a subdirection in Γ, then there exists an element $\gamma \in \Gamma_1$ such that $\gamma \geq \beta$. Then $\psi_\alpha = f_{\alpha,\beta} \circ f'_\beta = f_{\alpha,\beta} \circ f_{\beta,\gamma} \circ f'_\gamma = f_{\alpha,\beta} \circ \psi_\beta$. Therefore, according to Proposition 4.2.8, there exists a continuous homomorphism $\psi : \varprojlim T \to \varprojlim S$ such that $f_\alpha \circ \psi = \psi_\alpha$ for any $\alpha \in \Gamma$.

Now we prove that $\varphi \circ \psi$ is a mapping identical on $\varprojlim T$. Indeed, let $t \in \varprojlim T$, then for any $\alpha \in \Gamma_1$ we get

$$f'_\alpha \big((\varphi \circ \psi)(t) \big) = \big(f'_\alpha \circ \varphi \big) \big(\psi(t) \big) = f_\alpha \big(\psi(t) \big)$$
$$= (f_\alpha \circ \psi)(t) = \psi_\alpha(t) = (f_{\alpha,\alpha} \circ f'_\alpha)(t) = f'_\alpha(t).$$

Therefore, $(\varphi \circ \psi)(t) = t$ fo any $t \in T$, and, hence, the mapping $\varphi \circ \psi$ is identical on $\varprojlim T$.

Now we prove that $\psi \circ \varphi$ is a mapping identical on $\varprojlim S$. Indeed, let $s \in \varprojlim S$. Since for any $\alpha \in \Gamma$ there exists an element $\beta \in \Gamma_1$ such that $\beta \geq \alpha$, then we have:

$$f_\alpha \big((\psi \circ \varphi)(s) \big) = \big(f_\alpha \circ \psi \big) \big(\varphi(s) \big) = \psi_\alpha \big(\varphi(s) \big) = \big(f_{\alpha,\beta} \circ f'_\beta \big) \big(\varphi(s) \big)$$
$$= f_{\alpha,\beta} \big((f'_\beta \circ \varphi)(s) \big) = \big(f_{\alpha,\beta} \circ f_\beta \big)(s) = f_\alpha(s).$$

Therefore, $(\psi \circ \varphi)(s) = s$ for any $s \in S$, and, hence, the mapping $\psi \circ \varphi$ is identical on $\varprojlim S$.

Thus, φ and ψ are mutually inverse continuous homomorphisms. It means that φ and ψ are topological isomorphisms. The proposition is proved.

4.2.13. COROLLARY. *Let $S = \{X_\alpha, f_{\alpha,\beta} | \alpha, \beta \in \Gamma\}$ be an inverse spectrum of topological Abelian groups and the direction Γ has a maximal element α_0. Then the projection $f_{\alpha_0} : \varprojlim S \to X_{\alpha_0}$ is a topological isomorphism between the topological groups $\varprojlim S$ and X_{α_0}.*

PROOF. The subset $\{\alpha_0\}$ is a subdirection of the direction Γ. The further proof follows from Proposition 4.2.12.

4.2.14. REMARK. Let $S = \{X_\alpha, f_{\alpha,\beta} | \alpha, \beta \in \Gamma\}$ be an inverse spectrum of topological Abelian groups and for any $\alpha \in \Gamma$ in the group X_α is defined a subgroup Y_α such that $f_{\alpha,\beta}(Y_\beta) \subseteq Y_\alpha$ for any $\alpha, \beta \in \Gamma$, $\beta \geq \alpha$. Consider the inverse spectrum $T = \{Y_\alpha, h_{\alpha,\beta} | \alpha, \beta \in \Gamma\}$, where the homomorphism $h_{\alpha,\beta} : Y_\beta \to Y_\alpha$ is a restriction of the homomorphism $f_{\alpha,\beta}$ on Y_β for each $\alpha, \beta \in \Gamma$. Since, according to Proposition 4.1.40, $(\prod_{\gamma \in \Gamma} Y_\gamma, \tau_T)$ is a subgroup of the topological group $(\prod_{\gamma \in \Gamma} X_\gamma, \tau_T)$ and the threads of the spectrum T are threads of the spectrum S, then $\varprojlim T$ is a subgroup of the topological group $\varprojlim S$.

4.2.15. PROPOSITION. *Let A be a closed subgroup in the limit of an inverse spectrum $S = \{X_\alpha, f_{\alpha,\beta} | \alpha, \beta \in \Gamma\}$ of topological Abelian groups. Since for any $\alpha, \beta \in \Gamma$, $\beta \geq \alpha$, we have $f_{\alpha,\beta}(f_\beta(x)) = f_\alpha(x)$ for any $x \in \varprojlim S$, then $f_{\alpha,\beta}(f_\beta(A)) = f_\alpha(A)$. Consider, as in Remark 4.2.14, the inverse spectrum $T = \{f_\alpha(A), h_{\alpha,\beta} | \alpha, \beta \in \Gamma\}$, where the homomorphism $h_{\alpha,\beta} : f_\beta(A) \to f_\alpha(A)$ is a restriction of the homomorphisms $f_{\alpha,\beta}$ on $f_\beta(A)$ for each $\alpha, \beta \in \Gamma$. Then $A = \varprojlim T$.*

PROOF. Since, according to Remark 4.2.14, $\varprojlim T$ is a subgroup of the topological group $\varprojlim S$, then it is enough to show that the subgroups A and $\varprojlim T$ coincide.

Indeed, let $a \in A$, then $a \in \varprojlim S$ and $a \in \prod_{\alpha \in \Gamma} f_\alpha(A)$. Thus, $a \in \varprojlim T$ and, hence, $A \subseteq \varprojlim T$.

Vice versa, let $a \in \varprojlim T$ and $a \notin A$. Since A is a closed subgroup in $\varprojlim S$, then, according to Proposition 4.2.4, there exist $\alpha_0 \in \Gamma$ and a neighborhood U_{α_0} of zero in X_{α_0} such that $(a + f_{\alpha_0}^{-1}(U_{\alpha_0})) \bigcap A = \emptyset$. Since $a \in \varprojlim T$, then $a \in \prod_{\alpha \in \Gamma} f_\alpha(A)$, hence, there exists $b \in A$ such that $f_{\alpha_0}(a) = f_{\alpha_0}(b)$, i.e. $f_{\alpha_0}(b - a) = 0 \in U_{\alpha_0}$. Then $b - a \in f_{\alpha_0}^{-1}(U_{\alpha_0})$, hence, $b \in (a + f_{\alpha_0}^{-1}(U_{\alpha_0})) \bigcap A$, which contradicts the choice of the neighborhood U_{α_0} of zero in X_{α_0}. Hence, $a \in A$. Thus, $A = \varprojlim T$.

4.2.16. REMARK. According to Proposition 4.2.15, it is possible to transfer from the consideration of the limit $X = \varprojlim S$ of an inverse spectrum $S = \{X_\alpha, f_{\alpha,\beta} | \alpha, \beta \in \Gamma\}$, whose projections $f_\alpha : X \to X_\alpha$ are not surjective, to the limit $V = \varprojlim T$ of an inverse spectrum $T = \{Y_\alpha, f'_{\alpha,\beta} | \alpha, \beta \in \Gamma\}$, the projections $f'_\alpha : Y \to Y_\alpha$ of which are surjective homomorphisms. To do so it is enough to consider the inverse spectrum T, where $Y_\alpha = f_\alpha(X)$, and $f'_{\alpha,\beta}$ is the restriction of $f_{\alpha,\beta}$ on $f_\beta(X) = Y_\beta$ for $\alpha, \beta \in \Gamma$, $\beta \geq \alpha$.

4.2.17. **PROPOSITION.** *Let Γ be a countable direction, $S = \{X_\alpha, f_{\alpha,\beta} | \alpha, \beta \in \Gamma\}$ be an inverse spectrum of topological Abelian groups, where all the homomorphisms $f_{\alpha,\beta}$ are surjective. Then $f_\alpha(\varprojlim S) = X_\alpha$ for any $\alpha \in \Gamma$.*

PROOF. If the direction Γ has a maximal element α_0, then, according to Corollary 4.2.13, $f_{\alpha_0} : \varprojlim S \to X_{\alpha_0}$ is a topological isomorphism. Since for any $\alpha \in \Gamma$ the homomorphism f_{α,α_0} is surjective, then $f_\alpha = f_{\alpha,\alpha_0} \circ f_{\alpha_0}$ is a surjective homomorphism.

Now let the direction Γ not have a maximal element, and let $f_{\alpha_1}(\varprojlim S) \neq X_{\alpha_1}$ for a certain $\alpha_1 \in \Gamma$. Let $x_{\alpha_1} \in X_{\alpha_1} \setminus f_{\alpha_1}(\varprojlim S)$.

According to Proposition 3.1.8, there exists a subdirection Γ_1 in Γ, similar to $\mathbb{N}$. Without loss of generality we can consider that α_1 is the first element in Γ_1, and $\Gamma_1 = \{\alpha_k | k \in \mathbb{N}\}$, where $\alpha_{k+1} > \alpha_k$ for any $k \in \mathbb{N}$. Since $f_{\alpha_l, \alpha_{l+1}} : X_{\alpha_{l+1}} \to X_{\alpha_l}$ is a surjective mapping, then for any $k \in \mathbb{N}$ by induction, beginning with x_{α_1}, we can construct a sequence of elements $x_{\alpha_k} \in X_{\alpha_k}$ such that $x_{\alpha_k} = f_{\alpha_k, \alpha_{k+1}}(x_{\alpha_{k+1}})$. It is clear that in that case for any $n, m \in \mathbb{N}$, where $n \geq m$, we get $x_{\alpha_m} = f_{\alpha_m, \alpha_n}(x_{\alpha_n})$.

For any $\alpha \in \Gamma$ we define the element $x_\alpha \in X$ the following way:

since Γ_1 is a subdirection in Γ, then there exists $\alpha_k \in \Gamma_1$ such that $\alpha_k \geq \alpha$. Put $x_\alpha = f_{\alpha,\alpha_k}(x_{\alpha_k})$.

Now we verify that the definition of the element x_α is correct. Indeed, let $\alpha_k, \alpha_n \in \Gamma_1$, where $\alpha_k \geq \alpha$ and $\alpha_n \geq \alpha$. To be definite suppose that $n > k$, then $\alpha_n > \alpha_k$. Then

$$f_{\alpha,\alpha_k}(x_{\alpha_k}) = f_{\alpha,\alpha_k} \circ f_{\alpha_k,\alpha_n}(x_{\alpha_n}) = f_{\alpha,\alpha_n}(x_{\alpha_n}),$$

which proves that the definition of the element $x_\alpha \in X_\alpha$ is correct for any $\alpha \in \Gamma$.

Let now x be the element of $\prod_{\alpha \in \Gamma} X_\alpha$ such that $x(\alpha) = x_\alpha$ for any $\alpha \in \Gamma$. We'll prove that $x \in \varprojlim S$. Indeed, let $\alpha, \beta \in \Gamma$, where $\beta \geq \alpha$. Since Γ_1 is a subdirection in Γ, then there exists $\alpha_n \in \Gamma$ such that $\alpha_n \geq \beta \geq \alpha$. Then $x_\beta = f_{\beta,\alpha_n}(x_{\alpha_n})$ and $x_\alpha = f_{\alpha,\alpha_n}(x_{\alpha_n}) = f_{\alpha,\beta} \circ f_{\beta,\alpha_n}(x_{\alpha_n}) = f_{\alpha,\beta}(x_\beta)$. Therefore, x is a thread of the spectrum S.

Since $f_{\alpha_1}(x) = x_{\alpha_1}$, then $x_{\alpha_1} \in f_{\alpha_1}(\underleftarrow{\lim}S)$, which contradicts the choice of the element $x_{\alpha_1} \in X_{\alpha_1}$. It shows that $f_\alpha(\underleftarrow{\lim}S) = X_\alpha$ for any $\alpha \in \Gamma$.

4.2.18. THEOREM. *Let $S = \{X_\alpha, f_{\alpha,\beta}|\alpha,\beta \in \Gamma\}$ be an inverse spectrum of compact Hausdorff Abelian groups and $X = \underleftarrow{\lim}S$. Then $f_\sigma(X) = \bigcap_{\gamma \geq \sigma} f_{\sigma,\gamma}(X_\gamma)$ for any $\sigma \in \Gamma$.*

PROOF. Let $\mathfrak{N}$ be the set of such families $\mathfrak{A} = \{A_\alpha|\alpha \in \Gamma\}$, that:

$\emptyset \notin \mathfrak{A}$;

A_α is a closed subset in X_α;

$f_{\alpha,\beta}(A_\beta) \subseteq A_\alpha$ for any α, $\beta \in \Gamma$, $\alpha \leq \beta$.

For any $\mathfrak{A} = \{A_\alpha|\alpha \in \Gamma\}$ and $\mathfrak{A}' = \{A'_\alpha|\alpha \in \Gamma\}$ from $\mathfrak{N}$ we put:

$\mathfrak{A} \geq \mathfrak{A}'$ if $A_\alpha \subseteq A'_\alpha$ for any $\alpha \in \Gamma$. It is clear that this is a partial order on $\mathfrak{N}$.

Now we prove that $\mathfrak{N}$ is an inductive set (see A.11). Indeed, let $\mathcal{L} = \{\mathfrak{A}_\lambda | \mathfrak{A}_\lambda = \{A_{\lambda,\alpha}|\alpha \in \Gamma\}, \lambda \in \Lambda\}$ be a linearly ordered subset in $\mathfrak{N}$. Let $\mathfrak{A} = \{A_\alpha|\alpha \in \Gamma\}$, where $A_\alpha = \bigcap_{\lambda \in \Lambda} A_{\lambda,\alpha}$. Since for any $\alpha \in \Gamma$ the family $\{A_{\lambda,\alpha}|\lambda \in \Lambda\}$ is linearly ordered, then it forms a basis of a filter of the compact set X_α, then $A_\alpha \neq \emptyset$. Besides, $f_{\alpha,\beta}(A_\beta) = f_{\alpha,\beta}(\bigcap_{\lambda \in \Lambda} A_{\lambda,\beta}) \subseteq \bigcap_{\lambda \in \Lambda} f_{\alpha,\beta}(A_{\lambda,\beta}) \subseteq \bigcap_{\lambda \in \Lambda} A_{\lambda,\alpha} = A_\alpha$ for any $\alpha,\beta \in \Gamma$, $\alpha \leq \beta$. Thus, $\mathfrak{A} \in \mathfrak{N}$.

It is clear that $\mathfrak{A} \geq \mathfrak{A}_\lambda$ for any $\lambda \in \Lambda$. Thus, $\mathfrak{N}$ is an inductive set. Due to Zorn lemma, the set $\mathfrak{N}$ contains maximal elements. Let $\bar{\mathfrak{A}} = \{\bar{A}_\alpha|\alpha \in \Gamma\}$ be a maximal element in $\mathfrak{N}$.

Now we show that in that case $\bar{A}_\alpha = f_{\alpha,\beta}(\bar{A}_\beta)$ for any α, $\beta \in \Gamma$, $\beta \geq \alpha$. Indeed, for any $\alpha \in \Gamma$ let $A'_\alpha = \bigcap_{\beta \geq \alpha} f_{\alpha,\beta}(\bar{A}_\beta)$.

At first we show that $\mathfrak{A}' = \{A'_\alpha|\alpha \in \Gamma\} \in \mathfrak{N}$. The set $f_{\alpha,\beta}(\bar{A}_\beta)$ is compact, hence, it is a closed subset in X_α for any $\alpha,\beta \in \Gamma$, $\alpha \leq \beta$. Besides, for any $\beta,\gamma \in \Gamma$, where $\beta \geq \alpha$ and $\gamma \geq \alpha$, there exists $\delta \in \Gamma$ such that $\delta \geq \beta$ and $\delta \geq \gamma$. Then $f_{\alpha,\delta}(\bar{A}_\delta) = f_{\alpha,\beta} \circ f_{\beta,\delta}(\bar{A}_\delta) \subseteq f_{\alpha,\beta}(\bar{A}_\beta)$ and $f_{\alpha,\delta}(\bar{A}_\delta) = f_{\alpha,\gamma} \circ f_{\gamma,\delta}(\bar{A}_\delta) \subseteq f_{\alpha,\gamma}(\bar{A}_\gamma)$, i.e. $f_{\alpha,\delta}(\bar{A}_\delta) \subseteq f_{\alpha,\beta}(\bar{A}_\beta) \bigcap f_{\alpha,\gamma}(\bar{A}_\gamma)$. It means that the family $\{f_{\alpha,\beta}(\bar{A}_\beta)|\beta \in \Gamma, \beta \geq \alpha\}$ forms a basis of a filter of the compact set X_α, hence, $A'_\alpha \neq \emptyset$ for any $\alpha \in \Gamma$. Let $\alpha,\beta \in \Gamma$ and $\beta \geq \alpha$, then $f_{\alpha,\beta}(\bar{A}'_\beta) = f_{\alpha,\beta}(\bigcap_{\gamma \geq \beta} f_{\beta,\gamma}(\bar{A}_\gamma)) \subseteq \bigcap_{\gamma \geq \beta} f_{\alpha,\beta} \circ f_{\beta,\gamma}(\bar{A}_\gamma) = \bigcap_{\gamma \geq \beta} f_{\alpha,\gamma}(\bar{A}_\gamma)$. Since for any $\delta \in \Gamma$, where $\delta \geq \alpha$, there exists $\rho \in \Gamma$ such that $\rho \geq \delta$ and $\rho \geq \beta$, then $f_{\alpha,\rho}(\bar{A}_\rho) = f_{\alpha,\delta}(f_{\delta,\rho}(\bar{A}_\rho)) \subseteq f_{\alpha,\delta}(\bar{A}_\delta)$. Therefore, $\bigcap_{\gamma \geq \beta} f_{\alpha,\gamma}(\bar{A}_\gamma) \subseteq \bigcap_{\delta \geq \alpha} f_{\alpha,\delta}(\bar{A}_\delta) = A'_\alpha$, hence, $f_{\alpha,\beta}(A'_\beta) \subseteq A'_\alpha$. Hence, we've proved that $\mathfrak{A}' \in \mathfrak{N}$ and $\mathfrak{A}' \geq \bar{\mathfrak{A}}$.

Since the element $\bar{\mathfrak{A}}$ is maximal in $\mathfrak{N}$, then $\mathfrak{A}' = \mathfrak{A}$, i.e. $\bar{A}_\alpha = A'_\alpha$ for any $\alpha \in \Gamma$. If we assume that $f_{\alpha,\beta}(\bar{A}_\beta) \neq \bar{A}_\alpha$ for some $\alpha, \beta \in \Gamma$, $\beta \geq \alpha$, then $A'_\alpha = \bigcap_{\beta \geq \alpha} f_{\alpha,\beta}(\bar{A}_\beta) \neq \bar{A}_\alpha$, which contradicts the proved above. Thus, $f_{\alpha,\beta}(\bar{A}_\beta) = \bar{A}_\alpha$ for any α, $\beta \in \Gamma$, $\beta \geq \Gamma$.

Now we show that $\bar{A}_\alpha$ is a one-element subset in X_α for any $\alpha \in \Gamma$. Indeed, let $\alpha_0 \in \Gamma$ and $x_{\alpha_0} \in \bar{A}_{\alpha_0}$. For any $\beta \in \Gamma$, $\beta \geq \alpha_0$ put $B_\beta = \bar{A}_\beta \bigcap f_{\alpha_0,\beta}^{-1}(x_{\alpha_0})$, and for all other $\beta \in \Gamma$ put $B_\beta = \bar{A}_\beta$.

Now we prove that $\mathcal{L} = \{B_\alpha | \alpha \in \Gamma\} \in \mathfrak{N}$. Indeed, if $\beta \not\geq \alpha_0$, then for any $\gamma \in \Gamma$, where $\gamma \geq \beta$, we get that $f_{\beta,\gamma}(B_\gamma) \subseteq f_{\beta,\gamma}(\bar{A}_\gamma) \subseteq \bar{A}_\beta = B_\gamma$. If $\beta \geq \alpha_0$, then for any $\gamma \in \Gamma$, where $\gamma \geq \beta$, we get that $f_{\beta,\gamma}(f_{\alpha_0,\gamma}^{-1}(x_{\alpha_0})) = f_{\beta,\gamma}(f_{\beta,\gamma}^{-1}(f_{\alpha_0,\beta}^{-1}(x_{\alpha_0}))) \subseteq f_{\alpha_0,\beta}^{-1}(x_{\alpha_0})$, and since $f_{\beta,\gamma}(\bar{A}_\gamma) \subseteq \bar{A}_\beta$, then $f_{\beta,\gamma}(B_\gamma) \subseteq B_\beta$. Since $\bar{A}_{\alpha_0} = f_{\alpha_0,\beta}(\bar{A}_\beta)$ for any $\beta \in \Gamma$, $\beta \geq \alpha_0$, then $f_{\alpha_0,\beta}^{-1}(x_{\alpha_0}) \bigcap \bar{A}_\beta \neq \emptyset$. Therefore, $B_\beta \neq \emptyset$ for any $\beta \in \Gamma$. It means that $\mathcal{L} \in \mathfrak{N}$. Since $\mathcal{L} \geq \bar{\mathfrak{A}}$, then $B_\beta = \bar{A}_\beta$ for all $\beta \in \Gamma$, since $\bar{\mathfrak{A}}$ is maximal. In particular, $\bar{A}_{\alpha_0} = B_{\alpha_0} = \{x_{\alpha_0}\}$. Thus, $\bar{A}_\alpha$ is a one-element subset in X_α for any $\alpha \in \Gamma$.

Pass to completion of the proof of the theorem, i.e. prove that $f_\sigma(X) = \bigcap_{\beta \geq \sigma} f_{\sigma,\beta}(X_\beta)$ for any $\sigma \in \Gamma$. Assume the contrary, i.e. that for a certain $\sigma \in \Gamma$ we have $f_\sigma(X) \neq \bigcap_{\beta \geq \sigma} f_{\sigma,\beta}(X_\beta)$. Since $f_\sigma(X) = f_{\sigma,\beta}(f_\beta(X)) \subseteq f_{\sigma,\beta}(X_\beta)$ for any $\beta \in \Gamma$, $\beta \geq \sigma$, then $f_\sigma(X) \subseteq \bigcap_{\beta \geq \sigma} f_{\sigma,\beta}(X_\beta)$. Then there exists $x_\sigma \in \bigcap_{\beta \geq \sigma} f_{\sigma,\beta}(X_\beta)$ such that $x_\sigma \notin f_\sigma(X)$. For $\gamma \in \Gamma$ put $B_\gamma = f_{\sigma,\gamma}^{-1}(x_\sigma)$ for $\gamma \geq \sigma$ and $B_\gamma = X_\gamma$ for all other $\gamma \in \Gamma$. Then $B_\gamma \neq \emptyset$ for any $\gamma \in \Gamma$. Then for any $\beta, \gamma \in \Gamma$, $\gamma \geq \beta$ we have:

a) if $\beta \not\geq \sigma$, then $f_{\beta,\gamma}(B_\gamma) \subseteq f_{\beta,\gamma}(X_\gamma) \subseteq X_\beta = B_\beta$;

b) if $\beta \geq \sigma$, then $f_{\beta,\gamma}(B_\gamma) = f_{\beta,\gamma}(f_{\sigma,\gamma}^{-1}(x_\gamma)) = f_{\beta,\gamma}\left(f_{\beta,\gamma}^{-1}(f_{\sigma,\beta}^{-1}(x_\sigma))\right) \subseteq f_{\sigma,\beta}^{-1}(x_\sigma) = B_\beta$. Thus, $\mathcal{L} = \{B_\gamma | \gamma \in \Gamma\} \in \mathfrak{N}$.

Let $\tilde{\mathfrak{A}} = \{\tilde{A}_\gamma | \gamma \in \Gamma\}$ be such maximal element in $\mathfrak{N}$ that $\tilde{\mathfrak{A}} \geq \mathcal{L}$. Then, as we have shown above, for any $\gamma \in \Gamma$ we get that $\tilde{A}_\gamma = \{a_\gamma\}$ for a certain element $a_\gamma \in X_\gamma$. Since $\tilde{A}_\sigma \subseteq B_\sigma$, then $a_\sigma = x_\sigma$. Since $f_{\beta,\gamma}(\tilde{A}_\gamma) \subseteq \tilde{A}_\beta$, then $f_{\beta,\gamma}(a_\gamma) = a_\beta$ for any $\beta, \gamma \in \Gamma$, $\beta \leq \gamma$. Let a be the element of $\prod_{\gamma \in \Gamma} X_\gamma$ such that $a(\gamma) = a_\gamma$. Then $a \in \varprojlim S = X$ and $f_\sigma(a) = a_\sigma = x_\sigma$. Thus, $x_\sigma \in f_\sigma(X)$, which contradicts the choice of the element $x_\sigma \in X_\sigma$. The obtained contradiction shows that $f_\sigma(X) = \bigcap_{\beta \geq \sigma} f_{\sigma,\beta}(X_\beta)$ for any $\sigma \in \Gamma$. The theorem is proved.

4.2.19. COROLLARY. *Let $S = \{X_\alpha, f_{\alpha,\beta} | \alpha, \beta \in \Gamma\}$ be an inverse spectrum of*

*compact Hausdorff Abelian groups. If the homomorphism $f_{\alpha,\beta}$ is surjective for all $\alpha, \beta \in$
Γ, $\alpha \leq \beta$, then the projection $f_\alpha : \varprojlim S \to X_\alpha$ is a surjective homomorphism for any
$\alpha \in \Gamma$.*

PROOF. Indeed, according to Theorem 4.2.18, $f_\alpha(\varprojlim S) = \bigcap_{\beta \geq \alpha} f_{\alpha,\beta}(X_\beta) = X_\alpha$ for
any $\alpha \in I$.

4.2.20. PROPOSITION. *Let X be a topological Abelian group, $\mathcal{L} = \{Y_\alpha | \alpha \in \Gamma\}$ be a
basis of a certain filter of the group X consisting of subgroups, and Γ be the direction defined
by the basis $\mathcal{L}$ (see Remark 3.1.3), also $X_\alpha = X/Y_\alpha$ for any $\alpha \in \Gamma$. For any $\alpha, \beta \in \Gamma$,
$\beta \geq \alpha$, consider the natural homomorphisms $f_{\alpha,\beta} : X_\beta \to X_\alpha$ and $\phi_\alpha : X \to X_\alpha$. It is
clear that $S = \{X_\alpha, f_{\alpha,\beta} | \alpha, \beta \in \Gamma\}$ is an inverse spectrum of topological Abelian groups.*

*Let $\phi : X \to \varprojlim S$ be the continuous homomorphism constructed in Proposition 4.2.8.
Then:*

a) $\ker \phi = \bigcap_{\alpha \in \Gamma} Y_\alpha$;

b) $\phi(X)$ is a dense subgroup in $\varprojlim S$.

PROOF. Let $x \in X$, then $x \in \ker \phi$ if and only if $f_\alpha(\phi(x)) = 0$ for any $\alpha \in \Gamma$, where
$f_\alpha : \varprojlim S \to X_\alpha$ is the natural projection. Since $f_\alpha(\phi(x)) = \phi_\alpha(x)$, then $x \in \ker \phi$ if and
only if $\phi_\alpha(x) = 0$ for any $\alpha \in \Gamma$, i.e. $x \in Y_\alpha$ for any $\alpha \in \Gamma$. Hence, $\ker \phi = \bigcap_{\alpha \in \Gamma} Y_\alpha$.

Now we prove that $\phi(X)$ is a dense subgroup in $\varprojlim S$. Let $y \in \varprojlim S$ and U be a
neighborhood of zero in $\varprojlim S$. According to Proposition 4.2.4, there exist $\alpha_0 \in \Gamma$ and a
neighborhood U_{α_0} of zero in X_{α_0} such that $f_{\alpha_0}^{-1}(U_{\alpha_0}) \subseteq U$. Let $x \in \phi_{\alpha_0}^{-1}(y(\alpha_0))$, then
$f_{\alpha_0}(\phi(x) - y) = \phi_{\alpha_0}(x) - y(\alpha_0) = 0$. Thus, $\phi(x) - y \in f_{\alpha_0}^{-1}(0) \subseteq f_{\alpha_0}^{-1}(U_{\alpha_0}) \subseteq U$, i.e.
$\phi(x) \in y + U$. It means that $\phi(X)$ is a dense subgroup in $\varprojlim S$.

4.2.21. THEOREM. *Let X be a Hausdorff topological Abelian group that has a basis
$\mathcal{L} = \{U_\gamma | \gamma \in \Gamma\}$ of neighborhoods of zero consisting of subgroups, Γ be the direction, defined
by the basis $\mathcal{L}$, and $S = \{X/U_\alpha, f_{\alpha,\beta} | \alpha, \beta \in \Gamma\}$ be the inverse spectrum (see Proposition
4.2.20) of the discrete factor-groups X/U_α. Then the completion $\widehat{X}$ of the group X is
topologically isomorphic to $\varprojlim S$.*

PROOF. According to Corollary 4.2.7, $\varprojlim S$ is a Hausdorff complete group. Let
$\phi_\alpha : X \to X/U_\alpha$ be the canonical homomorphism and $\phi : X \to \varprojlim S$ be the continu-

ous homomorphism, constructed in Proposition 4.2.8.

According to Proposition 4.2.20 and Theorem 1.3.2, $\ker \phi = \bigcap_{\alpha \in \Gamma} U_\alpha = \{0\}$, hence, $\ker \phi = \{0\}$. According to Proposition 4.2.20, $\phi(X)$ is dense subgroup in $\varprojlim S$.

Now we prove that ϕ is an open mapping of X onto $\phi(X)$. Indeed, let $U_\gamma \in \mathcal{L}$. Then $V = f_\gamma^{-1}(\{0\})$ is a neighborhood of zero in $\varprojlim S$, hence $V \cap \phi(X)$ is a neighborhood of zero in $\phi(X)$. Let $y \in V \cap \phi(X)$, then there exists $x \in X$ such that $\phi(x) = y$, hence $\phi(x) = f_\alpha(\phi(x)) = f_\alpha(y)$ for any $\alpha \in \Gamma$. In particular, $\phi_\gamma(x) = f_\gamma(y) \in f_\gamma(V) = \{0\}$, i.e. $\phi(x) = 0$, hence, $x \in \ker \phi_\gamma = U_\gamma$. Then $V \cap \phi(X) \subseteq \phi(U_\gamma)$. Since $V \cap \phi(X)$ is a neighborhood of zero in $\phi(X)$, then $\phi(U_\gamma)$ is a neighborhood of zero in $\phi(X)$. Thus, ϕ is open mapping from X onto $\phi(X)$. Hence, ϕ is a topological isomorphism of the group X onto the dense subgroup $\phi(X)$ of the complete Hausdorff group $\varprojlim S$. According to Theorem 3.3.2, the group $\varprojlim S$ is topologically isomorphic to the completion $\widehat{X}$ of the topological group X.

4.2.22. **COROLLARY.** *Let X be a compact Hausdorff group with a basis $\mathcal{L} = \{U_\gamma \mid \gamma \in \Gamma\}$ of neighborhoods of zero consisting of subgroups, $S = \{X/U_\alpha, f_{\alpha,\beta} \mid \alpha, \beta \in \Gamma\}$ be the inverse spectrum (see Proposition 4.2.20) of the compact discrete (hence, finite) factor-groups X/U_γ of the group X. Then the group X is topologically isomorphic to the limit of the inverse spectrum of the finite groups X/U_γ (i.e. is a profinite group).*

PROOF. According to Corollary 3.2.4, the group X is complete. According to Theorem 3.3.2, it coincides with its completion $\widehat{X}$, therefore, according to Theorem 4.2.21, the group X is topologically isomorphic to $\varprojlim S$.

4.2.23. **REMARK.** It is easy to note that if in the statements 4.2.12–4.2.22 we consider an inverse spectrum of topological rings (modules), then the homomorphisms constructed in these statements will also be ring (module) homomorphisms. And, since the topological properties of the limits of inverse spectra of topological rings (modules) are defined by the properties of the limits of the inverse spectra of its additive groups, then the analogs of the statements 4.2.12 - 4.2.22 are also valid for limits of inverse spectra of topological rings (modules).

4.2.24. COROLLARY. *A compact totally disconnected ring R (in particular, compact ring with the unitary element) is topologically isomorphic to the limit of an inverse spectrum of discrete compact (hence, finite) factor-rings of the ring R by open two-sided ideals.*

PROOF. Due to Corollary 1.6.37 (Corollary 1.6.68), a compact totally disconnected ring (in particular, compact ring with the unitary element) has a basis of neighborhoods of zero consisting of two-sided ideals. The further proof follows from Corollary 4.2.22 and Remark 4.2.23.

4.2.25. COROLLARY. *A Hausdorff linearly compact ring (module) X, that has a basis of neighborhoods of zero consisting of two-sided ideals (submodules) is topologically isomorphic to the limit of an inverse spectrum of discrete factor-rings (factor-modules) by open two-sided ideals (submodules) of the ring (module) X.*

PROOF. According to Corollary 3.2.11 (Proposition 3.2.10), the ring (module) X is complete. According to Theorem 3.3.5 (Proposition 3.3.8), it coincides with its completion $\widehat{X}$. The further proof follows from corollary 4.2.22 and Remark 4.2.23.

§ 4.3. Subdirect Products

4.3.1. DEFINITION. Let Y be a topological Abelian group (ring, module), $\{X_i \mid i \in I\}$ a family of topological Abelian groups (rings, modules), and for any $i \in I$ is defined a continuous open surjective homomorphism $f_i : Y \to X_i$. If $\bigcap_{i \in I} \ker f_i = \{0\}$, then the group (ring, module) Y is called a subdirect product of the family $\{X_i \mid i \in I\}$ of the groups (rings, modules), and the homomorphisms f_i are called the natural homomorphisms, defined on the subdirect product.

4.3.2. PROPOSITION. *Let a topological Abelian group Y be a subdirect product of a family $\{X_i \mid i \in I\}$ of a Hausdorff Abelian group. Then Y is a Hausdorff group.*

PROOF. Since X_i is a Hausdorff group for any $i \in I$, then the kernel $\ker f_i$ of the natural homomorphism $f_i : Y \to X_i$ is closed in Y. Then the set $\{0\} = \bigcap_{i \in I} \ker f_i$ is closed. Hence, due to Theorem 1.3.2, the group Y is Hausdorff.

4.3.3. PROPOSITION. *Let $\mathfrak{m}$ be a cardinal number, Y a topological Abelian group, the topology of which is a $\mathfrak{m}$-topology. If the group Y is a subdirect product of a family $\{X_i \mid i \in I\}$ of topological Abelian groups and X is the $\mathfrak{m}$-product of the family $\{X_i \mid i \in I\}$, then there exists a continuous monomorphism $f : Y \to X$.*

PROOF. Let $f_i : Y \to X_i$ be the natural homomorphism, $i \in I$. Construct the mapping $f : Y \to X$, putting: $f(y)(i) = f_i(y)$ for $i \in I, y \in Y$.

It is clear that f is a group homomorphism. If $f(y) = 0$, then $f_i(y) = 0$ for any $i \in I$, i.e. $y \in \bigcap_{i \in I} \ker f_i = \{0\}$. Thus, f is a monomorphism from the group Y to the group X.

Now we prove that f is a continuous homomorphism. Indeed, let U be a neighborhood of zero in X. Then there exists a family $\{U_i \subseteq X_i \mid i \in I\}$, where U_i is a neighborhood of zero in X_i, such that $\left| \{i \in I \mid U_i \neq X_i\} \right| < \mathfrak{m}$, and $\prod_{i \in I} U_i \subseteq U$. For any $i \in I$ let $V_i = f_i^{-1}(U_i)$. Let $J = \{i \in I \mid V_i \neq Y\}$, then $\left| J \right| = \left| \{i \in I \mid U_i \neq X_i\} \right| < \mathfrak{m}$. Since the topology on Y is a $\mathfrak{m}$-topology, then $V = \bigcap_{i \in I} V_i = \bigcap_{i \in J} V_i$ is a neighborhood of zero in Y, and

$$f(V) \subseteq \prod_{i \in I} f_i(V) \subseteq \prod_{i \in I} f_i(V_i) \subseteq \prod_{i \in I} U_i \subseteq U.$$

Thus, f is a continuous homomorphism. The proposition is proved.

4.3.4. COROLLARY. *Let a topological Abelian group* Y *be a subdirect product of a family* $\{X_i \mid i \in I\}$ *of topological Abelian groups. Let* X *be the Tychonoff product of the family* $\{X_i \mid i \in I\}$. *Then there exists a continuous monomorphism* $f : Y \to X$.

PROOF. Consequently applying Remark 4.1.13, Definition 4.1.3 and Proposition 4.3.3, we can prove the corollary.

4.3.5. COROLLARY. *Let* (Y, τ) *be a Hausdorff topological Abelian group endowed with a non-weakable Hausdorff topology. Let* Y *be a subdirect product of a family* $\{X_i \mid i \in I\}$ *of Hausdorff topological Abelian groups. Then the group* Y *is topologically isomorphic to a certain subgroup of the Tychonoff product* X *of the family* $\{X_i \mid i \in I\}$.

PROOF. According to Corollary 4.3.4, there exists a continuous monomorphism $f : Y \to X$. Consider the subgroup $f(Y)$ of the topological group X. According to Corollary 4.1.44, the group $f(Y)$ is Hausdorff. Let τ_1 be the prototype on X of the group topology on $f(Y)$ with respect to the monomorphism f. Since f is a continuous homomorphism, then $\tau \geq \tau_1$, and since $f^{-1}(\{0\}) = \{0\} = [\{0\}]_{f(Y)}$, then, according to Theorem 1.3.2 , (Y, τ_1) is a Hausdorff group. Since the topology τ is a non-weakable Hausdorff topology, then $\tau = \tau_1$. Hence, $f : Y \to f(Y)$ is an open homomorphism.

Thus, we have shown that f is a topological isomorphism of the group Y to the subgroup $f(Y)$ of the topological group X.

4.3.6. PROPOSITION. *Let* $\{X_i \mid i \in I\}$ *be a family of topological Abelian groups,* $X = \prod_{i \in I} X_i$ *be their* $\mathfrak{m}$*-product. Let* Y *be a subgroup of the topological Abelian group* X, *and* Y *contains the subgroup* $\sum_{i \in I} X_i$. *Then the topological group* Y *is a subdirect product of the family* $\{X_i \mid i \in I\}$.

PROOF. Denote over f_i the restriction on Y of the canonical projections $\pi_i : X \to X_i$. Then f_i is a continuous homomorphism for any $i \in I$ and

$$\bigcap_{i \in I} \ker f_i \subseteq \bigcap_{i \in I} \ker \pi_i = \{0\}.$$

Since

$$f_j(Y) \supseteq f_j\left(\sum_{i \in I} X_i\right) = \pi_j\left(\sum_{i \in I} X_i\right) = X_j,$$

then $\hat{f}_j$ is a surjective homomorphism for any $j \in I$.

Now we show that f_i is an open homomorphism for any $i \in I$. Indeed, let U be a neighborhood of zero in Y, then $U = W \cap Y$, where W is a certain neighborhood of zero in X. For any $j \in I$ there exists a neighborhood W_j of zero in X_j such that $\prod_{j \in I} W_j \subseteq W$. Then $W_i' = \{x \in W \mid x(j) = 0 \text{ for } j \neq i\} \subseteq W \cap Y$, hence $f_i(U) \supseteq f_i(W_i') = W_i$. Therefore, $f_i(U)$ is a neighborhood of zero in X_i, i.e. f_i is open homomorphism.

Thus, Y is a subdirect product of the family $\{X_i \mid i \in I\}$.

4.3.7. REMARK. As we've shown in Corollary 4.3.4, a subdirect product of a family $\{X_i \mid i \in I\}$ of topological Abelian groups is a subgroup of their direct product $\prod_{i \in I} X_i$, that, may have a stronger topology than the topology, induced from the Tychonoff product (this clarifies the meaning of the term "subdirect product").

4.3.8. PROPOSITION. *Let R be a topological ring, $\{M_i \mid i \in I\}$ be a family of topological right R-modules, where for any $i \in I$ in M_i is defined an open submodule N_i, and $N = \prod_{i \in I} N_i$ is the Tychonoff product of the family $\{N_i \mid i \in I\}$ of the topological modules. Let $Z = \{f \in \prod_{i \in I} M_i \mid \{i \in I \mid f(i) \notin N_i\} \text{ is a finite subset}\}$. Then Z is a submodule in $\prod_{i \in I} M_i$ such that $\prod_{i \in I} N_i \subseteq Z$ and on the module Z exists the unique topology τ such that (Z, τ) is topological right R-module, and the family $\mathfrak{A} = \{W \mid W \text{ is a neighborhood of zero in } N\}$ is a basis of neighborhoods of zero in (Z, τ).*

PROOF. Directly from the definition of the subset Z follows that Z is a submodule in $\prod_{i \in I} M_i$ and $Z \supseteq \prod_{i \in I} N_i$.

To complete the proof (see Theorem 1.2.6) it is enough to verify that the family $\mathfrak{A}$ satisfies the conditions (BN1)–(BN4), (BN5'), (BN6') and (BN6''). Since N is a topological R-module, then the family $\mathfrak{A}$ satisfies the conditions (BN1) - (BN4), (BN5'),(BN6'). Let $W \in \mathfrak{A}$ and $f \in Z$. Without loss of generality we can consider that $W = \prod_{i \in I} W_i$, where W_i is a neighborhood of zero in N_i and $J = \{i \in I \mid W_i \neq N_i\}$ is a finite set. Since $f \in Z$, then $J_1 = \{i \in I \mid f(i) \notin N_i\}$ is a finite set. Since N_i is an open submodule in M_i, then W_i is a neighborhood of zero in M_i for any $i \in I$. Then for any $i \in J \cup J_1$ there exists a neighborhood V_i of zero in R such that $f(i) \cdot V_i \subseteq W_i$. Let $V = \bigcap_{i \in J \cup J_1} V_i$, then V is a neighborhood of zero in R. Then $f(i) \cdot V \subseteq N_i = W_i$, for $i \notin J \cup J_1$ and $f(i) \cdot V \subseteq W_i$ for

$i \in J \cup J_1$. Hence, $f \cdot W \subseteq \prod_{i \in I} W_i = W$, i.e. the condition (BN6″) is fulfilled.

The proposition is proved.

4.3.9. DEFINITION. Let R be a topological ring, $\{M_i \,|\, i \in I\}$ be a family of topological R-modules and for any $i \in I$ in the module M_i is determined an open submodule N_i. We call the submodule $Z = \{f \in \prod_{i \in I} M_i \,|\, \{i \in I \,|\, f(i) \notin N_i\}$ is a finite set $\}$ of the module $\prod_{i \in I} M_i$, endowed with the topology, constructed in Proposition 4.3.8, a local product of the family $\{M_i \,|\, i \in I\}$ of the modules with determined open submodules $N_i \subseteq M_i$, $i \in I$, and denote it over $\prod_{i \in I}(M_i : N_i)$.

4.3.10. PROPOSITION. *Let R be a topological ring, $\{M_i \,|\, i \in I\}$ be a family of topological R-modules and for any $i \in I$ in the module M_i is determined an open submodule N_i. Let $Z = \prod_{i \in I}(M_i : N_i)$ be the local product of the family $\{M_i \,|\, i \in I\}$ with determined open submodules $N_i \subseteq M_i$, N be the Tychonoff product of the family $\{N_i \,|\, i \in I\}$. Denote over $\rho_i : Z \to M_i$ the restriction on Z of the canonical projection $\pi_i : \prod_{j \in I} M_j \to M_i, i \in I$. Then:*

1) for any $i \in I$ the homomorphism $\rho_i : Z \to M_i$ is a continuous and open surjective homomorphism;

2) the module Z is a subdirect product of the family $\{M_i \,|\, i \in I\}$;

3) the submodule $N = \prod_{i \in I} N_i$ is open in Z;

4) if for any $i \in I$ the module M_i is complete, then the module Z is complete, too;

5) if for any $i \in I$ the module M_i is Hausdorff, then the module Z is Hausdorff, too;

6) if for any $i \in I$ the module M_i has a basis of neighborhoods of zero consisting of submodules, then the module Z has a basis of neighborhoods of zero consisting of submodules, too;

7) if for any $i \in I$ the submodule N_i is compact, then the module Z is a locally compact module.

PROOF. Let $i \in I$. Since $Y_i = \{f \in \prod_{j \in J} M_i \,|\, f(j) = 0 \text{ for } j \neq i\} \subseteq Z$, then $\rho_i(Z) \supseteq \rho_i(Y_i) = X_i$, i.e. ρ_i is a surjective homomorphism. Let U_i be a neighborhood of zero in M_i, then $U_i \cap N_i$ is a neighborhood of zero in N_i. Since $\pi_i\big|_N : N \to N_i$ is continuous mapping, then there exists a neighborhood V of zero in N such that $\pi_i(V) \subseteq U_i \cap N_i$. Then

V is a neighborhood of zero in Z, and $\rho_i(V) = \pi_i(V) \subseteq U_i$. Hence, we've proved that ρ_i is a continuous homomorphism.

Now let U be a neighborhood of zero in Z. Then $U \cap N$ is a neighborhood of zero in N. Since $\pi_i|_N : N \to N_i$ is an open mapping, then $\pi_i(U \cap N)$ is a neighborhood of zero in N_i, and since N_i is an open submodule in M_i, then $\pi_i(U \cap N)$ is a neighborhood of zero in M_i. Then $\rho_i(U) \supseteq \rho_i(U \cap N) = \pi_i(U \cap N)$, hence $\rho_i(U)$ is a neighborhood of zero in M_i. Thus, we've proved the validity of the item 1) of the proposition.

Since ρ_i is a continuous and open surjective homomorphisms for each $i \in I$, then to prove the item 2) it is enough to note that

$$\bigcap_{i \in I} \ker \rho_i \subseteq \prod_{i \in I} \ker \pi_i = \{0\}.$$

Since $N \in \mathfrak{A}$ (see Proposition 4.3.8), then N is an open submodule in Z. Thus, the item 3) is proved.

Let for any $i \in I$ the module M_i be complete. Since the open submodule N_i is closed in M_i, then, according to Theorem 3.2.16, N_i is a complete submodule for any $i \in I$. Due to Proposition 4.1.47, the module N is complete. Since N is an open submodule in Z, then, according to Corollary 3.2.23, the module Z is complete, too. It proves the validity of the item 4).

Let for any $i \in I$ the module M_i be Hausdorff. Then for any $i \in I$ the module N_i is Hausdorff, too. According to Corollary 4.1.44, the module N is Hausdorff. Since N is an open, hence, closed submodule in Z, then, due to Proposition 1.4.35, Z is a Hausdorff module. Thus, the item 5) is proved.

Let for any $i \in I$ the module M_i have a basis of neighborhoods of zero consisting of submodules. Then for any $i \in I$ the module N_i has a basis of neighborhoods of zero consisting of submodules, too. According to Proposition 4.1.19, the module $N = (\prod_{i \in I} N_i, \tau_T)$ has a basis of neighborhoods of zero consisting of submodules. Hence, Z has a basis of neighborhoods of zero consisting of submodules. It proves the validity of the item 6).

Let now for any $i \in I$ the submodule N_i be compact. Then, according to Corollary 4.1.53, the module N is compact, too. Since N is an open submodule in Z, then N is a compact neighborhood of zero in Z. Thus, Z is a locally compact module. Hence, we have proved the item 7), which completes the proof of the proposition.

4.3.11. REMARK. If in the conditions of Proposition 4.3.10 for any $i \in I$ the equation $N_i = M_i$ is fulfilled, then $\prod_{i \in I}(M_i : N_i) = \left(\prod_{i \in I} M_i, \tau_\tau\right)$, where τ_τ is the Tychonoff topology on $\prod_{i \in I} M_i$.

4.3.12. PROPOSITION. *Let R be a topological ring, $\{M_i \,|\, i \in I\}$ be a family of topological right R-modules, and for any $i \in I$ in the module M_i is determined an open submodule N_i, and $Z = \prod_{i \in I}(M_i : N_i)$ be the local product of the family $\{M_i \,|\, i \in I\}$. Let $N = \prod_{i \in I} N_i$, then the factor-module Z/N is discrete and isomorphic to the module $\sum_{i \in I}(M_i/N_i)$ endowed with discrete topology.*

PROOF. Since N is an open submodule in Z, then the factor-module Z/N is discrete. For any $i \in I$ denote over $\gamma_i : M_i \to M_i/N_i$ the natural homomorphism and over $\gamma : Z \to Z/N$ the natural homomorphism of the module Z to Z/N. Consider the direct product $\bar{M} = \prod_{i \in I}(M_i/N_i)$ and define the mapping $\varphi : Z/N \to \bar{M}$ as the following:

for $\bar{f} \in Z/N$ let f be an element of Z such that $\bar{f} = \gamma(f)$, then put $\varphi(\bar{f}) = g \in \bar{M}$, where $g(i) = \gamma_i(f(i))$ for any $i \in I$.

Now we prove that this definition is correct. Indeed, let $\bar{f} \in Z/N$ and $f, h \in Z$ be such that $\gamma(f) = \gamma(h) = \bar{f}$. Then $\varphi(\bar{f}) = g$ and $\varphi(\bar{f}) = g'$, where $g(i) = \gamma_i(f(i))$ and $g'(i) = \gamma_i(h(i))$ for any $i \in I$. Since $\gamma(f) = \gamma(h)$, then $\gamma(f - h) = 0$. Hence, $f - h \in N$. Therefore, $(f - h)(i) = f(i) - h(i) \in N_i$ for any $i \in I$, hence, $\gamma_i(f(i) - h(i)) = 0$, i.e. $\gamma_i(f(i)) = \gamma_i(h(i))$ for any $i \in I$. Thus, $g = g'$, which proves the correctness of the definition.

Let's prove that φ is a homomorphism of R-modules. Indeed, let $\bar{f}, \bar{h} \in Z/N$, and f and h be elements of Z such that $\gamma(f) = \bar{f}$ and $\gamma(h) = \bar{h}$. Then $\gamma(f - h) = \bar{f} - \bar{h}$. Let $\varphi(\bar{f}) = g$, $\varphi(\bar{h}) = t$ and $\varphi(\bar{f} - \bar{h}) = s$, then

$$s(i) = \gamma_i((f - h)(i)) = \gamma_i(f(i) - h(i)) = \gamma_i(f(i)) - \gamma_i(h(i)) = g(i) - t(i).$$

Thus, $s = g - t$, i.e. $\varphi(\bar{f} - \bar{h}) = \varphi(\bar{f}) - \varphi(\bar{h})$.

Now let $r \in R$ and $\bar{f} \in Z/N$. Let f be an element of Z such that $\gamma(f) = \bar{f}$, then $\gamma(f \cdot r) = \gamma(f) \cdot r = \bar{f} \cdot r$. Let $g = \varphi(\bar{f})$ and $t = \varphi(\bar{f} \cdot r)$. Then $t(i) = \gamma_i((f \cdot r)(i)) =$

$\gamma_i(f(i) \cdot r) = \gamma_i(f(i)) \cdot r = g \cdot r$. Therefore, $t = g \cdot r$, i.e. $\varphi(\bar{f} \cdot r) = \varphi(\bar{f}) \cdot r$. Thus, we've proved that φ is a homomorphism of R-modules.

Let's prove that $\varphi(Z/N) = \sum_{i \in I} M_i/N_i \subseteq \bar{M}$. Indeed, let $\bar{f} \in Z/N$ and f be an element of Z such that $\gamma(f) = \bar{f}$. Then, due to the definition of the module Z, the set $J = \{i \in I \mid f(i) \notin \bar{N}_i\}$ is finite. Then if $\varphi(\bar{f}) = g$, hence, $g(i) = \gamma_i(f(i)) = 0_i$ for any $i \in I \backslash J$, where 0_i is the neutral element of the additive group M_i/N_i. Therefore, $g \in \sum_{i \in I}(M_i/N_i)$, i.e. $\varphi(\bar{f}) \in \sum_{i \in I}(M_i/N_i)$. Thus, we've proved that $\varphi(Z/N) \subseteq \sum_{i \in I}(M_i/N_i)$.

Now let $g \in \sum_{i \in I}(M_i/N_i)$. Then $\{i \in I \mid g(i) \neq 0\}$ is a finite set. We choose for any $i \in I$ a certain preimage f_i of the element $g(i)$ in M_i in correspondence to the natural homomorphism γ_i, i.e. $\gamma_i(f_i) = g(i)$. Then $\{i \in I \mid f_i \notin N_i\} = \{i \in I \mid g(i) \neq 0_i\}$. Let $f \in \prod_{i \in I} M_i$ be such that $f(i) = f_i$ for any $i \in I$. Since $\{i \in I \mid f(i) \notin N_i\}$ is a finite set, then $f \in Z$. Let $\bar{f} = \gamma(f) \in Z/N$ and $\varphi(\bar{f}) = h$. Then $h(i) = \gamma_i(f(i)) = \gamma_i(f_i) = g(i)$ for any $i \in I$, i.e. $\varphi(\bar{f}) = g$. Thus, we've proved that $\varphi(Z/N) = \sum_{i \in I} M_i/N_i$.

Let's prove that φ is a monomorphism. Indeed, let $\bar{f} \in Z/N$ and $\varphi(\bar{f}) = 0$. Choose an element $f \in Z$ such that $\gamma(f) = \bar{f}$. Then $\gamma_i(f(i)) = 0_i$ for any $i \in I$. Therefore, $f \in \prod_{i \in I} N_i = N$, hence $\bar{f} = \gamma(f) = 0$. It means that $\ker \varphi = \{0\}$, i.e. φ is a monomorphism. Thus, we've proved that φ is an isomorphic mapping of the module Z/N to the module $\sum_{i \in I}(M_i/N_i)$. This completes the proof of the proposition.

4.3.13. REMARK. Let $\{X_i \mid i \in I\}$ be a family of topological Abelian groups, where for any $i \in I$ in the group X_i is determined an open subgroup H_i. Consider the groups X_i and H_i, $i \in I$, as modules over the ring of integers $\mathbb{Z}$ with discrete topology. Then their local product $\prod_{i \in I}(X_i : H_i)$ is called a local product of the family $\{X_i \mid i \in I\}$ of groups with determined open subgroups H_i.*

4.3.14. PROPOSITION. *Let $\{R_i \mid i \in I\}$ be a family of topological rings and for any $i \in I$ in the ring R_i is determined an open subring S_i. Consider the local product $Z = \prod_{i \in I}(R_i : S_i)$ of the additive topological groups $R_i(+)$ with determined open subgroups S_i. Then Z is a topological ring with respect to the ring operations defined on $\prod_{i \in I} R_i$.*

---

* In [432] is considered a more general construction of the local product of topological groups with determined closed subgroups.

PROOF. Let's prove that Z is a ring. Indeed, since Z is a subgroup in $\prod_{i \in I} R_i$, then it is necessary to prove that the product of any two elements of Z belongs to Z. Let $f, g \in Z$, then the sets $J_1 = \{i \in I \mid f(i) \notin S_i\}$ and $J_2 = \{i \in I \mid g(i) \notin S_i\}$ are finite. If $i \in I \setminus (J_1 \cup J_2)$, then $f(i) \in S_i$ and $g(i) \in S_i$, hence $f(i) \cdot g(i) \in S_i$. Since $J_1 \cup J_2$ is a finite set, then $f \cdot g \in Z$. Thus, Z is a subring in $\prod_{i \in I} R_i$.

Now we show that Z is a topological ring. According to Remark 4.3.13, $Z(+)$ is a topological Abelian group. Let's prove that the multiplication operation on Z is continuous. Indeed, let $f, g \in Z$ and U be a neighborhood of zero in Z. Without loss of generality we can consider that $U = \prod_{i \in I} U_i$, where U_i is a neighborhood of zero in S_i, $i \in I$, and the set $J_1 = \{i \in I \mid U_i \neq S_i\}$ is finite. Since $f, g \in S$, then the sets $J_2 = \{i \in I \mid f(i) \notin S_i\}$ and $J_3 = \{i \in I \mid g(i) \notin S_i\}$ are finite, too. For any $i \in I$ we construct a neighborhood V_i of zero in R_i as the following:

if $i \in I \setminus (J_1 \cup J_2 \cup J_3)$, then put $V_i = S_i$;

if $i \in J_1 \cup J_2 \cup J_3$, then since S_i is open in R_i, then W_i is a neighborhood of zero in R_i, and we choose neighborhoods W_i and V_i of zero in R_i such that $W_i + W_i + W_i \subseteq U_i$, $f(i) \cdot V_i \subseteq W_i$, $V_i \cdot g(i) \subseteq W_i$ and $V_i \cdot V_i \subseteq W_i$.

Then $f(i) \cdot V_i + V_i \cdot g(i) + V_i \cdot V_i \subseteq W_i + W_i + W_i \subseteq U_i$ for any $i \in J_1 \cup J_2 \cup J_3$. Let $V = \prod_{i \in I} V_i$, then V is a neighborhood of zero in Z (since the set $\{i \in I \mid V_i \neq S_i\}$ is finite), and

$$(f + V) \cdot (g + V) \subseteq f \cdot g + V \cdot g + f \cdot V + V \cdot V$$

$$= f \cdot g + \prod_{i \in I} (V_i \cdot g(i)) + \prod_{i \in I} (f(i) \cdot V_i) + \prod_{i \in I} (V_i \cdot V_i)$$

$$\subseteq f \cdot g + \prod_{i \in I} (V_i \cdot g(i) + f(i) \cdot V_i + V_i \cdot V_i) \subseteq f \cdot g + \prod_{i \in I} U_i = f \cdot g + U.$$

It proves that the multiplication operation on the ring Z is continuous.

The proposition is proved.

4.3.15. DEFINITION. Let $\{R_i \mid i \in I\}$ be a family of topological rings, and for any $i \in I$ in R_i is determined an open subring S_i. The topological ring $\prod_{i \in I} (R_i : S_i)$ from the Proposition 4.3.14 we call a *local product* of the family $\{R_i \mid i \in I\}$ of topological rings with determined open subrings S_i.

4.3.16. REMARK. Since the considered topological properties of rings and modules are the topological properties of their additive groups, then for the local products of topological rings the analogs of the statements from Proposition 4.3.10 are fulfilled.

4.3.17. DEFINITION. Let R be a topological ring, M_0 a topological R-module, $\{M_i \mid i \in I\}$ be a family of topological R-modules and for any $i \in I$ is determined a continuous homomorphism $f_i : M_i \to M_0$ of R-modules. Consider the Tychonoff product $M = (\prod_{i \in I} M_i, \tau_T)$. Denote over F the subset in M, consisting of all such elements $g \in M$, that there exists $m_0 \in M_0$ such that $f_i(g(i)) = m_0$ for all $i \in I$. It is clear that F is a submodule in M. The module F endowed with the topology, induced from M, we call a fan product of the family $\{M_i \mid i \in I\}$ with respect to the family $\{f_i : M_i \to M_0 \mid i \in I\}$ of the homomorphisms, and denote it over $\prod_{i \in I}(f_i : M_i \to M_0)$. The restrictions of the canonical projections $\pi_i : M \to M_i$ on F we call the projections of the fan product F to M_i and denote over $\rho_i : F \to M_i$.

Let $i_0 \in I$, then the homomorphism $\rho_0 = f_{i_0} \circ \rho_{i_0} : F \to M_0$ is continuous, since it is a composition of continuous homomorphisms f_{i_0} and ρ_{i_0}. Since $(f_{i_1} \circ \rho_{i_1})(g) = f_{i_1}(g(i_1)) = f_{i_0}(g(i_0)) = (f_{i_0} \circ \rho_{i_0})(g)$ for any $i_1 \in I$ and any $g \in F$, then the definition of the homomorphism ρ_0 does not depend on the choice of the element i_0. We call this homomorphism the canonical homomorphism of the fan product to M_0.

4.3.18. REMARK. Let R be a topological ring, M_0 be a topological R-module, $\{M_i \mid i \in I\}$ be a family of topological R-modules, and for any $i \in I$ is determined a continuous homomorphism $f_i : M_i \to M_0$ of R-modules. Consider the fan product $F = \prod_{i \in I}(f_i : M_i \to M_0)$. It is clear that if $g \in F$, then for any $j \in I$ we have $f_j(g(j)) = \bigcap_{i \in I} f_i(M_i) = N_0 \subseteq M_0$, hence $F \subseteq \prod_{i \in I} f_i^{-1}(N_0)$. Therefore, considering a fan product, we can suppose that all homomorphisms f_j are surjective.

4.3.19. PROPOSITION. *Let R be a topological ring, (M_0, τ_0) be a topological R-module, $(\{M_i, \tau_i\} | i \in I)$ be a family of topological R-modules, where for any $i \in I$ is determined a continuous surjective homomorphism $f_i : (M_i, \tau_i) \to (M_0, \tau_0)$ of R-modules. Let for any $i \in I$ the homomorphism f_i be a isomorphism. Then the fan product $F = \prod_{i \in I}(f_i : M_i \to M_0)$ is topologically isomorphic to the module (M_0, τ) with a topology*

$\tau = \sup\{\tau_i' \,|\, i \in I\}$, *where for any* $i \in I$ *the topology* τ_i' *on the module* M_0 *is the prototype of the topology* τ_i *with respect to the isomorphism* $f_i^{-1} : M_0 \to M_i$.

PROOF. Let $M = (\prod_{i \in I} M_i, \tau_{\scriptscriptstyle T})$ be the Tychonoff product of the family $\{M_i | i \in I\}$. Consider the mapping $f : M_0 \to M$, where $f(m)(i) = f_i^{-1}(m)$ for any $i \in I$ and for any $m \in M_0$. Due to Corollary 4.1.39 and Remark 4.1.54, f is a continuous homomorphism of R–modules. It is clear that $\ker f = \{0\}$. For any $i, j \in I$ and $m \in M_0$ we get:

$$f_i(f(m)(i)) = f_i(f_i^{-1}(m)) = m = f_j(f_j^{-1}(m)) = f_j(f(m)(j)).$$ Therefore, $f(M_0) \subseteq F$. On the other hand, let $p \in F$, then $f_i(p(i)) = f_j(p(j)) \in M_0$ for any $i, j \in I$. Let $f_{i_0}(p(i_0)) = m_0$ for certain $i_0 \in I$ and $m_0 \in M_0$. Then $f(m_0)(i) = f_i^{-1}(f_{i_0}(p_{i_0}))) = f_i^{-1}(f_i(p(i))) = p(i)$ for any $i \in I$. Hence, $f(m_0) = p$. Thus, we have proved that the mapping f is a continuous isomorphism between the modules (M_0, τ) and F.

Since f_i^{-1} is a topological isomorphism between the modules (M_0, τ_i') and (M_i, τ_i) for any $i \in I$, then the module M is topologically isomorphic to the module $N = (\prod_{i \in I} N_i, \tau_{\scriptscriptstyle T})$, where $N_i = (M_0, \tau_i')$ for any $i \in I$, and this isomorphism is realized by the mapping $\varphi : M \to N$, where $\varphi(p)(i) = f_i^{-1}(p(i))$ for any $i \in I$ and $p \in M$. The mapping $\gamma : (M_0, \tau) \to N$, where $(\gamma(m))(i) = m$ for any $i \in I$ and $m \in M_0$, is, according to Proposition 4.1.37 and Remark 4.1.54, the topological isomorphism between the modules (M_0, τ) and $(\gamma(M_0), \tau_{\scriptscriptstyle T}|_{\gamma(M_0)})$. Since $f = \varphi \circ \gamma : M_0 \to F$, then f is a topological isomorphism of the module (M_0, τ) on to F.

Thus, the proposition is proved.

4.3.20. PROPOSITION. *Let R be a topological ring, M_0 a topological R-module, $\{M_i | i \in I\}$ be a family of topological R-modules, and for any $i \in I$ is determined a continuous homomorphism $f_i : M_i \to M_0$. Let $F = \prod_{i \in I}(f_i : M_i \to M_0)$ be the fan product of the family $\{M_i | i \in I\}$. If for any $i \in I$ the homomorphism f_i is an open and surjective mapping, then the canonical homomorphism $p_0 : F \to M_0$ as well as the projections $\pi_i : F \to M_i$, $i \in I$, are open surjective mappings, too.*

PROOF. Let's prove that the homomorphism $p_0 : F \to M_0$ is surjective. Indeed, let $m_0 \in M_0$. Since the homomorphism f_i is surjective for any $i \in I$, then there exists $m_i \in M_i$ such that $f_i(m_i) = m_0$. Then let $g \in \prod_{i \in I} M_i$ be such that $g(j) = m_j$ for $j \in I$. Then $g \in F$ and $p_0(g) = m_0$. Thus, p_0 is a surjective mapping.

Now we prove that the projection $\pi_{i_0} : F \to M_{i_0}$ is surjective mapping for any $i_0 \in I$. Indeed, let $m_{i_0} \in M_{i_0}$, then, due to the proved above, there exists an element $g \in F$ such that $p_0(g) = f_{i_0}(m_{i_0})$. The element $g \in F$ can be chosen so that $g(i_0) = m_{i_0}$. Then $\pi_{i_0}(g) = m_{i_0}$. Therefore, $\pi_{i_0} : F \to M_{i_0}$ is a surjective mapping.

Finally we'll prove that the homomorphisms p_0 and π_i, $i \in I$, are open mappings. Indeed, let U be a neighborhood of zero in F. Fix an arbitrary element $i_0 \in I$. According to the definition of the topology on F and Remark 4.1.4, there exist indexes $i_1, \ldots, i_n \in I$ and neighborhoods U_{i_k} of zero in M_{i_k}, for $k = 0, 1, \ldots, n$, such that $\bigcap_{k=0}^n \pi_{i_k}^{-1}(U_{i_k}) \subseteq U$. Let $V_0 = \bigcap_{k=0}^n f_{i_k}(U_{i_k})$. Since all the homomorphisms f_i, $i \in I$, are open, then V_0 is a neighborhood of zero in M_0. Let $V_{i_0} = U_{i_0} \cap f_{i_0}^{-1}(V_0)$. It is clear that V_{i_0} is a neighborhood of zero in M_{i_0}.

Now we prove that $\pi_{i_0}(U) \supseteq V_{i_0}$ and $p_0(U) \supseteq V_0$. Indeed, let $m_{i_0} \in V_{i_0}$ and $m_0 = f_{i_0}(m_{i_0})$. Since $f_{i_k}(U_{i_k}) \supseteq V_0$, then for any $k = 0, 1, \ldots, n$ in the neighborhood U_{i_k} exists an element m_{i_k} such that $f_{i_k}(m_{i_k}) = m_0$. Since all the homomorphisms f_i, $i \in I$, are surjective, then for any $j \in I \backslash \{i_0, i_1, \ldots, i_n\}$ in the module M exists an element m_j such that $f_j(m_j) = m_0$. Let $g \in \prod_{i \in I} M_i$ be such that $g(j) = m_j$ for any $j \in I$. Then $g \in U \subseteq F$ and $\pi_{i_0}(g) = g(i_0) = m_{i_0}$. Therefore, $\pi_{i_0}(U) \supseteq V_{i_0}$ and

$$p_0(U) = f_{i_0}\left(\pi_{i_0}(U)\right) \supseteq f_{i_0}(V_{i_0}) = f_{i_0}(U_{i_0} \cap f_{i_0}^{-1}(V_0)) = f_{i_0}(U_{i_0}) \cap V_0 = V_0.$$

The proposition is proved.

4.3.21. COROLLARY. *Under the conditions of Proposition 4.3.20, the fan product $F = \prod_{i \in I}(f_i : M_i \to M_0)$ is a subdirect product of the family $\{M_i | i \in I\}$.*

PROOF. According to Proposition 4.3.20, for any $i \in I$ the homomorphism $\pi_i : F \to M_i$ is an open and continuous surjective homomorphism. Since $\bigcap_{i \in I} \ker \pi_i = 0$, then F is a subdirect product of the family $\{M_i | i \in I\}$.

4.3.22. PROPOSITION. *Let R be a topological ring, M_0 and M_i, $i \in I$, be topological R-modules and for any $i \in I$ is determined a continuous homomorphism $f_i : M_i \to M_0$. Let $F = \prod_{i \in I}(f_i : M_i \to M_0)$ be the fan product of the family $\{M_i | i \in I\}$, N be a topological R-module and be defined a family $\{g_i : N \to M_i | i \in I\}$ of continuous homomorphisms such* that $f_{i_1} \circ g_{i_1} = f_{i_0} \circ g_{i_0}$ for any pair of elements $i_0, i_1 \in I$. Then there exists a continuous homomorphism $h : N \to F$ such that $p_i \circ h = g_i$ for any $i \in I$.

PROOF. We define the homomorphism $h : N \to \prod_{i \in I} M_i$ as the following:

for $n \in N$ put $h(n) = t$, where $t(i) = g_i(n)$ for any $i \in I$.

Since $f_{i_1}(t(i_1)) = f_{i_1}(g_{i_1}(n)) = (f_{i_1} \circ g_{i_1})(n) = (f_{i_0} \circ g_{i_0})(n) = f_{i_0}(g_{i_0}(n)) = f_{i_0}(t(i_0))$ for any $i_0, i_1 \in I$, then $h(n) \in F$ for any $n \in N$. Hence, we can consider the homomorphism h as a mapping $h : N \to F$. It is clear that $(p_i \circ h)(n) = g_i(n)$ for any $i \in I$ and any $n \in N$. The continuity of h follows from Corollary 4.1.39.

4.3.23. PROPOSITION. *Let R be a topological ring, M_0 be a topological R-module, $\{M_i | i \in I\}$ be a family of topological R-modules, where for any $i \in I$ is determined a continuous homomorphism $f_i : M_i \to M_0$, and let $F = \prod_{i \in I}(f_i : M_i \to M_0)$ be the fan product of the family $\{M_i | i \in I\}$. Then if the module M_0 is Hausdorff, then F is a closed submodule of the Tychonoff product $M = (\prod_{i \in I} M_i, \tau_T)$.*

PROOF. Assume the contrary, then there exists $g \in [F]_M \backslash F$. Hence, there exist $i_0, i_1 \in I$ such that $m_0 = f_{i_0}(g(i_0)) \neq f_{i_1}(g(i_1)) = m_1$. Since M_0 is a Hausdorff module, then there exists a neighborhood U of zero in M_0 such that $(m_0 + U) \bigcap (m_1 + U) = \emptyset$. Since f_{i_0} and f_{i_1} are continuous homomorphisms, then $f_{i_0}^{-1}(U)$ and $f_{i_1}^{-1}(U)$ are neighborhoods of zero in M_{i_0} and M_{i_1} correspondingly. Since the projections $\pi_{i_0} : M \to M_{i_0}$ and $\pi_{i_1} : M \to M_{i_1}$ are continuous, then $V_0 = \pi_{i_0}^{-1}(f_{i_0}^{-1}(U))$ and $V_1 = \pi_{i_1}^{-1}(f_{i_1}^{-1}(U))$ are neighborhoods of zero in M, hence $V_0 \bigcap V_1$ is a neighborhood of zero in M. Since $g \in [F]_M$, then $(g + (V_0 \bigcap V_1)) \bigcap F \neq \emptyset$. Therefore, there exists $h \in (g + (V_0 \bigcap V_1)) \bigcap F$. Then $f_{i_1}(h(i_1)) \in f_{i_1}(\pi_{i_1}(g + V_1)) \subseteq f_{i_1}(g(i_1)) + f_{i_1}(f_{i_1}^{-1}(U)) = f_{i_1}(g(i_1)) + U = m_1 + U$ and $f_{i_0}(h(i_0)) \in f_{i_0}(\pi_{i_0}(g + V_0)) \subseteq f_{i_0}(g(i_0)) + f_{i_0}(f_{i_0}^{-1}(U)) = f_{i_0}(g(i_0)) + U = m_0 + U$. Since $(m_0 + U) \bigcap (m_1 + U) = \emptyset$, then $f_{i_0}(h(i_0)) \neq f_{i_1}(h(i_1))$, which contradicts the choice of the element $h \in F$. It means that F is a closed submodule in $M = (\prod_{i \in I} M_i, \tau_T)$.

4.3.24. PROPOSITION. *Let R be a topological ring, I be a direction that contains the smallest element i_0 and $S = \{M_\lambda, f_{\lambda,\beta} \,|\, \lambda, \beta \in I\}$ be an inverse spectrum of topological modules. Then $\varprojlim S$ is a submodule of the fan product $F = \prod_{i \in I \backslash \{i_0\}} (f_i : M_i \to M_{i_0})$, where $f_i = f_{i_0,i}$ for any $i \in I \backslash \{i_0\}$.*

PROOF. Let $J = I\backslash\{i_0\}$. Then the elements of the fan product F are exactly such elements $h \in \prod_{j \in J} M_j$ that $f_{i_1}(h(i_1)) = f_{i_2}(h(i_2))$ for any $i_1, i_2 \in J$. Due to Proposition 4.1.36, we can consider that $\prod_{i \in I} M_i = (\prod_{i \in J} M_i) \times M_{i_0}$. Let $\tilde{\pi} : \prod_{i \in I} M_i \to \prod_{i \in J} M_i$ be the natural projection, then $f_{i_1}((\tilde{\pi}(g)(i_1)) = f_{i_0,i_1}(g(i_1)) = g(i_0) = f_{i_0,i_2}(g(i_2)) = f_{i_2}((\tilde{\pi}(g))(i_2))$ for any $g \in \lim\limits_{\leftarrow} S$ and any $i_1, i_2 \in J$, i.e. $\tilde{\pi}(g) \in F$.

Let $\varphi : F \to \prod_{i \in J} M_i$ be the natural embedding and $\rho_0 : F \to M_{i_0}$ be the canonical homomorphism of a fan product (see Definition 4.3.17). Let $\psi : F \to (\prod_{i \in J} M_i) \times M_{i_0}$ be the continuous homomorphism, constructed according to Corollary 4.1.39 for the homomorphisms φ and ρ_0, then $\psi \circ \tilde{\pi}$ is identical on $\lim\limits_{\leftarrow} S$. Then, since $\tilde{\pi}$ and ψ are continuous homomorphisms, then the restriction of the mapping $\tilde{\pi}$ on $\lim\limits_{\leftarrow} S$ is a topological isomorphism of the module $\lim\limits_{\leftarrow} S$ onto $\tilde{\pi}(\lim\limits_{\leftarrow} S)$.

4.3.25. DEFINITION. Let X_0 be a topological Abelian group, $\{X_i | i \in I\}$ a family of topological Abelian groups, and for any $i \in I$ is determined a continuous homomorphism $f_i : X_i \to X_0$. Consider the Abelian groups X_0 and X_i, $i \in I$, as topological $\mathbb{Z}$-modules over the ring $\mathbb{Z}$ of integers with the discrete topology. The fan product $F = \prod_{i \in I}(f_i : X \to X_0)$ of the family $\{X_i | i \in I\}$ of topological $\mathbb{Z}$-modules we call a fan product of the family $\{X_i | i \in I\}$ of topological Abelian groups.

4.3.26. REMARK. Let R_0 be a topological ring, $\{R_i | i \in I\}$ be a family of topological rings and for any $i \in I$ is determined a continuous ring homomorphism $f_i : R_i \to R_0$. Consider the fan product $F = \prod_{i \in I}(f_i : R_i \to R_0)$ of the family $\{R_i(+) | i \in I\}$ of additive topological groups $R_i(+)$. It is easy to see that F is a subring of the topological ring $(\prod_{i \in I} R_i, \tau_T)$. The topological ring F with the topology induced from the $(\prod_{i \in I} R_i, \tau_T)$ we call a fan product of the family $\{R_i | i \in I\}$ of topological rings.

Since the considered topological properties of the ring F are the topological properties of its additive groups, then for a fan product of topological rings the analogs of the statements from 4.3.19 - 4.3.24 are valid.

4.3.27. DEFINITION. Let R be a topological ring, M a topological right R-module. We say that on the module M is defined a $\mathcal{D}$-boundary, if there is defined a family $\mathcal{D}$ of subsets of the module M (that are called $\mathcal{D}$-bounded sets) and:

1) if $A \in \mathcal{D}$, then $-A = \{-a | a \in A\} \in \mathcal{D}$;

2) if $A \in \mathcal{D}$ and $B \subseteq A$, then $B \in \mathcal{D}$;

3) if $A, B \in \mathcal{D}$, then $A \cup B \in D$ and $A + B = \{a + b | a \in A, b \in B\} \in \mathcal{D}$;

4) if $m \in M$, then $\{m\} \in \mathcal{D}$;

5) if $A \in D$ and $r \in R$, then $A \cdot r = \{a \cdot r | a \in A\} \in \mathcal{D}$.

4.3.28. REMARK. The examples of a $\mathcal{D}$-boundary on the module M could be:

the family of all finite subsets of M;

the family of all precompact subsets of M;

the family of all bounded subsets of M (see Definition 1.6.1);

the family of all subsets of M.

Indeed, the fulfillment of the properties 1)–5) for the family of all finite subsets and for the family of all subsets is evident. The fulfillment of these properties for the family of all bounded subsets follows from the statements 1.6.4, 1.6.19, 1.1.42, 1.6.27. The fulfillment of these properties for the family of all precompact subsets follows from Propositions 3.4.5 and 3.4.7 and from the fact that the multiplication by an element of the ring is a continuous homomorphism of the Abelian group $M(+)$ (see 1.1.42).

4.3.29. REMARK. Let X be a topological Abelian group. Consider X as a $\mathbb{Z}$-module over the ring $\mathbb{Z}$ of integers with the discrete topology. Let on $\mathbb{Z}$-module X be defined a family $\mathcal{L}$ of subsets, then for this family to define a $\mathcal{L}$-boundary on X it is enough to fulfill the conditions 1) - 4) of Definition 4.3.27, since in this case the condition 5) follows from the condition 3).

4.3.30. DEFINITION. Let R be a topological ring, M be a topological R-module with $\mathcal{L}$-boundary, I be a set of indexes. Consider $\prod_{i \in I} M_i$, where $M_i = M$ for each $i \in I$. The subset

$$M_{\mathcal{L}}^I = \left\{ f \in \prod_{i \in I} M_i \,\middle|\, \{f(i) | i \in I\} \in \mathcal{L} \right\}$$

we call a $\mathcal{L}$-bounded power of the module M. Due to Definition 4.3.27, $M_{\mathcal{L}}^I$ is a submodule in $\prod_{i \in I} M_i$.

4.3.31. REMARK. Let R be a topological ring, M be a topological R-module and I be a set of indexes. Then:

if $\mathcal{L}$ is a family of all finite subsets of M, then $M_{\mathcal{L}}^I = \sum_{i \in I} M_i$, where $M_i = M$ for each $i \in I$;

if $\mathcal{L}$ is a family of all subsets of M, then $M_{\mathcal{L}}^I = \prod_{i \in I} M_i$, where $M_i = M$ for each $i \in I$.

4.3.32. PROPOSITION. *Let R be a topological ring, M a topological R-module, $\mathcal{L}$ the family of all precompact subsets of M and I be a set of indexes. Let τ be the topology on $\mathcal{L}$-bounded power $M_{\mathcal{L}}^I$ of the module M, induced by the cubic topology τ_c on Abelian group $\prod_{i \in I} M_i$, where $M_i = M$ for each $i \in I$. Then $(M_{\mathcal{L}}^I, \tau)$ is a topological right R-module that is a subdirect product of the family $\{M_i | M_i = M, \ i \in I\}$ of modules.*

PROOF. According to Definition 4.1.68, $(\prod_{i \in I} M_i, \ \tau_c)$ is a topological Abelian group, hence, $M_{\mathcal{L}}^I$ is a topological Abelian group in the topology τ.

Now we prove that the multiplication operation by elements of the ring R is continuous on $(M_{\mathcal{L}}^I, \tau)$. Indeed, let U be a neighborhood of zero in $(M_{\mathcal{L}}^I, \tau)$. Without loss of generality we can consider that $U = (\prod_{i \in I} U_i) \cap M_{\mathcal{L}}^I$, where $U_i = U_0$ for a certain neighborhood U_0 of zero in M and each $i \in I$. There exist a neighborhood U_1 of zero in M such that $U_1 + U_1 + U_1 \subseteq U_0$. Let $f \in M_{\mathcal{L}}^I$ and $r \in R$. Since M is a topological right R-module, then there exists a neighborhood V_0 of zero in M and a neighborhood W_0 of zero in R such that $V_0 \cdot r \subseteq U_1$ and $V_0 \cdot W_0 \subseteq U_1$. Let $\widehat{U} = \prod_{i \in I} \widehat{U}_i$, where $\widehat{U}_i = U_1$ for any $i \in I$. Due to Proposition 3.4.24, the precompact set $\{f(i) | i \in I\}$ is bounded in M, hence, there exists a neighborhood W_1 of zero in R such that $\{f(i) | i \in I\} \cdot W_1 \subseteq U_1$. Let $W_2 = W_0 \cap W_1$, then W_2 is a neighborhood of zero in R. Let $\widehat{V} = \prod_{i \in I} V_i$, where $V_i = V_0$ for any $i \in I$. According to the definition of the topology τ, $V = \widehat{V} \cap M_{\mathcal{L}}^I$ is a neighborhood of zero in $(M_{\mathcal{L}}^I, \tau)$. Then,

$$(f + V) \cdot (r + W_2) \subseteq f \cdot r + V \cdot r + f \cdot W_2 + V \cdot W_2 \subseteq f \cdot r$$
$$+ \left(\left(\prod_{i \in I} (V_i \cdot r) \right) \cap M_{\mathcal{L}}^I \right) + \left(\left(\prod_{i \in I} (f(i) \cdot W_2) \right) \cap M_{\mathcal{L}}^I \right) + \left(\left(\prod_{i \in I} (V_i \cdot W_2) \right) \cap M_{\mathcal{L}}^I \right)$$
$$\subseteq f \cdot r + \left((\prod_{i \in I} \widehat{U}_i) \cap M_{\mathcal{L}}^I \right) + \left((\prod_{i \in I} \widehat{U}_i) \cap M_{\mathcal{L}}^I \right) + \left((\prod_{i \in I} \widehat{U}_i) \cap M_{\mathcal{L}}^I \right)$$

$$\subseteq f \cdot r + \left(\left(\prod_{i \in I} (\widehat{U}_i + \widehat{U}_i + \widehat{U}_i) \right) \cap M_{\mathcal{L}}^I \right) \subseteq f \cdot r + \left(\left(\prod_{i \in I} U_i \right) \cap M_{\mathcal{L}}^I \right) = f \cdot r + U.$$

Thus, the multiplication operation on $M_{\mathcal{L}}^I$ is continuous. Hence, we have proved that $M_{\mathcal{L}}^I$ is a topological right R-module.

Due to Remark 4.1.69 and Proposition 4.1.2, the restrictions on the subgroup $M_{\mathcal{L}}^I$ of the natural projections $\pi_j : \prod_{i \in I} M_i \to M_j$ of $(\prod_{i \in I} M_i, \tau_c)$ onto M_j are continuous homomorphisms of Abelian groups for all $j \in I$. It is clear that for any $i \in I$ the homomorphism $q_i = \pi_i|_{M_{\mathcal{L}}^I} : M_{\mathcal{L}}^I \to M_i = M$ is a homomorphism of R-modules.

Now we prove that for any $i \in I$ the mapping $q_i : M_{\mathcal{L}}^I \to M$ is surjective and open. Indeed, for any $i_0 \in I$, the subgroup $Y_{i_0} = \{ f \in \prod_{i \in I} M_i \,|\, f(i) = 0 \text{ for any } i \in I \backslash \{i_0\} \}$ of the topological group $(\prod_{i \in I} M_i, \tau_c)$ is a subgroup of $M_{\mathcal{L}}^I$. Due to Proposition 4.1.5, the restriction of the canonical projection π_{i_0} on the subgroup Y_{i_0} is an open and surjective mapping. Therefore, the homomorphism $q_{i_0} : M_{\mathcal{L}}^I \to M$ is an open mapping for any $i_0 \in I$. Since $\bigcap_{i \in I} \ker q_i \subseteq \bigcap_{i \in I} \ker \pi_i = \{0\}$, then $M_{\mathcal{L}}^I$ is a subdirect product of the family $\{ M_i \,|\, M_i = M, \ i \in I \}$ of modules. Thus, the proposition is proved.

4.3.33. DEFINITION. Let R be a topological ring, M a topological R-module $\mathcal{L}$ the family of all precompact sets in M and I be a set of indexes. The topological R-module $M_{\mathcal{L}}^I$ that has a topology induced from $(\prod_{i \in I} M_i, \tau_c)$, where $M_i = M$ for each $i \in I$, is called a $\mathcal{L}$-bounded power of the module M.

4.3.34. DEFINITION. Let R be a topological ring. We say that on the ring R is defined a $\mathcal{L}$-boundary, if there is defined a family $\mathcal{L}$ of its subsets that defines $\mathcal{L}$-boundary on the topological Abelian group $R(+)$ (i.e. satisfies the conditions 1) - 4) of Definition 4.3.27) and satisfies also the condition:

6) if $A, B \in \mathcal{L}$, then $A \cdot B = \{ a \cdot b \,|\, a \in A, b \in B \} \in \mathcal{L}$.

4.3.35. REMARK. The examples of $\mathcal{L}$-boundaries on a topological ring R could be ones enumerated in Remark 4.3.28.

Indeed, the fulfillment of the properties 1)–4) and 6) for for the family of all finite subsets and the family of all subsets is evident. The fulfillment of these properties for the family of all bounded subsets follows from Corollary 1.6.5, Proposition 1.6.19 and Corollary 1.6.22.

The fulfillment of the properties 1)–4) for the family of all precompact subsets follows from Remark 4.3.28, and the fulfillment of the property 6) from Propositions 4.1.52 and 3.4.7 and the fact that the multiplication operation is a continuous homomorphism from the Tychonoff product $R \times R$ onto R (see Definition 1.1.6).

4.3.36. DEFINITION. Let R be a topological ring with $\mathcal{L}$-boundary and I be a set of indexes. The subset $R_{\mathcal{L}}^I = \{f \in \prod_{i \in I} R_i \mid \{f(i) | i \in I\} \in \mathcal{L}\}$, where $R_i = R$ for each $i \in I$, we call a $\mathcal{L}$-bounded power of the ring R. Due to Definition 4.3.34, $R_{\mathcal{L}}^I$ is a subring of $\prod_{i \in I} R_i$.

4.3.37. PROPOSITION. *Let R be a topological ring, $\mathcal{L}$ be the family of all precompact sets in R and I be a set of indexes. Let τ be the topology on the $\mathcal{L}$-bounded power $R_{\mathcal{L}}^I$ of the ring R, induced by the cubic topology τ_c on the group $\prod_{i \in I} R_i$, where $R_i = R$ for each $i \in I$. Then $(R_{\mathcal{L}}^I, \tau)$ is a topological ring that is a subdirect product of the family $\{R_i | R_i = R, \ i \in I\}$ of rings.*

PROOF. According to Definition 4.1.68, $(\prod_{i \in I} R_i, \ \tau_c)$ is a topological Abelian group, hence $R_{\mathcal{L}}^I$ is a topological Abelian group in the topology τ.

Now we prove that the multiplication operation is continuous on $(R_{\mathcal{L}}^I, \tau)$. Indeed, let U be a neighborhood of zero in $(R_{\mathcal{L}}^I, \tau)$. Without loss of generality we can consider that $U = (\prod_{i \in I} U_i) \cap R_{\mathcal{L}}^I$, where $U_i = U_0$ for a certain neighborhood U_0 of zero in R and each $i \in I$. There exists a neighborhood U_1 of zero in R such that $U_1 + U_1 + U_1 \subseteq U_0$. Let $\widehat{U} = \prod_{i \in I} \widehat{U}_i$, where $\widehat{U}_i = U_1$ for any $i \in I$. Since R is a topological ring, then there exists a neighborhood V_1 of zero in R such that $V_1 \cdot V_1 \subseteq U_1$. Let $r, s \in R_{\mathcal{L}}^I$. Since the precompact sets $\{r(i) | i \in I\}$ and $\{s(i) | i \in I\}$ are bounded in R, (see Proposition 3.4.24) then there exist neighborhoods V_2 and V_3 of zero in R such that $\{r(i) | i \in I\} \cdot V_2 \subseteq U_1$ and $V_3 \cdot \{s(i) | i \in I\} \subseteq U_1$. Let $V_0 = V_1 \cap V_2 \cap V_3$, then V_0 is a neighborhood of zero in R. Let $V = (\prod_{i \in I} V_i) \cap R_{\mathcal{L}}^I$, where $V_i = V_0$ for any $i \in I$. Then V is a neighborhood of zero in $R_{\mathcal{L}}^I$ and

$$(r + V) \cdot (s + V) \subseteq r \cdot s + V \cdot s + r \cdot V + V \cdot V \subseteq r \cdot s$$
$$+ \left(\left(\prod_{i \in I} (V_i \cdot s(i)) \right) \cap R_{\mathcal{L}}^I \right) + \left(\left(\prod_{i \in I} (r(i) \cdot V_i) \right) \cap R_{\mathcal{L}}^I \right)$$

$$+\left(\left(\prod_{i\in I}(V_i\cdot V_i)\right)\cap R_{\mathcal{L}}^I\right)\subseteq r\cdot s+\left(\left(\prod_{i\in I}\widehat{U}_i+\prod_{i\in I}\widehat{U}_i+\prod_{i\in I}\widehat{U}_i\right)\cap R_{\mathcal{L}}^I\right)$$

$$\subseteq r\cdot s+\left(\left(\prod_{i\in I}U_i\right)\cap R_{\mathcal{L}}^I\right)=r\cdot s+U.$$

Hence we've proved that the multiplication operation on $R_{\mathcal{L}}^I$ is continuous. Thus, $R_{\mathcal{L}}^I$ is a topological ring.

The proof of the statement that $R_{\mathcal{L}}^I$ is a subdirect product of the family $\{R_i|R_i=R,\ i\in I\}$ of topological rings is realized analogously to the proof of the corresponding statement of Proposition 4.3.32. Thus, the proposition is proved.

4.3.38. DEFINITION. Let R be a topological ring, $\mathcal{L}$ be the family of all precompact sets in R and I be a set of indexes. The topological ring $R_{\mathcal{L}}^I$, which has a topology induced from $(\prod_{i\in I}R_i,\ \tau_c)$, where $R_i=R$ for each $i\in I$, we call a $\mathcal{L}$–bounded power of the ring R.

4.3.39. PROPOSITION. *Let R be a topological ring, M a topological R-module and I be a set of indexes that is a topological space in a certain topology τ_0. The set of all mappings from I to M is naturally identified with the set $\prod_{i\in I}M_i$, where $M_i=M$ for any $i\in I$. Consider $\prod_{i\in I}M_i$ as a Tychonoff product of the family $\{M_i|M_i=M,\ i\in I\}$ of topological modules. Then any submodule F of the topological module $\left(\prod_{i\in I}M_i,\ \tau_\tau\right)$, where F contains all the constant mappings from I to M (for instance, the submodule of all continuous mappings), is a subdirect product of the family $\{M_i|M_i=M,\ i\in I\}$ of topological modules.*

PROOF. For any $i\in I$ let $\gamma_i=\pi_i|_F:F\to M_i$ be a restriction on F of the canonical projection $\pi_i:\prod_{j\in I}M_j\to M_i=M$.

Let's prove that the homomorphism γ_i is surjective for any $i\in I$. Indeed, let $m_0\in M$ and g be the element of $\prod_{i\in I}M_i$ such that $g(i)=m_0$ for any $i\in I$. Then $g\in F$ and $\gamma_j(g)=\pi_j(g)=m_0$ for any $j\in I$, i.e. the homomorphism γ_j is surjective.

Besides, $\bigcap_{i\in I}\ker\gamma_i\subseteq\bigcap_{i\in I}\ker\pi_i=\{0\}$. Since for any $i\in I$ the homomorphism π_i is continuous, then γ_i is a continuous homomorphism for any $i\in I$.

Now we prove that for any $i_0\in I$ the homomorphism γ_{i_0} is open. Indeed, let U be a neighborhood of zero in F. Due to Remark 4.1.4 and the definition of the topology on

F, we can consider that $U = \left(\bigcap_{k=1}^{n} \pi_{i_k}^{-1}(U_{i_k})\right) \bigcap F$, where U_{i_k} is a neighborhood of zero in $M_{i_k} = M$, for a certain finite subset $\{i_1, \ldots i_n\} \subseteq I$. Let $V = \bigcap_{k=1}^{n} U_{i_k}$, then V is a neighborhood of zero in M. Let $m_0 \in V$ and $g \in \prod_{i \in I} M_i$ be such that $g(i) = m_0$ for any $i \in I$. Then $g \in F$ and $g \in \pi_{i_k}^{-1}(U_{i_k})$ for any $k = 1, \ldots, n$. Hence, $g \in U$ and $\gamma_{i_0}(g) = g(i_0) = m_0$. It means that $\gamma_{i_0}(U) \supseteq V$, i.e. $\gamma_{i_0}(U)$ is a neighborhood of zero in M. Hence, γ_{i_0} is an open homomorphism for any $i_0 \in I$. Therefore, the family $\{\gamma_i | i \in I\}$ of the homomorphisms satisfies Definition 4.3.1.

Thus, the proposition is proved.

§ 4.4. Semidirect Products

4.4.1. PROPOSITION. *Let R and S be topological rings and on S be defined a multiplication operation of elements of S by elements of R from the left and the right such that the group $S(+)$ becomes a right and a left topological R-module. Let Q be the direct sum of the topological groups $R(+)$ and $S(+)$, then:*

1) Q transforms into a topological ring (possibly, non-associative), if we define a multiplication operation on it in the following way:

$(s_1, r_1) \cdot (s_2, r_2) = (s_1 \cdot s_2 + r_1 \cdot s_2 + s_1 \cdot r_2, \; r_1 \cdot r_2)$, *for all $s_1, s_2 \in S$ and $r_1, r_2 \in R$;*

2) if rings R and S are associative and $(r_1 \cdot s) \cdot r_2 = r_1 \cdot (s \cdot r_2)$ and $(r \cdot s_1) \cdot s_2 = r \cdot (s_1 \cdot s_2)$ for any $r, r_1, r_2 \in R$ and $s, s_1, s_2 \in S$, then the ring Q is associative, too.

PROOF. Since the ring S is a right and left R-module, then Q is a ring with respect to the operation "+" and the introduced operation " $\cdot$ ". Since $Q(+)$ is a topological group, then to complete the proof it is enough to verify that the operation " $\cdot$ ", introduced on Q, is continuous.

Let $(s_1, r_1), (s_2, r_2) \in Q$ and W be a neighborhood of the element $(s_1, r_1) \cdot (s_2, r_2)$. Since $Q(+)$ is a topological group, then there exists a neighborhood W_0 of zero in Q such that $(s_1, r_1) \cdot (s_2, r_2) + W_0 \subseteq W$, and $W_0 = V_0 \times U_0$, where U_0 and V_0 are certain neighborhoods of zero in R and S correspondingly. Since R is a topological ring, then there exists a neighborhood U_1 of zero in R such that $r_1 \cdot U_1 + U_1 \cdot r_2 + U_1 \cdot U_1 \subseteq U_0$. Let V_1 be a neighborhood of zero in S such that $V_1 + V_1 + V_1 \subseteq V_0$. Since S is a topological ring, then there exists a neighborhood V_2 of zero in S such that $V_2 \cdot V_2 + s_1 \cdot V_2 + V_2 \cdot s_2 \subseteq V_1$. Since S is a left and right topological R-module, then there exist neighborhoods V_3 and U_2 of zero in S and R correspondingly such that $V_3 \cdot U_2 + V_3 \cdot r_2 + s_1 \cdot U_2 \subseteq V_1$ and $U_2 \cdot V_3 + r_1 \cdot V_3 + U_2 \cdot s_2 \subseteq V_1$. Let $V = V_1 \cap V_2 \cap V_3$ and $U = U_1 \cap U_2$, then $W' = U \times V$ is a neighborhood of zero in Q and, hence, $(s_1, r_1) + W'$ and $(s_2, r_2) + W'$ are neighborhoods of the elements (s_1, r_1) and (s_2, r_2) in Q correspondingly. Moreover, for any two elements

(s, r), $(s', r') \in W'$ we get

$$((s_1, r_1) + (s, r)) \cdot ((s_2, r_2) + (s', r')) = (s_1 + s, r_1 + r) \cdot (s_2 + s', r_2 + r')$$

$$= (s_1 \cdot s_2 + s_1 \cdot s' + s \cdot s_2 + s \cdot s' + r_1 \cdot s_2 + r_1 \cdot s' + r \cdot s_2 + r \cdot s'$$

$$+ s_1 \cdot r_2 + s_1 \cdot r' + s \cdot r_2 + s \cdot r', \; r_1 \cdot r_2 + r_1 \cdot r' + r \cdot r_2 + r \cdot r').$$

Also:

$$r_1 \cdot r' + r \cdot r_2 + r \cdot r' \in r_1 \cdot U_1 + U_1 \cdot r_2 + U_1 \cdot U_1 \subseteq U_0;$$

$$s_1 \cdot s' + s \cdot s_2 + s \cdot s' \in s_1 \cdot V_2 + V_2 \cdot s_2 + V_2 \cdot V_2 \subseteq V_1;$$

$$r_1 \cdot s' + r \cdot s_2 + r \cdot s' \in r_1 \cdot V_3 + U_2 \cdot s_2 + U_2 \cdot V_3 \subseteq V_1;$$

$$s_1 \cdot r' + s \cdot r_2 + s \cdot r' \in s_1 \cdot U_2 + V_3 \cdot r_2 + V_3 \cdot U_2 \subseteq V_1 .$$

Then

$$((s_1, r_1) + (s, r)) \cdot ((s_2, r_2) + (s', r'))$$

$$\in (r_1 \cdot s_2 + s_1 \cdot r_2 + s_1 \cdot s_2, r_1 \cdot r_2) + (V_1 + V_1 + V_1) \cdot U_0$$

$$\subseteq (s_1, r_1) \cdot (s_2, r_2) + (V_0 \cdot U_0) \subseteq W.$$

Thus, $((s_1, r_1) + W') \cdot ((s_2, r_2) + W') \subseteq W$. Hence, 1) is proved.

It is easy to see, that if rings R and S are associative and $(r_1 \cdot s) \cdot r_2 = r_1 \cdot (s \cdot r_2)$, $r \cdot (s_1 \cdot s_2) = (r \cdot s_1) \cdot s_2$ for any $r, r_1, r_2 \in R$ and $s, s_1, s_2 \in S$, then the ring Q is associative. The proof is completed.

4.4.2. DEFINITION. Let R and S be topological rings from Proposition 4.4.1, then the topological ring Q, constructed there, is called a semidirect product of the topological rings S and R, and is denoted over $S \rtimes R$.

4.4.3. REMARK. Let R and S be topological rings. We define the multiplication operation of the elements of S by the elements of R as the following:

$r \cdot s = s \cdot r = 0$ for any $r \in R$ and $s \in S$.

It is clear that with respect to this operation S is a right and a left topological R-module. Then for the rings R and S can be considered a semidirect product $S \rtimes R$. Since

$$(s_1, r_1) \cdot (s_2, r_2) = (s_1 \cdot s_2 + r_1 \cdot s_2 + s_1 \cdot r_2, r_1 \cdot r_2) = (s_1 \cdot s_2, r_1 \cdot r_2)$$

for any (s_1, r_1), $(s_2, r_2) \in S \times R$, then the ring $S \times R$ is topologically isomorphic to the direct product $S \times R$ of the topological rings R and S. Thus, the direct product of the topological rings R and S can be considered as a particular case of a semidirect product.

4.4.4. EXAMPLE. Let $\mathbb{Z}$ be a ring of integers with the discrete topology, S be a topological ring. The group $S(+)$ is a topological right and left $\mathbb{Z}$-module with respect to the following natural multiplication operation of elements of S by the elements of $\mathbb{Z}$:

$$
s \cdot n = n \cdot s = \begin{cases} \underbrace{s + \ldots + s}_{n \text{ times}}, & \text{if } n > 0; \\ 0, & \text{if } n = 0; \\ \underbrace{(-s) + \ldots + (-s)}_{|n| \text{ times}}, & \text{if } n < 0. \end{cases}
$$

Then the corresponding semidirect product $S \times \mathbb{Z}$ is a topological ring. If S is an associative ring, then $S \times \mathbb{Z}$ is an associative ring, since $(n_1 \cdot s) \cdot n_2 = n_1 \cdot (s \cdot n_2)$ and $n \cdot (s_1 \cdot s_2) = (n \cdot s_1) \cdot s_2$ for any $n, n_1, n_2 \in \mathbb{Z}$ and $s, s_1, s_2 \in S$.

4.4.5. EXAMPLE. Let R be a topological ring. Since R can be in the natural way considered as a left and right topological R-module, then the semidirect product $R \times R$ is a topological ring. Since for the associative ring R we have that $(r_1 \cdot r_2) \cdot r_3 = r_1 \cdot (r_2 \cdot r_3)$ for any $r_1, r_2, r_3 \in R$, then $R \times R$ is an associative ring.

4.4.6. PROPOSITION. *Let $S \times R$ be a semidirect product of topological rings S and R, then the subrings $S' = \{(s, 0) \mid s \in S\}$ and $R' = \{(0, r) \mid r \in R\}$ of the topological ring $S \times R$ are isomorphic to the topological rings S and R correspondingly.*

PROOF. Let $\pi_1 : S \times R \to S$ and $\pi_2 : S \times R \to R$ be the canonical projections, then it is easy to verify that their restrictions on S' and R' are topological isomorphisms.

4.4.7. PROPOSITION. *Let $S \times R$ be a semidirect product of topological rings S and R, then $S' = \{(s, 0) \mid s \in S\}$ is an ideal in $S \times R$ and the factor–ring $(S \times R)/S'$ is isomorphic to the topological ring R.*

PROOF. From the definition of the operation " $\cdot$ " on $S \times R$ it follows that S' is a two-sided ideal in $S \times R$. Let $\pi_2 : S \times R \to R$ be the natural projection, then π_2 is an open, continuous and surjective homomorphism of Abelian groups, besides, $\ker \pi_2 = S'$.

From the definition of the operation " · " on $S \lambda R$ follows that π_2 is a ring homomorphism. Then the ring R is topologically isomorphic to the ring $(S \lambda R)/S'$.

4.4.8. LEMMA. *Let* $\mathbb{Z}$ *be the ring of integers with a topology, where all non-zero ideals form a basis of neighborhoods of zero, and* X *be a precompact topological Abelian group with a basis of neighborhoods of zero consisting of subgroups. Then with respect to the multiplication operations of elements of* X *by elements of* $\mathbb{Z}$, *defined in Remark 4.4.4,* X *is a right and left topological* $\mathbb{Z}$-*module.*

PROOF. Since X is a topological Abelian group and X is a right and left $\mathbb{Z}$-module, then it is enough to verify that the multiplication operation of elements of X by elements of $\mathbb{Z}$ is continuous.

Let $n \in \mathbb{Z}$ and $x \in X$, and W be a neighborhood of the element $n \cdot x$ in X. Then there exists an open subgroup U in X such that $n \cdot x + U \subseteq W$. Since, according to Proposition 3.4.5, the subgroup $< x >$, generated by the element x, is precompact in the topological group X, then it is either finite or non-discrete. Hence, in both cases there exists a natural number m such that $m \cdot x \in U$. Then $V = m \cdot \mathbb{Z}$ is a neighborhood of zero in $\mathbb{Z}$, and
$$(n+V) \cdot (x+U) \subseteq n \cdot x + V \cdot x + n \cdot U + V \cdot U \subseteq n \cdot x + \mathbb{Z} \cdot (m \cdot x) + \mathbb{Z} \cdot U + \mathbb{Z} \cdot U \subseteq n \cdot x + U,$$
since U is a subgroup. Therefore, X is a topological left $\mathbb{Z}$-module.

Analogously we verify that $(x + U) \cdot (n + V) \subseteq x \cdot n + U$, i.e. that X is a topological right $\mathbb{Z}$-module.

The proof is completed.

4.4.9. THEOREM. *A Hausdorff topological ring* R *is embeddable into a compact topological ring* Q *with the unitary element if and only if* R *is precompact and it has a basis of neighborhoods of zero consisting of subgroups of its additive group.*

PROOF. Let the ring R be embedded into a compact ring Q with the unitary element. Since the ring Q has a basis of neighborhoods of zero consisting of subgroups (see Corollaries 1.6.68 and 1.4.47), then the ring R, as a subring of compact topological ring Q, is a precompact ring and has a basis of neighborhoods of zero consisting of subgroups.

Now let R be a precompact ring with a basis of neighborhoods of zero consisting of subgroups. Let $\mathbb{Z}$ be the ring of integers with the topology, where all non-zero ideals form a basis of neighborhoods of zero. Then, according to Lemma 4.4.8, R is a right and left topological $\mathbb{Z}$-module. Hence, there exists a semidirect product $R \rtimes \mathbb{Z}$ that is an associative ring. It is clear that the element (0,1) is the unitary element of the ring $R \rtimes \mathbb{Z}$.

Since the topological groups $R(+)$ and $\mathbb{Z}(+)$ are Hausdorff and precompact, then the additive group of the ring $R \rtimes \mathbb{Z}$ is a Hausdorff and precompact group (see Corollary 4.1.44 and Proposition 4.1.52).

Let Q be the completion of the topological ring $R \rtimes \mathbb{Z}$, then Q is a compact ring. Due to Proposition 4.4.6, its subring $R' = \{(r,0) \,|\, r \in R\}$ is isomorphic to the topological ring R. Identifying the ring R' with the ring R, we get that R is a subring of the compact ring Q. Since in the ring $R \rtimes \mathbb{Z}$ the element (0,1) is unitary, then, as the ring Q is Hausdorff and the ring $R \rtimes \mathbb{Z}$ is dense in Q, then the element (0,1) is unitary in the ring Q, too. The theorem is proved.

4.4.10. COROLLARY. A Hausdorff compact ring R is embeddable into a compact ring Q with the unitary element if and only if R is a totally disconnected ring.

5. Non-discrete Topologizations of Rings and Modules

One of the first arising problems in the theory of topological groups, rings, and modules is the problem of endowing groups, rings, and modules with topologies satisfying certain conditions. This chapter is about methods of constructing such topologies.

The proof of some results included in this chapter could also be found in [26, 27, 29-34, 132, 182, 184-186, 214, 215, 509] and several closed results in [2, 76, 101, 102, 112, 123, 130, 142, 162-165, 173-175, 189-192, 218-226, 228, 265-270, 276, 288, 294, 298, 310].

The question of the existence of non-discrete topologies (or the topologies satifying certain conditions) on other algebraic systems is considered in [4, 73, 98, 117, 167, 188, 262, 273, 314, 317, 345, 348, 357, 389, 390, 396-399, 505, 514-516]. The question of estimation of the number of different topologies is considered in [55, 74, 92, 171, 172, 197, 198, 217, 227, 267, 268, 284, 290, 306, 323, 325, 340-343, 353, 410-412, 419, 420, 425, 430, 431].

§ 5.1. Non-Discrete Topologizability of Infinite Modules

5.1.1. LEMMA. *Let G be an infinite Abelian group, then there exists a non-discrete Hausdorff group topology τ such that (G, τ) is a precompact group.*

PROOF. Let $\{\xi_\alpha \mid \alpha \in A\}$ be the set of all homomorphisms of the group G into the group $\mathbb{R}/\mathbb{Z}$.

Let τ_0 be the topology on the group $\mathbb{R}/\mathbb{Z}$ induced by the interval topology on the ring R. Then $(\mathbb{R}/\mathbb{Z}, \tau_0)$ is a Hausdorff compact group (see Example 1.5.39). For any $\alpha \in A$ let τ_α be the prototype on G of the topology τ_0 with respect to the homomorphism ξ_α. According to Proposition to 3.4.9, (G, τ_α) is a precompact group. Then, following Proposition 3.4.11, $\tau = \sup\{\tau_\alpha \mid \alpha \in A\}$ is a precompact group topology on G. Since G is an infinite group, then (G, τ) cannot be discrete.

Let's verify that τ is a Hausdorff topology on G. Let a be an arbitrary non-zero element in G. According to the theorem from B.11, there exists $\alpha_0 \in A$ such that $\xi_{\alpha_0}(a) \neq 0$. Since

$(\mathbb{R}/\mathbb{Z}, \tau_0)$ is a Hausdorff group, then there exists a neighborhood V of zero in $(\mathbb{R}/\mathbb{Z}, \tau_0)$ such that $\xi_{\alpha_0}(a) \notin V$. Then $\xi_{\alpha_0}^{-1}(V)$ is a neighborhood of zero in (G, τ_{α_0}). Thus, $\xi_{\alpha_0}^{-1}(V)$ is a neghborhood of zero in (G, τ) and $a \notin \xi_{\alpha_0}^{-1}(V)$.

Hence, we've proved that the group (G, τ) is Hausdorff.

5.1.2. THEOREM. *Let (R, τ_0) be a ring with the discrete topology, M be an infinite R-module. Then there exists a non-discrete Hausdorff topology τ on M such that (M, τ) is a topological (R, τ_0)-module.*

PROOF. Let $\{\tau_\gamma \mid \gamma \in \Gamma\}$ be the set of all (not necessarily Hausdorff) precompact group topologies on the group M and $\tau = \sup\{\tau_\gamma \mid \gamma \in \Gamma\}$. Then, according to Proposition 3.4.11, (M, τ) is a precompact group. Since M is an infinite group, then τ is a non-discrete topology. From Lemma 5.1.1 it follows that τ is a Hausdorff topology as far as there are Hausdorff topologies among τ_γ.

To complete the proof, we must verify that (M, τ) is a topological (R, τ_0)-module, i.e., according to Theorem 1.2.6, it remains to verify that the family of all neighborhoods of zero in (M, τ) satisfies the conditions (BN1)–(BN4), (BN5'), (BN6') and (BN6'').

Since (M, τ) is a topological group, then the conditions (BN1)–(BN4) are fulfilled.

The fulfillment of the conditions (BN5') and (BN6'') follows from the discreteness of the topology τ_0 on the ring R.

Let now $r \in R$ and U be a neighborhood of zero in (M, τ). Consider the mapping $\xi : M \longrightarrow M$ acting by the rule: $\xi(m) = r \cdot m$ for every $m \in M$. It is clear that ξ is an endomorphism of the group M. Let τ' be the prototype of the topology τ with respect to the homomorphism ξ, then (M, τ') is a precompact group. From the definition of the topology τ it follows that $\tau \geq \tau'$. Then $\xi^{-1}(U)$ is a neighborhood of zero in (M, τ'), hence, is the same in (M, τ), and $r \cdot \xi^{-1}(U) = \xi(\xi^{-1}(U)) = U$. Thus, it is verified that the condition (BN6') is fulfilled.

The proof of the theorem is completed.

The following example shows that there exists a topological ring (R, τ_0) and an R-module such that the latter cannot be considered as a topological (R, τ_0)–module in any Hausdorff topology.

5.1.3. EXAMPLE. Let R be the ring of polynomials of argument x over the field of rational numbers and M be a one-dimensional left vector space over the field of rational numbers.

For each positive integer number n denote over V_n the ideal in R generated by the element x^n. It is easy to verify that the family $\{V_n \mid n = 1, 2, \dots\}$ is a basis of neighborhoods of zero of a certain topology τ_0 that performs R into a topological ring (R, τ_0).

Fix some non-zero element $z \in M$, then $M = \{r \cdot z \mid r$ is a rational number$\}$.

Put $\left(\sum_{i=0}^k r_i \cdot x^i\right) \cdot (r \cdot z) = \sum_{i=0}^k (r_i \cdot r) \cdot z$ for any $\sum_{i=0}^k r_i \cdot x^i \in R$ and any element $r \cdot z \in M$. Then M becomes an left R–module.

Let's show that M cannot be considered as a topological (R, τ_0)-module in any Hausdorff topology.

Assume the contrary, i.e. that M is a topological (R, τ_0)-module in some Hausdorff topology τ. Then there exists a neighborhood U of zero in (M, τ) such that $z \notin U$ and a neighborhood V_n of zero in (R, τ_0) such that $V_n \cdot z \subseteq U$. Then $z = x^n \cdot z \in V_n \cdot z \subseteq U$ (as $x \cdot z = z$). We have obtaned a contradiction. Thus, (M, τ) is not a topological (R, τ_0)-module.

§ 5.2 Additional Properties of Polynomials over Rings

5.2.1. DEFINITION. Let R be a ring and x an argument. Over $R[x]$ we denote a free ring generated by the set $R \bigcup \{x\}$, hence the elements of the ring $R[x]$ are the finite sums of associative words composed from the elements of R as well as x.

The ring $R[x]$ is called a ring of generalized polynomials of x over R, and its elements generalized polynomials*.

If R is a commutative ring, then over $\overline{R[x]}$ we denote a free commutative ring generated by the set $R \bigcup \{x\}$.

The elements of the ring $\overline{R[x]}$ are finite sums of the associative-commutative words composed from the elements of R as well as x. The ring $\overline{R[x]}$ is called a ring of polynomials, and its elements polynomials of x over R**.

5.2.2. DEFINITION. A degree of the word composed from elements of R and the argument x is the number of the occurrences of x in this word.

A degree of a polynomial (generalized polynomial) of x over the ring R is the maximal degree of the words its summands.

5.2.3. DEFINITION. For any element $a \in R$ consider the mappings

$\varphi_a : R \bigcup \{x\} \longrightarrow R$ and $\psi_a : R \bigcup \{x\} \longrightarrow R[x]$ acting as follows:

$\varphi_a(b) = \psi_a(b) = b$ for any $b \in R$;

$\varphi_a(x) = a$ and $\psi_a(x) = a + x$.

Since $R[x]$ is a free ring generated by the set $R \bigcup \{x\}$, then the mappings φ_a and ψ_a can be extended to the homomorphic mappings of the ring $R[x]$ onto the rings R and $R[x]$,

---

* In the literature on algebra another entity is also considered as a generalized polynomial. However, it is easy to show that there exists a homomorphism ξ of $R[x]$ onto the ring of generalized polynomials in that terminology such that $f(a) = (\xi(f))(a)$ for any $f(x) \in R[x]$ and $a \in R$ (for the value of $f(a)$, see Definition 5.2.3). Taking all these into account, we decide to name the elements of $R[x]$ generalized polynomials.

** As in the case of generalized polynomials, over polynomials another object is denoted too; we, however, will name elements of $\overline{R[x]}$ polynomials.

respectively. We also denote these homomorphisms over φ_a and ψ_a.

Similarly, in the case when R is a commutative ring, the mappings φ_a and ψ_a can be extended to the homomorpic mappings of the ring $\overline{R[x]}$ onto the rings R and $\overline{R[x]}$. These homomorphisms will be also denoted by φ_a and ψ_a.

Now let $f(x)$ be a generalized polynomial (polynomial) over the ring (commutative ring) R and a be a certain element of R. Put $f(a) = \varphi_a(f(x))$ and $f(a + x) = \psi_a(f(x))$.

An element $a \in R$ is called a root of generalized polynomial (polynomial) $f(x)$ if $f(a) = 0$.

For any $f(x) \in R[x]$ $(f(x) \in \overline{R[x]})$, the element $f(0)$ is called a free member of a generalized polynomial (polynomial) $f(x)$.

5.2.4. REMARK. It is easy to observe that an element $b \in R$ is a root of a generalized polynomial $f(x)$ if and only if the element $b - a$ is a root of the generalized polynomial $f(x + a)$.

5.2.5. LEMMA. *Let $f(x)$ be a generalized polynomial of degree m over a ring R and $f(0) = 0$. Then for any element $a \in R$ there exists a generalized polynomial $g(x)$ of degree not more than $m - 1$ such that:*

1) $g(0) = 0$;

2) $f(a + x) = f(x) + f(a) + g(x)$.

PROOF. The proof of the lemma follows from the fact that for any word $z(x) \in R[x]$ of degree $k \geq 1$, and any element $a \in R$ the equation $z(a + x) = z(x) + z(a) + q(x)$ is true, where $q(x)$ is a generalized polynomial of degree $k - 1$, and $q(0) = 0$.

5.2.6. THEOREM. *Let, for a generalized polynomial $f(x)$ of degree m over a ring R, there be elements $a_1, \dots, a_{m+1} \in R$ such that for all pairwise distinct numbers $i_1, \dots, i_r \in \{1, \dots, m + 1\}$ a sum of the form $\sum_{k=1}^{r} a_{i_k}$ is a root of generalized polynomial $f(x)$. Then $f(0) = 0$.*

PROOF. Proof will be based on induction by the degree of generalized polynomial $f(x)$.

If $f(x)$ is a generalized polynomial of zero degree, then $f(x) = b$, where b belongs to the subring in $R[x]$ generated by R. Since a_1 is a root of the generalized polynomial $f(x)$, then $f(a_1) = 0$. Then (see Denotation 5.2.3)

$$f(0) = \varphi_0(f(x)) = \varphi_0(b) = \varphi_{a_1}(b) = \varphi_{a_1}(f(x)) = f(a_1) = 0.$$

Let the theorem be proved for every generalized polynomial of degree not more than $m - 1$, and $f(x)$ be of degree m. Let $a_1, \ldots, a_{m+1}$ be elements of the ring R such that the condition of the theorem is satisfied. Consider the generalized polynomial $\varphi(x) = f(x) - f(0)$, whose degree is equal to m, too. Since $\varphi(0) = 0$, then, according to Lemma 5.2.5, $\varphi(x + a_{m+1}) = \varphi(x) + \varphi(a_{m+1}) + g(x)$, where $g(x)$ is a generalized polynomal of degree not more than $m - 1$, and $g(0) = 0$. Then

$$f(x + a_{m+1}) = \varphi(x + a_{m+1}) + f(0)$$
$$= \varphi(x) + \varphi(a_{m+1}) + g(x) + f(0) = f(x) + f(a_{m+1}) - f(0) + g(x).$$

Since $f(a_{m+1}) = 0$, then $g(x) - f(0) = f(x + a_{m+1}) - f(x)$.

We'll show now that the generalized polynomial $g(x) - f(0)$ and the elements $a_1, \ldots, a_m$ satisfy the condition of the theorem. Indeed, for any sum of the form $\sum_{k=1}^{r} a_{i_k}$ we have:

$$g\left(\sum_{k=1}^{r} a_{i_k}\right) - f(0) = f\left(\sum_{k=1}^{r} a_{i_k} + a_{m+1}\right) - f\left(\sum_{k=1}^{r} a_{i_k}\right) = 0.$$

Then $g(0) - f(0) = 0$ according to the assumption, and it means that $f(0) = g(0) = 0$.

5.2.7. REMARK. Let R be a commutative ring and $f(x) \in R[x]$. Let $\overline{f(x)} \in \overline{R[x]}$ be the image of the element $f(x)$ by the natural homomorphism of $R[x]$ onto $\overline{R[x]}$. Then $f(a) = \overline{f(a)}$ for any element $a \in R$.

From Remark 5.2.7, we also get the following Corollary.

5.2.8. COROLLARY. For any generalized polynomial $f(x)$ over a commutative ring R there exists a polynomial $\overline{f(x)}$ such that in the ring R the sets of roots of $f(x)$ and of $\overline{f(x)}$ coincide.

5.2.9. THEOREM. *Let R be a commutative ring without divisors of zero and $\overline{f(x)} \in \overline{R[x]}$. If $f(a) \neq 0$ for some $a \in R$, then $\overline{f(x)}$ can have only the finite set of roots in the ring R.*

PROOF. We shall use the induction by the degree of the polynomial $\overline{f(x)}$. If $\overline{f(x)}$ has zero degree, then $\overline{f(x)} = b$, where b belongs to the subring in $R[x]$ generated by the set R. Since $\overline{f(a)} \neq 0$, then (see Denotation 5.2.3)

$$\overline{f(c)} = \varphi_c\left(\overline{f(x)}\right) = \varphi_c(b) = \varphi_a(b) = \varphi_a\left(\overline{f(x)}\right) = \overline{f(a)} \neq 0$$

for any $c \in R$ and, consequently, $\overline{f(x)}$ has no roots in R.

Suppose now that the theorem is proved for all polynomials of degree not more than k.

Let the degree of the polynomial $\overline{f(x)}$ be equal to $k+1$ and c be a root of $\overline{f(x)}$, then there exists a polynomial $\overline{\varphi(x)}$ of degree k such that $\overline{f(x)} = (x - c) \cdot \overline{\varphi(x)}$.

Since $\overline{f(a)} \neq 0$, then $\overline{\varphi(a)} \neq 0$. According to our supposition, $\overline{\varphi(x)}$ can have only a finite number of roots in R. Since R is a ring without divisors of zero, then an element $d \in R$ is a root of the polynomial $\overline{f(x)}$ if and only if $d - c = 0$ or d is a root of $\overline{\varphi(x)}$. Hence, $\overline{f(x)}$ can have only a finite number of roots in R.

From Corollary 5.2.8 and Theorem 5.2.9 we obtain the following corollary.

5.2.10. COROLLARY. Let R be a commutative ring without divisors of zero and $f(x) \in R[x]$. If $f(a) \neq 0$ for some $a \in R$, then the generalized polynomial $f(x)$ can have in R only a finite number of roots.

5.2.11. DEFINITION. Let Φ be a family of generalized polynomials over a ring R. Over $\Phi^{-1}(0)$ we denote the set of all such elements of the ring R that each of them is a root of at least one generalized polynomial from Φ.

A subset A of the ring R is called algebraic over R with respect to a subset B of the ring R if there exists a set Φ of generalized polynomials over R such that:

1) $|\Phi| < |R|^*$;

2) $A \subseteq \Phi^{-1}(0)$;

3) $B \not\subseteq \Phi^{-1}(0)$.

If Φ is a finite set, then A is called finite-algebraic over R with respect to B. If $B = R$, then A is called algebraic, or finite-algebraic over R, respectively.

---

* Here $|S|$ is a cardinality of the set S (see A.7).

5.2.12. REMARK. It is obvious that for countable rings the conception of algebraity and finite algebraity coincide.

5.2.13. THEOREM. *Let A be a subgroup of the additive group of a ring R and S be a subset in A. If S is not algebraic over R with respect to A, then $b + S$ is not algebraic over R with respect to A for any element $b \in A$.*

PROOF. Let Φ be a set of generalized polynomials over R such that $|\Phi| < |R|$ and $A \not\subseteq \Phi^{-1}(0)$. Consider the set $\Phi_1 = \{f(x + b) \,|\, f(x) \in \Phi\}$ of generalized polynomials over R. It is easy to observe (see Remark 5.2.4) that $\Phi^{-1}(0) - b = \Phi_1^{-1}(0)$ and, hence $A = A - b \not\subseteq \Phi^{-1}(0) - b = \Phi_1^{-1}(0)$, since S is not algebraic over R with respect to A and $|\Phi_1| = |\Phi| < |R|$, then $S \not\subseteq \Phi_1^{-1}(0)$. Therefore, $b + S \not\subseteq b + \Phi_1^{-1}(0) = \Phi^{-1}(0)$. Thus, $b + S$ is not algebraic over R with respect to A.

5.2.14. PROPOSITION. *Let G be an infinite discrete multiplicative (not necessarily Abelian) group and βG be the set of all free ultrafilters of the set G. Then βG is a semigroup with respect to binary operation " $\circ$ " defined as follows:*

for $\Phi, \mathcal{F} \in \beta G$ put $\Phi \circ \mathcal{F} = \big\{A \subseteq G \,\big|\, \{g \in G \,|\, A \cdot g^{-1} \in \Phi\} \in \mathcal{F}\big\}$ (i.e.) so, $\Phi \circ \mathcal{F}$ consists of all subsets $A \subseteq G$ such that for each of them exists a subset $F_A \in \mathcal{F}$ such that $A \cdot f^{-1} \in \Phi$ for any $f \in F_A$.

PROOF. At first verify that the operation " $\circ$ " is correctly defined on βG, i.e. that $\Phi \circ \mathcal{F} \in \beta G$ holds for any $\Phi, \mathcal{F} \in \beta G$. It is sufficient to verify the fulfillment of the ultrafilter axioms for $\Phi \circ \mathcal{F}$.

From the definition of the operation " $\circ$ " it follows that $\emptyset \notin \Phi \circ \mathcal{F}$. Let $A, B \in \Phi \circ \mathcal{F}$. Then $\{g \in G \,|\, A \cdot g^{-1} \in \Phi\} \in \mathcal{F}$ and $\{g \in G \,|\, B \cdot g^{-1} \in \Phi\} \in \mathcal{F}$. Since

$$\big\{g \in G \,\big|\, (A \textstyle\bigcap B) \cdot g^{-1} \in \Phi\big\} = \{g \in G \,|\, A \cdot g^{-1} \in \Phi\} \textstyle\bigcap \{g \in G \,|\, B \cdot g^{-1} \in \Phi\},$$

then $\{g \in G \,|\, (A \bigcap B) \cdot g^{-1} \in \Phi\} \in \mathcal{F}$ and, hence $A \bigcap B \in \Phi \circ \mathcal{F}$.

Let $A \in \Phi \circ \mathcal{F}$ and $A \subseteq B$, then

$$\big\{g \in G \,\big|\, A \cdot g^{-1} \in \Phi\big\} \subseteq \big\{g \in G \,\big|\, B \cdot g^{-1} \in \Phi\big\}.$$

Hence, $\{g \in G \,|\, B \cdot g^{-1}\} \in \mathcal{F}$, i.e. $B \in \Phi \circ \mathcal{F}$.

Let's show that $\Phi \circ \mathcal{F}$ is a free ultrafilter. Let $A \subseteq G$ and $A \notin \Phi \circ \mathcal{F}$. Then $\{g \in G \mid A \cdot g^{-1} \in \Phi\} \notin \mathcal{F}$. Since $\mathcal{F}$ is an ultrafilter, then $G \backslash \{g \in G \mid A \cdot g^{-1} \in \Phi\} \in \mathcal{F}$; i.e. $G \backslash A \in \Phi \circ \mathcal{F}$. Hence, $\Phi \circ \mathcal{F}$ is an ultrafilter (see A.13).

It remains to verify that $\Phi \circ \mathcal{F}$ is a free ultrafilter. Assume the contrary, i.e. that $\{g\} \in \Phi \circ \mathcal{F}$ for some $g \in G$. Then there exists $F \in \mathcal{F}$ such that $\{g\} \cdot b^{-1} \in \Phi$ for any $b \in F$. We have a contradiction with the fact that Φ is a free ultrafilter. Thus, the operation " $\circ$ " is defined correctly.

Verify now that "$\circ$" is an associative operation. Let $\mathcal{A}, \mathcal{E}, \mathcal{F} \in \beta G$ and X be a subset in G such that $X \in \mathcal{A} \circ (\mathcal{E} \circ \mathcal{F})$. Then $\{g \in G \mid X \cdot g^{-1} \in \mathcal{A}\} \in \mathcal{E} \circ \mathcal{F}$. Therefore, there esists $F \in \mathcal{F}$ such that

$$\{g \in G \mid X \cdot g^{-1} \in \mathcal{A}\} \cdot f^{-1} \in \mathcal{E}$$

for any $f \in F$, i.e. for any $f \in F$ there exists $E_f \in \mathcal{E}$ such that

$$E_f \cdot f \subseteq \{g \in G \mid X \cdot g^{-1} \in \mathcal{A}\}.$$

The last inclusion means that for any $d \in E_f \cdot f$ there exists $A_{d,f} \in \mathcal{A}$ such that $A_{d,f} \subseteq X \cdot d^{-1}$. Since $\mathcal{A}$ is a filter, then $X \cdot d^{-1} \in \mathcal{A}$ for any $d \in E_f \cdot f$.

Thus, we have shown that there exists $F \in \mathcal{F}$ such that for every $f \in F$ exists $E_f \in \mathcal{E}$ such that $X \cdot (d \cdot f)^{-1} \in \mathcal{A}$ for any $d \in E_f$, i.e. $(X \cdot f^{-1}) \cdot d^{-1} \in \mathcal{A}$ for any $d \in E_f$. Then

$$E_f \subseteq \{g \in G \mid (X \cdot f^{-1}) \cdot g^{-1} \in \mathcal{A}\},$$

therefore,

$$\{g \in G \mid (X \cdot f^{-1}) \cdot g^{-1} \in \mathcal{A}\} \in \mathcal{E},$$

i.e. $X \cdot f^{-1} \in \mathcal{A} \circ \mathcal{E}$ for any $f \in F$. Hence,

$$F \subseteq \{g \in G \mid X \cdot g^{-1} \in \mathcal{A} \circ \mathcal{E}\},$$

and it means that

$$\{g \in G \mid X \cdot g^{-1} \in \mathcal{A} \circ \mathcal{E}\} \in \mathcal{F}.$$

Therefore, $X \in (\mathcal{A} \circ \mathcal{E}) \circ \mathcal{F}$. Thus, $\mathcal{F} \circ (\mathcal{E} \circ \mathcal{F}) \subseteq (\mathcal{A} \circ \mathcal{E}) \circ \mathcal{F}$. Since $\mathcal{A} \circ (\mathcal{E} \circ \mathcal{F})$ and $(\mathcal{A} \circ \mathcal{E}) \circ \mathcal{F}$ are ultrafilters, then $\mathcal{A} \circ (\mathcal{E} \circ \mathcal{F}) = (\mathcal{A} \circ \mathcal{E}) \circ \mathcal{F}$.

This completes the proof of the proposition.

5.2.15. PROPOSITION. *For any $Z \subseteq G$ put $U_Z = \{\mathcal{A} \in \beta G \mid Z \in \mathcal{A}\}$. Then:*

1) the family $\{U_Z \mid Z \subseteq G\}$ is a basis consisting of open subsets for a certain Hausdorff compact topology on βG;

2) for any $\mathcal{F} \in \beta G$ the mapping $f_{\mathcal{F}} : \beta G \longrightarrow \beta G$ acting by the rule $f_{\mathcal{F}}(\Phi) = \mathcal{F} \circ \Phi$, is continuous in this topology.

PROOF. Since $\beta G = U_G$, then $\beta G = \bigcup_{Z \subseteq G} U_Z$.

Let now $Z_1, Z_2 \subseteq G$ and $\mathcal{D} \in U_{Z_1} \bigcap U_{Z_2}$. Then $Z_1, Z_2 \in \mathcal{D}$ and, hence $Z_1 \bigcap Z_2 \in \mathcal{D}$, i.e. $\mathcal{D} \in U_{Z_1 \cup Z_2} \subseteq U_{Z_1 \bigcap Z_2}$. Since $U_{Z_1 \bigcap Z_2} \subseteq U_{Z_1} \bigcap U_{Z_2}$, then $\mathcal{D} \in U_{Z_1} \bigcap U_{Z_2}$. Thus, we've shown that the family $\{U_Z \mid Z \subseteq G\}$ is a basis consisting of open subsets of a certain topology τ on βG (see C.2.).

We show now that the topological space $(\beta G, \tau)$ is Hausdorff. Let $\mathcal{A}, \mathcal{D} \in \beta G$ and $\mathcal{A} \neq \mathcal{D}$. Then there exists a subset $A \subseteq G$ such that $A \in \mathcal{A}$ and $\mathcal{A} \notin \mathcal{D}$. Since $\mathcal{D}$ is an ultrafilter of G then $B = G \backslash A \in \mathcal{D}$. From the definition of sets U_Z follows that $\mathcal{A} \in U_A$ and $\mathcal{D} \in U_B$, i.e. that U_A and U_B are neighborhoods of the elements $\mathcal{A}$ and $\mathcal{D}$ in $(\beta G, \tau)$, respectively. Since for any filter, and, hence, for any ultrafilter $\mathcal{F}$, the subsets A and B cannot belong simultaneously to $\mathcal{F}$, then $U_A \bigcap U_B = \emptyset$. This shows that $(\beta G, \tau)$ is a Hausdorff space.

Let's show that $(\beta G, \tau)$ is a compact space. Assume the contrary. Let $\{U_\gamma \mid \gamma \in \Gamma\}$ be a cover of the space βG by open sets U_γ, from which it is impossible to choose a finite subcover. Without loss of generality we can consider that each $U\gamma = U_{Z_\gamma}$, where $Z_\gamma \subseteq G$ for any $\gamma \in \Gamma$.

For each $\gamma \in \Gamma$ consider the subset $F_\gamma = G \backslash Z_\gamma$. Since $\beta G \neq \bigcup_{i=1}^{n} U_{\gamma_i}$ for any finite subset $\{\gamma_1, \dots, \gamma_n\} \subseteq \Gamma$, then there exists an element $\mathcal{F} \in \beta G$ such that $\mathcal{F} \notin \bigcup_{i=1}^{n} U_{\gamma_i}$ and, therefore, $Z_{\gamma_i} \notin \mathcal{F}$ for $i = 1, \dots, n$. Since $\mathcal{F}$ is an ultrafilter, then $F_{\gamma_i} = G \backslash Z_{\gamma_i} \in \mathcal{F}$ for $i = 1, \dots, n$ (see A.13) and, therefore, $\bigcap_{i=1}^{n} F_{\gamma_i} \neq \emptyset$. Hence, the family $\{F_\gamma \mid \gamma \in \Gamma\}$ is a basis of some filter. Extend this filter to an ultrafilter Φ. Since $G = \bigcup_{\gamma \in \Gamma} U_\gamma$, then $\bigcap_{\gamma \in \Gamma} F_\gamma = \emptyset$ and therefore, Φ is a free ultrafilter. Since $F_\gamma \in \Phi$ for any $\gamma \in \Gamma$, then $Z_\gamma = G \backslash F_\gamma \notin \Phi$ for any $\gamma \in \Gamma$. From the definition of the sets U_Z it follows that $\Phi \notin U_{Z_\gamma} = U_\gamma$ for any $\gamma \in \Gamma$. We've got a contradiction with the fact that $\{U_\gamma \mid \gamma \in \Gamma\}$ is a cover of the space $(\beta G, \tau)$. Therefore, $(\beta G, \tau)$ is a compact space. Thus, the statement 1) is verified.

To prove the statement 2), we should verify that the mapping $f_{\mathcal{F}}$ is continuous for every $\mathcal{F} \in \beta G$. Let $\Phi \in \beta G$, and U_Z be a neighborhood of the element $\mathcal{F} \circ \Phi$ in βG. Then $Z \in \mathcal{F} \circ \Phi$ and (see Proposition 5.2.14) there is $F \in \Phi$ such that $Z \cdot g^{-1} \in \mathcal{F}$ for any $g \in F$. Since $\Phi \in U_F$ (see the definition of the sets U_X) and U_F is an open subset, then U_F is a neighborhood of the element Φ in $(\beta G, \tau)$.

Let now $\mathcal{D} \in U_F$, then $F \in \mathcal{D}$. Since $Z \cdot g^{-1} \in \mathcal{F}$ for any $g \in F$, then $Z \in \mathcal{F} \circ \mathcal{D}$. Hence, $\mathcal{F} \circ \mathcal{D} \in U_Z$. From this follows that $f_{\mathcal{F}}(U_F) \subseteq U_Z$. This completes the proof of Proposition.

5.2.16. PROPOSITION. *Let (G, τ) be a compact Hausdorff space. Let on G be defined a semigroup operation " $\circ$ " such that for every $g \in G$ the mapping $f_g : G \longrightarrow G$ acting by the rule $f_g(x) = g \circ x$, is continious. Then the semigroup G contains an idempotent.*

PROOF. Consider the family

$$\mathfrak{A} = \left\{ A \subseteq G \,\middle|\, A \neq \emptyset, \, A = [A]_G, \, A \circ A \subseteq A \right\}.$$

Since $G \in \mathfrak{A}$, then this family is not empty. Since (G, τ) is a compact space, then for any system $\{A_i \,|\, i \in I\}$ of subsets $A_i \in \mathfrak{A}$ which is linearly ordered with respect to inclusion, the subset $A = \bigcap_{i=I} A_i$ is not empty and, as an intersection of closed subsets, is closed itself. From $A \circ A \subseteq A_i \circ A_i \subseteq A_i$ for any $i \in I$ follows that $A \circ A \subseteq A$ and, therefore, $A \in \mathfrak{A}$.

Thus, the family $\mathfrak{A}$ is inductive (see A.11) with respect to the order defined by the inclusion of sets in G. Hence, $\mathfrak{A}$ contains minimal elements (see A.11, Zorn's Lemma). Let X be some minimal element in $\mathfrak{A}$ and $\rho \in X$. Since X is a closed subset in G (see definition of the family $\mathfrak{A}$), then X is a compact subset. Hence, $\rho \circ X = f_\rho(X)$ (as a continuous image of the compact subset) is compact. And, since (G, τ) is a Hausdorff space, then $\rho \circ X$ is a closed subset. From $\rho \circ X \neq \emptyset$ and

$$(\rho \circ X) \circ (\rho \circ X) \subseteq \rho \circ (X \circ X \circ X) \subseteq \rho \circ X$$

follows that $\rho \circ X \in \mathfrak{A}$. Since X is a minimal element in $\mathfrak{A}$ and $\rho \circ X \subseteq X \circ X \subseteq X$, then $\rho \circ X = X$. Thus, $B = \{g \in X \mid \rho \circ g = \rho\} \neq \emptyset$. Then $B = X \bigcap f_\rho^{-1}(\rho)$ is a closed subset (as an intersection of two closed subsets), and

$$(d_1 \circ d_2) \circ \rho = d_1 \circ (d_2 \circ \rho) = d_1 \circ \rho = \rho$$

for any $d_1, d_2 \in B$. Then $B \circ B \subseteq B$. Therefore, $B \in \mathfrak{A}$. Since X is a minimal element in $\mathfrak{A}$, then $B = X$. Then $\rho \in B$, and, hence $\rho \circ \rho = \rho$.

This completes the proof of the proposition.

5.2.17. PROPOSITION. *Let G be an infinite multiplicative (not necessary Abelian) group, e be the unitary element in G and $A_1, \ldots, A_n$ be subsets in G such that $G \backslash \{e\} = \bigcup_{i=1}^{n} A_i$. Then there exist $1 \leq i_0 \leq n$ and a countable sequence $g_1, g_2, \ldots$ of elements of A_{i_0} such that $g_{k_1} \cdot g_{k_2} \ldots g_{k_m} \in A_{i_0}$ for any $m \in \mathbb{N}$ and $k_1 > k_2 > \ldots > k_m$.*

PROOF. From Propositions 5.2.14, 5.2.15, and 5.2.16 follows that there exists an idempotent $\mathcal{E}$ in the semigroup $(\beta G, \circ)$. Since $\mathcal{E}$ is a free ultrafilter in G, and $G = \{e\} \bigcup (\bigcup_{i=1}^{n} A_i)$, then there exists $1 \leq i_0 \leq n$ such that $A_{i_0} \in \mathcal{E}$ (see A.13).

By the induction we can construct the sequences $E_1, E_2, \ldots$ of elements of $\mathcal{E}$ and $g_1, g_2, \ldots$ of elements of G as follows: put $E_1 = A_{i_0}$. Since $\mathcal{E} \circ \mathcal{E} = \mathcal{E}$, then $\{g \in G \mid E \cdot g^{-1} \in \mathcal{E}\} \in \mathcal{E}$ for any $E \in \mathcal{E}$, and, hence, $\{g \in G \mid E \cdot g^{-1} \in \mathcal{E}\} \bigcap E_1 \neq \emptyset$. As g_1 take an arbitrary element from $\{g \in G \mid E \cdot g^{-1} \in \mathcal{E}\} \bigcap E_1$.

Suppose that $E_1, \ldots, E_k$, $g_1, \ldots, g_k$ are already defined, and the following conditions are fulfilled: $g_i \in E_i$; $E_i \cdot g_i^{-1} \in \mathcal{E}$ for $1 \leq i \leq k$, and $E_r \subseteq E_j \cdot g_j^{-1}$ for $1 \leq j < r \leq k$. Put $E_{k+1} = A_{i_0} \bigcap (\bigcap_{j=1}^{k} E_j \cdot g_j^{-1})$, and as g_{k+1} choose an arbitrary element from

$$E_{k+1} \bigcap \left\{ g \in G \mid E_{k+1} \cdot g^{-1} \in \mathcal{E} \right\}.$$

Then $E_{k+1} \in \mathcal{E}$ (since $\mathcal{E}$ is a filter), $g_i \in E_i$, $E_i \cdot g_i^{-1} \in \mathcal{E}$ for $1 \leq i \leq k+1$, and $E_r \subseteq E_j \cdot g_j^{-1}$ for $1 \leq j < r \leq k+1$.

To complete the proof we must verify that the constructed sequence $g_1, g_2, \ldots$ is required. So, let $m \in \mathbb{N}$ and $i_1 > i_2 \ldots > i_m$.

If $m = 1$, then $g_{i_1} \in E_{i_0} \subseteq A_{i_0}$.

If $m > 1$, then

$$g_{i_1} \cdot g_{i_2} \cdot \ldots \cdot g_{i_m} \in E_{i_1} \cdot g_{i_2} \cdot \ldots \cdot g_{i_m} \subseteq (E_{i_2} \cdot g_{i_2}^{-1}) \cdot g_{i_2} \cdot \ldots \cdot g_{i_m}$$

$$= E_{i_2} \cdot g_{i_3} \cdot \ldots \cdot g_{i_m} \subseteq \ldots \subseteq E_{i_{m-1}} \cdot g_{i_m} \subseteq (E_{i_m} \cdot g_{i_m}^{-1}) \cdot g_{i_m} = E_{i_m} \subseteq A_{i_0}.$$

The proof is completed.

5.2.18. THEOREM. *Let R be an infinite ring and G be an infinite subgroup of the additive group of the ring R. Then $G \backslash \{0\}$ is not finite-algebraic over the ring R with respect to $\{0\}$ (see Definition 5.2.11).*

PROOF. Let's assume the contrary. Let $\Phi = \{f_1(x), \ldots, f_n(x)\}$ be a finite family of generalized polynomials over R such that $G \backslash \{0\} \subseteq \Phi^{-1}(0)$ but $0 \notin \Phi^{-1}(0)$.

For every $1 \leq i \leq n$ consider the subset $A_i = \{g \in G \mid f_i(g) = 0\}$. Then $G \backslash \{0\} = \bigcup_{i=1}^n A_i$. Since G is a commutative additive group, then from Proposition 5.2.17 follows the existence of such $1 \leq i_0 \leq n$ and such sequence $g_1, g_2, \ldots$ of elements of G that $\sum_{i=1}^m g_{i_k} \in A_{i_0}$ for any $m \in \mathbb{N}$ and any pairwise distinct numbers $i_1, i_2, \ldots, i_m$. Thus, $\sum_{k=1}^m g_{i_k}$ is a root of the generalized polynomial $f_{i_0}(x)$. According to Theorem 5.2.6, $f_{i_0}(0) = 0$, i.e. $0 \in \Phi^{-1}(0)$. We have a contradiction with the choice of the family Φ.

This completes the proof of Theorem.

§ 5.3. Non-Discrete Topologization of Countable Rings

5.3.1. DEFINITION. Let $V_1, V_2, \ldots$ and $S_1, S_2 \ldots$ be some symmetric subsets of a ring R, i.e. $-V_i = V_i$ and $-S_i = S_i$ for any $i \in \mathbb{N}$.

For every natural number k let $F_k(S_1, S_2, \ldots, S_k; V_1, V_2, \ldots, V_k)$ be such set that:

$$F_1(S_1; V_1) = S_1 \cdot V_1 + V_1 \cdot S_1 + V_1 \cdot V_1 + V_1 + V_1$$

$$= \left\{ a_1 \cdot b_1 + b_2 \cdot a_2 + b_3 \cdot b_4 + b_5 + b_6 \,\middle|\, a_1, a_2 \in S_1; b_1, b_2, b_3, b_4, b_5, b_6 \in V_1 \right\}$$

and $F_{k+1}(S_1, \ldots, S_{k+1}; V_1, \ldots, V_{k+1}) = F_1\big(S_1; V_1 \cup F_k(S_2, \ldots, S_{k+1}; V_2, \ldots, V_{k+1})\big)$.

5.3.2. PROPOSITION. *Subsets* $F_k(S_1, \ldots S_k; V_1, \ldots, V_k)$ *satisfy the following conditions:*

I)   *if* $0 \in V_1$, *then* $V_1 - V_1 \subseteq F_1(S_1; V_1)$;

II) *if* $0 \in V_1$, *then* $V_1 \cdot V_1 \subseteq F_1(S_1; V_1)$;

III) *if* $0 \in V_1$, *then* $S_1 \cdot V_1 \subseteq F_1(S_1; V_1)$ *and* $V_1 \cdot S_1 \subseteq F_1(S_1; V_1)$;

IV) $F_k(S_1, \ldots, S_k; \{0\}, \ldots, \{0\}) = \{0\}$ *for* $k \in \mathbb{N}$;

V)   *if* $V_i' \subseteq V_i$ *and* $S_i' \subseteq S_i$ *for* $i = 1, \ldots, k$, *then*

$$F_k(S_1', \ldots, S_k'; V_1', \ldots V_k') \subseteq F_k(S_1, \ldots, S_k; V_1, \ldots, V_k);$$

VI) *if* S_i *and* V_i *are finite sets for all* $i \in \mathbb{N}$, *then the set* $F_k(S_1, \ldots, S_k; V_1, \ldots, V_k)$ *is finite and symmetric for any* $k \in \mathbb{N}$;

VII) *if* $0 \in V_i$ *for all* $i \in \mathbb{N}$, *then* $V_j \subseteq F_k(S_1, \ldots, S_k; V_1, \ldots, V_k)$ *for all* $k \in \mathbb{N}$ *and* $1 \leq j \leq k$;

VIII) *if* $0 \in V_i$ *for all* $i \in \mathbb{N}$, *then*

$$F_{k+p}(S_1, \ldots, S_{k+p}; V_1, \ldots, V_k, \{0\}, \ldots, \{0\}) = F_k(S_1, \ldots, S_k; V_1, \ldots, V_k)$$

for all $k, p \in \mathbb{N}$;

IX) *if* J *is a two-sided ideal of the ring* R, *then* $F_k(S_1, \ldots, S_k; J, J, \ldots, J) \subseteq J$ *for every* $k \in \mathbb{N}$;

X) $F_k\big(S_1, \ldots, S_k; V_1 \bigcup F_{k-1}(S_1, \ldots, S_k; V_2, \ldots, V_k), \ldots, V_{k-1} \bigcup F_1(S_k; V_k), V_k\big) \quad =$ $F_k(S_1, \ldots, S_k; V_1, \ldots, V_k)$ *for every* $k \geq 2$.

PROOF. The fulfillment of the conditions I)-III) follows from the definition of the set $F_1(S_1; V_1)$; the properties IV)-IX) are easily proven by induction.

Now let's verify the fulfillment of the condition X). Indeed,

$$F_2(S_1, S_2; V_1, V_2) = F_1\left(S_1; V_1 \cup F_1(S_2; V_2)\right)$$
$$= F_1\left(S_1; V_1 \cup F_1(S_2; V_2) \cup F_1(S_2; V_2)\right) = F_2\left(S_1, S_2; V_1 \cup F_1(S_2; V_2), V_2\right).$$

Suppose that the required equality is proved for $k \leq r$. Then

$$F_{r+1}(S_1, \ldots, S_{r+1}; V_1, \ldots, V_{r+1})$$
$$= F_1\left(S_1; V_1 \cup F_r(S_2, \ldots, S_{r+1}; (V_2, \ldots, V_{r+1}))\right)$$
$$= F_1\left(S_1; V_1 \cup F_r(S_2, \ldots, S_{r+1}; (V_2 \cup F_{r-1}(S_3, \ldots, S_{r+1}; V_3, \ldots, V_{r+1})), \ldots, \right.$$
$$\left. (V_r \cup F_1(S_{r+1}, V_{r+1})), V_{r+1})\right) = F_r\left(S_1, \ldots, S_{r+1}; (V_1 \bigcup F_r(S_2, \ldots, S_{r+1}; \right.$$
$$\left. V_2, \ldots, V_{r+1})), \ldots, (V_r \cup F_1(S_{r+1}; V_{r+1})), V_{r+1}\right).$$

The proposition is proved.

5.3.3. DENOTATION. Let now R be some countable ring. We enumerate the set of its non-zero elements by natural numbers, i.e. $R = \{0, a_1, a_2, \ldots\}$.

For every non-negative number k put

$$S_k = \{\pm a_1, \pm a_2, \ldots, \pm a_k\}^*$$

and for every pair of non-negative integers i, j define subsets $V_{i,j}$ and $S_{i,j}$ in the ring R and a subset $\Phi_{i,j}$ in $R[x]$ (see 5.2.1) as follows:

$$V_{0,j} = \{0\}; \quad S_{0,j} = S_j; \quad \Phi_{0,j} = \{x - a \mid a \in S_j\}, \quad \text{for all} \quad j \in \mathbb{N}.$$

Suppose that the subsets $V_{i,j}$, $S_{i,j}$ in R and $\Phi_{i,j}$ in $R[x]$ are already defined for all $j \in \mathbb{N}$ and all $i \leq k$.

---

* We suppose that $S_0 = \emptyset$.

If $k+1$ is an even number, then put $S_{k+1,j} = S_{k,j}$ and $\Phi_{k+1,j} = \Phi_{k,j}$ for all $j \in \mathbb{N}$, put also

$$V_{k+1,j} = V_{k,j} \cup F_{k-j}(S_{j+1}, \dots, S_k; V_{k,j+1}, \dots, V_{k,k}) \text{ for } j < k,$$

and $V_{k+1,j} = \{0\}$ for $j \geq k$.

If $k+1$ is an odd number, then put $S_{k+1,0} = \emptyset$, $\Phi_{k+1,0} = \emptyset$ and $V_{k+1,0} = V_{k,0} \cup \{\pm a\}$, where a is an arbitrary element of R.

Suppose that the sets $V_{k+1,j}$, $S_{k+1,j}$ and $\Phi_{k+1,j}$ for $j \leq p$ are already defined. For any non-negative integer $i \leq p$ define a set of generalized polynomials $\Phi_{k+i,i,p+1} \subseteq R[x]$. Taking into account that $R \subseteq R[x]$ and $x \in R[x]$, put:

$$\Phi_{k+1,i,p+1} = F_{p-i+1}\left(S_{i+1}, \dots, S_{p+1}; V_{k+1,i+1}, \dots, V_{k+1,p}, V_{k,p+1} \cup \{\pm x\}\right).$$

Now put:

$$\Phi_{k+1,p+1} = \bigcup_{i=0}^{p} \left(\Phi_{k+1,i,p+1} - S_i\right) = \bigcup_{i=0}^{p} \{f(x) - c \mid f(x) \in \Phi_{k+1,i,p+1}, c \in S_i\}$$

and

$$S_{k+1,p+1} = S_{p+1} \cup \Phi_{k+1,p+1}^{-1}(0).$$

(see Definition 5.2.11).

Let a be an element of R and a does not belong to the set $S_{k+1,p+1}$. Then put $V_{k+1,p+1} = V_{k,p+1} \cup \{\pm a\}$ for $p \leq k$ (if, however, we have that $S_{k+1,p+1} = R$ for some $p \leq k$, then put $V_{k+1,p+1} = V_{k,p+1}$) and $V_{k+1,p+1} = \{0\}$ for $p \geq k$.

5.3.4. PROPOSITION. *The sets $V_{i,j}$ and $\Phi_{i,j}$, $i, j \in \mathbb{N}$, satisfy the following properties:*

I) $V_{i,j} = \{0\}$, *for $i, j \in \mathbb{N}$ if $j \geq i$;*

II) $V_{i,j} \subseteq V_{k,j}$, *for $i, j, k \in \mathbb{N}$ if $k \geq i$;*

III) $V_{i,j}$ *are finite symmetric sets for $i, j \in \mathbb{N}$;*

IV) $F_p(S_{j+1}, \dots, S_{j+p}; V_{i,j+1}, \dots, V_{i,j+p}) \subseteq F_{i-j}(S_{j+1}, \dots, S_i; V_{i,j+1}, \dots, V_{i,i})$ *for all $i, j, p \in \mathbb{N}$, $j \leq i$;*

V) $F_p(S_{j+1}, \dots, S_{j+p}; V_{i,j+1}, \dots, V_{i,j+p}) \subseteq V_{i+1,j}$ *for any $i, j, p \in \mathbb{N}$;*

VI) $V_{i,j} \subseteq V_{i+1,k}$ *for any* $i, j, k \in \mathbb{N}$, $j \geq k$;

VII) $F_p(S_{j+1}, \dots, S_{j+p}; V_{k+1,j+1}, \dots, V_{k+1,j+p}) =$

$\quad F_{p+1}(S_{j+1}, \dots, S_{j+p+1}; V_{k+1,j+1}, \dots, V_{k+1,j+p}, V_{k,j+p+1})$ *for any* $j, k, p \in \mathbb{N}$;

VIII) $S_j \bigcap V_{i,j} = \emptyset$, *for any* $i, j \in \mathbb{N}$;

IX) $\Phi_{i,j}$ *is a finite set for any* $i, j \in \mathbb{N}$;

X) $0 \notin \Phi_{i,j}^{-1}(0)$ *for any* $i, j \in \mathbb{N}$.

PROOF. The validity of the properties I) and II) directly follows from the definition of the sets $V_{i,j}$.

The property III) can be proven by induction using the property VI) of Proposition 5.3.2.

Now verify the validity of the condition IV). Indeed, let $p \leq i - j$, then, applying consequently the properties VIII) and V) of Proposition 5.3.2, we obtain:

$$F_p(S_{j+1}, \dots, S_{j+p}; V_{i,j+1}, \dots, V_{i,j+p})$$
$$= F_{i-j}(S_{j+1}, \dots, S_i; V_{i,j+1}, \dots, V_{i,j+p}, \{0\}, \dots, \{0\})$$
$$\subseteq F_{i-j}(S_{j+1}, \dots, S_i; V_{i,j+1}, \dots, V_{i,i}).$$

Let now $p > i - j$. Since $V_{i,k} = \{0\}$ for $k > i$, then, applying the property V) of Proposition 5.3.2, we get

$$F_p(S_{j+1}, \dots, S_{j+p}; V_{i,j+1}, \dots, V_{i,j+p})$$
$$= F_p(S_{j+1}, \dots, S_{j+p}; V_{i,j+1}, \dots, V_{i,i}, \{0\}, \dots, \{0\})$$
$$= F_{i-j}(S_{j+1}, \dots, S_i; V_{i,j+1}, \dots, V_{i,i}).$$

Verify the validity of the property V). Let $j \geq i - 1$, then, according to already proved property I), we get:

$$V_{i,j+1} = V_{i,j+2} = \dots = V_{i,j+p} = \{0\}.$$

Then, by the property IV) of Proposition 5.3.2:

$$F_p(S_{j+1}, \dots, S_{j+p}; V_{i,j+1}, \dots, V_{i,j+p})$$
$$= F_p(S_{j+1}, \dots, S_{j+p}; \{0\}, \dots, \{0\}) = \{0\} \subseteq V_{i+1,j}.$$

Let now $j < i - 1$. Consider two cases:

1) $i + 1$ is an even number. Then, according to already proven property IV) and the definition of the set $V_{i+1,j}$ for the case when $i + 1$ is an even number, we have:

$$F_p(S_{j+1}, \dots, S_{j+p}; V_{i,j+1}, \dots, V_{i,j+p})$$
$$\subseteq F_{i-j}(S_{j+1}, \dots, S_i, V_{i,j+1}, \dots, V_{i,i}) \subseteq V_{i+1,j}.$$

2) $i + 1$ is an odd number. Then i is even and, hence,

$$V_{i,k} = V_{i-1,k} \cup F_{i-k-1}(S_{k+1}, \dots, S_{i-1}; V_{i-1,k+1}, \dots, V_{i-1,i-1}) \text{ for } k < i - 1$$

and $V_{i,k} = \{0\}$ for $k \geq i - 1$.

Applying consequently the already proven property IV) as well as properties VIII), X) of Proposition 5.3.2 and the property I) of this proposition, we have:

$$F_p(S_{j+1}, \dots, S_{j+p}; V_{i,j+1}, \dots, V_{i,j+p}) \subseteq F_{i-j}(S_{j+1}, \dots, S_i, V_{i,j+1}, \dots, V_{i,i})$$
$$= F_{i-j}(S_{j+1}, \dots, S_i; (V_{i-1,j+1} \cup F_{i-j-2}(S_{j+2}, \dots, S_{i-1};$$
$$V_{i-1,j+2}, \dots, V_{i-1,i-1})), \dots, (V_{i-1,i-2} \cup F_1(S_{i-1}; V_{i-1,i-1}))\{0\}, \{0\})$$
$$= F_{i-j}(S_{j+1}, \dots, S_i; (V_{i-1,j+1} \cup F_{i-j-1}(S_{j+2}, \dots, S_i; V_{i-1,j+2}, \dots, V_{i-1,i-1}, \{0\})),$$
$$\dots, (V_{i-1,i-2} \cup F_2(S_{i-1}, S_i; V_{i-1,i-1}, \{0\})), (V_{i-1,i-1} \cup F_1(S_1; \{0\})), \{0\})$$
$$= F_{i-j}(S_{j+1}, \dots, S_i; V_{i-1,j+1}, \dots, V_{i-1,i-1}, \{0\})$$
$$= F_{i-j-1}(S_{j+1}, \dots, S_{i-1}; V_{i-1,j+1}, \dots, V_{i-1,i-1}) = V_{i,j} \subseteq V_{i+1,j}.$$

The validity of the property VI) follows from the properties VII) of Proposition 5.3.2, and V) of this proposition.

Examine the property VII). Let $p = 1$, then, according to the property V) of Denotation 5.3.3, we get:

$$F_2(S_{j+1}, S_{j+2}; V_{k+1,j+1}, V_{k,j+2}) = F_1(S_{j+1}; V_{k+1,j+1} \cup F_1(S_{j+2}, V_{k,j+2}))$$
$$= F_1(S_{j+1}; V_{k+1,j+1} \cup V_{k+1,j+1}) = F_1(S_{j+1}; V_{k+1,j+1}).$$

Suppose that the required equality is proven for $p \leq r$ and for any j and k. Then

$$F_{r+2}(S_{j+1}, \dots, S_{j+r+2}; V_{k+1,j+1}, \dots, V_{k+1,j+r+1}, V_{k,j+r+2})$$

$$F_1(S_{j+1}; V_{k+1,j+1} \cup F_{r+1}(S_{j+2}, \dots, S_{j+r+2}; V_{k+1,j+2}, \dots, V_{k+1,j+r+1}, V_{k,j+r+2}))$$

$$= F_1(S_{j+1}; V_{k+1,j+1} \cup F_r(S_{j+2}, \dots, S_{j+r+1}; V_{k+1,j+2}, \dots, V_{k+1,j+r+1}))$$

$$F_{r+1}(S_{j+1}, \dots, S_{j+r+1}; V_{k+1,j+1}, \dots, V_{k+1,j+r+1}).$$

Thus, the property VII) is proven for any $j, k, r \in \mathbb{N}$.

Examine the validity of the property VIII). Indeed, $S_j \cap V_{0,j} = S_j \cap \{0\} = \emptyset$ for any $j \in \mathbb{N}$.

Suppose that $S_j \cap V_{i,j} = \emptyset$ for all $i \leq k$ and all $j \in \mathbb{N}$. Consider two cases:

1) $k + 1$ is an odd number. Then either $V_{k+1,j} = V_{k,j} \cup \{\pm a\}$, where $a \notin S_{k+1,j}$, or $V_{k+1,j} = V_{k,j}$. Since $S_j \subseteq S_{k+1,j}$, then either $V_{k+1,j} \cap S_j = (V_{k,j} \cup \{\pm a\}) \cap S_j = V_{k,j} \cap S_j = \emptyset$ or $V_{k+1,j} \cap S_j = V_{k,j} \cap S_j = \emptyset$;

2) $k + 1$ is an even number. Assume that $V_{k+1,j_0} \cap S_{j_0} \neq \emptyset$ for some j_0. Then

$$S_{j_0} \cap F_{k-j_0}(S_{j_0+1}, \dots, S_k; V_{k,j_0+1}, \dots, V_{k,k})$$

$$= S_{j_0} \cap (V_{k,j_0} \cup F_{k-j_0}(S_{j_0+1}, \dots, S_k; V_{k,j_0+1}, \dots, V_{k,k})) = S_{j_0} \cap V_{k+1,j_0} \neq \emptyset.$$

Let $p_0 = \min\{p \in \mathbb{N} \,|\, S_{j_0} \cap F_p(S_{j_0+1}, \dots, S_{j_0+p}; V_{k,j_0+1}, \dots, V_{k,j_0+p}) \neq \emptyset\}$. Since k is an odd number, then either $V_{k,j_0+p} = V_{k-1,j_0+p} \cup \{\pm a\}$, where $a \notin S_{k,j_0+p}$, or $V_{k,j_0+p_0} = V_{k-1,j_0+p_0}$.

If $V_{k,j_0+p_0} = V_{k-1,j_0+p_0} \cup \{\pm a\}$, then

$$S_{j_0} \cap F_{p_0}(S_{j_0+1}, \dots, S_{j_0+p_0}; V_{k,j_0+1}, \dots, V_{k,j_0+p_0-1}, V_{k-1,j_0+p_0} \cup \{\pm a\}) \neq \emptyset.$$

Then there exists a generalized polynomial $f(x) \in \Phi_{k,j_0,j_0+p_0}$ such that $f(a) \in S_{j_0}$ (see the definition of the set $\Phi_{k,j,p}$). Hence, $f(a) - c = 0$ for some $c \in S_{j_0}$. Thus, we have shown that the element a is a root of some generalized polynomial from $\Phi_{k,j_0,j_0+p_0} - S_{j_0}$ and, therefore, $a \in S_{k,j_0+p_0}$. We have obtained a contradiction. Hence, $V_{k,j_0+p_0} = V_{k-1,j_0+p_0}$.

Consider now two subcases:

a) $p_0 > 1$. Then, according to already proven property VII), we get:

$$S_{j_0} \cap F_{p_0-1}(S_{j_0+1}, \dots, S_{j_0+p_0-1}; V_{k,j_0+1}, \dots, V_{k,j_0+p_0-1})$$

$$= S_{j_0} \cap F_{p_0}(S_{j_0+1}, \dots, S_{j_0+p_0}; V_{k,j_0+1}, \dots, V_{k,j_0+p_0-1}, V_{k-1,j_0+p_0})$$

$$= S_{j_0} \cap F_{p_0}(S_{j_0+1}, \dots, S_{j_0+p_0}; V_{k,j_0+1}, \dots, V_{k,j_0+p_0-1}, V_{k,j_0+p_0}) \neq \emptyset.$$

We have obtained a contradiction with the definition of p_0;

b) $p_0 = 1$. Then, according to already proven property V),

$$S_{j_0} \cap V_{k,j_0} \supseteq S_{j_0} \bigcap F_1(S_{j_0+1}; V_{k-1,j_0+1}) = S_{j_0} \cap F_1(S_{j_0+1}; V_{k,j_0+1}) \neq \emptyset.$$

We have obtained a contradiction with our assumption.

Hence, $V_{k+1,j_0} \bigcap S_{j_0} = \emptyset$ in the case when $k+1$ is an even number, too, i.e. property VIII) is fulfilled.

Since $R \subseteq R[x]$, then from the property VI) of Proposition 5.3.2 and already proven property III) it easily follows that for any $i, j \in \mathbb{N}$ the set $\Phi_{i,j}$ is finite, i.e. the property IX) is fulfilled.

Verify the property X). Assume the contrary, i.e. that there exists a generalized polynomial $f(x) \in \Phi_{i_0,j_0}$, for some i_0 and j_0, such that $f(0) = 0$.

From the construction of the set $\Phi_{i,j}$ it follows that there exists a generalized polynomial $\varphi(x) \in \Phi_{i_0,k,j_0}$, for some $k \leq j_0$, and an element $c \in S_k$ such that $f(x) = \varphi(x) - c$.

It can be easily observed that

$$\varphi(0) \in F_{j_0-k}\left(S_{k+1}, \dots, S_{j_0}; V_{i_0,k+1} \dots, V_{i_0,j_0-1}, V_{i_0-1,j_0}\right).$$

Applying consequently the properties II) and V) to the last inclusion, we obtain that $\varphi(0) \in V_{i_0+1,k}$.

Since $c = \varphi(0) - f(0) = \varphi(0)$, then $c \in S_k \bigcap V_{i_0+1,k}$.

We have obtained a contradiction with the property VIII). Hence, $0 \notin \Phi_{i,j}^{-1}(0)$ for any i and j. This completes the proof of the proposition.

5.3.5. REMARK. From the method of construction of the set $S_{i,j}$ it follows that any element of $S_{i,j}$ is a root for at least one generalized polynomial (see Definition 5.2.3) from the set $\Phi_{i,j} \bigcup \{x - c \mid c \in S_j\}$. Since $0 \notin S_n$ for $n \in \mathbb{N}$, then from the properties IX) and X) of Proposition 5.3.3 follows that $S_{i,j}$ is a finite-algebraic set over the ring R with respect to $\{0\}$ (see Definition 5.2.11) for any i and j.

5.3.6. THEOREM. *The family* $\{W_j = \bigcup_{i=1}^{\infty} V_{i,j} \mid j = 1, 2, \dots\}$ *of subsets (definition of* $V_{i,j}$ *see in Denotation 5.3.3) of a countable ring* R *can be considered as a basis of neighborhoods of zero in a certain topology on* R *such that* R *is a Hausdorff topological ring.*

PROOF. According to Theorem 1.2.5 and Corollary 1.3.7, it is enough to verify that the family $\{W_j \mid j \in \mathbb{N}\}$ satisfies the conditions (BN1)-(BN6) and (BN1').

Validity of the condition (BN1) follows from the property I) of Proposition 5.3.4, and validity of (BN2) follows from the property VI) of Proposition 5.3.4 and the definition of W_j.

Validity of conditions (BN3) and (BN5) follows from the property V) of Proposition 5.3.4, and validity of (BN4) results from the property III) of Proposition 5.3.4.

Let now $a \in R$ and $j \in \mathbb{N}$. If $a = 0$, then $a \cdot W_j = W_j \cdot a = \{0\} \subseteq W_j$. If $a \neq 0$, then $a \in S_k$ for some k. Let $p = \max\{k, j\}$. Since $S_k \subseteq S_{p+1}$, then applying consequently properties III) of Proposition 5.3.2, and properties V) and VI) of Proposition 5.3.3, we obtain:

$$a \cdot W_{p+1} = a \cdot \left(\bigcup_{i=0}^{\infty} V_{i,p+1} \right) = \bigcup_{i=0}^{\infty} (a \cdot V_{i,p+1}) \subseteq \bigcup_{i=0}^{\infty} (S_{p+1} \cdot V_{i,p+1})$$

$$\subseteq \bigcup_{i=0}^{\infty} F_1(S_{p+1}; V_{i,p+1}) \subseteq \bigcup_{i=0}^{\infty} V_{i+1,p} = W_p \subseteq W_j.$$

Analogously, $W_{p+1} \cdot a \subseteq W_j$. Hence, the family $\{W_k \mid k \in \mathbb{N}\}$ satisfies the condition (BN6), too.

From the property VIII) of Proposition 5.3.4 follows that $S_j \bigcap W_j = \emptyset$ for any $j \in \mathbb{N}$. Since $\bigcup_{j=1}^{\infty} S_j = R \backslash \{0\}$, then $\bigcap_{j=1}^{\infty} W_j = \{0\}$. Hence, the family $\{W_k \mid k \in \mathbb{N}\}$ satisfies the condition (BN1'), too. It completes the proof of the theorem.

5.3.7. THEOREM. *Any non-discrete Hausdorff ring topology on a countable ring* R *satisfying the first countability axiom can be obtained by the method above.*

symmetric neighborhoods of zero in R such that $S_j \bigcap U_i = \emptyset$ and

$$F_1(S_i; U_i) = S_i \cdot U_i + U_i \cdot S_i + U_i \cdot U_i + U_i + U_i \subseteq U_{i-1}$$

for $i = 2, 3, \dots$.

By induction by p it can be verified that for any p and j we have:

$$F_p(S_{j+1}, \dots, S_{j+p}; U_{j+1}, \dots, U_{j+p}) \subseteq U_j. \qquad (*)$$

When building the sets $V_{k+1,j}$ as in Proposition 5.3.3 and in the case of the odd $k+1$, let's take as an element a the element a_j with the smallest index (see numbering in the beginning of the proof) from the set $(U_j \backslash U_{k,j}) \bigcap (R \backslash S_{k+1,j})$.

Then, using the inclusion $(*)$ and carrying out the induction by j, it is easy to verify that $V_{j,n} \subseteq U_n$ for any $j, n \in \mathbb{N}$. Therefore, $W_j = \bigcup_{k=0}^{\infty} V_{k,j} \subseteq U_j$ for any $j \in \mathbb{N}$.

Now we show that $(U_j \backslash V_{k,j}) \bigcap (R \backslash S_{k+1,j}) = U_j \backslash V_{k,j}$ for any $k, j \in \mathbb{N}$.

Since $V_{k,p} \subseteq V_{k+1,p} \subseteq U_p$ for all $k, p \in \mathbb{N}$, then for any natural j and arbitrary element $b \in U_j$ we have:

$$S_n \cap F_{j-n}(S_{n+1}, \dots, S_j; V_{k+1,n+1}, \dots, V_{k+1,j-1}, V_{k,j} \cup \{\pm b\})$$
$$\subseteq S_n \cap F_{j-n}(S_{n+1}, \dots, S_j; U_{n+1}, \dots, U_j) \subseteq S_n \cap U_n = \emptyset$$

for any $n < j$. Hence, for any $n < j$ and arbitrary generalized polynomial

$$f(x) \in F_{j-n}(S_{n+1}, \dots, S_j; V_{k+1,n+1}, \dots, V_{k+1,j-1}, V_{k,j} \cup \{\pm x\})$$

we have $f(b) \notin S_n$. Therefore, any $b \in U_j$ cannot be a root of a polynomial from $\Phi_{k+1,n,j} - S_n$ for $n, j \in \mathbb{N}$. Moreover, since $b \notin S_j$ (because of $S_j \bigcap U_j = \emptyset$), then $b \notin S_j \bigcup \Phi_{k+1,j}^{-1}(0) = S_{k+1,j}$, i.e. $(U_j \backslash V_{k,j}) \bigcap (R \backslash S_{k+1,j}) = U_j \backslash V_{k,j}$.

Further, when building the set $V_{k+1,j}$ and in the case of the odd $k+1$, let's take as an element a an element with the smallest index (see the numbering of elements of the ring R) from the set $U_j \backslash V_{k,j}$.

Then for any element $a_r \in U_j$ there exists an odd $k+1$ such that either $a_r \in V_{k,j}$, or a_r is an element with the smallest index in the set $U_j \backslash V_{k,j}$, hence $a_r \in V_{k,j} \bigcup \{\pm a_r\} = V_{k+1,j}$. Thus, $U_j \subseteq W_j$ and therefore, $U_j = W_j$.

Hence, we have shown that the initial topology and the topology given by the family $\{W_j \mid i \in \mathbb{N}\}$ are identical. This completes the proof.

5.3.8. THEOREM. *Any countable ring R permits a non-discrete Hausdorff topology**.

PROOF. According to Remark 5.3.5, for any $i, j \in \mathbb{N}$ a set $S_{i,j}$ is finite-algebraic over the ring R with respect to $\{0\}$. According to Theorem 5.2.18, the set $R\backslash\{0\}$ is not finite-algebraic over R with respect to $\{0\}$. Therefore, $R\backslash\{0\} \not\subseteq S_{i,j}$ for any $i, j \in \mathbb{N}$. For any $t \in \mathbb{N}$ in the constructing sets $V_{k+1,t}$ in the case of the odd $k + 1$ and $k + 1 > t$, some non-zero element from $(R\backslash\{0\})\backslash S_{k+1,t}$ can be chosen as element a. Hence, $W_t \neq \{0\}$ for any $t \in \mathbb{N}$.

This completes the proof.

---

* While proving the theorem, the associativity of R will not be used anywhere. Therefore, considering $R[x]$ as a free (not necessarily associative) ring, we get that this proof is valid also for an arbitrary (not necessarily associative) ring R.

§ 5.4. Method of Topologization of
Infinite Uncountable Rings

5.4.1. DENOTATION. Let S and V be subsets of a ring R. Over $F_R(S; V)$ we denote the ideal, generated by the set V in the subring, generated by the subset $S \bigcup V$ in the ring R.

5.4.2. DEFINITION. A non-limit transfinite number α is called even (odd), if $\alpha = \xi + n$, where ξ is either zero or a limit transfinite number, and n is an even (odd) natural number.

5.4.3. DENOTATION. Let R be a infinite uncountable ring and $\mathfrak{m}$ be its cardinality. Then the set of all non-zero elements of the ring R can be enumerated by the transfinite numbers less than $\omega(\mathfrak{m})^*$ (i.e. $R = \{0,\} \cup \{a_\alpha \mid 1 \le \alpha < \omega(\mathfrak{m})\}$). For any transfinite number $\alpha < \omega(\mathfrak{m})$ put $S_\alpha = \{a_\gamma \mid 1 \le \gamma \le \alpha\}$. It is obvious that $|S_\alpha| = |\alpha| < \mathfrak{m}^{**}$ and $R \backslash \{0\} = \bigcup_{\alpha < \omega(\mathfrak{m})} S_\alpha$. For every transfinite number $\alpha < \omega(\mathfrak{m})$ we'll define by induction the sets V_α in R and subsets Φ_α in $R[x]$ (see Definition 5.2.1). Namely, put

$$\Phi_1 = \{f(x) - a_1 \mid f(x) \in F_{R[x]}(S_1; \{x\})\}$$

(definition of a_1 uses the numbering of the ring R). If now $c_1 \in R \backslash \Phi_1^{-1}(0)$ (see Definition 5.2.11), then put $V_1 = F_R(S_1; \{c_1\})$ (if $R = \Phi_1^{-1}(0)$, then put $V_1 = \{0\}$).

Let $\beta < \omega(\mathfrak{m})$. Suppose that V_α and Φ_α for $\alpha < \beta$ are already defined. Taking into account that $R \subseteq R[x]$ and $\{x\} \subseteq R[x]$ for every $\gamma < \beta$, put:

$$\Phi_{\gamma,\beta} = F_{R[x]}\left(S_\gamma \cup \left(\bigcup_{\rho < \gamma} V_\rho\right); \left(\bigcup_{\gamma \le \rho < \beta} V_\rho\right) \cup F_{R[x]}\left(S_\beta \cup \left(\bigcup_{\rho < \beta} V\rho\right); \{x\}\right)\right) - S_\gamma$$

and $\Phi_\beta = \bigcup_{\gamma < \beta} \Phi_{\gamma,\beta}$. If $c_\beta \in R \backslash \Phi_\beta^{-1}(0)$, then put $V_\beta = F_R(S_\beta \bigcup (\bigcup_{\gamma < \beta} V_\gamma); \{c_\beta\})$. As in defining V_1, put $V_\beta = \{0\}$, if $\Phi_\beta^{-1}(0) = R$. Hence, for every $\beta < \omega(\mathfrak{m})$ we have constructed the subset V_β in R and the subset Φ_β in $R[x]$.

---

* Here $\omega(\mathfrak{m})$ is the smallest transfinite number of the cardinality $\mathfrak{m}$ (see A.7).

** $|S|$ and $|\alpha|$ are the cardinalities of the set S and of the transfinite number α, respectively (see A.7).

For every $\alpha < \omega(\mathfrak{m})$ put $W_\alpha = \sum_{\alpha \leq \gamma < \omega(\mathfrak{m})} V_\gamma$.

5.4.4. PROPOSITION. The subsets V_α and Φ_α for $\alpha < \omega(\mathfrak{m})$ satisfy the following properties:

1) $(V_\alpha \cdot V_\beta) \bigcup (V_\beta \cdot V_\alpha) \subseteq V_\beta$ for $\alpha \leq \beta < \omega(\mathfrak{m})$;

2) $|V_\alpha|, |\Phi_\alpha| \leq \max\{|\alpha|, \aleph_0\}$;

3) $0 \notin \Phi_\alpha^{-1}(0)$ for $\alpha < \omega(\mathfrak{m})$.

PROOF. The fulfillment of the condition 1) easily follows from the method of constructing of the set V_γ for $\gamma < \omega(\mathfrak{m})$.

The verification of property 2) we'll carry out by the induction by α. Let $\alpha = 1$. Then the cardinalities of the subrings, generated by the subsets $\{a_1, x\} \subseteq R[x]$ and $\{a_1, c_1\} \subseteq R$, are not greater than $\aleph_0$. Hence, Φ_1 and V_1 as ideals of these subrings have the cardinalities not greater than $\aleph_0$.

Let $\beta < \omega(\mathfrak{m})$ and suppose that property 2) is already satisfied for all $\alpha < \beta$. Then the cardinalities of the subrings in $R[x]$ and R, generated by the subsets

$$\{x\} \cup \Big(\bigcup_{\alpha < \beta} V_\alpha \Big) \cup S_\beta \subseteq R[x]$$

and

$$\{c_\beta\} \cup \Big(\bigcup_{\alpha < \beta} V_\alpha \Big) \cup S_\beta \subseteq R$$

respectively are not greater than $\max\{|\beta|, \aleph_0\}$. It means that the cardinalities of the sets Φ_β and V_β as subsets of these rings are not greater than $\max\{|\beta|, \aleph_0\}$. This completes the verification of the property 2).

Let us verify the validity of property 3). Assume the contrary, i.e. that $0 \in \Phi_\beta^{-1}(0)$ for some $\beta < \omega(\mathfrak{m})$. Since $f(0) = 0$ for any generalized polynomial $f(x) \in F_{R[x]}(S_1; \{x\})$, then $\varphi(0) = a_1 \neq 0$ for any $\varphi(x) \in \Phi_1$, i.e. $0 \notin \Phi_1^{-1}(0)$. Hence, $\beta > 1$. From the definition of the set Φ_β follows that $0 \in \Phi_{\alpha,\beta}^{-1}(0)$ for some $\alpha < \beta$.

Let φ by a generalized polynomial from $\Phi_{\alpha,\beta}$ such that $\varphi(0) = 0$. Then there exists a polynomial

$$f(x) \in F_{R[x]}\Big(S_\alpha \cup \big(\bigcup_{\gamma < \alpha} V_\gamma \big); \big(\bigcup_{\alpha \leq \gamma < \beta} V_\gamma \big) \cup F_{R[x]} \big(S_\beta \cup \big(\bigcup_{\gamma < \beta} V_\gamma \big); \{x\} \big) \Big)$$

and an element $s \in S_\alpha$ such that $\varphi(x) = f(x) - s$. Since $\varphi(0) = 0$, then

$$s = f(0) \in F_R\Big(S_\alpha \cup \big(\bigcup_{\gamma < \alpha} V_\gamma\big); \big(\bigcup_{\alpha \leq \gamma < \beta} V_\gamma\big) \cup F_R\big(S_\beta \cup \big(\bigcup_{\gamma < \beta} V_\gamma\big); \{0\}\big)\Big) =$$

$$F_R\Big(S_\alpha \cup \big(\bigcup_{\gamma < \alpha} V_\gamma\big); \big(\bigcup_{\alpha \leq \gamma < \beta} V_\gamma\big)\Big)$$

because of $F_R(S; \{0\}) = 0$ for any $S \subseteq R$. Since s could be represented in terms of finite number of elements of the set $S_\alpha \cup (\bigcup_{\alpha < \beta} V_\gamma)$ (as an element of the subring, generated by this set), then

$$s \in F_R\Big(S_\alpha \cup \big(\bigcup_{\gamma < \alpha} V_\gamma\big); \bigcup_{\alpha \leq \gamma \leq \beta_1} V_\gamma\Big)$$

for some $\alpha \leq \beta_1 < \beta$. Then $\beta_1 > 1$. We can suppose that β_1 is the smallest transfinite number between such transfinite numbers, i.e. that

$$s \notin F_R\Big(S_\alpha \cup \big(\bigcup_{\gamma < \alpha} V_\gamma\big); \bigcup_{\alpha \leq \gamma < \beta_1} V_\gamma\Big).$$

Taking into account the definition of the set V_{β_1}, we derive that

$$s \in F_R\Big(S_\alpha \cup \big(\bigcup_{\gamma < \alpha} V_\gamma\big); \big(\bigcup_{\alpha \leq \gamma < \beta_1} V_\gamma\big) \cup F_R\big(S_{\beta_1} \cup \big(\bigcup_{\gamma < \beta_1} V_\gamma\big); \{c_{\beta_1}\}\big)\Big).$$

Then there exists a generalized polynomial

$$f_1(x) \in F_{R[x]}\Big(S_\alpha \cup \big(\bigcup_{\gamma < \alpha} V_\gamma\big); \big(\bigcup_{\alpha \leq \gamma < \beta_1} V_\gamma\big) \cup F_{R[x]}\big(S_{\beta_1} \cup \big(\bigcup_{\gamma < \beta_1} V_\gamma\big); \{x\}\big)\Big)$$

for which $f_1(c_{\beta_1}) = s$. Since

$$\varphi_1(x) = f_1(x) - s \in \Phi_{\alpha, \beta_1} \subseteq \Phi_{\beta_1}$$

and $\varphi_1(c_{\beta_1}) = f_1(c_{\beta_1}) - s = 0$, then $c_{\beta_1} \in \Phi_{\beta_1}^{-1}(0)$. We have the contradiction with the choice of the element c_{β_1} (see the construction of the sets V_γ above). This completes the proof of the proposition.

5.4.5. THEOREM. *Let* $R = \{0\} \cup \{a_\alpha \mid \alpha < \omega(\mathfrak{m})\}$ *be an infinite ring with the uncountable cardinality* $\mathfrak{m}$ *and* W_α, $\alpha < \omega(\mathfrak{m})$, *be the subsets, defined in Denotation 5.4.3.*

Then the family $\{W_\alpha \mid \alpha < \omega(\mathfrak{m})\}$ can be taken as a basis of neighborhoods of zero for a certain topology on R, and in this topology R is a Hausdorff topological ring*.

PROOF. As in proving Theorem 5.3.6, it is sufficient to verify that the family $\{W_\alpha \mid \alpha < \omega(\mathfrak{m})\}$ satisfies properties (BN1)–(BN6) and (BN1') (see Theorem 1.2.5 and Corollary 1.3.7). From the method of construction of the sets V_α and W_α for $\alpha < \omega(\mathfrak{m})$ it easily follows that the family $\{W_\alpha \mid \alpha < \omega(\mathfrak{m})\}$ satisfies the conditions (BN1) and (BN2).

Since W_α for $\alpha < \omega(\mathfrak{m})$ is a subgroup of the additive group of the ring R, then the conditions (BN3) and (BN4) are fulfilled.

Taking into account the property 1) of Proposition 5.4.4, for any $\alpha \leq \beta < \omega(\mathfrak{m})$ we have:

$$W_\alpha \cdot W_\beta = \left(\sum_{\alpha \leq \gamma} V_\gamma\right) \cdot \left(\sum_{\beta \leq \rho} V_\rho\right) = \sum_{\alpha \leq \gamma, \beta \leq \rho} (V_\gamma \cdot V_\rho)$$

$$\subseteq \sum_{\alpha \leq \gamma, \beta \leq \rho} V_{\max\{\gamma, \rho\}} \subseteq \sum_{\max\{\alpha, \beta\} \leq \xi} V_\xi = W_{\max\{\alpha, \beta\}}.$$

From this inclusion follows the fulfillment of the condition (BN5).

Now let $a \in R$ and $\beta < \omega(\mathfrak{m})$. If $a = 0$, then $a \cdot W_\alpha = 0 \cdot W_\alpha = \{0\} \subseteq W_\beta$ and $W_\alpha \cdot a = W_\alpha \cdot 0 = \{0\} \subseteq W_\beta$ for any $\alpha < \omega(\mathfrak{m})$. Therefore, in the case when $a = 0$ the condition (BN6) is fulfilled. If $a \neq 0$, then $a = a_\gamma$ (see the numbering of the ring R in Denotation 5.4.3) for some $\gamma < \omega(\mathfrak{m})$. Then $a = a_\gamma \in S_\rho$ for any $\gamma \leq \rho < \omega(\mathfrak{m})$ and therefore,

$$a \cdot W_{\max\{\beta, \gamma\}} = a \cdot \left(\sum_{\max\{\beta, \gamma\} < \rho} V_\rho\right) \subseteq \sum_{\max\{\beta, \gamma\} \leq \rho} (S_\rho \cdot V_\rho) \subseteq \sum_{\max\{\beta, \gamma\} \leq \rho} V_\rho \subseteq W_\beta.$$

Similarly it is verified that $W_{\max\{\beta, \gamma\}} \cdot a \subseteq W_\beta$. By means of this the fulfillment of the condition (BN6) is verified.

To complete the proof it remains to verify the fulfillment of the condition (BN1'). Assume the contrary, i.e. that $0 \neq a \in \bigcap_{\gamma < \omega(\mathfrak{m})} W_\gamma$. Let $a = a_\rho$ (see Denonation 5.4.3), then, according to the assumption, $a_\rho \in W_\rho$ and therefore, there exist transfinite numbers $\rho < \rho_1 < \rho_2 \cdots < \rho_n < \omega(\mathfrak{m})$ and non-zero elements $b_i \in V_{\rho_i}$, for $i = 1, \ldots, n$, such that

---

* In general case the constructed topology could be discrete.

$a_\rho = \sum_{i=1}^{n} b_i$. Since $b_n \in V_{\rho_n}$, then from the construction of the set V_{ρ_n} follows that $b_n = f(c_{\rho_n})$ for some generalized polynomial

$$f(x) \in F_{R[x]}\left(S_{\rho_n} \cup \left(\bigcup_{\gamma < \rho_n} V_\gamma\right); \{x\}\right).$$

Since $\sum_{i=1}^{n-1} b_i \in \sum_{\rho < \gamma < \rho_n} V_\gamma$, then from the definition of the set Φ_{ρ,ρ_n} follows that

$$\varphi(x) = f(x) + \sum_{i=1}^{n-1} b_i \in$$

$$F_{R[x]}\left(S_\rho \cup \left(\bigcup_{\gamma < \rho} V_\gamma\right); \left(\bigcup_{\rho \leq \gamma < \rho_n} V_\gamma\right) \cup F_{R[x]}\left(S_{\rho_n} \cup \left(\bigcup_{\gamma < \rho_n} V_\gamma\right); \{x\}\right)\right).$$

Hence, $\varphi(x) - a_\rho \in \Phi_{\rho,\rho_n} - S_\rho \subseteq \Phi_{\rho_n}$. In addition,

$$\varphi(c_{\rho_n}) - a_\rho = f(c_{\rho_n}) + \sum_{n=1}^{n-1} b_i - a_\rho = b_n + \sum_{n=1}^{n-1} b_i - a_\rho = 0.$$

We have the contradiction with the fact that the element c_{ρ_n} is not a root of any generalized polynomial from Φ_{ρ_n} (see Denotation 5.4.3, the construction of the sets V_β). This completes the proof of the theorem.

5.4.6. THEOREM. *Let R be an infinite ring with uncountable cardinality $\mathfrak{m}$ such that for any set Φ of generalized polynomials with non-zero free members (see Definition 5.2.2) over the ring R, which has the cardinality less than the cardinality of the ring R, (i.e. $|\Phi| < |R|$) in the ring R exists an element which is not a root of any generalized polynomial from Φ. Then the ring R admits a non-discrete Hausdorff ring topology.*

PROOF. Let τ be a ring topology, constructed on the ring R in Theorem 5.4.5. It is clear that for proving the theorem it is sufficient to verify that for every $\alpha < \omega(\mathfrak{m})$ the set V_α can be chosen different from $\{0\}$, i.e. that $\Phi_\alpha^{-1}(0) \not\supseteq R \backslash \{0\}$ for every $\alpha < \omega(\mathfrak{m})$.

According to Proposition 5.4.4,

$$|\Phi_\alpha| \leq \max\{|\alpha|, \aleph_0\} < \mathfrak{m} = |R|$$

and $0 \notin \Phi_\alpha^{-1}(0)$ for every $\alpha < \omega(\mathfrak{m})$. Since $f(0) \neq 0$ for any generalized polynomial $f(x) \in \Phi_\alpha$, i.e. the free member of the generalized polynomial $f(x) \in \Phi_\alpha$ is different from zero, then, by the condition of the theorem, in the ring R exists a non-zero element c_α, which is not a root of any generalized polynomial from Φ_α. It means that the set V_α can be chosen different from $\{0\}$.

This completes the proof of the theorem.

5.4.7. COROLLARY. *Let R be an infinite uncountable commutative ring without divisors of zero, then R admits a non-discrete Hausdorff ring topology.*

PROOF. Indeed, according to Theorem 5.4.6, it is sufficient to prove that $\Phi^{-1}(0) \not\supseteq R\backslash\{0\}$ for any family Φ of generalized polynomials with non-zero free members over the ring R, such that $|\Phi| < |R|$. Let Φ be the family mentioned above, then $f(0) \neq 0$ for any generalized polynomial $f(x) \in \Phi$. Then from Corollary 5.2.10 follows that $|\Phi^{-1}(0)| < |R|$ and therefore, $R\backslash\{0\} \not\subseteq \Phi^{-1}(0)$.

§ 5.5. Non-Discrete Topologization of Some Infinite Rings

5.5.1. LEMMA. *Let A and B be finite subsets of a ring R and let G be an infinite subgroup of the additive group of the ring R. Then:*

1) if $G \cdot A = \{g \cdot a \,|\, g \in G, \, \alpha \in A\} \subseteq B$, then the set $A_l^* = \{g \in G \,|\, g \cdot A = 0\}$ is infinite;

2) if $A \cdot G = \{a \cdot g \,|\, a \in A, \, g \in G\} \subseteq B$, then the set $A_r^* = \{g \in G \,|\, A \cdot g = 0\}$ is infinite;

3) if $A \cdot G \subseteq B$ and $G \cdot A \subseteq B$, then the set $A^* = \{g \in G \,|\, A \cdot g = g \cdot A = 0\}$ is infinite.

PROOF. Let $A = \{a_1, \dots, a_n\}$ and $B = \{b_1, \dots, b_m\}$. Consider the set M of all sequences of the length n, composed from natural numbers from 1 to m, i.e.

$$M = \left\{ \bar{k} = (k_1, k_2, \dots, k_n) \,\middle|\, 1 \le k_i \le m \quad \text{for} \quad 1 \le i \le n \right\}.$$

For every sequence $\bar{k} = (k_1, k_2, \dots, k_n) \in M$ put $G_{\bar{k}} = \{g \in G \,|\, g \cdot a_i = b_{k_i} \text{ for } i = 1, 2, \dots, n\}$. Since $G = \bigcup_{\bar{k} \in M} G_{\bar{k}}$ and M is a finite set, then for some $\bar{k}_0 \in M$ the subset $G_{\bar{k}_o}$ is infinite. Then the set $\{g_1 - g_2 \,|\, g_1, g_2 \in G_{\bar{k}_0}\}$ is infinite, too.

In this case, if $\bar{k}_0 = (k_1, \dots, k_n)$, then $(g_1 - g_2) \cdot a_j = b_{k_j} - b_{k_j} = 0$ for any $g_1, g_2 \in G_{\bar{k}_0}$ and any $j = 1, \dots, m$. It means that the subset $A_l^* = \{g \in G \,|\, g \cdot A = 0\}$ is infinite. This proves the statement 1).

The proof of statement 2) is carried out analogously.

The statement 3) is proved by consequent applying of the statements 1) and 2) at first to G and then to A_l^*.

5.5.2. PROPOSITION. *Let R be a ring and N be the sum of all nilpotent ideals of the ring R. If N is infinite, then there exists an infinite ideal J_0 in R such that $J_0^3 = 0$.*

PROOF. Let $\mathfrak{M}$ be the set of all ideals J of the ring R, such that $J^3 = 0$. Since the union of an ascending sequence of ideals from $\mathfrak{M}$ belongs to $\mathfrak{M}$, then, by Zorn's Lemma (see A.11), there exist maximal elements in $\mathfrak{M}$.

Let J_0 be some ideal, being a maximal element in $\mathfrak{M}$.

We'll show that it is an infinite ideal. Assume the contrary, i.e. that J_0 is finite. Then, putting $A = B = J_0$ in Lemma 5.5.1, we get, according to the statement 3) of this lemma, that

$$J_0^* = \{g \in N \,|\, g \cdot J_0 = J_0 \cdot g = 0\}$$

is an infinite ideal. Since $J_0^* \subseteq N$, then there exists a nilpotent ideal J of the ring R such that $J \subseteq J_0^*$ but $J \not\subseteq J_0$.

Let $m = \min\{k \in \mathbb{N} \,|\, \text{there exists an ideal } J_k \text{ in } R \text{ such that } J_k \subseteq J_0^*, J_k \not\subseteq J_0, (J_k)^k = 0\}$ and J_m be the ideal, corresponding to this number m. Since $J_m^2 \subseteq J_0^*$ and $(J_m^2)^{m-1} \subseteq J_m^m = 0$, then from the fact that m is minimal number follows that $J_m^2 \subseteq J_0$. Then

$$(J_0 + J_m)^3 \subseteq J_0^3 + J_0 J_m + J_m J_0 + J_m^3 \subseteq \{0\} + J_0 J_0^* + J_0^* J_0 + J_0 J_0^* = \{0\}.$$

Therefore, by virtue of the maximality J_0 in $\mathfrak{M}$, we get that $J_m \subseteq J_0$. Thus, we have the contradiction with the choice of ideal J_m. The proposition is proved.

5.5.3. PROPOSITION. *Let R be a commutative ring and N be the set of its nilpotent elements. If N is infinite, then in R exists an infinite ideal J such that $J^2 = 0$.*

PROOF. Assume the contrary, i.e. that any ideal of the ring R, for which $J^2 = 0$, is finite. Since an ideal generated by a nilpotent element is also nilpotent in the commutative ring, then N is the sum of all nilpotent ideals of the ring R. Then, according to Proposition 5.5.2, there exists an infinite ideal J_0 in the ring R such that $J_0^3 = 0$. Since $(J_0^2)^2 = 0$, then according to the assumption, J_0^2 is a finite ideal.

By induction we'll construct the sequence $a_1, a_2, \ldots$ of pairwise distinct elements of J such that $a_i \cdot a_j = 0$ for $i, j \in \mathbb{N}$, $i \neq j$.

Put $a_1 = 0$ and suppose that the elements $a_1, \ldots, a_m \in J_0$ are already determined, moreover that $a_i \cdot a_j = 0$ for $1 \leq i, j \leq m$, $i \neq j$.

Put $A = \{a_1, \ldots, a_m\}$, $B = J_0^2$ and $G = J_0$. Then $G \cdot A \subseteq J_0 \cdot J_0 \subseteq J_0^2 = B$ and $A \cdot G = J_0 \cdot J_0 \subseteq J_0^2 = B$. According to Lemma 5.5.1, the set $A^* = \{g \in G \,|\, A \cdot g = g \cdot A = 0\}$ is infinite. As a_{m+1} we take an arbitrary element from $A^* \setminus \{a_1, \ldots, a_m\}$.

Since $a_{m+1} \cdot a_i = a_i \cdot a_{m+1} = 0$ for any $1 \leq i \leq m$, then taking into account the inductive supposition, we obtain that $a_i \cdot a_j = 0$ for $1 \leq i, j \leq m+1$ and $i \neq j$. Therefore, we've constructed the required sequence $a_1, a_2, \ldots$

Since $a_i^2 \in J_0^2$ for any $i \in \mathbb{N}$ and J_0^2 is a finite ideal, then there exists an element $a \in J_0^2$ such that the set $C = \{i \in \mathbb{N} \,|\, a_i^2 = a\}$ is infinite. Let $|J_0^2| = n_0$ (see A.7), then $n_0 \cdot a = 0$.

Consider the splitting of the set C into a countable number of pairwise disjoint subsets C_k for $k \in \mathbb{N}$, each of which contains exactly n_0 numbers.

Put $b_k = \sum_{i \in C_k} a_i$, $k \in \mathbb{N}$. Then from the fact that $a_i \cdot a_j = 0$ for $i \neq j$ follows that

$$b_k^2 = \left(\sum_{i \in C_k} a_i \right)^2 = \sum_{i,j \in C_k} a_i \cdot a_j = \sum_{i \in C_k} a_i^2 = n_0 \cdot a = 0$$

for any $k \in \mathbb{N}$ and $b_i \cdot b_j = 0$ for any $i, j \in \mathbb{N}$, $i \neq j$.

Since R is a commutative ring, then the ideal J, generated by the set $\{b_k \mid k \in \mathbb{N}\}$, is infinite and, moreover $J^2 = 0$. We have the contradiction with our assumption. That completes the proof of the proposition.

5.5.4. PROPOSITION. *Let R be a commutative ring in which any non-zero ideal contains a non-zero nilpotent element. If the set N of all nilpotent elements of the ring R is finite, then the ring R is finite, too.*

PROOF. Assume the contrary, i.e. that R is an infinite ring. Putting in Lemma 5.5.1 $A = B = N$ and $G = R$, we obtain that the set $N^* = \{r \in R \mid r \cdot N = 0\}$ is infinite. Therefore, there exists a non-nilpotent element $a \in N^*$. Then $a^2 \cdot R$ is a non-zero ideal in the ring R. According to our assumption, it contains some non-zero nilpotent element, i.e. there exists an element $b \in R$ such that $a^2 \cdot b$ is a non-zero nilpotent element. Since $(a \cdot b)^{2k} = (a^2 \cdot b)^k \cdot b^k$ for any $k \in \mathbb{N}$, then the element $a \cdot b$ is nilpotent, too. Then $0 \neq a^2 \cdot b = a \cdot (a \cdot b) \in N^* \cdot N = \{0\}$. We have a contradiction. This completes the proof of the proposition.

5.5.5. PROPOSITION. *Let in an infinite commutative ring R the set of all nilpotent elements be finite. Then R contains an infinite ideal, which contains no non-zero nilpotent elements.*

PROOF. Assume the contrary, i.e. that every ideal of the ring R, which contains no non-zero nilpotent elements, is finite. Let $\mathfrak{M}$ be the set of all ideals of the ring R, containing no non-zero nilpotent elements. Then, since a union of an ascending chain of ideals from $\mathfrak{M}$ is an ideal from $\mathfrak{M}$, then in $\mathfrak{M}$ exist maximal elements (see A.11, Zorn's lemma).

Let J_1 be some maximal element in $\mathfrak{M}$. Then, according to our assumption, the ideal J_1 is finite and, therefore, is a direct sum of a finite number of finite fields (see B.13). Then J_1 possesses the unitary elements e. Let $R = R \cdot e + R \cdot (1-e)$, where $R \cdot (1-e) = \{a - a \cdot e \mid a \in R\}$. Since $R \cdot e = J_1$, then from the maximality of the ideal J_1 follows that in the ring $R \cdot (1-e)$

any non-zero ideal contains a non-zero nilpotent element. Then, according to Proposition 5.5.4, $R \cdot (1 - e)$ is a finite ring. Therefore, the ring R, as the direct sum of two finite rings, is finite. We have the contradiction. This completes the proof of the proposition.

5.5.6. THEOREM. *Let a ring R contain an infinite ideal J such that $J^2 = 0$. Then R admits a non-discrete Hausdorff ring topology in which J is an open ideal.*

PROOF. It is clear that J is a right as well as left R-module. Let τ be the largest precompact group topology on J. Then from the proof of Theorem 5.1.2 follows that (J, τ) is a topological left module over the ring R with the discrete topology. Analogously to Theorem 5.1.2 it could be shown that (J, τ) is a topological right module over the ring R with the discrete topology.

Let $\mathcal{B}$ be the family of all neighborhoods of zero in (J, τ). We show now that the family $\mathcal{B}$ considered as a family of the subsets of the ring R is a basis of neighborhoods of zero in a certain Hausdorff ring topology on R. To do this, according to Corollary 1.3.7, it is sufficient to verify that $\mathcal{B}$ satisfies the conditions (BN1)-(BN6) and (BN1').

Since τ is a Hausdorff group topology on J, then the family $\mathcal{B}$ fulfills conditions (BN1)-(BN4) and (BN1').

Since $V \subseteq J$ for any $V \in \mathcal{B}$ and $J^2 = 0$, then $V^2 = \{0\} \subseteq V$ for any $V \in \mathcal{B}$. Thus, we have verified the fulfillment of the condition (BN5).

Let now $V \in \mathcal{B}$ and $r \in R$. Since (J, τ) is both a left and right topological R-module, then there are neighborhoods V_1 and V_2 of zero in (J, τ) such that $r_1 \cdot V_1 \subseteq V$ and $V_2 \cdot r \subseteq V$. Then $(V_1 \cap V_2) \in \mathcal{B}$, moreover, $r \cdot (V_1 \cap V_2) \subseteq V$ and $(V_1 \cap V_2) \cdot r \subseteq V$. By this we have verified also the fulfillment of the condition (BN6), hence, the theorem is completely proved.

5.5.7. COROLLARY. If a commutative ring R contains an infinite number of nilpotent elements, then it admits a non-discrete Hausdorff ring topology.

5.5.8. THEOREM. *Let a commutative ring R have an infinite ideal J, which contains no non-zero nilpotent elements. Then R admits a non-discrete Hausdorff ring topology in which J is an open ideal.*

PROOF. Denote over $\mathfrak{A}$ the family of all ideals of the ring R, each of which is itself a field and is contained in J. The following 3 cases are possible:

1) $\mathfrak{A}$ is an infinite set;

2) an infinite ideal $A \in \mathfrak{A}$ exists;

3) $\mathfrak{A}$ is a finite set, and any ideal $A \in \mathfrak{A}$ is finite.

Consider each case separately.

First case. Let $\{A_i \mid i \in \mathbb{N}\}$ be some countable subset in $\mathfrak{A}$. For any natural number n consider the ideal $V_n = \sum_{i=n}^{\infty} A_i$ in the ring R and show that the family $\{V_n \mid n \in \mathbb{N}\}$ is a basis of neighborhoods of zero in a certain Hausdorff non-discrete topology on R.

Since V_n is an ideal in the ring R for every $n \in \mathbb{N}$, then the family $\{V_n \mid n \in \mathbb{N}\}$ satisfies the conditions (BN1)-(BN6).

Let now $a \in \bigcap_{n=1}^{\infty} V_n$. There exists $m \in \mathbb{N}$ such that $a \in \sum_{i=1}^{m} A_i$. Since each of A_k, $k \in \mathbb{N}$, is a minimal ideal in R, then $A_i \cdot A_j = 0$ for $i, j \in \mathbb{N}$, $i \neq j$, and, hence

$$a^2 \in \left(\sum_{n=1}^{m} A_i \right) \cdot V_{m+1} = \left(\sum_{i=1}^{m} A_i \right) \cdot \left(\sum_{i=m+1}^{\infty} A_i \right) = \{0\}.$$

Since J contains no non-zero nilpotent elements, then $a = 0$. It means that the family $\{V_n \mid n \in \mathbb{N}\}$ satisfies the condition (BN1$'$), too.

Since $V_n \neq \{0\}$ for any $n \in \mathbb{N}$, then the topology, defined by the family $\{V_n \mid n \in \mathbb{N}\}$ as a basis of neighborhoods of zero, is non-discrete.

Second case. Let $A \in \mathfrak{A}$ be an infinite ideal of the ring R. Since A is an infinite field, then from Theorem 5.3.8 and Corollary 5.4.7 follows that A admits a non-discrete Hausdorff ring topology τ. Let $\{V_\gamma \mid \gamma \in \Gamma\}$ be the family of all neighborhoods of zero in (A, τ). Then, according to Proposition 1.2.2, this family satisfies the conditions (BN1)-(BN6). Then, considering each of V_γ, $\gamma \in \Gamma$, as a subset in R too, we obtain that the family $\{V_\gamma \mid \gamma \in \Gamma\}$ of subsets in R satisfies the conditions (BN1)-(BN5).

Now let $r \in R$ and $\gamma \in \Gamma$. Let e be the unitary element of the field A, then $r \cdot e \in A$ and it means that there exists $\gamma_1 \in \Gamma$ such that $r \cdot e \cdot V_{\gamma_1} \subseteq V_\gamma$. Then $r \cdot V_{\gamma_1} = r \cdot (e \cdot V_{\gamma_1}) = (r \cdot e) \cdot V_{\gamma_1} \subseteq V_\gamma$. Since R is a commutative ring, then from the arbitrariness of the choice of $r \in R$ and $\gamma \in \Gamma$ follows the fulfillment of the condition (BN6).

Furthermore, since the topological ring (A, τ) is Hausdorff, then $\bigcap_{\gamma \in \Gamma} V_\gamma = \{0\}$, hence the condition (BN1$'$) is fulfilled, too. Thus, we have shown that the family $\{V_\gamma \mid \gamma \in \Gamma\}$

defines on R a non-discrete Hausdorff ring topology, being a basis of the neighborhoods of zero.

Third case. Let $\mathfrak{A} = \{A_1, \ldots, A_n\}$, then $A = \sum_{i=1}^{n} A_i$ is a finite ideal in R. Then from Lemma 5.5.1 it follows that the set $A^* = \{r \in R \mid r \cdot A = 0\}$ is an infinite ideal in R.

By transfinite induction we'll construct a descending transfinite chain $J_1 \supseteq J_2 \ldots$ of non-zero ideals J_α in the ring R such that $\bigcap_\alpha J_\alpha = \{0\}$.

Put $J_1 = A^*$ and suppose that the ideals are constructed J_α for $\alpha < \beta$.

If $\bigcap_{\alpha < \beta} J_\alpha = \{0\}$, then the required chain is constructed.

Let now $B = \bigcap_{\alpha < \beta} J_\alpha \neq 0$. Since $B \cdot (\sum_{i=1}^{n} A_i) = B \cdot A \subseteq A^* \cdot A = \{0\}$ and $\{A_1, \ldots, A_n\}$ is the family of all ideals of the ring R, each of which is a field and is contained in J, then B is not a field. Since J does not contain nilpotent ideals and $B \subseteq J$, then $B^2 \neq 0$. Hence (see B.10), the ring B possesses some proper ideal B_1. Then BB_1 is a non-zero ideal in R, moreover, $BB_1 \subseteq B = \bigcap_{\alpha < \beta} J_\alpha$. Hence, we can take $J_\beta = BB_1$.

Thus, we have constructed the descending transfinite sequence $J_1 \supseteq J_2 \supseteq \ldots$ of non-zero ideals J_α of the ring R such that $\bigcap_\alpha J_\alpha = \{0\}$.

It is clear that the family of all constructed ideals J_α satisfies the conditions (BN1)-(BN6) and (BN1'). It means that it defines on R a non-discrete Hausdorff ring topology, being a basis of neighborhoods of zero.

It is obvious that in any of considered cases J is an open ideal. This proves completely the theorem.

5.5.9. THEOREM. *Any infinite commutative ring R admits a non-discrete Hausdorff ring topology.*

PROOF. Let N be the set of all nilpotent elements of the ring R.

If N is an infinite ideal in R, then, according to Proposition 5.5.3, the ring R possesses an infinite ideal J such that $J^2 = 0$. Then by Theorem 5.5.6 the ring R admits a non-discrete Hausdorff ring topology.

If N is a finite ideal, then by the virtue of the infinity of the ring R, from Proposition 5.5.5 follows that R contains an infinite ideal J containing no non-zero nilpotent elements. Then, according to Theorem 5.5.8, the ring R admits a non-discrete Hausdorff ring topology. This proves completely the theorem.

5.5.10. THEOREM. *Let* $\{A_\gamma \mid \gamma \in \Gamma\}$ *be a family of subgroups of the additive group of a ring* R, *and in every subgroup* A_γ *is determined a subset* C_γ, *which is not algebraic over the ring* R *with respect to* A_γ. *If* $|\Gamma| \leq |R|$ *(see A.7), then the ring* R *admits a Hausdorff ring topology such that for every* $\gamma \in \Gamma$ *the set* C_γ *is dense in* A_γ.

PROOF. Let $\mathfrak{m}$ be the cardinality of the ring R. Since the set of all pairs of the type (γ, a), where $\gamma \in \Gamma$ and $a \in R$, has the cardinality $\mathfrak{m}$, then the set $M = \{(\gamma, a) \mid \gamma \in \Gamma, a \in A_\gamma\}$ has a cardinality not more than $\mathfrak{m}$. Then the set $B = \{\alpha \mid \alpha < \omega(\mathfrak{m})\}$ can be split into a family of pairwise disjoint subsets $B(\gamma, a)$, $(\gamma, a) \in M$ such that $|B(\gamma, a)| = \mathfrak{m}$.

Let α_0 be a transfinite number less than $\omega(\mathfrak{m})^*$. Then there exists, being unique, a pair $(\gamma_0, b_0) \in B$ such that $\alpha_0 \in B(\gamma_0, b_0)$. By Theorem 5.2.13, the set $C_\gamma - b_0$ is not algebraic over the ring R with respect to A_{γ_0}. Since $0 \in A_\gamma$, then the set $C_{\gamma_0} - b_0$ is in particular not algebraic over the ring R with respect to $\{0\}$.

Further, if $\mathfrak{m} = \aleph_0$, then from Proposition 5.3.4 (items IX and X) follows that $C_{\gamma_0} - b_0 \not\subseteq \Phi_{i,j}^{-1}(0) = S_{i,j}$ for any $i, j \in \mathbb{N}$, hence $C_{\gamma_0} - b_0 \not\subseteq S_{2\alpha_0+1,\alpha}$ for $\alpha_0, \alpha \in \mathbb{N}$. Therefore, when constructing the sets $V_{2\alpha_0+1,\alpha_0}$, for $\alpha_0 \in \mathbb{N}$ as a we take some element from $(C_{\gamma_0} - b_0)\backslash S_{2\alpha_0+1,\alpha-0}$.

Similarly, if $\mathfrak{m} > \aleph_0$, then from Proposition 5.4.4 (items 2 and 3) follows that $C_\gamma - b_0 \not\subseteq \Phi_{\alpha_0}^{-1}(0)$ for $\alpha_0 < w(\mathfrak{m})$. Therefore, when constructing the sets V_{α_0}, as c_{α_0} we take an element from $(C_{\gamma_0} - b_0)\backslash\Phi_{\alpha_0}^{-1}(0)$ for $\alpha_0 < w(\mathfrak{m})$.

If now W_{α_0} is the set, constructed according to Theorem 5.3.6 for the case $\mathfrak{m} = \aleph_0$ and according to Theorem 5.4.5 for the case $\mathfrak{m} > \aleph_0$, then $W_{\alpha_0} \cap (C_{\gamma_0} - b_0) \neq 0$, i.e. in both cases in the neighborhood W_{α_0} is contained some element from $C_{\gamma_0} - b_0$.

Now we'll show that in the ring topology τ, defined on R by the family $\{W_\alpha \mid \alpha < \omega(\mathfrak{m})\}$, (see Theorems 5.3.6 and 5.4.5), the subset C_γ is dense in A_γ for $\gamma \in \Gamma$.

Indeed, let $\gamma \in \Gamma$ and b be an element from A_γ. Let W_α be some neighborhood of zero. Since $|\alpha| < \mathfrak{m}$, then in the set $B(\gamma, b)$ exists some transfinite number $\xi > \alpha$ (otherwise, the cardinality of the set $B(\gamma, b)$ could be not greater than $|\alpha|$).

As it has been shown above, $W_\xi \cap (C_\gamma - b_0) \neq 0$. From Proposition 5.3.4 (item VI) for the case when $\mathfrak{m} = \aleph_0$, and from the definition of W_ρ (see Denotation 5.4.3) for

---

* If $\mathfrak{m} = \aleph_0$, then α_0 is a natural number.

the case when $\mathfrak{m} > \aleph_0$ it follows that $W_\xi \subseteq W_\alpha$. Then $W_\alpha \cap (C_\gamma - b) \neq 0$ and, hence $(W_\alpha + b) \cap C_\gamma \neq 0$. From the arbitrariness of the choice of the element b and of the neighborhood W_α follows the density of the set C_γ in the subgroup A_γ. This completely proves the theorem.

If in Theorem 5.5.10 as $\{C_\gamma \mid \gamma \in \Gamma\}$ we take the set of all principal non-zero (left, right, one-sided, two-sided) ideals of the ring R and put $A_\gamma = R$ for all $\gamma \in \Gamma$, then we get:

5.5.11. COROLLARY. Let in a ring R any non-zero ideal (left, right, one-sided, two-sided) is not algebraic over R. Then the ring R admits a Hausdorff ring topology, in which it does not contain closed proper ideals (left, right, one-sided, two-sided).

From Theorems 5.2.9 and 5.5.10 we get:

5.5.12. COROLLARY. Let $\{C_\gamma \mid \gamma \in \Gamma\}$ be a family of subsets of an infinite commutative ring R without divisors of zero. If $|\Gamma| \leq |R|$ and $|C_\gamma| = |R|$ for every $\gamma \in \Gamma$, then the ring R admits a Hausdorff ring topology τ such that every C_γ is totally dense in (R, τ).

In particular, taking $C_0 = R \backslash \{0\}$ and $C_a = a \cdot R$ for every non-zero element $a \in R$, we get:

5.5.13. COROLLARY. Any infinite commutative ring R without divisors of zero admits a non-discrete Hausdorff ring topology in which the ring R does not contain closed proper ideals.

5.5.14. COROLLARY. Let R be a commutative ring without divisors of zero and let $\widehat{R}$ be its quotient field (see B.18). Then on $\widehat{R}$ exists a Hausdorff non-discrete topology such that $(\widehat{R}, \tau)$ is a topological field and, moreover, R is a totally dense subring in $(\widehat{R}, \tau)$.

Indeed, $|\widehat{R}| = |R|$. According to Corollary 5.5.12, on $\widehat{R}$ exists a ring topology τ such that R is a dense subset in $(\widehat{R}, \tau)$. According to Theorem 1.7.8, the topology τ can be weakened to such topology $\widehat{\tau}$ that $(\widehat{R}, \widehat{\tau})$ becomes a topological field. The density of the ring R in $(\widehat{R}, \tau)$ is obvious.

§ 5.6. Example of Infinite Non-Associative Commutative
Ring Not Admitting Non-Discrete Topologies*

5.6.1. DENOTATIONS. Let $\mathfrak{m}$ be a cardinality greater than the cardinality of the continuum $\mathfrak{c}$, and let B be a set of the cardinality $\mathfrak{m}$. Then B can be enumerated by transfinite numbers less than $\omega(\mathfrak{m})$, i.e. $B = \{x_\alpha \,|\, \alpha < \omega(\mathfrak{m})\}$.

Let $\mathfrak{M}$ be the set of all countable sequences of natural numbers, then $\mathfrak{M}$ can be enumerated by transfinite numbers greater than ω_0** and less than $\omega(\mathfrak{c})$, i.e.

$$\mathfrak{M} = \{\mathfrak{A}_\alpha \,|\, \omega_0 < \alpha < \omega(\mathfrak{c})\}.$$

For any natural numbers i and k we define the set $B_{i,k}$ of all elements x_α, such that in the sequence $\mathfrak{A}_\alpha$ (with the same index) on the place i is the number k , i.e.

$$B_{i,k} = \{x_\alpha \,|\, \mathfrak{A}_\alpha = n_1, n_2, \dots , \text{ and } n_i = k\}.$$

It is obvious that for any natural number i we have:

1) $B_{i,k} \cap B_{i,k_2} = \emptyset$, if $k_1 \neq k_2$;

2) $\bigcup_{k=1}^{\infty} B_{k,i} = \{x_\alpha | \omega_0 < \alpha < \omega(\mathfrak{c})\}$.

We enumerate the set of all positive rational numbers by natural numbers $r_1, r_2, \dots$, and for any natural numbers i and k put $r_{i,k} = \frac{1}{r_i \cdot r_k}$.

Now let A be an algebra over the field $\mathbb{Q}$ of rational numbers with the set of generators $B \bigcup \{e\}$ and the following rule of multiplication for components:

$$e \cdot x_\alpha = x_\alpha \cdot e = x_\alpha \quad \text{for any} \quad x_\alpha \in B \quad \text{and}$$

$$x_\alpha \cdot x_\beta = \begin{cases} e, & \text{if } \alpha = \beta; \\ r_{i,k} \cdot x_{\beta+\alpha}, & \text{if } \alpha = i \text{ and } x_\beta \in B_{i,k}; \\ r_{i,k} \cdot x_{\alpha+\beta}, & \text{if } \beta = i \text{ and } x_\alpha \in B_{i,k} \\ 0, & \text{in other cases.} \end{cases}$$

---

* The first example of the ring not admitting non-discrete Hausdorff topologies was constructed in 1970 (see [29]). Then in 1990 in [132] was noted that insignificant change in this example permits one to obtain a commutative ring not admitting non-discrete Hausdorff topologies.

** ω_0 is the smallest countable transfinite number.

From properties 1) and 2) it follows that the introduced operation of multiplication of generators is correct.

5.6.2. REMARK. Since $x_\alpha \cdot x_\beta = x_\beta \cdot x_\alpha$ for any $\alpha, \beta < \omega(\mathfrak{m})$, then the ring A is commutative. Since $(x_1 \cdot x_1) \cdot x_2 = x_2 \neq 0 = x_1 \cdot (x_1 \cdot x_2)$, then the ring A is not associative.

5.6.3. THEOREM. *The ring A, constructed in Denotation 5.6.1, does not admit a non-discrete Hausdorff ring topology.*

PROOF. Assume the contrary and let τ be a non-discrete Hausdorff ring topology on A.

Let us show first that for any neighborhood V_0 of zero in (A, τ) there exists a non-zero rational number b such that $b \cdot e \in V_0$. Let $V_1 \supseteq V_2 \supseteq V_3 \supseteq V_4 \supseteq V_5$ be neighborhoods of zero in (A, τ) such that: $V_5 \cdot V_5 \subseteq V_4$; $x_{\omega_0} \cdot V_4 \subseteq V_3$; $V_3 \cdot V_3 \subseteq V_2$; $x_1 \cdot V_2 \subseteq V_1$; $V_1 \cdot V_1 \subseteq V_0$. Let now $0 \neq r_0 \cdot e + \sum_{i=1}^{n} r_i \cdot x_{\alpha_i} \in V_5$, then

$$V_4 \supseteq V_5 \cdot V_5 \ni \left(r_0 \cdot e + \sum_{i=1}^{n} r_i \cdot x_{\alpha_i} \right)^2$$

$$= \sum_{i=0}^{n} r_i^2 \cdot e + \sum_{i=1}^{n} 2 r_i \cdot r_0 \cdot x_{\alpha_i} + \sum_{i \neq j} r_i \cdot r_j \cdot x_{\alpha_i} \cdot x_{\alpha_j} = a_0 \cdot e + \sum_{i=1}^{k} a_i \cdot x_{\gamma_i},$$

where $a_i \in \mathbb{Q}$ and $\gamma_i \leq \omega(\mathfrak{m})$ for $1 \leq k$, moreover, $a_0 \neq 0$. Since $x_{\omega_0} \cdot x_\alpha = 0$ for any $\alpha \neq \omega_0$ and $x_{\omega_0}^2 = e$, then $V_3 \supseteq x_{\omega_0} \cdot V_4 = a_0 \cdot x_{\omega_0} + a \cdot e$ (where $a \in \mathbb{Q}$ and could also be equal to zero, too). Further,

$$V_2 \supseteq V_3 \cdot V_3 \ni (a_0 \cdot x_{\omega_0} + a \cdot e)^2 = a_0^2 \cdot e + a^2 \cdot e + 2 a_0 \cdot a \cdot x_{\omega_0} = b_0 \cdot e + b_1 \cdot x_{\omega_0},$$

moreover, $b_0 \neq 0$. Since $x_1 \cdot x_{\omega_0} = 0$, then

$$V_1 \supseteq x_1 \cdot V_2 \ni x_1 \cdot (b_0 + b_1 \cdot x_{\omega_0}) = b_0 \cdot x_1.$$

Then $V_0 \supseteq V_1 \cdot V_1 \ni (b_0 \cdot x_1)^2 = b_0^2 \cdot e$, moreover, as it was mentioned above, $b_0 \neq 0$.

In summary, we have proved that for any neighborhood V_0 of zero in (A, τ) there exists a positive rational number b such that $b \cdot e \in V_0$.

Then for any $\alpha < \omega(\mathfrak{m})$ and any neighborhood U of zero in (A, τ) there exists a positive rational number r such that $r \cdot x_\alpha \in U$.

Since (A, τ) is a Hausdorff ring, then there exists a neighborhood U_0 of zero such that $e \notin U_0$. Let U_1 be a neighborhood of zero in (A, τ) such that $(U_1 \cdot U_1) \cdot (U_1 \cdot U_1) \subseteq U_0$. Then, as was noted above, for every $i \in \mathbb{N}$ exists a positive rational number r_{n_i} (see the numbering of rational numbers in Denotation 5.6.1) such that $r_{n_i} \cdot x_i \in U_1$. Then the sequence $n_1, n_2 \ldots$ of natural numbers belongs to $\mathfrak{M}$ and, hence $\mathfrak{A}_{\alpha_0} = n_1, n_2 \ldots$ for some $\omega_0 < \alpha_0 < \omega(\mathfrak{c})$.

Consider the element x_{α_o}. As it was noted above, there exists a positive rational number r_k such that $r_k \cdot x_{\alpha_0} \in U_1$. Since in the sequence $\mathfrak{A}_{\alpha_0}$ on the place k is the number n_k, then $x_{\alpha_0} \in B_{k,n_k}$. Then (see Definition of $r_{i,k}$)

$$e = (x_{\alpha_0+k})^2 = (r_{n_k} \cdot r_k \cdot r_{k,n_k} \cdot x_{\alpha_0+k})^2$$
$$= ((r_{n_k} \cdot x_k) \cdot (r_k \cdot x_{\alpha_0}))^2 \in (U_1 \cdot U_1) \cdot (U_1 \cdot U_1) \subseteq U_0.$$

We have the contradiction with the choice of the neighborhood U_0.

Thus, the theorem is completely proved.

6. Extension of Topologies

In this chapter is considered a question of the extension of a topology (pseudonorm) on a ring onto certain of its overrings. Besides, the results on extension of a ring topology and a topology on a completely regular space onto the ring of polynomials imply the decision of the question of existence of free topological rings and modules generated by a completely regular space.

The proof of some propositions of this chapter could be also found in [8-11, 35, 40, 45-54, 72, 500-504]. The question of the extension of a topology on a ring onto its semigroup ring is considered also in [336-339], and onto several other its quotient rings in [20, 127, 128, 200, 230, 250, 251, 255-257, 346, 352, 361, 446, 448, 457, 458]. The questions of the existence of free topological groups, modules and other topological systems, as well as some of their properties are considered in [1, 8-14, 41, 57, 60, 79, 86, 88, 124, 135-137, 158, 168, 169, 203, 204, 211, 212, 263, 272, 274, 307, 318-321, 327, 328, 400-409].

Close to this chapter are results on the extension of a topology on a field onto several of its extensions (see [36, 56, 37, 58, 59, 97, 240, 324, 349, 436-439]) and the results on embedding of topological rings into topological rings with certain additional properties (see [21, 109, 170, 183, 256, 295, 304, 374]).

§ 6.1. Extension of Topologies on the Rings of Polynomials

6.1.1. DEFINITION. Let $\mathfrak{M}$ be the variety of all semigroups or the variety of all commutative semigroups. Over F_X denote the free semigroup in the variety $\mathfrak{M}$, generated by a set X, with externally joint unitary element.

Let now R be a ring with the unitary element, then the semigroup ring RF_X (see B.17) we shall call a ring of polynomials over the ring R with the set of variables X. We shall denote it over $R[X]$.

6.1.2. DEFINITION. A degree $\deg z$ of the element $z = x_1^{k_1} \cdot x_2^{k_2} \cdot \ldots \cdot x_n^{k_n}$ of F_X is the number $k_1 + k_2 + \ldots + k_n$. Also we consider that a degree of the unitary element 1 is equal to zero.

6.1.3. DENOTATIONS. Let:

(R, τ_0) be a Hausdorff topological ring with the unitary element 1, and $\mathcal{B}_0(R)$ be a basis of neighborhoods of zero in (R, τ_0);

(X, τ_1) be a completely regular topological space and Δ be the family of continuous real-valued functions on X which is a semigroup with respect to the following multiplicative operation: $(\varphi \cdot \psi)(x) = \varphi(x) \cdot \psi(x)$ for $\varphi, \psi \in \Delta$, $x \in X$ and such that for any closed subset F in (X, τ_1) and any element $x \in X \backslash F$ there could be found a function $\varphi \in \Delta$ for which $1 - \varphi \in \Delta$, $\varphi(x) = 1$ and $\varphi(y) = 0$ for any $y \in F$;

$\mathcal{P}_0(X)$ is the set of such mappings $\gamma : X \bigcup \{0\} \to X \bigcup \{0\}$ that $\gamma(0) = 0$, where 0 is the zero element of the ring R;

for every $\gamma \in \mathcal{P}_0(x)$ over $\tilde{\gamma}$ we denote the endomorphism of the subsemigroup $\{0\} \bigcup F_X$ of multiplicative semigroup of the ring $R[X]$, which extends the mapping γ, i.e.

$$\tilde{\gamma}\left(x_1^{k_1} \cdot x_2^{k_2} \cdot \ldots \cdot x_n^{k_n}\right) = \left(\gamma(x_1)\right)^{k_1} \cdot \left(\gamma(x_2)\right)^{k_2} \cdot \ldots \cdot \left(\gamma(x_n)\right)^{k_n},$$

and let $\Gamma = \{\tilde{\gamma} \,|\, \gamma \in \mathcal{P}_0(x)\}$.

6.1.4. DENOTATIONS. Let $V \in \mathcal{B}_0(R)$, $\Phi = \{\varphi_1, \ldots \varphi_n\} \subseteq \Delta$, $n \in \mathbb{N}$, $0 < \epsilon \in \mathbb{R}$,

then over $U(V, \Phi, n, \epsilon)$ we denote the set of all such elements $u \in R[X]$, which can be presented in the form:

$$u = \sum_{k=1}^{t} v_k \cdot z_k + \sum_{i=1}^{s} f_i \cdot (x_i - y_i) \cdot h_i + g$$

where:

$$v_k \in R; \quad z_k \in F_X \quad \text{and} \quad \deg z_k < n, \quad \text{for} \quad 1 \leq k \leq t;$$

$$x_i, y_i \in X \quad \text{and} \quad g, f_i, h_i \in R[X], \quad \text{for} \quad 1 \leq i \leq s,$$

moreover,

$\sum_{\tilde{\gamma}(z_k)=z} v_k \in V$ for any $\tilde{\gamma} \in \Gamma$ and $z \in F_X$, $\deg z < n$;

$\sum_{i=1}^{s} |\varphi(x_i) - \varphi(y_i)| < \epsilon$ for any $\varphi \in \Phi$;

$g = 0$ or it is a sum of monomials from $R[X]$, the degree of each of which is greater or equal to n.

6.1.5. PROPOSITION. *Let (R, r_0) be a Hausdorff topological ring, (X, τ_1) a completely regular topological space. Then the family $\mathcal{U}$ of all subsets of type $U(V, \Phi, n, \epsilon)$ (see Denotations 6.1.4) defines on $R[X]$ a certain Hausdorff ring topology τ, being a basis of neighborhoods of zero, moreover $\tau|_R = \tau_0$, $\tau|_X = \tau_1$, and X is a closed subset in $(R[X], \tau)$.*

PROOF. Let us check first that the family $\mathcal{U}$, considered as a basis of neighborhoods of zero, defines on $R[X]$ some ring topology. For doing this, according to Theorem 1.2.5, it is sufficient to verify that the family of subsets of type $U(V, \Phi, n, \epsilon)$ fulfills conditions (BN1) – (BN6).

It is obvious that $0 \in U(V, \Phi, n, \epsilon)$ for any $V \in \mathcal{B}_0$, $\Phi \subseteq \Delta$, $n \in \mathbb{N}$, $0 < \epsilon \in \mathbb{R}$, and it means that the condition (BN1) is satisfied.

Since $U(V'', \Phi'', n'', \epsilon'') \subseteq U(V, \Phi, n, \epsilon) \cap U(V', \Phi', n', \epsilon')$, then we get that if $V'' \subseteq V \cap V'$, $\Phi'' \supseteq \Phi \bigcup \Phi$, $n'' > \max\{n, n'\}$, $\epsilon'' \leq \min\{\epsilon, \epsilon'\}$, hence, the condition (BN2) is satisfied.

Let now $V, V' \in \mathcal{B}_0(R)$ and $V' + V' \subseteq V$; $\Phi \subseteq \Delta$; $n \in \mathbb{N}$; $\epsilon \in \mathbb{R}$. Let $u_1, u_2 \in U(V', \Phi, n, \epsilon/2)$ and

$$u_j = \sum_{k=1}^{t_j} v_{k,j} \cdot z_{k,j} + \sum_{i=1}^{s_j} f_{i,j} \cdot (x_{i,j} - y_{i,j}) \cdot h_{i,j} + g_j$$

are the presentations of the elements u_1 and u_2, constructed in accordance with Denotations 6.1.4 for $j = 1, 2$. Then

$$u_1 + u_2 = \sum_{k=1}^{t_1+t_2} v_k \cdot z_k + \sum_{i=1}^{s_1+s_2} f_i \cdot (x_i - y_i) \cdot h_i + g,$$

where:

$$v_k = \begin{cases} v_{k,1}, & \text{if } 1 \leq k \leq t_1; \\ v_{k-t_1,2}, & \text{if } t_1 < k \leq t_1 + t_2; \end{cases}$$

$$z_k = \begin{cases} z_{k,1}, & \text{if } 1 \leq k \leq t_1; \\ z_{k-t_1,2}, & \text{if } t_1 < k \leq t_1 + t_2; \end{cases}$$

$$x_k = \begin{cases} x_{k,1}, & \text{if } 1 \leq k \leq s_1; \\ x_{k-s_1,2}, & \text{if } s_1 < k \leq s_1 + s_2; \end{cases}$$

$$y_k = \begin{cases} y_{k,1}, & \text{if } 1 \leq k \leq s_1; \\ y_{k-s_1,2}, & \text{if } s_1 < k \leq s_1 + s_2; \end{cases}$$

$$f_k = \begin{cases} f_{k,1}, & \text{if} \quad 1 \le k \le s_1; \\ f_{k-s_1,2}, & \text{if} \quad s_1 < k \le s_1 + s_2; \end{cases}$$

$$h_k = \begin{cases} h_{k,1}, & \text{if} \quad 1 \le k \le s_1; \\ h_{k-s_1,2}, & \text{if} \quad s_1 < k \le s_1 + s_2; \end{cases}$$

and $g = g_1 + g_2$.

Moreover, for any $\tilde{\gamma} \in \Gamma$ and $z \in F_X$, $\deg z < n$, (see Definition 6.1.1) we have

$$\sum_{\tilde{\gamma}(z_k)=z} v_k = \sum_{\tilde{\gamma}(z_{i,1})=z} v_{i,1} + \sum_{\tilde{\gamma}(z_{i,2})=z} v_{i,2} \in V' + V' \subseteq V$$

and

$$\sum_{k=1}^{s_1+s_2} |\varphi(x_k) - \varphi(y_k)| = \sum_{i=1}^{s_1} |\varphi(x_{i,1}) - \varphi(y_{i,1})| + \sum_{i=1}^{s_2} |\varphi(x_{i,2}) - \varphi(y_{i,2})| \le \frac{\epsilon}{2} + \frac{\epsilon}{2} = \epsilon$$

for any $\varphi \in \Phi$.

Since $g = g_1 + g_2$ is a sum of monomials, the degree of each of which is more or equal to n, then $u_1 + u_2 \in U(V, \Phi, n, \epsilon)$. From the arbitrariness of the choice of elements u_1 and u_2 the fulfilment of the condition (BN3) follows.

Since $U(V, \Phi, n, \epsilon) = -U(-V, \Phi, n, \epsilon)$ for any $V \in \mathcal{B}_0(R)$, $\Phi \subseteq \Delta$, $n \in \mathbb{N}$, $\epsilon \in \mathbb{R}$, then the condition (BN4) is satisfied.

Let $n \in \mathbb{N}$, and $V, V' \in \mathcal{B}_0(R)$, where

$$\underbrace{V' \cdot V' + V' \cdot V' + \ldots + V' \cdot V'}_{2^n \text{ summands}} \subseteq V.$$

Let $u_1, u_2 \in U(V', \Phi, n, \epsilon/2)$ and

$$u_j = \sum_{k=1}^{t_j} v_{k,j} \cdot z_{k,j} + \sum_{k=1}^{s_j} f_{k,j} \cdot (x_{k,j} - y_{k,j}) \cdot h_{k,j} + g_j$$

are the presentations of the elements u_1 and u_2, constructed in accordance with Denotations 6.1.4 for $j = 1, 2$. Then

$$u_1 \cdot u_2 = \sum_{k=1}^{t_1 \cdot t_2} v_k \cdot z_k + \sum_{k=1}^{s_1+s_2} f_k \cdot (x_k - y_k) \cdot h_k + g,$$

where:

$$z_k = \begin{cases} z_{i,1} \cdot z_{j,2}, & \text{if } k = (i-1) \cdot t_2 + j, \quad 1 \le i \le t_1, \\ & \qquad 1 \le j \le t_2, \quad \deg(z_{i,1} \cdot z_{j,2}) < n; \\ 1, & \text{in other cases;} \end{cases}$$

$$v_k = \begin{cases} v_{i,1} \cdot v_{j,2}, & \text{if } k = (i-1) \cdot t_2 + j, \quad 1 \le i \le t_1, \\ & \qquad 1 \le j \le t_2, \quad \deg(z_{i,1} \cdot z_{j,2}) < n; \\ 0, & \text{in other cases;} \end{cases}$$

$$f_k = \begin{cases} u_1 \cdot f_{i,2}, & \text{if } k = i, \quad 1 \le i \le s_2; \\ f_{i,1}, & \text{if } k = s_2 + i, \quad 1 \le i \le s_1; \end{cases}$$

$$x_k = \begin{cases} x_{i,2}, & \text{if } k = i, \quad 1 \le i \le s_2; \\ x_{i,1}, & \text{if } k = s_2 + i, \quad 1 \le i \le s_1; \end{cases}$$

$$y_k = \begin{cases} y_{i,2}, & \text{if } k = i, \quad 1 \le i \le s_2; \\ y_{i,1}, & \text{if } k = s_2 + i, \quad 1 \le i \le s_1; \end{cases}$$

$$h_k = \begin{cases} h_{i,2}, & \text{if } k = i, \quad 1 \le i \le s_2; \\ h_{i,1} \cdot (\sum_{j=1}^{t_2} v_{j,2} \cdot z_{j,2} + g_2), & \text{if } k = s_2 + i, \quad 1 \le i \le s_1, \end{cases}$$

and

$$g = u_1 \cdot g_2 + g_1 \cdot \left(\sum_{k=1}^{t_2} v_{k,2} \cdot z_{k,2} \right) + \sum_{\deg(z_{i,1} \cdot z_{j,2}) \ge n} v_{i,1} \cdot v_{j,2} \cdot z_{i,1} \cdot z_{j,2} \,.$$

Moreover,

$$\sum_{i=1}^{s_1+s_2} |\varphi(x_i) - \varphi(y_i)| = \sum_{i=1}^{s_2} |\varphi(x_{i,2}) - \varphi(y_{i,2})|$$

$$+ \sum_{i=1}^{s_1} |\varphi(x_{i,1}) - \varphi(y_{i,1})| \le \frac{\epsilon}{2} + \frac{\epsilon}{2} = \epsilon$$

and g is a sum of monomials, the degree of each of which is more or equal to n.

Let now $\widetilde{\gamma} \in \Gamma$, $z \in F_X$, $\deg z < n$ and $\{(z_1', z_1''), \dots, (z_t', z_t'')\}$ is the family of all pairs of elements of F_X such that $z = z_i' \cdot z_i''$ for $1 \le i \le t$. Since the number of various subwords of the word z is not more than 2^n, then $t \le 2^n$. Then

$$\sum_{\widetilde{\gamma}(z_k)=z} v_k = \sum_{j=1}^{t} \left(\left(\sum_{\widetilde{\gamma}(z_{i,1})=z_j'} v_{i,1} \right) \cdot \left(\sum_{\widetilde{\gamma}(z_{i,2})=z_j''} v_{i,2} \right) \right) \in \underbrace{V' \cdot V' + \dots + V' \cdot V'}_{2^n \text{ summands}} \subseteq V.$$

Thus, we have shown, that $u_1 \cdot u_2 \in V$. From the arbitrariness of the choice of elements u_1 and u_2 the fulfilment of the condition (BN5) follows.

Let $f = \sum_{i=1}^{m} r_i \cdot g_i \in R[X]$, where $r_i \in R$, $g_i \in F_X$ for $1 \le i \le m$ and $V \in \mathcal{B}_0(R)$, $\Phi \subseteq \Delta$, $n \in \mathbb{N}$, $0 < \epsilon \in \mathbb{R}$. There exists a neighborhood $V' \in \mathcal{B}_0(R)$ such that $\sum_{i=1}^{m} r_i \cdot V' \subseteq V$ and $\sum_{i=1}^{m} V' \cdot r_i \subseteq V$.

Let $u \in U(V', \Phi, n, \epsilon)$ and

$$u = \sum_{k=1}^{t} v_k \cdot z_k + \sum_{i=1}^{s} f_i \cdot (x_i - y_i) \cdot h_i + g$$

is the presentation of the element u, constructed in accordance with Denotations 6.1.4. Then

$$f \cdot u = \sum_{i=1}^{m \cdot t} v_i' \cdot z_i' + \sum_{i=1}^{s} f_i' \cdot (x_i - y_i) \cdot h_i + g'$$

where:

$v_k' = r_i \cdot v_j$ and $z_k' = g_i \cdot z_j$ if $k = (i-1)t + j$, for $1 \le i \le m$, $1 \le j \le t$, and $\deg g_i \cdot z_i \le n$;

$v_k' = 0$ and $z_k' = 1$ if $k = (i-1) \cdot t + j$, for $1 \le i \le m$, $1 \le j \le t$, and $\deg g_i \cdot z_j \ge n$;

$f_t' = f \cdot f_t$ for $1 \le t \le s$;

$g' = f \cdot g + \sum_{\deg g_i \cdot z_j \ge n} (r_i \cdot v_j) \cdot (g_i \cdot z_j)$.

Moreover, $\sum_{i=1}^{s} |\varphi(x_i) - \varphi(y_i)| < \epsilon$ for any $\varphi \in \Phi$, and g' is a sum of monomials of $R[X]$, the degree of each of which is not less than n .

Let now $\tilde{\gamma} \in \Gamma$ and $z \in F_X$, $\deg z < n$. For each $1 \le i \le t$ there exists, moreover unique, element $z_i'' \in F_X$ such that $g_i \cdot z_i'' = z$. Then

$$\sum_{\tilde{\gamma}(z_i')=z} v_i' = \sum_{i=1}^{m} r_i \cdot \left(\sum_{\tilde{\gamma}(z_j)=z_i'} v_j \right) \in \sum_{i=1}^{m} r_i \cdot V' \subseteq V.$$

Therefore, $f \cdot u \in U(V, \Phi, n, \epsilon)$.

In a similar way it is proved that $u \cdot f \in U(V, \Phi, n, \epsilon)$. From the arbitrariness of the choice of element u the fulfillment of the condition (BN6) follows.

Thus, the family of all subsets of type $U(V, \Phi, n, \epsilon)$ defines on $R[X]$ a ring topology τ, being a basis of neighborhoods of zero.

To complete the proof of the theorem, it remains to verify that:

τ is a Hausdorff topology; $\tau|_R = \tau_0$; $\tau|_X = \tau_1$ and X is a closed subset in $(R[X], \tau)$.

Before completing the proof of theorem we shall prove the following lemma.

6.1.6. LEMMA. Let: $m, p \in \mathbb{N}, p \geq 2; S = \{\widetilde{x}_1, \ldots, \widetilde{x}_p\} \subseteq X; \widetilde{z}_1, \ldots, \widetilde{z}_m \in F_S \subseteq F_X; \widetilde{r}_1, \ldots, \widetilde{r}_m \in R; V_0 \in \mathcal{B}_0(R), r_1 \notin V_0; \Phi = \{\varphi_1, \ldots, \varphi_p\} \subseteq \Delta, \varphi_i(\widetilde{x}_i) = 1$ and $\varphi_i(\widetilde{x}_j) = 0$ for $j \neq i, 1 \leq i < j \leq p; n_0 \in \mathbb{N}, n_0 > \deg \widetilde{z}_1$. If $f = \sum_{i=1}^{m} \widetilde{r}_i \cdot \widetilde{z}_i$, then $f \notin U(V_0, \Phi, n_0, \frac{1}{2})$.

PROOF. Assume the contrary, i.e. $f \in U(V_0, \Phi, n_0, \frac{1}{2})$ and

$$f = \sum_{k=1}^{t} v_k \cdot z_k + \sum_{k=1}^{s} f_k \cdot (x_k - y_k) \cdot h_k + g$$

is the presentation of the element f, constructed in accordance with Denotations 6.1.4.

Let us define on the set X the relation δ, defined as follows: $x \delta y$ if and only if $x = y$ or there exists a finite collection of two-element sets $\{x_{i_1}, y_{i_1}\}, \ldots, \{x_{i_l}, y_{i_l}\}$ (the definition of the elements x_i and y_i see in Denotations 6.1.4) such that: $x \in \{x_{i_1}, y_{i_1}\}; y \in \{x_{i_l}, y_{i_l}\}$ and $\{x_{i_{j-1}}, y_{i_{j-1}}\} \bigcap \{x_{i_j}, y_{i_j}\} \neq \emptyset$, for $1 < j \leq l$. It is clear that δ is the equivalence relation on the set X.

We shall show that each class of δ-equivalent elements contains not more than one element from the set $\{\widetilde{x}_j | 1 \leq j \leq p\}$. Assume the contrary, i.e. $\widetilde{x}_{j_1} \delta \widetilde{x}_{j_2}$ for $1 \leq j_1 < j_2 \leq p$ and let $\{x_{i_1}, y_{i_1}\}, \ldots \{x_{i_l}, y_{i_l}\}$ be a collection of two-element sets, establishing the equivalence of the elements $\widetilde{x}_{j_1}$ and $\widetilde{x}_{j_2}$. Without loss of generality we can consider that this is a collection of a minimal length. Therefore, among the numbers $i_1, \ldots, i_l$ there are no coinciding ones, i.e. they are pairwise distinct.

Then

$$1 = |\varphi_{j_1}(\widetilde{x}_{j_1}) - \varphi_{j_1}(\widetilde{x}_{j_2})| \leq \sum_{k=1}^{l} |\varphi_{j_1}(x_{i_k}) - \varphi_{j_1}(y_{i_k})| \leq \sum_{i=1}^{s} |\varphi_{j_1}(x_i) - \varphi_{j_1}(y_i)| < \frac{1}{2}.$$

We have a contradiction.

Hence, various elements from the set $\{x_1, \ldots, x_p\}$ cannot be equivalent with respect to δ. Consider the mapping $\gamma : X \bigcup\{0\} \rightarrow X \bigcup\{0\}$, defined in the following manner:

$$\gamma(x) = \begin{cases} \widetilde{x}_i, & \text{if } x \delta \widetilde{x}_i, \quad 1 \leq i \leq p; \\ 0, & \text{in other cases.} \end{cases}$$

We shall extend the mapping γ to an endomorphism $\bar{\gamma}$ of the ring $R[X]$, putting
$\bar{\gamma}(\varphi(x'_1, \ldots, x'_q)) = \varphi(\gamma(x'_1), \ldots \gamma(x'_q))$ for any $\varphi(x'_1, \ldots, x'_q) \in R[X]$.

It is clear that the reduction $\tilde{\gamma}$ of the endomorphism $\bar{\gamma}$ on $F_X \bigcup\{0\}$ belongs to Γ (see Denotations 6.1.3). Then $\sum_{\tilde{\gamma}(z_k)=\tilde{z}_1} v_k \in V_0$.

Since $x_i \delta y_i$ for $1 \leq i \leq s$, and in the set $\{\tilde{x}_1, \ldots \tilde{x}_p\}$ there are no two equivalent elements with respect to δ, then $\tilde{\gamma}(x_i - y_i) = 0$ for $1 \leq i \leq s$ and $\tilde{\gamma}(\tilde{x}_j) = \tilde{x}_j$ for $1 \leq j \leq p$, and, hence, $(\tilde{z}_k) = \tilde{z}_k$ for $1 \leq k \leq m$. Then

$$\sum_{i=1}^{m} \tilde{r}_i \cdot \tilde{z}_i = \bar{\gamma}(\sum_{i=1}^{m} \tilde{r}_i \cdot \tilde{z}_i) = \bar{\gamma}(f) = \sum_{i=1}^{t} v_i \cdot \tilde{\gamma}(z_i) + \bar{\gamma}(g).$$

Since g and, hence, $\bar{\gamma}(g)$ is a sum of monomials the degree of each of which is more than $n_0 > \deg \tilde{z}_1$, then among the summands of g there are no members similar to the element $\tilde{r}_i \cdot \tilde{z}_1$. Then, taking into account the definition of the set $U(V, \Phi, n, \epsilon)$ and the fact that the reduction $\tilde{\gamma}$ on F_X of the endomorphism $\bar{\gamma}$ belongs to Γ, we obtain

$$\tilde{r}_1 = \sum_{\tilde{\gamma}(z_i)=\tilde{z}_1} v_i \in V_0.$$

We have the contradiction to the lemma condition. This completes the proof of the lemma.

Let us pass to completing the proof of the theorem.

Let $0 \neq f = \sum_{i=1}^{m} \tilde{r}_i \cdot \tilde{z}_i$, where $0 \neq \tilde{r}_i \in R$ and $\tilde{z}_i \neq \tilde{z}_j$ for $1 \leq i < j \leq m$. There exist $V_0 \in \mathcal{B}_0(R)$ and $S = \{\tilde{x}_1, \ldots, \tilde{x}_p\} \subseteq X$ such that $\tilde{r}_1 \notin V_0$ and $\tilde{z}_1, \ldots, \tilde{z}_m \in F_S$. For each $1 \leq i \leq p$ there exists a function $\varphi \in \Delta$ such that $\varphi_i(\tilde{x}_i) = 1$, and $\varphi_i(\tilde{x}_j) = 0$ for $1 \leq j \leq p$, and $j \neq i$ (such function exists because the subset $\{\tilde{x}_j \,|\, 1 \leq j \leq p, \; j \neq i\}$ is closed in (X, τ_1)). If now $n_0 > \deg \tilde{z}_1$, then, according to Lemma 6.1.6,

$$f \notin U(V_0, \Phi, n_0, \frac{1}{2}).$$

We have verified that the constructed topology τ is Hausdorff.

Let now $V \in \mathcal{B}_0(R)$, $\Phi \subseteq \Delta$; $n \in \mathbb{N}$; $0 < \epsilon \in \mathbb{R}$.

Since $1 \in F_X$, and the endomorphism $\tilde{\gamma}: F_X \bigcup\{0\} \to F_X \bigcup\{0\}$, acting by the rule:

$$\tilde{\gamma}(z) = \begin{cases} 1, & \text{if} \quad z = 1; \\ 0, & \text{in other cases}; \end{cases}$$

belongs to Γ, then from the definition of the set $U(V, \Phi, n, \epsilon)$ and from $R \subseteq R[X]$ it easily follows that $V = R \bigcap U(V, \Phi, n, \epsilon)$ for any: $V \in \mathcal{B}_0(R)$; $\Phi \subseteq \Gamma$; $n \in \mathbb{N}$; $0 \leq \epsilon \in \mathbb{R}$. Therefore, $\tau\big|_R = \tau_0$.

Let now $\widetilde{x}_1 \in X$ and W be a neighborhood of the element $\widetilde{x}_1$ in (X, τ_1). Let W_0 be an open subset in (X, τ_1) such that $x_1 \in W_0 \subseteq W$. Then there exists a function $\varphi_1 \in \Delta$ such that: $\varphi_2 = 1 - \varphi_1 \in \Delta$; $\varphi_1(\widetilde{x}_1) = 1$ and $\varphi_1(x) = 0$ for $x \in X \backslash W_0$ (see Denotations 6.1.3). We choose a neighborhood $V_0 \in \mathcal{B}_0(R)$ such that $1 \notin V_0$. Then $\Phi_0 = \{\varphi_1, \varphi_2\} \subseteq \Delta$.

If now $\widetilde{y}_1 \notin W$, then $\widetilde{y}_1 \in X \backslash W_0$. Since: $1 \notin V_0$; $\varphi_1(\widetilde{x}_1) = 1$; $\varphi_1(\widetilde{y}_1) = 0$; $\varphi_2(\widetilde{y}_1) = 1$ and $2 = \deg \widetilde{y}_1 + 1$, then by Lemma 6.1.6, $\widetilde{y}_1 - \widetilde{x}_1 \notin U(V_0, \Phi_0, 2, \frac{1}{2})$ and, hence, $\widetilde{y}_1 \notin x_1 + U(V_0, \Phi_0, 2, \frac{1}{2})$. From the arbitrariness of the choice of the element $\widetilde{y}_1$ it follows that

$$W \supseteq X \cap \left(\widetilde{x}_1 + U\left(V_0, \Phi_0, 2, \frac{1}{2}\right)\right).$$

Thus, we have shown that $\tau\big|_X \geq \tau_1$.

Let now $U(V, \Phi, n, \epsilon)$ be a neighborhood of zero in $(R[X], \tau)$ and $x_0 \in X$. Since Φ consists of the finite number of continuous functions, then

$$W = \bigcap_{\varphi \in \Phi} \{x \in X \,|\, |\varphi(x) - \varphi(x_0)| < \epsilon\}$$

is an open neighborhood of the element x_0 in (X, τ_1). Since $x - x_0 = v_1 \cdot z_1 + (x - x_0) + g$, where: $v_1 = 0$; $z_1 = 1$ and $g = 0$, and moreover, $|\varphi(x) - \varphi(x_0)| < \epsilon$ for $x \in W$, $\varphi \in \Phi$, then $x - x_0 \in U(V, \Phi, n, \epsilon)$ for any $x \in W$. Then $W \subseteq (x_0 + U(V, \Phi, n, \epsilon)) \cap X$. In summary, we have shown that $\tau\big|_X \leq \tau_1$ and therefore, $\tau\big|_X = \tau_1$.

It remains to check the closedness of X in $(R[X], \tau)$. Let $f \in R[x] \backslash X$. Then there can be found non-zero elements $r_1, \ldots, r_m \in R$, pairwise distinct elements $z_1, \ldots, z_m \in F_X$ and $x_1, \ldots, x_p \in X$ such that $z_1, \ldots, z_m \in F_S$, where $S = \{x_1, \ldots, x_p\}$ and $f = \sum_{i=1}^m r_i \cdot z_i$. Put $n_0 = \max\{\deg z_i \,|\, 1 \leq j \leq m\} + 1$. Choose neighborhoods $V_0, V_1 \in \mathcal{B}_0(R)$ such that $\{1, r_i, 1 - r_i \,|\, r_i \neq 1, \ 1 \leq i \leq m\} \cap V_0 = \emptyset$ and $V_1 - V_1 \subseteq V_0$. We can choose open sets $W_1, \ldots, W_p$ in (X, τ_1) such that $x_i \in W_i$ and $W_i \bigcap W_j = \emptyset$ for $1 \leq i < j \leq p$. There exist functions $\varphi_1, \ldots, \varphi_p \in \Delta$ such that: $1 - \varphi_i \in \Delta$; $\varphi_i(x_i) = 1$ and $\varphi_i(x) = 0$ for $x \in X \backslash W_i$ for $1 \leq i \leq p$. Let $\Phi = \{\varphi_1, \ldots, \varphi_p\}$. Put $W_i' = \{x \in X \,|\, |\varphi_i(x) - 1| < \frac{1}{4}\}$ for $1 \leq i \leq p$. Then $x_i \in W_i' \subseteq W_i$, and W_i' are open sets in (X, τ_1) for $1 \leq i \leq p$.

There exist $\varphi_1', \ldots, \varphi_p' \in \Delta$ such that: $1 - \varphi_i \in \Delta$; $\varphi_i'(x_i) = 1$ and $\varphi_i'(x) = 0$ for $x \in X \backslash W_i'$, $1 \le i \le p$. Since Δ is a semigroup, then $\varphi = (1-\varphi_1') \cdot (1-\varphi_2') \cdot \ldots \cdot (1-\varphi_p') \in \Delta$. Note that $\varphi(x_i) = 0$ for $i = 1, \ldots, p$ and $\varphi(x) = 1$ for $x \notin \bigcup_{i=1}^p W_i'$.

Put $\Phi' = \{\varphi_1', \ldots, \varphi_p', \varphi\}$ and show that

$$\left(f + U(V_1, \Phi \cup \Phi', n_0, \frac{1}{4}) \right) \cap X = \emptyset.$$

Assume the contrary. Let

$$x_{p+1} \in X \cap \left(f + U(V_1, \Phi \cup \Phi', n_0, \frac{1}{4}) \right).$$

Then $x_{p+1} - f \in U(V_1, \Phi \cup \Phi', n_0, \frac{1}{4})$.

If $x_{p+1} \notin \bigcup_{i=1}^p W_i'$, then $x_{p+1} \ne x_i$ for $1 \le i \le p$ and $\varphi_j'(x_{p+1}) = 0$ for $1 \le j \le p$. Putting in Lemma 6.1.6 $\tilde{x}_i = x_i$ for $1 \le i \le p+1$; $\tilde{z}_j = z_j$ for $1 \le j \le m$; $\tilde{z}_{m+1} = x_{p+1}$; $\tilde{r}_i = -r_i$ for $1 \le i \le m$ and $\tilde{r}_{m+1} = 1$, we obtain

$$x_{p+1} - f = \sum -r_i z_i + x_{p+1} \notin U\left(V_0, \Phi', n_0, \frac{1}{2} \right)$$

and moreover, $x_{p+1} - f \notin U(V_1, \Phi \bigcup \Phi', n_0, \frac{1}{4})$. We have the contradiction to the choice of the element x_{p+1}.

Hence, $x_{p+1} \in \bigcup_{i=1}^p W_i'$, i.e. $x_{p+1} \in W_{i_0}'$ for some, uniquely defined, number $1 \le i_0 \le p$ (as $W_k' \subseteq W_k$ and $W_i \cap W_j = \emptyset$ for $i \ne j$).

Then $|\varphi_{i_0}(x_{p+1}) - 1| < \frac{1}{4}$. Furthermore, since $W_{i_0}' \subseteq W_{i_0}$ and $W_{i_0} \cap W_i = \emptyset$ for $i \ne i_0$ and $1 \le i \le p$, then $x_{i_0} \notin W_i$, and $x_{p+1} \notin W_i$ for $i \ne i_0$, $1 \le i \le p$. Hence, $|\varphi_i(x_{p+1}) - \varphi_i(x_{i_0})| = 0 - 0 = 0 < \frac{1}{4}$ for $i \ne i_0$, $1 \le i \le p$. Therefore (see Denotations 6.1.4), $x_{i_0} - x_{p+1} \in U(V_1, \Phi, n_0, \frac{1}{4})$.

Since $V_1 + V_1 \subseteq V_0$, (the proof of validity of the condition (BN3), see above), then

$$x_{i_0} - f = (x_{i_0} - x_{p+1}) + (x_{p+1} - f) \in U\left(V_1, \Phi, n_0, \frac{1}{4} \right) + U\left(V_1, \Phi \cup \Phi', n_0, \frac{1}{4} \right)$$
$$\subseteq U\left(V_1, \Phi, n_0, \frac{1}{4} \right) + U\left(V_1, \Phi, n_0, \frac{1}{4} \right) \subseteq U\left(V_0, \Phi, n_0, \frac{1}{2} \right).$$

On the other hand, since $f \notin X$, then

$$0 \ne x_{i_0} - f = x_{i_0} - \sum_{i=1}^m r_i \cdot z_i = \sum_{j=1}^k \hat{r}_j \cdot \hat{z}_j,$$

where

$$\{\widehat{z}_j \,|\, 1 \le j \le k\} \subseteq \{x_{i_0}, z_i \,|\, 1 \le i \le m\} \subseteq F_S,$$
$$\{\widehat{r}_j \,|\, 1 \le j \le k\} \subseteq \{1, r_j, 1 - r_j \,|\, 1 \le j \le m \text{ and } r_j \ne 1\}.$$

Then for the element $x_{i_0} - f$ and the neighborhood $U(V_0, \Phi, n_0, \frac{1}{2})$ the conditions of Lemma 6.1.6 are satisfied and therefore, $x_{i_0} - f \notin U\left(V_0, \Phi, n_0, \frac{1}{2}\right)$. But there was proved above that $x_{i_0} - f \in U\left(V_0, \Phi, n_0, \frac{1}{2}\right)$. The contradiction completes the proof of the theorem.

6.1.7. THEOREM. *Let* (R, τ_0) *be a Hausdorff topological ring with the unitary element,* (X, τ_1) *be a completely regular topological space. Then on the ring* $R[X]$ *exists a Hausdorff ring topology* τ *such that* $\tau|_R = \tau_0$ *and* $\tau|_X = \tau_1$. *Moreover,* X *is a closed subset in* $(R[X], \tau)$.

PROOF. Indeed, since (X, τ_1) is a completely regular space, then the family Δ of all continuous mappings $\varphi : (X, \tau_1) \to [0, 1]$ fulfills the conditions of Denotation 6.1.3. Then, according to Theorem 6.1.5, on the ring $R[X]$ exists a required topology.

6.1.8. THEOREM. *Let a Hausdorff topological ring* (R, τ_0) *with the unitary element possess a basis of neighborhoods of zero consisting of subgroups (subrings), and* (X, τ_1) *be a zero-dimensional (see C.14) completely regular topological space. Then on the ring* $R[X]$ *exists a ring topology* τ *such that* $\tau|_R = \tau_0$, $\tau|_X = \tau_1$, *and* $(R[X], \tau)$ *possesses a basis of neighborhoods of zero consisting of subgroups (subrings).*

PROOF. Take as Δ the set of all continuous real-valued functions on (X, τ_1), which get only the values 0 and 1. Since (X, τ_1) is a zero-dimensional space, then Δ fulfills the condition, mentioned in Denotations 6.1.3.

Let $\mathcal{B}_0(R)$ be a basis of neighborhoods of zero in (R, τ_0) consisting of subgroups (subrings), then, as it was shown in Proposition 6.1.5, the family of all subsets of type $U(V, \Phi, n, \epsilon)$, where $V \in \mathcal{B}_0(R)$, $\Phi \in \Delta$ and $|\Phi| < \infty$, $n \in \mathbb{N}$, $0 < \epsilon \in \mathbb{R}$, defines on $R[X]$ a ring topology τ such that $\tau|_R = \tau_0$ and $\tau|_X = \tau_1$.

It remains to verify that $U(V, \Phi, n, \epsilon)$ is a subgroup (subring) if V is one and $0 < \epsilon < 1$. From the proof of the validity of the conditions (BN3) and (BN4) within Proposition 6.1.5

it follow that it is sufficient to verify that in this case

$$U(V, \Phi, n, \epsilon) = U\left(V, \Phi, n, \frac{\epsilon}{2}\right).$$

Let $f \in U(V, \Phi, n, \epsilon)$ and

$$f = \sum_{i=1}^{r} v_i \cdot z_i + \sum_{i=1}^{s} f_i \cdot (x_i - y_i) \cdot h_i + g$$

is the presentation, constructed in accordance with Denotations 6.1.4. Since $\varphi(x) \in \{0, 1\}$ for any $x \in X$, $\varphi \in \Phi$ and $|\varphi(x_i) - \varphi(y_i)| < \epsilon < 1$ for $1 \leq i \leq s$, then $\varphi(x_i) - \varphi(y_i) = 0$ for any $\varphi \in \Phi$ and $1 \leq i \leq s$. Then

$$\sum_{i=1}^{s} |\varphi(x_i) - \varphi(y_i)| = 0 < \frac{\epsilon}{2}$$

for any $\varphi \in \Phi$. Furthermore, $\sum_{\tilde{\gamma}(z_i)=z} v_i \in V$ for any $\tilde{\gamma} \in \Gamma$ and $z \in F_X$, $\deg z < n$. Since $\deg g \geq n$, then $f \in U(V, \Phi, n, \frac{\epsilon}{2})$, i.e. $U(V, \Phi, n, \epsilon) \subseteq U\left(V, \Phi, n, \frac{\epsilon}{2}\right)$. Since the inverse inclusion is obvious, then

$$U(V, \Phi, n, \epsilon) = U\left(V, \Phi, n, \frac{\epsilon}{2}\right).$$

This completely proves the theorem.

§ 6.2. Free Topological Rings and Modules

6.2.1. DEFINITION. Let (R, τ_0) be a Hausdorff topological ring with the unitary element and (X, τ_1) be a completely regular topological space. We say that a topological (R, τ_0)-module (M, τ) is a free topological module over the topological ring (R, τ_0), generated by (X, τ_1), if:

a) (X, τ_1) is a subspace of the topological space (M, τ);

b) M does not contain any proper R-submodules, which contain the set X;

c) for any topological (R, τ_0)-module (M', τ') every continuous mapping $\xi : (X, \tau_1) \to (M, \tau')$ has an extension $\widehat{\xi} : (M, \tau) \to (M', \tau')$, which is a continuous homomorphism of modules.

6.2.2. THEOREM. *For any Hausdorff topological ring (R, τ_0) with the unitary element and for arbitrary completely regular space (X, τ_1) there exists a free topological module $(M_X, \widehat{\tau})$ over the ring (R, τ_0), generated by the topological space (X, τ_1), moreover, X is a closed subset in $(M_X, \widehat{\tau})$.*

PROOF. Let M_X be a free R-module, generated by the set X. Consider the family $\mathcal{T}$ of all such topologies τ' on M_X (not necessarily Hausdorff) that (M_X, τ') is a topological (R, τ_0)-module, moreover, $\tau'|_X \leq \tau_1$. Let $\widehat{\tau} = \sup\{\tau' \mid \tau' \in \mathcal{T}\}$. We shall show that $(M_X, \widehat{\tau})$ is the required module, i.e. $\widehat{\tau}$ is a Hausdorff topology and conditions a), b), c) of Definition 6.2.1 are satisfied.

Let τ be the ring topology on the ring $R[X]$, constructed according to Theorem 6.1.7. Then, since $\tau|_R = \tau_0$, $(R[x], \tau)$ can be considered as a topological (R, τ_0)-module.

Identifying the element $\sum_{i=1}^{n} r_i \cdot x_i$ from M_x with the corresponding element of $R[X]$ we can consider that M_X is a submodule of R-module $R[X](+)$. Then $\tau|_{M_X} \in \mathcal{T}$ and, hence, $\widehat{\tau} \geq \tau|_{M_X}$. Since τ is a Hausdorff topology and X is a closed subset in $(R[X], \tau)$, then $\widehat{\tau}$ is also a Hausdorff topology, moreover, X is a closed subset in $(M_X, \widehat{\tau})$.

Furthermore,

$$\tau_1 = \tau|_X = (\tau|_{M_X})|_X \leq \widehat{\tau}|_X = \sup\{\tau|_X \mid \tau \in \mathcal{T}\} \leq \tau_1,$$

i.e. $\widehat{\tau}|_X = \tau_1$. Thus, we have verified that the condition a) of Definition 6.2.1 is satisfied.

From the fact that M_X is a free R-module, generated by the set X, the fulfillment of the condition b) follows.

Let now (M', τ_1') be a topological (R, τ_0)-module and $\xi : (X, \tau_1) \to (M', \tau_1')$ be a continuous mapping. Since M_X is a free R-module, the mapping ξ can be extended to a homomorphism $\widehat{\xi} : M_X \to M'$ of modules.

According to Theorem 1.5.33, (M_X, τ_ξ) is a topological (R, τ_0)-module, where τ_ξ is the prototype in M_X of the topology τ_1' with respect to the homomorphism $\widehat{\xi}$, moreover $\widehat{\xi}$ is a continuous homomorphism.

Let us show that $\tau_\xi \in \mathcal{T}$, i.e. $\tau_\xi|_X \leq \tau_1$. Indeed, let W be an open set in τ_ξ, then $W = \widehat{\xi}^{-1}(U)$, where U is an open set in (M', τ'). Since $\xi : (X, \tau_1) \to (M', \tau_1')$ is a continuous mapping, then $\xi^{-1}(U)$ is an open set in (X, τ_1). But

$$\xi^{-1}(U) = \{x \in X \mid \xi(x) \in U\} = \widehat{\xi}^{-1}(U) \cap X = W \cap X$$

and, hence, $W \bigcap X$ is an open set in (X, τ_1). From the arbitrariness of the choice of the set W it follows that $\tau_\xi|_X \leq \tau_1$, i.e. $\tau_\xi \in \mathcal{T}$. Then $\tau_\xi \leq \widehat{\tau}$ and therefore, the homomorphism $\widehat{\xi} : (M_X, \widehat{\tau}) \to (M', \tau_1')$ is continuous. Thus, the theorem is completely proved.

6.2.3. REMARK. Since any topological Abelian group is a topological module over the ring of integers endowed with discrete topology, then from Theorem 6.2.2 it follows that for any completely regular space (X, τ_1) there exists a free topological Abelian group, i.e. there exists a Hausdorff topological group (G_X, τ) such that the following conditions are satisfied:

a) (X, τ_1) is a subspace of the topological space (G_X, τ);

b) G_X does not contain any proper subgroups containing the set X ;

c) for any topological Abelian group (G', τ') an arbitrary continuous mapping $\xi : (X, \tau_1) \to (G', \tau')$ has an extension $\widehat{\xi} : (G_X, \tau) \to (G', \tau')$ which is a continuous group homomorphism.

6.2.4. DEFINITION. Let (X, τ_1) be a completely regular topological space. A Hausdorff topological ring (R, τ_0) is called a free topological ring generated by the topological space (X, τ_1) if the following conditions are satisfied:

a) (X, τ_1) is a subspace of the topological space (R, τ_0);

b) R does not contain any proper subrings containing the set X;

c) for any topological ring (R', τ') an arbitrary continuous mapping $\xi : (X, \tau_1) \to (R', \tau')$ has an extension $\widehat{\xi} : (R, \tau_0) \to (R', \tau')$, which is a continuous ring homomorphism.

6.2.5. THEOREM. *For any completely regular topological space (X, τ_1) there exists a free topological ring (R_X, τ_0) generated by the space (X, τ_1), moreover X is a closed subset in (R_X, τ_0).*

PROOF. Let the variety $\mathfrak{M}$ mentioned in Definition 6.1.1, be the variety of all semi-groups, $\mathbb{Z}[X]$ be the ring of polynomials over the ring $\mathbb{Z}$ of integers, and let R_X be the subring in $\mathbb{Z}[X]$ consisting of all polynomials without free members, i.e. $R_X = \mathbb{Z}[X] \cdot X$.

Consider the family $\mathcal{T}$ of all ring topologies τ on R_X (not necessarily Hausdorff) such that $\tau|_X \leq \tau_1$. Let $\tau_0 = \sup\{\tau \mid \tau \in \mathcal{T}\}$, then (R_X, τ_0) is a topological ring.

We shall show that it is a Hausdorff ring and it fulfills conditions a), b), c) of Definition 6.2.4.

Considering on $\mathbb{Z}$ the discrete topology, we shall define on the ring $\mathbb{Z}[X]$ according to Proposition 6.1.7 a Hausdorff ring topology $\widehat{\tau}_0$ such that $\widehat{\tau}_0|_X = \tau_1$ and X is a closed subset in $(\mathbb{Z}[X], \tau_0')$. Then $\tau_0' = \widehat{\tau}_0|_{R_X} \in \mathcal{T}$ and, therefore, τ_0 being stronger than the topology τ_0', is a Hausdorff topology, moreover, X is a closed subset in (R_X, τ_0).

Furthermore,

$$\tau_1 = \widehat{\tau}_0|_X = \tau_0'|_X \leq \tau_0|_X = \sup\{\tau|_X \mid \tau \in \mathcal{T}\} \leq \tau_1,$$

i.e. $\tau_1 = \tau_0|_X$.

Thus, we have verified that for the ring (R_X, τ_0) the condition a) is satisfied.

From the definition of the ring of polynomials (see Definition 6.1.1) it follows that R_X is a free ring generated by the set X and therefore, the condition b) is satisfied.

Let now (R', τ') be a topological ring and $\xi : (X, \tau_1) \to (R', \tau')$ be a continuous mapping. Since, as was mentioned above, R_X is a free ring, generated by the set X, the mapping ξ can be extended to the ring homomorphism $\widehat{\xi} : R_X \to R'$.

Let τ_ξ be the prototype of the topology τ' on R_X with respect to the homomorphism $\widehat{\xi}$, then, according to Theorem 1.5.35, (R_X, τ_ξ) is a topological ring, moreover, $\widehat{\xi}$ is a continuous ring homomorphism of (R_X, τ_ξ) to (R', τ').

As in the proving of Theorem 6.2.2, it is verified that $\tau_\xi \in \mathcal{T}$. Then $\tau_\xi \leq \tau_0$ and, therefore, $\widehat{\xi} : (R_X, \tau_0) \to (R', \tau')$ is a continuous ring homomorphism. Thus, the theorem is completely proved.

6.2.6. REMARK. Considering the family of all commutative topological rings by analogy with Definition 6.2.4, the definition of a free topological commutative ring, generated by a completely regular space (X, τ_1) could be given. Considering in the proving of Theorem 6.2.5 the variety of all commutative semigroups instead of the variety of all semigroups, the existence of a free topological commutative ring generated by a completely regular topological space could be proved.

6.2.7. REMARK. Since there exist completely regular spaces, which are not normal, then from Theorem 6.2.5 it follows that there exist Hausdorff topological rings and, hence, Abelian groups whose spaces are not normal.

§ 6.3. Extension of Norms on Semigroup Rings

6.3.1. DEFINITION. A real-valued function η, defined on a semigroup G, is called a pseudo-seminorm if: $\eta(g) > 0$ and $\eta(a \cdot b) \leq \eta(a) \cdot \eta(b)$ for any $g, a, b \in G$. If moreover, $\eta(a \cdot b) = \eta(a) \cdot \eta(b)$ for any $a, b \in G$, then η is called a seminorm.

6.3.2. THEOREM. *Let η be a pseudo-seminorm on a monoid G, and (R, ξ) be a pseudonormed ring with the unitary element 1. If $\xi(1) = \eta(e) = 1$, where e is the unitary element in G, then on the semigroup ring RG exists a pseudonorm $\widehat{\xi}$ such that $\widehat{\xi}(r) = \xi(r)$ and $\widehat{\xi}(g) = \eta(g)$ for any $\in R$ and $g \in G$.*

PROOF. Let us define on RG a real-valued function $\widehat{\xi}$ as follows:

for $u = \sum_{i=1}^{n} r_i \cdot g_i$, where $r_i \in R$, $g_i \in G$ and $g_i \neq g_j$ for $i, j \leq n$ and $i \neq j$, put $\widehat{\xi}(u) = \sum_{i=1}^{n} \xi(r_i) \cdot \eta(g_i)$.

It is clear that the mapping $\widehat{\xi}$ is defined correctly, moreover, $\widehat{\xi}(r \cdot e) = \xi(r)$ and $\widehat{\xi}(1 \cdot g) = r(g)$ for any $r \in R$ and $g \in G$.

We shall verify that $\widehat{\xi}$ is a pseudonorm on RG, i.e. it satisfies the conditions (NR1) – (NR3) of Definition 1.1.12.

Since $\xi(0) = 0$, then from the definition of the function $\widehat{\xi}$ it follows that $\widehat{\xi}(0) = 0$. Let now $0 \neq u = \sum_{i=1}^{n} r_i \cdot g_i \in RG$, moreover $g_i \neq g_j$ for $1 \leq i, j \leq n$. There exists $i_0 \leq n$ such that $r_{i_0} \neq 0$. Since $\eta(g_i) > 0$ for $1 \leq i \leq n$, then

$$\widehat{\xi}(u) = \sum_{i=1}^{n} \xi(r_i) \cdot \eta(g_i) \geq \xi(r_{i_0}) \cdot \eta(g_{i_0}) > 0.$$

The fulfillment of condition (NR1) is verified.

Let now $u = \sum_{i=1}^{n} r_i \cdot g_i$ and $u' = \sum_{i=1}^{m} r'_i \cdot g'_i \in RG$, where $g_i \neq g_j$ and $g'_i \neq g'_j$ for $i \neq j$. If necessary, adding the summands of the form of $0 \cdot g$ and substituting the summands, we can consider that $m = n$ and $g'_i = g_i$ for $1 \leq i \leq n$. Then $u - u' = \sum_{i=1}^{n} (r_i - r'_i) \cdot g_i$ and, hence,

$$\widehat{\xi}(u - u') = \sum_{i=1}^{n} \xi(r_i - r'_i) \cdot \eta(g_i) \leq \sum_{i=1}^{n} (\xi(r_i) + \xi(r'_i)) \cdot \eta(g_i)$$

$$= \sum_{i=1}^{n} \xi(r_i) \cdot \eta(g_i) + \sum_{i=1}^{n} \xi(r'_i) \cdot \eta(g_i) = \widehat{\xi}(u) + \widehat{\xi}(u').$$

The fulfillment of the condition (NR2) is verified.

As above, let $u = \sum_{i=1}^{n} r_i \cdot g_i$, $u' = \sum_{i=1}^{m} r'_i \cdot g'_i \in RG$, where $g_i \neq g_j$ and $g'_i \neq g'_j$ for $i \neq j$.

Let $\{h_k \mid k = 1, \ldots, t\} = \{g_i \cdot g_j \mid 1 \leq i \leq n, 1 \leq j \leq m\}$, where the elements h_k are selected from the set $\{g_i \cdot g_j \mid 1 \leq i \leq n, 1 \leq j \leq m\}$ without repetitions, then $u \cdot u' = \sum_{k=1}^{t} (\sum_{g_i \cdot g_j = h_k} r_i \cdot r'_j) \cdot h_k$, moreover, $h_k \neq h_p$ for $1 \leq k < p \leq t$. Then

$$\widehat{\xi}(u \cdot u') = \sum_{k=1}^{t} \xi \Big(\sum_{g_i \cdot g_j = h_k} r_i \cdot r_j \Big) \cdot \eta(h_k)$$

$$\leq \sum_{k=1}^{t} \Big(\sum_{g_i \cdot g_j = h_k} \xi(r_i \cdot r'_j) \Big) \cdot \eta(h_k)$$

$$\leq \sum_{i=1}^{n} \sum_{j=1}^{m} \xi(r_j) \cdot \xi(r'_i) \cdot \eta(g_i) \cdot \eta(g'_j)$$

$$= \Big(\sum_{i=1}^{n} \cdot \xi(r_i) \cdot \eta(g_i) \Big) \cdot \Big(\sum_{j=1}^{m} \xi(r'_j) \cdot \eta(g'_j) \Big) = \widehat{\xi}(u) \cdot \widehat{\xi}(u').$$

Thus, the theorem is completely proved.

6.3.3. DEFINITION. A pseudonorm (norm) ξ on a ring R is called non-Archimedean if $\xi(a + b) \leq \max\{\xi(a), \xi(b)\}$ for any $a, b \in R$.

6.3.4. THEOREM. *Let η be a pseudo-seminorm on a monoid G and let ξ be a non-Archimedean pseudonorm ξ on a ring R with the unitary element 1. If $\xi(1) = \eta(e) = 1$ where e is the unitary element in G, then on the semigroup ring RG there exists a non-Archimedean pseudonorm $\widehat{\xi}$ such that $\widehat{\xi}(r) = \xi(r)$ and $\widehat{\xi}(g) = \eta(g)$ for any $r \in R$ and $g \in G$.*

PROOF. Let us define on RG a real-valued function $\widehat{\xi}$ as follows:

for $u = \sum_{i=1}^{n} r_i \cdot g_i \in RG$, where $r_i \in R$, $g_i \in G$ and $g_i \neq g_j$ for $1 \leq i, j \leq n$ and $i \neq j$, put $\widehat{\xi}(u) = \max\{\xi(r_i) \cdot \eta(g_i) \mid 1 \leq i \leq n\}$.

It is clear that the function $\widehat{\xi}$ is defined correctly, moreover $\widehat{\xi}(r \cdot e) = \xi(r)$ and $\widehat{\xi}(1 \cdot g) = \eta(g)$ for any $r \in R$ and $g \in G$.

We shall verify that $\widehat{\xi}$ is non-Archimedean pseudonorm on RG. As in Theorem 6.3.2, it is verified that $\widehat{\xi}(u) = 0$ if and only if $u = 0$.

Since $\xi(-r) = \xi(r)$ for any $r \in R$, then $\widehat{\xi}(-u) = \widehat{\xi}(u)$ for any $u \in RG$.

Let $u = \sum_{i=1}^{n} r_i \cdot g_i$, $u' = \sum_{i=1}^{m} r'_i \cdot g'_i \in RG$, where $g_i \neq g_j$ and $g'_i \neq g'_j$ for $i \neq j$. As in the proving of Theorem 6.3.2, we can consider that $n = m$ and $g_i = g'_i$ for $1 \leq i \leq n$. Then $u + u' = \sum_{i=1}^{n} (r_i + r'_i) \cdot g_i$ and, hence,

$$\xi(u + u') = \max\{\xi(r_i + r'_i) \cdot \eta(g_i) \,|\, 1 \leq i \leq n\}$$
$$\leq \max\{\xi(r_i) \cdot \eta(g_i), \xi(r'_i) \cdot \eta(g_i) \,|\, 1 \leq i \leq n\} \leq \max\{\widehat{\xi}(u), \widehat{\xi}(u')\}.$$

To complete the proof of the theorem it remains to verify the fulfillment of condition (NR3). Let $u = \sum_{i=1}^{n} r_i \cdot g_i$, $u' = \sum_{i=1}^{m} r'_i \cdot g'_i \in RG$, where $g_i \neq g_j$ and $g'_i \neq g'_j$ for $i \neq j$. As in the proof of Theorem 6.3.2, put $\{h_k \,|\, k = 1, \ldots, t\} = \{g_i \cdot g'_j \,|\, 1 \leq i < n, \, 1 \leq j \leq m\}$, where $h_k \neq h_p$ for $1 \leq k < p \leq t$. Then $u \cdot u' = \sum_{k=1}^{t} (\sum_{g_i \cdot g_j = h_k} r_i \cdot r'_j) \cdot h_k$ and, hence,

$$\xi(u \cdot u') = \max\{\xi(\sum_{g_i \cdot g'_j = h_k} r_i \cdot r'_j) \cdot \eta(h_k) \,|\, 1 \leq i \leq n, \, 1 \leq j \leq m, \, 1 \leq k \leq t\}$$
$$\leq \max\{\max\{\widehat{\xi}(r_i \cdot r'_j) \cdot \eta(h_k) \,|\, g_i \cdot g'_j = h_k, \, 1 \leq i \leq n, \, 1 \leq j \leq m\}, \,|\, 1 \leq k \leq t\}$$
$$\leq \max\{\xi(r_i) \cdot \xi(r'_i) \cdot \eta(g_i) \cdot \eta(g'_j) \,|\, 1 \leq i \leq n, 1 \leq j \leq m\}$$
$$\leq \max\{\xi(r_i) \cdot \eta(g_i) \,|\, 1 \leq i \leq n\} \cdot \max\{\xi(r'_i) \cdot \eta(g'_j) \,|\, 1 \leq j \leq m\} = \widehat{\xi}(u) \cdot \widehat{\xi}(u').$$

Thus, the theorem is completely proved.

6.3.5. **THEOREM.** *Let (G, η) be a seminormed group and (R, ξ) be a non-Archimedean normed ring with the unitary element 1. If the group G is right-ordered (see B.15), then on the semigroup ring RG there exists a non-Archimedean norm $\widehat{\xi}$ such that $\widehat{\xi}(r) = \xi(r)$ and $\widehat{\xi}(g) = \eta(g)$ for any $r \in R$, $g \in G$.*

PROOF. Since η is a seminorm on G and ξ is a norm on R, then $\xi(1) = \eta(e) = 1$, where e is the unitary element of the group G. Then, according to Theorem 6.3.4, there exists a non-Archimedean pseudonorm $\widehat{\xi}$ on RG such that $\widehat{\xi}(r) = \xi(r)$ and $\widehat{\xi}(g) = \eta(g)$ for any $r \in R$ and $g \in G$. Thus, it remains to verify that the pseudonorm $\widehat{\xi}$ is a norm.

Let $u = \sum_{i=1}^{n} r_i \cdot g_i$ and $u' = \sum_{i=1}^{m} r'_i \cdot g'_i$ be arbitrary elements from RG. Without loss of generality we can consider that the elements $r_i \cdot g_i$ and $r'_j \cdot g'_j$ are numbered in such a way that $\widehat{\xi}(r_i \cdot g_i) \geq \widehat{\xi}(r_{i+1} \cdot g_{i+1})$ and $\widehat{\xi}(r'_j \cdot g'_j) \geq \widehat{\xi}(r'_{j+1} \cdot g'_{j+1})$ for $1 \leq i \leq n-1$, $1 \leq j \leq m-1$ (otherwise, the renumbering could be done).

Let $n' = \max\{i \,\big|\, \widehat{\xi}(r_i \cdot g_i) = \xi(r_1 \cdot g_1), \, 1 \le i \le n\}$ and $m' = \max\{j \,\big|\, \widehat{\xi}(r'_j \cdot g'_j) = \widehat{\xi}(r'_1 \cdot g'_1), \, 1 \le j \le m\}$.

According to the condition of the theorem, there exists a linear order " $\prec$ " on G such that $x \cdot y \prec z \cdot y$ for any $x, y, z \in G$ for which $x \prec y$. There exists also a natural number $1 \le i_1 \le n'$ such that $g_{i_1} \prec g_k$ for any $1 \le k \le n'$, $k \ne i_1$. Since G is a group and it is a linearly ordered set with respect to " $\prec$ ", there exists, being unique, a natural number $1 \le j_1 \le m'$ such that $g_{i_1} \cdot g_{j_1} \prec g_{i_1} \cdot g_k$ for $1 \le k \le m$ and $k \ne j_1$.

Then $\widehat{\xi}(u) = \xi(r_{i_1}) \cdot \eta(g_{i_1})$ and $\widehat{\xi}(u') = \xi(r'_{j_1}) \cdot \eta(g'_{j_1})$ (see the note about the numbering of elements $r_i \cdot g_i$, $r'_j \cdot g'_j$ and the choice of numbers n', m' above), moreover, $\xi(r_{i_1}) \cdot \eta(g_{i_1}) \ge \xi(r_i) \cdot \eta(g_i)$ for $n' < i \le n$ and $\xi(r_{j_1}) \cdot \eta(g_{j_1}) \ge \xi(r'_j) \cdot \eta(g'_j)$ for $m' < j \le m$.

From the fact that $(G, \prec)$ is a right-ordered group it follows that $g_{i_1} \cdot g_{j_1} \prec g_{i_1} \cdot g'_j \prec g_i \cdot g'_j$ for any $1 \le i \le n'$, $i \ne i_1$ and $1 \le j \le m'$, $j \ne j_1$. Hence, $g_{i_1} \cdot g_{j_1} \ne g_i \cdot g'_j$ for $1 \le i \le n'$, $i \ne i_1$ and $1 \le j \le m'$, $j \ne j_1$. If, however, $i = i_1$ and $j \ne j_1$, or $i \ne i_1$, and $j = j_1$, then, from the fact that G is a group, it follows that in these cases $g_{i_1} \cdot g'_{j_1} \ne g_i \cdot g'_j$, too. Thus, we have shown that $g_{i_1} \cdot g'_{j_1} \ne g_i \cdot g'_j$ for any $1 \le i \le n'$ and $1 \le j \le m'$ such that $(i_1, j_1) \ne (i, j)$.

Let now i_2, j_2 be such that $g_{i_2} \cdot g'_{i_2} = g_{i_1} \cdot g'_{i_1}$, then $i_2 > n'$ or $j_2 > m'$. Further, $\widehat{\xi}(r_{i_2} \cdot g_{i_2}) < (\widehat{\xi}(r_{i_1} \cdot g_{i_1})$ and $\widehat{\xi}(r'_{j_2} \cdot g'_{j_2}) \le \widehat{\xi}(r'_{j_1} \cdot g'_{j_1})$ or $\widehat{\xi}(r_{i_2} \cdot g_{i_2}) \le \widehat{\xi}(r_{i_1} \cdot g_{i_1})$ and $\widehat{\xi}(r'_{j_2} \cdot g'_{j_2}) < \widehat{\xi}(r'_{j_1} \cdot g'_{j_1})$. Since ξ is a norm on R, and η is a seminorm on G , then

$$\widehat{\xi}(r_{i_2} \cdot r'_{j_2} \cdot g_{i_2} \cdot g'_{j_2}) = \xi(r_{i_2} \cdot r'_{j_2}) \cdot \eta(g_{i_2} \cdot g'_{j_2}) = \xi(r_{i_2}) \cdot \xi(r'_{j_2}) \cdot \eta(g_{i_2}) \cdot \eta(g'_{j_2})$$
$$< \xi(r_{i_1}) \cdot \xi(r'_{j_1}) \cdot \eta(g_{i_1}) \cdot \eta(g_{j_1}) = \xi(r_{i_1} \cdot r'_{j_1}) \cdot \eta(g_{i_1} \cdot g'_{j_1}) = \widehat{\xi}(r_{i_1} \cdot r'_{j_1} \cdot g_{i_1} \cdot g'_{j_1}),$$

i.e. $\xi(r_{i_2} \cdot r'_{j_2} \cdot g_{i_2} \cdot g'_{j_2}) < \widehat{\xi}(r_{i_1} \cdot r'_{j_1} \cdot g_{i_1} \cdot g'_{j_1})$ for any $1 \le i_2 \le n$, $1 \le j_2 \le m$, $(i_2, j_2) \ne (i_1, j_1)$ and $g_{i_2} \cdot g'_{j_2} = g_{i_1} \cdot g'_{j_1}$

Therefore, from the fact that $\widehat{\xi}$ is a non-Archimedean pseudonorm, it follows that $\widehat{\xi}(r \cdot g_{i_1} \cdot g'_{j_1}) = \widehat{\xi}(r_{i_1} \cdot r'_{j_1} \cdot g_{i_1} \cdot g'_{j_1})$ where $r = \sum_{g_i \cdot g'_j = g_{i_1} \cdot g'_{j_1}} (r_i \cdot r'_j)$

Let now $u \cdot u' = \sum_{i=1}^{t} p_i \cdot h_i$, where $p_i \in R$, $h_i \in G$ and $h_i \ne h_j$ for $1 \le i \le t$, $1 \le j \le t$,

$i \neq j$, then $r \cdot g_{i_1} \cdot g_{j_1} \in \{p_1 \cdot h_1, \dots, p_t \cdot h_t\}$. Further,

$$\xi(u) \cdot \xi(u') = \xi(r_{i_1}) \cdot \eta(g_{i_1}) \cdot \xi(r'_{j_1}) \cdot \eta(g'_{j_1}) = \xi(r_{i_1} \cdot r'_{j_1}) \cdot \eta(g_{i_1} \cdot g'_{j_1})$$
$$= \widehat{\xi}(r_{i_1} \cdot r'_{j_1} \cdot g_{i_1} \cdot g'_{j_1}) = \widehat{\xi}(r \cdot g_{i_1} \cdot g'_{j_1}) = \xi(r) \cdot \eta(g_{i_1} \cdot g'_{j_1})$$
$$\leq \max\{\xi(p_i) \cdot \eta(h_i) \,|\, 1 \leq i \leq t\} = \widehat{\xi}(u \cdot u').$$

And since $\widehat{\xi}$ is a pseudonorm on RG, then $\widehat{\xi}(u) \cdot \widehat{\xi}(u') = \widehat{\xi}(u \cdot u')$. From the arbitrariness of the choice of the elements $u, u' \in RG$ it follows that $\widehat{\xi}$ is a norm on RG. Thus, the theorem is completely proved.

6.3.6. COROLLARY. *Let (G, η) be a seminormed group, and let (R, ξ) be a non-Archimedean normed ring with the unitary element. Then in each case below on the group ring RG there exists a non-Archimedean norm $\widehat{\xi}$ such that $\widehat{\xi}|_R = \xi$ and $\widehat{\xi}|_G = \eta$:*

1) G is a group of automorphisms of linearly ordered set;

2) G is an Abelian group without torsion;

3) G is a free solvable group;

4) G is a free nilpotent group;

5) G is a free group.

PROOF. Indeed, according to B.15, in each case above the group G admits a linear order with respect to which it is right-ordered.

Then, according to Theorem 6.3.5, on the ring RG exists a required norm .

6.3.7. REMARK. We shall show in Example 6.3.9 that the analogs of Theorem 6.3.5 and of Corollary 6.3.6 for Archimedean norms do not take place, i.e. the requirement of the norm ξ on the ring R to be non-Archimedean is essential and cannot be skipped. But for this we need a theorem, which gives the description of a class of normalized fields containing the field of real numbers. It is given below without proof. The proof of this theorem can be found, for instance, in [65, 145, 280].

6.3.8. THEOREM. (Gelfand-Mazur). *Any normalized field containing the field $\mathbb{R}$ of real numbers is izomorphic to a subfield of the field of complex numbers.*

6.3.9. EXAMPLE. Let: $(\mathbb{R}, \xi)$ be the field of real numbers with the natural norm, i.e. $\xi(a) = |a|$ for any $a \in \mathbb{R}$; G be an Abelian group without torsion, whose cardinality is greater than the continuum cardinality, and η be a real-valued function on G such that $\eta(g) = 1$ for any $g \in G$ (it is easy to see that η is a seminorm on G).

Let us show now that the norm ξ and the seminorm η cannot be extended onto the group ring $\mathbb{R}G$.

Assume the contrary, i.e. ξ and η can be extended to the norm $\widehat{\xi}$ on the ring $\mathbb{R}G$. Since $\mathbb{R}G$ is an associative-commutative ring without divisors of zero, then $\mathbb{R}G$ admits the quotient field $\widetilde{R}$ (see B.18). We extend the norm $\widehat{\xi}$ to the norm $\widetilde{\xi}$ on $\widetilde{R}$ putting $\widetilde{\xi}(\frac{a}{b}) = \frac{\xi(a)}{\xi(b)}$. It is easy to see that $\widetilde{\xi}$ is defined correctly, being the norm on $\widetilde{R}$. Since $\mathbb{R} \subseteq \widetilde{R}$ and $\widetilde{\xi}|_{\mathbb{R}} = \xi$, then by Theorem 6.3.8 we get that $\widetilde{R}$ is isomorphic to a subring of the ring of complex numbers and, hence, it has the cardinality not greater than the continuum cardinality. We have the contradiction to the choice of the group G .

§ 6.4. Extension of Topologies on
Semigroup Rings

6.4.1. DEFINITION. Let (R, τ_0) be a topological ring with the unitary element, and (G, τ_1) be a topological monoid*. If on the semigroup ring RG exists a Hausdorff ring topology τ extending the topologies τ_0 and τ_1 (i.e. $\tau|_R = \tau_0$ and $\tau|_G = \tau_1$), then we call the pair (τ_0, τ_1) extendable.

6.4.2. THEOREM. *Let τ_0 and τ_0' be ring topologies on a ring R, moreover $\tau_0' \geq \tau_0$, and (G, τ_1) be a topological monoid. If the pair (τ_0, τ_1) is extendable, then the pair (τ_0', τ_1) is extendable, too.*

PROOF. Let φ be a natural homomorphism of the ring RG onto the ring R (i.e. $\varphi(\sum_{i=1}^n r_i \cdot q_i) = \sum_{i=1}^n r_i)$ and let τ' be the prototype of the topology τ_0' in RG with respect to φ. According to Theorem 1.5.35, τ' is a ring non-Hausdorff topology on RG. It is easy to notice, that $\tau'|_R = \tau_0'$ and $\tau'|_G$ is an anti-discrete topology on G.

If now τ is ring Hausdorff topology on RG extending the topologies τ_0 and τ_1 then $\sup\{\tau', \tau\}$ is a required topology.

6.4.3. THEOREM. *Let (R, τ_0) be a topological ring with the unitary element and $\{\tau_\gamma | \gamma \in \Gamma\}$ be a family of semigroup topologies on a monoid G such that for each $\gamma \in \Gamma$ the pair (τ_0, τ_γ) is extendable. Then the pair $(\tau_0, \sup\{\tau_\gamma | \gamma \in \Gamma\})$ is extendable.*

PROOF. For each $\gamma \in \Gamma$ we denote over $\hat{\tau}_\gamma$ the ring topology on RG extending the topologies τ_0 and τ_γ. It is easy to verify that $\hat{\tau} = \sup\{\hat{\tau}_\gamma | \gamma \in \Gamma\}$ is a required topology.

6.4.4. THEOREM. *Let (R, τ_0) be a topological ring with the unitary element 1 and (G, τ_1) be a topological monoid. Let $\{(R_\alpha, \tau_\alpha) | \alpha \in A\}$ be a family of topological rings such that (R, τ_0) is a subdirect product of the family $\{(R_\alpha, \tau_\alpha) | \alpha \in A\}$ (see Definition 4.3.1) and for each $\alpha \in A$ the pair (τ_α, τ_1) is extendable. Then the pair (τ_0, τ_1) is extendable, too.*

PROOF. For each $\alpha \in A$ denote over $\hat{\tau}_\alpha$ a Hausdorff ring topology on RG extending the topologies τ_α and τ_1.

---

* Topological monoid is a monoid with such given topology that the existing binary algebraic operation is continuous.

Let $(\tilde{R}, \tilde{\tau}_0)$ and $(\hat{R}, \hat{\tau})$ be the direct products with Tychonoff topology of the families $\{(R_\alpha, \tau_\alpha) | \alpha \in A\}$ and $\{(R_\alpha G, \hat{\tau}_\alpha) | \alpha \in A\}$ respectively. Let $\varphi_\alpha : \tilde{R} \to R_\alpha$ and $\psi_\alpha : \hat{R} \to R_\alpha G$, for $\alpha \in A$, be the natural ring homomorphisms (projections). Consider the mapping $\xi : \tilde{R}G \to \hat{R}$ putting into correspondence to the element $\sum_{i=1}^k \tilde{r}_i \cdot g_i$ such element $f \in \hat{R}$ that $\psi_\alpha(f) = \sum_{i=1}^k \varphi_\alpha(\tilde{r}_i) \cdot g_i$ for any $\alpha \in A$. It is clear that ξ is the inclusion of the ring $\tilde{R}G$ into the ring $\hat{R}$.

Identifying the element $a \in \tilde{R}G$ with the element $\xi(a) \in \hat{R}$ we shall obtain the identification of the elements of the ring $\tilde{R}$ with the corresponding elements of the ring

$$\xi(\tilde{R}) = \{f \in \hat{R} | \psi_\alpha(f) = r_\alpha \cdot e, \ r_\alpha \in R_\alpha, \ \text{for} \ \alpha \in A\} = \prod_{\alpha \in A} R_\alpha$$

(e is the identity element of the monoid G), and the identification of the elements of the monoid $G = \{1 \cdot g | g \in G\}$ (1 is the identity element of the ring $\tilde{R}$) with the appropriate elements of the monoid

$$\xi(G) = \{f \in \hat{R} | \psi_\alpha(f) = 1 \cdot g, \ g \in G, \ \text{for all} \ \alpha \in A\}.$$

Then from Proposition 4.1.40 it follows that

$$\hat{\tau}\big|_{\xi(\tilde{R})} = \hat{\tau}\big|_{\prod_{\alpha \in A} R_\alpha} = \tilde{\tau}_0,$$

and from Proposition 4.1.37 it follows that

$$\hat{\tau}\big|_G = \sup\{\tilde{\tau}_\alpha\big|_G \big| \alpha \in A\} = \tau_1.$$

Hence, $\hat{\tau}\big|_{\tilde{R}G}$ is the extension of the topologies $\tilde{\tau}_0$ and τ_1 on $\tilde{R}G$.

Since $(\hat{R}, \hat{\tau})$ is a Hausdorff ring (as the direct product of the Hausdorff rings), then $\hat{\tau}\big|_{\tilde{R}G}$ is a Hausdorff topology and, hence, the pair $(\tilde{\tau}_0, \tau_1)$ is extendable. Since there exists a continuous inclusion (see Proposition 4.3.3) of the ring (R, τ_0) into

$$\prod_{\alpha \in A} (R_\alpha, \tau_\alpha) = (\tilde{R}, \tilde{\tau}),$$

then from Theorem 6.4.2 it follows that the pair (τ_0, τ_1) is extendable, too.

6.4.5. THEOREM.[*] *Let:* (R, τ_0) *be a topological ring with the unitary element;* Γ *be a directed set;* $\{(G_\gamma, \tau_\gamma) | \gamma \in \Gamma\}$ *be a family of topological Abelian groups. Let* (G, τ_1) *be a topological group, which is an inverse limit of the family* $\{(G_\gamma, \tau_\gamma) | \gamma \in \Gamma\}$, *and for each* $\gamma \in \Gamma$ *the pair* (τ_0, τ_γ) *be extendable. Then the pair* (τ_0, τ_1) *is extendable, too.*

PROOF. For each $\gamma \in \Gamma$ we denote over $\hat{\tau}_\gamma$ a Hausdorff ring topology on RG_γ which is an extension of the topologies τ_0 and τ_γ. Let $(\hat{R}, \hat{\tau}_0)$ be the direct product with the Tychonoff topology of the family $\{(RG_\gamma, \hat{\tau}_\gamma) | \gamma \in \Gamma\}$ of topological rings. Then $\hat{\tau}_0$ is a Hausdorff topology.

Consider the mapping $\xi : RG \to \hat{R}$ putting into correspondence to an element $\sum_{i=1}^{n} r_i \cdot g_i \in RG$ such element $f \in \hat{R}$ that $\hat{\pi}_\gamma(f) = \sum_{i=1}^{n} r_i \cdot \pi_\gamma(g_i)$ for any $\gamma \in \Gamma$, where $\hat{\pi}_\gamma : \hat{R} \to RG_\gamma$ and $\pi_\gamma : G \to G_\gamma$ are the canonical projections. It is clear that ξ is a ring homomorphism.

Let us show now that ξ is an injection, i.e. one-to-one mapping. Indeed, let $0 \neq \sum_{i=1}^{n} r_i \cdot g_i \in RG$. Without loss of generality we can consider that $g_i \neq g_j$ for $1 \leq i < j \leq n$.

For any pair (i, j), $i \neq j$, $1 \leq i, j \leq n$ there exists an element $\gamma_{i,j} \in \Gamma$ such that $\pi_{\gamma_{i,j}}(g_i) \neq \pi_{\gamma_{i,j}}(g_j)$. If now $\gamma_0 \geq \gamma_{i,j}$ for any $i, j = 1, \ldots, n$, $i \neq j$, then from Definition 4.2.1 it follows that $\pi_{\gamma_0}(g_i) \neq \pi_{\gamma_0}(g_j)$ for any $i, j = 1, \ldots, n$, $i \neq j$. Then

$$\hat{\pi}_{\gamma_0}\left(\xi\left(\sum_{i=1}^{n} r_i \cdot g_i\right)\right) = \sum_{i=1}^{n} r_i \cdot \pi_{\gamma_0} \cdot (g_i) \neq 0$$

and, hence, $\xi(\sum_{i=1}^{n} r_i \cdot g_i) \neq 0$, i.e. ξ is an injective mapping. Let's identify an element $a \in RG$ with the element $\xi(a) \in \hat{R}$.

Then from Proposition 4.1.37 it follows that $\hat{\tau}_0\big|_R = \sup\{\hat{\tau}_\gamma|_R | \gamma \in \Gamma\} = \tau_0$.

Since for each $\gamma \in \Gamma$ the mappings $\hat{\pi}_\gamma \xi : G \to G_\gamma$ and $\pi_\gamma \xi^{-1} : \xi(G) \to G_\gamma$ are continuous, then from Corollary 4.1.39 follows the continuity of the mappings $\xi : G \to \xi(G)$ and $\xi^{-1} : \xi(G) \to G$, i.e. ξ is a homeomorphic mapping of the topological space (G, τ_1) onto the topological space $(\xi(G), \hat{\tau}\big|_{\xi(G)})$. Then, taking into account the identification of

---

[*] This result is valid also for topological monoids (see [54]). The definition of a limit of inverse spectrum of monoids is analogous to the definition of a limit of inverse spectrum of Abelian groups (see Definition 4.2.1).

elements $g \in G$ and $\xi(g) \in \hat{R}$, we derive that $\tau_1 = \tau|_{\xi(G)}$. From the fact that the topology $\hat{\tau}$ is Hausdorff it follows that the pair (τ_0, τ_1) is extendable.

6.4.6. THEOREM. *Let (R, τ_0) be a Hausdorff topological ring with the unitary element, and (G, τ_1) be a finite monoid with the discrete topology. Then the pair (τ_0, τ_1) is extendable.*

PROOF. Let $G = \{g_1, \ldots, g_n\}$. For every neighborhood V of zero of the ring R we consider the subset $W_V = \{\sum_{i=1}^{n} r_i \cdot g_i | r_i \in V, \; i = 1, \ldots, n\}$ and verify that the family $\{W_V | V$ is a neighborhood of zero in $R\}$ fulfills the conditions (BN1) – (BN6) and, according to Theorem 1.2.5, it defines a certain ring topology τ on RG.

The fulfillment of conditions (BN1) – (BN4) is obvious.

Let W_V be a set from the constructed ones, and let $b \in RG$. Then $b = \sum_{i=1}^{n} r_i \cdot g_i$. Let V_1 and V_2 be neighborhoods of zero in (R, τ_0) such that

$$\underbrace{V_1 + \ldots + V_1}_{n^2 \text{ times}} \subseteq V,$$

and $r_i \cdot V_2 \subseteq V_1$ and $V_2 \cdot r_i \subseteq V_1$ for $i = 1, \ldots, n$, then for any element $d = \sum_{i=1}^{n} r_i' \cdot g_i \in W_{V_2}$ we have

$$b \cdot d = \sum_{i=1}^{n} \sum_{j=1}^{n} r_i \cdot r_j' \cdot g_i \cdot g_j = \sum_{k=1}^{n} \sum_{g_i \cdot g_j = g_k} (r_i \cdot r_j') \cdot g_k$$

$$\in \sum_{k=1}^{n} \underbrace{(V_1 + \ldots + V_1)}_{n^2 \text{ times}} \cdot g_k \subseteq \sum_{k=1}^{n} V \cdot g_k = W_V.$$

Analogously it is verified that $d \cdot b \in W_V$.

From the arbitrariness of the choice of element $d \in W_{V_2}$ the fulfillment of the condition (BN5) follows.

Finally, if W_V is a set from the constructed ones and V_1, V_2 be neighborhoods of zero in (R, τ_0) such that

$$\underbrace{V_1 + \ldots + V_1}_{n^2 \text{ times}} \subseteq V \quad \text{and} \quad V_2 \cdot V_2 \subseteq V_1,$$

then for any $\sum_{i=1}^{n} r_i \cdot g_i$, $\sum_{j=1}^{n} r'_j \cdot g_j \in W_{V_2}$ we have $r_i \cdot r_j \in V_2 \cdot V_2 \subseteq V_1$ and, hence

$$\left(\sum_{i=1}^{n} r_i \cdot g_i \right) \cdot \left(\sum_{j=1}^{n} r'_j \cdot g_j \right) = \sum_{k=1}^{n} \sum_{g_i \cdot g_j = g_k} (r_i \cdot r'_j) \cdot g_k$$

$$\in \sum_{k=1}^{n} \underbrace{(V_1 + \ldots + V_1)}_{n^2 \text{ times}} \cdot g_k \subseteq \sum_{k=1}^{n} V \cdot g_k = W_V,$$

i.e. $W_{V_2} \cdot W_{V_2} \subseteq W_V$. Thus, the fulfillment of the condition (BN6) is also verified.

Hence, the family $\{W_V | V \text{ is a neighborhood of zero in } R\}$ defines a certain ring topology τ on RG.

Since τ_0 is a Hausdorff topology, then from the constructing of the sets W_V it follows that τ is a Hausdorff topology, too.

As $W_V \bigcap R = V$ for any neighborhood V of zero in R, then $\tau|_R = \tau_0$.

Since τ_0 is a Hausdorff topology, there exists a neighborhood V of zero in R such that $-1 \notin V$. From the constructing of the set W_V it follows that $g - h \notin W_V$ for any elements g, $h \in G$, and $g \neq h$, i.e. $(g + W_V) \bigcap G = \{g\}$ for any $g \in G$. It means that $\tau|_G = \tau_1$.

This completely proves the Theorem.

From Corollary 4.2.22 and Theorems 6.4.5 and 6.4.6 the following corollary follows:

6.4.7. Corollary.* Let (R, τ_0) be a Hausdorff topological ring, and (G, τ_1) be a compact totally disconnected Abelian group. Then the pair (τ_0, τ_1) is extendable.

The further result shows that the requirement of total disconnectedness of a compact group in Corollary 6.4.7 is essential and cannot be simply skipped, not being replaced by another condition.

6.4.8. THEOREM.** *Let (G, τ_1) be a compact Abelian group, not being totally disconnected. Then there exists a Hausdorff locally bounded topological ring $(\hat{R}, \hat{\tau}_0)$, possessing*

---

* This result is also valid for an arbitrary compact group, not necessarily an Abelian one, since any compact totally disconnected group is an inverse limit of finite discrete groups (see [52]).

** The requirement of commutativity of the group is not essential. This result is true for an arbitrary compact not totally disconnected group (see [52]).

a basis of neighborhoods of zero consisting of subgroups, such that the pair $(\hat{\tau}_0, \hat{\tau}_1)$ *is not extendable.*

PROOF. Assume the contrary, i.e. that for any topological ring (R, τ_0) the pair (τ_0, τ_1) is extendable. Let C be the connected component of the identity element in the group (G, τ_1) and $\tau_1' = \tau_1|_C$, then the pair (τ_0, τ_1') is also extendable on the group ring RC. Therefore, without loss of generality, we can consider that G is a connected group.

Let $\hat{R}$ be a field of non-countable regular cardinality $\mathfrak{m}$, and let $\hat{\tau}_0$ be a ring topology on R, constructed according to Theorem 5.4.5. Since $W_1 \cdot W_\alpha \subseteq W_\alpha$ for any $\alpha < \omega(\mathfrak{m})$, $\hat{\tau}_0$ is a locally bounded topology. From the definition of the sets W_α it follows that W_α is a subgroup of the additive group of the ring $\hat{R}$ for any $\alpha < \omega(\mathfrak{m})$. According to Corollary 5.4.7, $\hat{\tau}_0$ is a non-discrete topology.

By the assumption on the ring $\hat{R}G$ exists a ring topology $\hat{\tau}$ such that $\hat{\tau}|_{\hat{R}} = \hat{\tau}_0$ and $\hat{\tau}|_G = \tau_1$. Then G is a compact subset in $(\hat{R}G, \hat{\tau})$ and, hence, according to Proposition 1.1.38, for any $n \in \mathbb{N}$ the subset $\hat{G}_n = 1 + \underbrace{G + \ldots + G}_{n \text{ times}}$ is compact in $(\hat{R}G, \hat{\tau})$. From Proposition 1.6.4 it follows that for any $n \in \mathbb{N}$ the subset $\hat{G}_n$ is bounded in the ring $(\hat{R}G, \hat{\tau})$.

As (G, τ) is a compact Abelian group, then, according to Theorem 1.5.40, there exists a non-zero character $\xi : G \longrightarrow \mathbb{R}/\mathbb{Z}$, i.e. a non-zero continuous homomorphism of the topological group (G, τ_1) to the topological group $\mathbb{R}/\mathbb{Z}$. Let $U = (-\frac{1}{4}; \frac{1}{4}) + \mathbb{Z}$, then U is a neighborhood of zero in $\mathbb{R}/\mathbb{Z}$ and, hence, $\xi^{-1}(U)$ is a neighborhood of the unitary element in (G, τ_1). From the fact that $\hat{\tau}|_G = \tau_1$ it follows that there exists a neighborhood $\hat{W}$ of zero in $\hat{R}G$ such that $(1 + \hat{W}) \cap G = \xi^{-1}(U)$.

Let $\hat{W}_0$ be a symmetric neighborhood of zero in $\hat{R}G$ such that $\hat{W}_0 \cdot \hat{W}_0 \subseteq \hat{W}$, then taking into account a boundedness of the sets $\hat{G}_n$ for $n \in \mathbb{N}$, we see that there exist neighborhoods $\hat{W}_1, \hat{W}_2, \ldots$ of zero in $(\hat{R}G, \hat{\tau})$ such that $\hat{W}_k \cdot \hat{G}_k \subseteq W_0$ for $k \in \mathbb{N}$. As $\hat{\tau}|_{\hat{R}} = \hat{\tau}_0$, for each $k \in \mathbb{N}$ there exists $\alpha_k < \omega(\frac{\mathfrak{m}}{})$ such that $W_{\alpha_k} \subseteq \hat{W}_k \cap \hat{R}$.

Since $\mathfrak{m}$ is a regular cardinal number, then $\sup\{\alpha_i | \alpha_i < \omega(\mathfrak{m}), i \in \mathbb{N}\} < \omega(\mathfrak{m})$ and, hence, the intersection $\bigcap_{i=1}^{\infty} W_{\alpha_i}$ of a countable number of neighborhoods W_{α_i} of zero in $\hat{R}$ contains some neighborhood W_α, where $\alpha = \sup\{\alpha_i | i \in \mathbb{N}\}$, i.e. $\bigcap_{i=1}^{\infty} W_{\alpha_i} \neq \{0\}$. Let $0 \neq r \in \bigcap_{k=1}^{\infty} W_{\alpha_k}$, then there exists a symmetric neighborhood $\hat{W}'$ of zero in $(\hat{R}G, \hat{\tau})$

such that $r^{-1} \cdot \hat{W}' \subseteq \hat{W}_0$.

As $U' = (1 + (\hat{W}' \cap \hat{W})) \cap G$ is a neighborhood of the unitary element in (G, τ_1), then from the connectedness of the group G it follows that G is generated by the set U'. Then from the fact that ξ is a non-zero homomorphism it follows that $\xi(U') \neq \{0\}$. Let a be an element from U' such that $\xi(a) \neq 0$, then $\xi(a) \in U = (-\frac{1}{4}; \frac{1}{4}) + \mathbb{Z}$ and, hence, $n\xi(a) \notin U$ for some $n \in \mathbb{N}$. Then

$$(1 - a^n) = r^{-1} \cdot (1 - a) \cdot r \cdot (1 + a + \ldots + a^{n-1})$$
$$\in r^{-1} \cdot (1 - (1 + \hat{W}')) \cdot \hat{W}_{\alpha_n} \cdot \hat{G}_n$$
$$\subseteq r^{-1} \cdot W' \cdot \hat{W}_n \cdot \hat{G}_n \subseteq \hat{W}_0 \cdot \hat{W}_0 \subseteq \hat{W}$$

and, hence, $a^n \in 1 + \hat{W}$. As $a^n \in G$, then $a^n \in (1 + \hat{W}) \cap G = \xi^{-1}(U)$ and, hence, $n\xi(a) = \xi(a^n) \in \xi(\xi^{-1}(U)) = U$. We have the contradiction to the choice of the number n. This completely proves the theorem.

6.4.9. THEOREM. *Let (R, τ_0) be a Hausdorff topological ring with the unitary element 1, which possesses a basis of neighborhoods of zero consisting of subgroups, and (G, τ_1) be a monoid with the discrete topology. Then the pair (τ_0, τ_1) is extendable.*

PROOF. Let $\mathcal{B}_0$ be a basis of zero of the ring (R, τ_0), consisting of subgroups. For each $V \in \mathcal{B}_0$ put

$$W_V = \left\{ \sum_{i=1}^{n} r_i \cdot g_i \middle| r_i \in V, g_i \in G, 1 \leq i \leq n, n \in \mathbb{N} \right\}^*.$$

Let's verify that the family $\{W_V | V \in \mathcal{B}_0\}$ fulfills the conditions (BN1) – (BN6) and, according to Theorem 1.2.5, it defines on RG a certain ring topology τ.

The fulfillment of the conditions (BN1) – (BN4) easily follows from the definition of the sets W_V.

Let now W_V be a set from the mentioned ones and let $b = \sum_{i=1}^{m} a_i \cdot h_i \in RG$. There exists a neighborhood $V' \in \mathcal{B}_0$ such that $a_i \cdot V' \subseteq V$ and $V' \cdot a_i \subseteq V$ for $i = 1, \ldots, m$. Let $d \in W_{V'}$, then $d = \sum_{i=1}^{n} r_i \cdot g_i$, where $n \in \mathbb{N}$, $r_i \in V'$, $g_i \in G$ for $1 \leq i \leq n$. Then $d \cdot b = \sum_{i=1}^{m} \sum_{j=1}^{n} (r_j \cdot a_i) \cdot (g_j \cdot h_i)$, where $a_i \cdot r_j \in a_i \cdot V' \subseteq V$, $r_j \cdot a_i \in V' \cdot a_i \subseteq V$ and

---

* We don't suppose that $g_i \neq g_j$ for $i \neq j$.

$h_i \cdot g_i, g_i \cdot h_i \in G$, for $1 \leq i \leq m$ and $1 \leq j \leq n$. Hence, $b \cdot d, d \cdot b \in W_V$. From the arbitrariness of the choice of element d it follows that $b \cdot W_{V'} \subseteq W_V$ and $W_{V'} \cdot b \subseteq W_V$, i.e. the condition (BN5) is satisfied.

Finally, let W_V be a set from the constructed ones, and $W_{V'}$ be such that $V' \cdot V' \subseteq V$. If b and d are elements of $W_{V'}$, then there exist their presentations in the form $b = \sum_{i=1}^{m} r_i \cdot g_i$, and $d = \sum_{j=1}^{n} p_i \cdot h_j$, where $r_i, p_i \in V'$, and $g_i, h_i \in G$ for $1 \leq i \leq m$ and $1 \leq j \leq n$. Then

$$b \cdot d = \sum_{i=1}^{m} \sum_{j=1}^{n} (r_i \cdot p_j) \cdot (g_i \cdot h_j),$$

moreover, $r_i \cdot p_j \in V' \cdot V' \subseteq V$ and $g_i \cdot h_i \in G$ for $1 \leq i \leq m$ and $1 \leq j \leq n$ i.e. $b \cdot d \in W_V$. From the arbitrariness of the choice of elements b and d it follows that $W_{V'} \cdot W_{V'} \subseteq W_V$. So, the fulfillment of the condition (BN6) is verified.

Hence, the family $\{W_V | V \in \mathcal{B}_0\}$ defines on RG a certain ring topology τ.

Let us verify that the topology τ is a Hausdorff, i.e. that the condition (BN1$'$) is satisfied. Assume the contrary, i.e. that $0 \neq \sum_{i=1}^{n} r_i \cdot g_i \in \bigcap_{V \in \mathcal{B}_0} W_V$. Without loss of generality we can consider, that $0 \neq r_i$ and $g_i \neq g_1$, for $1 < i \leq n$. There exist a neighborhood $V_0 \in \mathcal{B}_0$ of zero in (R, τ_0) such that $r_1 \notin V_0$. Since $\sum_{i=1}^{n} r_i \cdot g_i \in W_{V_0}$, then there exists s presentation of the element $\sum_{i=1}^{n} r_i \cdot g_i$ in the form $\sum_{j=1}^{m} a_j \cdot h_j$ such that $a_j \in V_0$, and $h_j \in G$ for $1 \leq j \leq m$. Then $r_1 = \sum_{h_j = g_1} a_j$, and since V_0 is a group, then $r_1 \in V_0$. We have the contradiction to the choice of the neighborhood V_0. Hence, τ is a Hausdorff topology.

From the definition of the sets W_V it easily follows that $V \subseteq W_V \cap R$ for any $V \in \mathcal{B}_0$. Let now $r \notin V$, and assume that $r \in W_V \cap R$. Then $r = \sum_{i=1}^{n} r_i \cdot g_i$, where $r_i \in V$ and $g_i \in G$ for $1 \leq i \leq n$. Since V is a group, then $r = \sum_{g_i = e} r_i \in V$, where e is the identity element of the monoid G. We have a contradiction. Hence, $r \notin W_V \cap R$. It means that $W_V \cap R = V$ for any $V \in \mathcal{B}_0$. Thus, we have shown that $\tau|_R = \tau_0$.

Let now $g \in G$ and V_0 be a neighborhood of zero in (R, τ_0) such that $-1 \notin V_0$. Then it is easy to notice that $g \in (g + W_{V_0}) \cap G$. Assume now that $h \neq g$ and $h \in (g + W_{V_0}) \cap G$. Then $g - h \in W_{V_0}$ and, hence, $h - g = \sum_{i=1}^{n} r_i \cdot g_i$, where $r_i \in V_0$ and $g_i \in G$ for $1 \leq i \leq n$. Since V_0 is a group, then $-1 = \sum_{g_i = g} r_i \in V_0$. We have a contradiction. Hence, $(g + W_{V_0}) \cap G = \{g\}$, i.e. $\tau|_G = \tau_1$. This completely proves the theorem.

6.4.10. Corollary.* Let (R, τ_0) be a Hausdorff topological ring with the unitary element, possessing a basis of neighborhoods of zero consisting of subgroups, and (G, τ_1) be a Hausdorff topological Abelian group, possessing a basis of neighborhoods of the unitary element consisting of subgroups. Then the pair (G, τ_1) is extendable.

PROOF. Indeed, if $(\hat{G}, \hat{\tau}_1)$ is the completion of the topological group (G, τ_1), then, according to Theorem 4.2.21, $(\hat{G}, \hat{\tau}_1)$ is a inverse limit of discrete groups. Then from Theorems 6.4.9 and 6.4.5 follows that the pair (τ_0, τ_1) is extendable. And since $\hat{\tau}_1|_G = \tau_1$, the pair (τ_0, τ_1) is extendable, too.

6.4.11. DEFINITION. Let R be a topological ring with the unitary element, and $\mathcal{U} = (V_1, V_2, \dots)$ be a sequence of symmetric neighborhoods of zero in R. A sequence $A = (a_1, a_2, \dots)$ of elements of the ring R is called admissible with respect to $\mathcal{U}$ or, briefly, $\mathcal{U}$–admissible, if there exists a sequence $B = (b_1, b_2, \dots)$ of elements of R such that:

a) $b_i \in V_i$ for all $i \in \mathbb{N}$;

b) for each $k \in \mathbb{N}$ there exists a number $s = s(k, A, B)$ such that $s \geq k$ and $a_k = c_1 \cdot c_2 \cdot \ldots \cdot c_s$, where $c_j \in \{0, 1, \pm b_j\}$ for $1 \leq j \leq s$ and moreover, $c_s \neq 1$.

The sequence B (defined, generally speaking, ambiguously) we call an appropriate one to the sequences $\mathcal{U}$ and A.

6.4.12. Proposition. Let R be a topological ring with the unitary element 1 and $\mathcal{U} = (V_1, V_2, \dots)$, $\mathcal{U}' = (V_1', V_2', \dots)$, $\mathcal{U}'' = (V_1'', V_2'', \dots)$ be sequences of symmetric neighborhoods of zero in R. If $A = (a_1, a_2, \dots)$ and $A' = (a_1', a_2', \dots)$ are admissible sequences with respect to $\mathcal{U}$ and $\mathcal{U}'$ respectively of elements of R with corresponding sequences $B = (b_1, b_2, \dots)$ and $B' = (b_1, b_2, \dots)$ of elements from R, then:

A) if $V_i \subseteq V_i'$ for all $i \in \mathbb{N}$, then the sequence A is $\mathcal{U}'$-admissible, too;

B) if $V_i \bigcup V_i' \subseteq V_{2i-1}'' \bigcap V_{2i}''$ for all $i \in \mathbb{N}$, then the sequence $A'' = (a_1'', a_2'', \dots)$ is $\mathcal{U}''$-

---

* The present result is also valid in the case when (G, τ_1) is non Abelian Hausdorff topological group, possessing the basis of neighborhoods of the unitary element consisting of normal subgroups. It is true because, as it is known in the theory of topological groups, such group has the completion, which is an inverse limit of discrete groups.

admissible, where:

$$a_k'' = \begin{cases} a_i, & \text{if} \quad k = 2i - 1, \quad \text{for} \quad i \in \mathbb{N}; \\ a_i', & \text{if} \quad k = 2i, \quad \text{for} \quad i \in \mathbb{N}; \end{cases}$$

C) if R is a commutative ring, $V_i \bigcup V_i' \subseteq \bigcap_{j=1}^{3i^2} V_j''$ for all $i \in \mathbb{N}$ and $n, m \in \mathbb{N}$, then the sequence $A'' = (a_1'', a_2'', \dots)$ is $\mathcal{U}''$-admissible, where:

$$a_k'' = \begin{cases} a_i \cdot a_j', & \text{if} \quad k = 3\rho(i,j) - 2, \quad \text{for} \quad 1 \le i \le m, \quad \text{and} \quad 1 \le j \le n; \\ 0, & \text{otherwise} \end{cases}$$

(see the definition of the function $\rho(i,j)$ in A.14);

D) if V_0 is a sub-semigroup of a multiplicative semigroup of the ring R, V_i, V_i', V_i'' are right ideals of the sub-semigroup V_0 and $V_i \bigcup V_i' \subseteq \bigcap_{j=1}^{3^{i+1}} V_j''$ for all $i \in \mathbb{N}$, then the sequence $A'' = (a_1'', a_2'', \dots)$ is $\mathcal{U}''$-admissible, where $a_k'' = a_i \cdot a_j'$ if $k = \rho(i,j)$, for $i, j \in \mathbb{N}$ (see the definition of the mapping $\rho : \mathbb{N} \times \mathbb{N} \to \mathbb{N}$ in A.14);

E) if: $n \in \mathbb{N}$; $d_1, \dots, d_n \in R$; $1 \notin V_i''$; $V_i \bigcup V_i \cdot d_k \bigcup d_k \cdot V_i \subseteq \bigcap_{j=1}^{(n+1) \cdot i} V_j''$ for all $i \in \mathbb{N}$ and $1 \le k \le n$, then each of the sequences $A'' = (a_1'', a_2'', \dots)$ and $A''' = (a_1''', a_2''', \dots)$ is $\mathcal{U}$-admissible, where:

$$a_k'' = \begin{cases} d_i \cdot a_j, & \text{if} \quad k = (n+1) \cdot (j-1) + i, \quad \text{for} \quad 1 \le i \le n \quad \text{and} \quad j \in \mathbb{N}; \\ 0, & \text{otherwise}, \end{cases}$$

and

$$a_k''' = \begin{cases} a_j \cdot d_i, & \text{if} \quad k = (n+1) \cdot (j-1) + i, \quad \text{for} \quad 1 \le i \le n, \quad \text{and} \quad j \in \mathbb{N}; \\ 0, & \text{otherwise}; \end{cases}$$

F) if V_0 and V_i for $i \in \mathbb{N}$ are neighborhoods of zero in R such that $V_i + V_i + V_i \cdot V_i + V_i \cdot V_i \subseteq V_{i-1}$ for $i \in \mathbb{N}$, then $a_{j_1} + a_{j_2} + \dots + a_{j_p} \in V_0$ for any finite set $\{j_1, \dots, j_p\}$ of pairwise distinct natural numbers;

G) let $n \in \mathbb{N}$, and $S = \{i_1, \dots, i_n\}$ and $S' = \{i_1', \dots, i_n'\}$ be such finite sequences of natural numbers that $i_j' \le i_j$, $i_l \le i_t$, $i_l' \le i_t'$ for $1 \le j \le n$ and $1 \le l < t \le n$. If $b_j = 0$ for $i \notin S$ and $V_1 \supseteq V_2 \supseteq \dots$, then the sequence $D = (d_1, d_2, \dots)$ is $\mathcal{U}$-admissible, where:

$$d_k = \begin{cases} a_{i_j}, & \text{if} \quad k = i_j' \quad \text{for} \quad 1 \le j \le n; \\ 0, & \text{otherwise}. \end{cases}$$

PROOF. Statement A) easily follows from the definition of an admissible sequence.

Let us verify statement B). Consider the sequence $B'' = (b_1'', b_2'', \dots)$ of elements of the ring R, where:

$$b_k'' = \begin{cases} b_i, & \text{if } k = 2i - 1 & \text{for } i \in \mathbb{N}; \\ b_i', & \text{if } k = 2i, & \text{for } i \in \mathbb{N}. \end{cases}$$

Since $b_{2i-1}'' = b_i \in V_i \subseteq V_{2i-1}''$ and $b_{2i}'' = b_i' \in V_i' \subseteq V_{2i}''$, it follows that the condition a) of Definition 6.4.11 is satisfied for the sequences B'' and $\mathcal{U}''$.

Let $k \in \mathbb{N}$. If $k = 2i - 1$, then $a_k'' = a_i$. For the sequences A and B and the number i there exists a number $s \geq i$ such that $a_i = c_1 \cdot c_2 \cdot \dots \cdot c_s$, where $c_j \in \{0, 1, \pm b_j\}$ for $1 \leq j \leq s$ and $c_s \neq 1$. Putting $c_{2j-1}'' = c_j$ and $c_{2j}'' = 1$ for $1 \leq j \leq s$, we get $a_k'' = c_1'' \cdot c_2'' \cdot \dots \cdot c_{2s-1}''$, where $c_j'' \in \{0, 1, \pm b_j\}$ for $1 \leq j \leq 2s - 1$ and $c_{2s-1}'' \neq 1$, i.e. the condition b) of Definition 6.4.11 holds for $k = 2i - 1$.

Analogously, by considering the sequences A'' and B'' we can verify the condition b) for $k = 2i$. Thus, the statement B) is proved.

Verification of the statement C). Consider the sequence $B'' = (b_1'', b_2'', \dots)$ of elements of the ring R, where:

$$b_k'' = \begin{cases} b_i, & \text{if } k = 3i^2 - 1, & \text{for } i \in \mathbb{N}; \\ b_i', & \text{if } k = 3i^2, & \text{for } i \in \mathbb{N}; \\ 0, & \text{otherwise}. \end{cases}$$

Since $b_{3i^2-1}'' = b_i \in V_i \subseteq V_{3i^2-1}''$, and $b_{3i^2}'' = b_i' \in V_i' \subseteq V_{3i^2}''$ for all $i \in \mathbb{N}$, and $b_r'' = 0 \in V_r''$ for $r \notin \{3i^2 - 1, 3i^2 \,|\, i \in \mathbb{N}\}$, then it follows that the condition a) of Definition 6.4.11 is satisfied for the sequenses B'' and $\mathcal{U}''$.

Let $k \in \mathbb{N}$. If $k \neq 3\rho(i,j) - 2$ for $1 \leq i \leq m$ and $1 \leq j \leq n$, then $a_k = 0 = c_1 \cdot c_2 \cdot \dots \cdot c_k$, where $c_l = 0 \in \{0, 1, \pm b_i\}$ for $1 \leq l \leq k$. Now let $k = 3\rho(i,j) - 2$ for $1 \leq i \leq m$ and $1 \leq j \leq n$. Then $a_k'' = a_i \cdot a_j'$. For the sequences A, B, A', B' and natural numbers i, j there exist natural numbers $s \geq i$ and $s' \geq j$ such that $a_i = c_1 \cdot c_2 \cdot \dots \cdot c_s$ and $a_j' = c_1' \cdot c_2' \cdot \dots \cdot c_{s'}'$. Besides, $c_l \in \{0, 1, \pm b_l\}$ and $c_l' \in \{0, 1, \pm b_l'\}$ with $c_s \neq 1$ and $c_{s'}' \neq 1$. Consider the number $s'' = \max\{3s^2 - 1, 3 \cdot (s')^2\}$ and the elements $c_1'', \dots, c_{s''}''$, where:

$$c_k'' = \begin{cases} c_l, & \text{if } k = 3l^2 - 1, & \text{for } l \leq s; \\ c_l', & \text{if } k = 3l^2, & \text{for } l \leq s'; \\ 1, & \text{otherwise}. \end{cases}$$

Then $s'' \geq \max\{3i^2, 3j^2\} \geq 3\rho(i,j) - 2$ and $a_k'' = a_i \cdot a_j = c_1'' \cdot c_2'' \cdot \ldots \cdot c_{s''}''$, where $c_k'' \in \{0, 1, \pm b_k''\}$ and $c_{s''}'' \neq 1$, i.e. and the condition b) of Definition 6.4.11 is satisfied. Thus, the statement C) is proved.

Let us verify the statement D). For each $n \in \mathbb{N}$ put (see B.12)

$$S_n = \{0, 1, \pm b_1'\} \cdot \{0, 1, \pm b_2'\} \cdot \ldots \cdot \{0, 1, \pm b_n'\}.$$

Since $b_i' \in V_i' \subseteq V_0$ for $1 \leq i \leq n$, then $S_n \subseteq V_0 \bigcup\{1\}$. Since $|S_n| \leq 3^n$ (see A.7), then recurring, if necessary, zero element, we consider that $S_n = \{d_{1,n}, \ldots, d_{3^n,n}\}$, moreover $d_{1,n} = 1$.

Let us construct the sequence $B'' = (b_1'', \ldots)$ of elements of the ring R as follows:

$$b_k'' = \begin{cases} b_i', & \text{if } k = 3^i, & \text{for } i \in \mathbb{N}; \\ b_i \cdot d_{t,i}, & \text{if } k = 3^i + t, & \text{for } 0 < t \leq 3^i, & \text{and } i \in \mathbb{N}; \\ 0, & \text{otherwise}. \end{cases}$$

We shall verify that B'' corresponds to the sequences $\mathcal{U}$ and A''. Indeed, if $k = 3^i$ for $i \in \mathbb{N}$, then

$$b_k'' = b_i' \in V_i' \subseteq \bigcap_{j=1}^{3^{i+1}} V_j'' \subseteq V_{3^i}'' = V_k''.$$

If, however, $k = 3^i + t$, where $0 < t \leq 3^i$, for $i \in \mathbb{N}$, then

$$b_k'' = b_i \cdot d_{t,i} \in V_i \cdot V_0 \subseteq V_i \subseteq \bigcap_{j=1}^{3^{i+1}} V_j'' \subseteq V_{3^i+t}'' = V_k''.$$

In other cases $b_k'' = 0 \in V_k''$. Thus, the condition a) of Definition 6.4.11 for the sequences B'' and $\mathcal{U}''$ is satisfied.

Let $k \in \mathbb{N}$. By virtue of A.14, $k = \rho(i,j)$ for some $i, j \in \mathbb{N}$, and hence, $a_k'' = a_i \cdot a_j'$. Consider the numbers $r = s(i, A, B)$ and $s = s(j, A', B')$ (see Definition 6.4.11, condition b). Then $r \geq i$ and $s \geq j$, moreover $a_i = c_1 \cdot c_2 \cdot \ldots \cdot c_r$ and $a_j' = c_1' \cdot c_2' \cdot \ldots \cdot c_s'$, where $c_l \in \{0, 1, \pm b_l\}$ and $c_m' \in \{0, 1, \pm b_m\}$ for $1 \leq l \leq r$, $1 \leq m \leq s$ and $c_r \neq 1$, $c_s' \neq 1$. Therefore,

$$a_k'' = a_i \cdot a_j' = c_1 \cdot c_2 \cdot \ldots \cdot c_r \cdot c_1' \cdot c_2' \cdot \ldots \cdot c_s'.$$

Let $p = \min\{r, s\}$. Then $c_1' \cdot c_2' \cdot \ldots \cdot c_p' \in S_p \subseteq S_r$ and, hence, $c_1' \cdot c_2' \cdot \ldots \cdot c_p' = d_{t,r}$ for some $1 \leq t \leq 3^r$. Therefore,

$$a_k = (c_1 \cdot c_2 \cdot \ldots \cdot c_{r-1}) \cdot (c_r \cdot c_1' \cdot c_2' \cdot \ldots \cdot c_p') \cdot (c_{p+1}' \cdot \ldots \cdot c_s')$$
$$= c_1 \cdot c_2 \cdot \ldots \cdot c_{r-1} \cdot (c_r \cdot d_{t,r}) \cdot c_{p+1}' \cdot \ldots \cdot c_s'.$$

Put:

$$c_l'' = \begin{cases} c_q, & \text{if } l = 3^q + 1, & \text{for } 1 \leq q \leq r-1; \\ c_r \cdot d_{t,r}, & \text{if } l = 3^r + t; \\ c_m', & \text{if } l = 3^m, & \text{for } p < m \leq s; \\ 1, & \text{otherwise}, \end{cases}$$

and

$$n = \begin{cases} 3^s, & \text{if } r < s; \\ 3^r + t, & \text{if } r \geq s. \end{cases}$$

Then $a_k = c_1'' \cdot c_2'' \cdot \ldots \cdot c_n''$, moreover, taking into account A.14, we get

$$n \geq 3^{\max\{r,s\}} \geq \max\{r^2, s^2\} \geq \rho(r,s) \geq \rho(i,j) = k.$$

Let us verify that $c_l'' \in \{0, 1, \pm b_l''\}$ for $1 \leq l \leq n$. Indeed, if $l = 3^m + 1$ for $1 \leq m \leq r-1$, then, since $d_{1,m} = 1$, we get

$$c_l'' = c_m \in \{0, 1, \pm b_m\} = \{0, 1, \pm b_m \cdot d_{1,m}\} = \{0, 1, \pm b_l''\}.$$

If $l = 3^m$ for $p < m \leq s$, then

$$c_l'' = c_m' \in \{0, 1 \pm b_m'\} = \{0, 1, \pm b_l''\}.$$

If $l = 3^r + t$, then
$$c_l'' = c_r \cdot d_{t,r} \in \{0, \pm b_r \cdot d_{t,r}\} \subseteq \{0, 1, b_l''\}.$$

In other cases $c_l'' = 1 \in \{0, 1, \pm b_l''\}$.

Let us verify that $c_n'' \neq 1$. Indeed, if $r < s$, then $n = 3^s$ and, hence, $c_n'' = c_{3^s}'' = c_s' \neq 1$ (see Definition 6.4.11, condition b). If, however, $r \geq s$, then $n = 3^r + t$ and, hence,

$$c_n'' = c_{3^r+t} = c_r \cdot d_{t,r} \in \{0, \pm b_r\} \cdot d_{t,r} \in V_r \cdot S_r \subseteq V_r \cdot V_0 \subseteq V_r$$

(since V_r is a right ideal in V_0). Since $1 \notin V_r$, then $c_n'' \neq 1$.

Thus, the statement D) is verified.

Verification of the statement E). Consider the sequences $B'' = (b_1'', b_2'', \dots)$ and $B''' = (b_1''', b_2''', \dots)$, where:

$$b_k'' = \begin{cases} d_i \cdot b_j, & \text{if } k = (n+1) \cdot (j-1) + i \quad \text{for } 1 \leq i \leq n \quad \text{and } j \in \mathbb{N}; \\ b_j, & \text{if } k = (n+1) \cdot j, \quad\quad\quad\quad\quad \text{for } j \in \mathbb{N}, \end{cases}$$

and

$$b_k''' = \begin{cases} b_j, & \text{if } k = (n+1) \cdot j, \quad\quad\quad\quad \text{for } j \in \mathbb{N}; \\ b_j \cdot d_i, & \text{if } k = (n+1) \cdot j + i, \quad \text{for } 1 \leq i \leq n, \quad \text{and } j \in \mathbb{N}. \end{cases}$$

We shall verify that the sequences B'' and B''' are appropriate to the sequences A'', $\mathcal{U}''$, and A''', $\mathcal{U}'''$, respectively.

If $k = (n+1) \cdot (j-1) + i$ for some $1 \leq i \leq n$ and $j \in \mathbb{N}$, then

$$b_k'' = d_i \cdot b_j \in d_i \cdot V_j \subseteq \bigcap_{l=1}^{(n+1)\cdot j} V_l'' \subseteq V_k''.$$

If $k = (n+1) \cdot j$, then

$$b_k'' = b_j \in V_j \subseteq \bigcap_{l=1}^{(n+1)\cdot j} V_j'' \subseteq V_k''.$$

Thus, the condition a) of Definition 6.4.11 for the sequences B'' and $\mathcal{U}''$ is satisfied.

Let now $k \in \mathbb{N}$. If $k = (n+1) \cdot m$ for some $m \in \mathbb{N}$, then $a_k'' = c_1'' \cdot c_2'' \cdot \dots \cdot c_k''$, where $c_l'' = 0 \in \{0, 1, \pm b_l''\}$ for $1 \leq l \leq k$, moreover, $c_k'' = 0 \neq 1$.

If $k = (n+1) \cdot (j-1) + i$ for some $1 \leq i \leq n$ and $j \in \mathbb{N}$, then $a_k'' = d_i \cdot a_j$. Consider the number $s = s(j, A, B)$ (see Definition 6.4.11, condition b). Then $s \geq j$, moreover $a_j = c_1 \cdot c_2 \cdot \dots \cdot c_s$, where $c_l \in \{0, 1, \pm b_l\}$ for $1 \leq l \leq s$ and $c_s \neq 1$. If $t = \min\{l \,|\, c_l \neq 1\}$, then $t \leq s$. Put

$$p = \begin{cases} (n+1) \cdot (t-1) + i, & \text{if } t = s; \\ (n+1) \cdot s, & \text{if } t < s, \end{cases}$$

and for $1 \leq l \leq p$ put:

$$c_l'' = \begin{cases} d_i \cdot c_t, & \text{if } l = (n+1) \cdot (t-1) + i; \\ c_m, & \text{if } l = (n+1) \cdot m \quad \text{for } t < m \leq s; \\ 1, & \text{otherwise.} \end{cases}$$

Then

$$a_k'' = d_i \cdot a_j = d_i \cdot c_1 \cdot c_2 \cdot \ldots \cdot c_s = (d_i \cdot c_t) \cdot c_{t+1} \cdot \ldots \cdot c_s = c_1'' \cdot c_2'' \cdot \ldots \cdot c_p'',$$

moreover $k = (n+1) \cdot (j-1) + i \leq (n+1) \cdot (s-1) + i \leq (n+1) \cdot s$ and, hence, in both cases (see the definition of the number p) we get $k \leq p$.

Let us verify now that $c_l'' \in \{0, 1, \pm b_l''\}$ for all $1 \leq l \leq p$. Indeed, if $l = (n+1) \cdot (t-1) + i$, then $c_l'' = d_i \cdot c_t \in d_i \cdot \{0, \pm b_t\} = \{0, \pm b_l''\}$. But if $l = (n+1) \cdot m$ for $t < m \leq s$, then $c_l'' = c_m \in \{0, 1, \pm b_m\} = \{0, 1, \pm b_l''\}$. In other cases $c_l'' = 1 \in \{0, 1, \pm b_l''\}$.

Let us verify now that $c_p'' \neq 1$. Indeed, if $t = s$, then $p = (n+1) \cdot (s-1) + i$. Therefore, $c_p'' = d_i \cdot c_s \in d_i \cdot V_s \subseteq V_1''$. Since, due to our condition, $1 \notin V_1''$, then $c_p'' \neq 1$. But if $t < s$, then $p = (n+1) \cdot s$ and therefore, $c_p'' = c_s \neq 1$.

Thus, it is verified that the sequence B'' is appropriate to the sequences A'' and $\mathcal{U}''$, i.e. A'' is $\mathcal{U}''$–admissible.

Let us verify now that the sequence B''' is appropriate to the sequences A''' and $\mathcal{U}'''$. Since for the sequence B''' it is verified, then $b_k''' \in V_k$ for all $k \in \mathbb{N}$, i.e. the condition a) of Definition 6.4.11 is valid for the sequences B''' and $\mathcal{U}'''$.

Let now $k \in \mathbb{N}$. If $k = (n+1) \cdot m$ for $m \in \mathbb{N}$, then $a_k''' = 0 = c_1''' \cdot c_2''' \cdot \ldots \cdot c_k'''$, where $c_l''' = 0 \in \{0, 1, \pm b_l\}$ for $1 \leq l \leq k$, moreover $c_k''' = 0 \neq 1$.

If $k = (n+1) \cdot (j-1) + i$ for some $1 \leq i \leq n$ and $j \in \mathbb{N}$, then $a_k''' = a_j \cdot d_i$. Consider the number $s = s(j, A, B)$ (see Definition 6.4.11, condition b). Then $s \geq j$, moreover $0 = a_j = c_1 \cdot c_2 \cdot \ldots \cdot c_s$, where $c_l \in \{0, 1, \pm b_l\}$ for $1 \leq l \leq s$ and $c_s \neq 1$.

Put $p = (n+1) \cdot (s-1) + i$, and for $1 \leq l \leq p$ put:

$$c_l''' = \begin{cases} c_m, & \text{if } l = (n+1) \cdot m \quad \text{for} \quad 1 \leq m \leq s-1; \\ c_s \cdot d_i, & \text{if } l = (n+1) \cdot (s-1) + i; \\ 1, & \text{otherwise}. \end{cases}$$

Then

$$a_k = a_j \cdot d_i = c_1 \cdot c_2 \cdot \ldots \cdot c_s \cdot d_i = c_1''' \cdot c_2''' \cdot \ldots \cdot c_p'''.$$

It is clear that $k = (n+1) \cdot (j-1) + i \leq (n+1) \cdot (s-1) + i = p$.

Let us verify now that $c_l''' \in \{0, 1, \pm b_l'''\}$ for all $1 \leq l \leq p$. Indeed, if $l = (n+1) \cdot m$ for $1 \leq m < s$, then

$$c_l''' = c_m \in \{0, 1, \pm b_m\} = \{0, 1, \pm b_l'''\}.$$

But if $l = (n+1) \cdot (s-1) + i$, then

$$c_l''' = c_s \cdot d_i \in \{0, \pm b_s\} \cdot d_i = \{0, \pm b_s \cdot d_i\} = \{0, \pm b_l'''\}.$$

In other cases $c_l''' = 1 \in \{0, 1, \pm b_l'''\}$.

We shall verify that $c_p''' \neq 1$. Indeed, $c_p''' = c_s \cdot d_i \in V_s \cdot d_i \subseteq V_1''$. Since, due to our condition, $1 \notin V_1''$, then $c_p''' \neq 1$. The statement E) is proved.

Verification of the statement F). For any natural numbers k and s with $k \leq s$ we consider the set $B_{k,s} = \{a \in R \,|\, a = d_k \cdot d_{k+1} \cdot \ldots \cdot d_t, \text{ where } t \geq s, \, d_i \in \{0, 1, \pm b_i\}, \text{ and } d_t \neq 1\}$.

At first we shall show that $B_{k,k} \subseteq V_k \bigcup (V_k \cdot V_k)$ for any $k \in \mathbb{N}$. Indeed, if $b \in B_{k,k}$, then $b = d_k \cdot d_{k+1} \cdot \ldots \cdot d_{k+m}$, where $m \geq 0$, $d_i \in \{0, 1, \pm b_i\}$, and $d_{k+m} \neq 1$. If $m = 0$, then $d_k \neq 1$, hence, $b = d_k \in V_k \subseteq V_k \bigcup V_k \cdot V_k$.

Suppose that the desired inclusion has been proved for $m \leq t$ and all k. Then

$$b = d_k \cdot d_{k+1} \cdot \ldots \cdot d_{k+t+1} = d_k \cdot (d_{k+1} \cdot \ldots \cdot d_{k+1+t})$$

$$\in \{1, \pm b_k\} \cdot \left(V_{k+1} \cup (V_{k+1} \cdot V_{k+1})\right) \subseteq \{1, \pm b_k\} \cdot V_k \subseteq V_k \cup (V_k \cdot V_k).$$

Thus, we have shown that $B_{k,k} \subseteq V_k \bigcup (V_k \cdot V_k)$.

We shall show now that $\sum_{i=0}^{n} B_{k,k+i} \subseteq V_{k-1}$ for any $k \in \mathbb{N}$ and $n \in \mathbb{N} \bigcup \{0\}$. Indeed, if $n = 0$, then $B_{k,k} \subseteq V_k \bigcup (V_k \cdot V_k) \subseteq V_{k-1}$.

Let us suppose that $\sum_{i=0}^{n} B_{k,k+i} \subseteq V_{k-1}$ for $n \leq m$ and any $k \in \mathbb{N}$, and establish this inclusion for $n = m + 1$. Since $B_{k,k+t} = B_{k+1,k+t} \bigcup (b_k \cdot B_{k+1,k+t})$ for any $k, t \in \mathbb{N}$, then it follows from the inductive supposition that

$$\sum_{i=0}^{m+1} B_{k,k+i} = B_{k,k} + \sum_{i=0}^{m} B_{k,k+1+i} \subseteq B_{k,k} + \sum_{i=0}^{m} B_{k+1,k+1+i} + b_k \cdot \left(\sum_{i=0}^{m} B_{k+1,k+1+i}\right)$$

$$\subseteq \left(V_k \cup (V_k \cdot V_k)\right) + V_k + b_k \cdot V_k \subseteq V_k + V_k + V_k \cdot V_k + V_k \cdot V_k \subseteq V_{k-1}.$$

Thus, we have shown that $\sum_{i=0}^{n} B_{k,k+i} \subseteq V_{k-1}$ for any $k \in \mathbb{N}$ and $n \in \mathbb{N} \bigcup \{0\}$.

It follows from the definition of an admissible sequence and the sets $B_{k,i}$ that $0, a_i \in B_{1,i}$ for all $i \in \mathbb{N}$. Then $\sum_{i=1}^{n} a_{ij} \in \sum_{i=0}^{l} B_{1,1+i} \subseteq V_0$ for any n such that $1 \leq n \leq p$, where $l = \max\{j_1, \ldots, j_n\}$. Thus, the statement F) is verified.

Verification of the statement G). Consider the sequence $C = (c_1, c_2, \dots)$ of elements of the ring R, where:

$$c_k = \begin{cases} b_{i_j}, & \text{if} \quad k = i'_j, \quad \text{for} \quad 1 \leq j \leq n; \\ 0, & \text{otherwise}. \end{cases}$$

Since $V_{i_j} \subseteq V_{i'_j}$ for all $j \in \mathbb{N}$, then $c_{i'_j} = b_{i_j} \in V_{i'_j}$ for $j = 1, \dots, n$ and $c_i = 0 \in V_i$ for $i \notin S'$, i.e. the condition a) of Definition 6.4.11 is satisfied for the sequences C and $\mathcal{U}$.

Let $l \in \mathbb{N}$. If $d_l = 0$, then $d_l = c'_1 \cdot c'_2 \cdot \dots \cdot c'_l$, where $c'_i = 0 \in \{0, 1, \pm c_i\}$ for $1 \leq i \leq l$. If $d_l \neq 0$, then $l = i'_j \in S'$ and $d_l = a_{i_j}$ (see the definition of d_k).

For the natural number i_j there exist a natural number $s \geq i_j$ and elements $c''_k \in \{0, 1, \pm b_k\}$ for $1 \leq k \leq s$ such that $a_{i_j} = c''_1 \cdot c''_2 \cdot \dots \cdot c''_s$. Since $a_{i_j} = d_l \neq 0$, it follows that $c''_k \neq 0$ for $1 \leq k \leq s$. Since $b_k = 0$ for $k \notin S$, then we have $c''_k = 1$ for $k \notin S$. Then, since $c''_s \neq 1$, we have $s = i_t$, where $1 \leq t \leq n$ and $t \geq j$. Let's put

$$c'_k = \begin{cases} c''_{i_r}, & \text{if} \quad k = i'_r, \quad \text{for} \quad 1 \leq r \leq n; \\ 1, & \text{if} \quad k \notin S', \end{cases}$$

then

$$d_l = a_{i_j} = c''_1 \cdot c''_2 \cdot \dots \cdot c''_s = c''_{i_1} \cdot c''_{i_2} \cdot \dots \cdot c''_{i_t} = c'_{i'_1} \cdot c'_{i'_2} \cdot \dots \cdot c'_{i'_t} = c'_1 \cdot c'_2 \cdot \dots \cdot c'_{i'_t}$$

and $c'_r \in \{0, 1, \pm c_r\}$ for $1 \leq r \leq i'_t$, moreover $c'_{i'_t} \neq 1$. Thus, the condition b) of Definition 6.4.11 is satisfied, i.e. the statement G) is proved.

Thus, the proposition is completely proved.

6.4.13. THEOREM. *Let (R, τ_0) be a Hausdorff topological ring with the unitary element 1 and (G, τ_1) be a monoid with the discrete topology. Then the pair (τ_0, τ_1) is extendable in each case below:*

a) R is a commutative ring ;

b) (R, τ_0) is a ring, locally bounded from left;

c) (R, τ_0) is a ring locally bounded from right.

PROOF. Let $\mathcal{B}_0$ be a basis of symmetrical neighborhoods of zero in (R, τ_0) and Δ be the family of all countable sequences $\mathcal{U} = (V_1, V_2, \dots)$ of neighborhoods from $\mathcal{B}_0$.

For $\mathcal{U} \in \Delta$ over $W_{\mathcal{U}}$, we denote the set of all $f \in RG$ such that for each of them there exist an $\mathcal{U}$-admissible sequence $A = (a_1, a_2, \dots)$ of elements of the ring R, a natural number n and a set $\{g_i, \dots, g_n\} \subseteq G$ (possibly, $g_i = g_j$ for some $i \neq j$) such that $f = \sum_{i=0}^{n} a_i \cdot g_i$.

Let us verify that the family $\{W_{\mathcal{U}} | \mathcal{U} \in \Delta\}$ fulfills conditions (BN1) – (BN6) and that, according to Theorem 1.2.5, it defines on RG a certain ring topology.

Indeed, $A = (0, 0, \dots)$ is a $\mathcal{U}$–admissible sequence for all $\mathcal{U} \in \Delta$, and therefore, $0 \in W_{\mathcal{U}}$ for any $\mathcal{U} \in \Delta$, i.e. the condition (BN1) is satisfied.

Let us verify the fulfillment of the condition (BN2). Let $\mathcal{U} = (V_1, V_2, \dots)$ and $\mathcal{U}' = (V_1', V_2', \dots) \in \Delta$, then for each $i \in \mathbb{N}$ we can find a neighborhood $V_i'' \in \mathcal{B}_0$ such that $V_i'' \subseteq V_i \cap V_i'$. Let $\mathcal{U}'' = (V_1'', V_2'', \dots)$ and $f \in W_{\mathcal{U}''}$. By virtue of the statement A) of Proposition 6.4.12 we derive that $f \in W_{\mathcal{U}}$ and $f \in W_{\mathcal{U}'}$, i.e. $W_{\mathcal{U}''} \subseteq W_{\mathcal{U}} \cap W_{\mathcal{U}'}$ and, hence, the condition (BN2) is satisfied.

Let us verify the fulfillment of the condition (BN3). Let $\mathcal{U} = (V_1, V_2, \dots) \in \Delta$, then we consider $\mathcal{U}' = (V_1', V_2', \dots) \in \Delta$, where $V_i' \subseteq V_{2i-1} \cap V_{2i}$ for $i \in \mathbb{N}$. Let $f, f' \in W_{\mathcal{U}'}$. Then there exist: $\mathcal{U}'$–admissible sequences $B = (b_1, b_2, \dots)$ and $B' = (b_1', b_2', \dots)$ of elements of the ring R; natural numbers n, m; sets $\{g_1, \dots, g_m\}$ and $\{g_1', \dots, g_m'\}$ of elements of G such that $f = \sum_{i=1}^{m} b_i \cdot g_i$ and $f' = \sum_{i=1}^{n} b_i' \cdot g_i'$.

From Definition 6.4.11 follows that the sequences $A = (a_1, a_2, \dots)$ and $A' = (a_1', a_2', \dots)$ are also $\mathcal{U}'$–admissible, where:

$$a_i = \begin{cases} b_i, & \text{if } 1 \leq i \leq m; \\ 0, & \text{if } i > m, \end{cases}$$

$$a_i' = \begin{cases} b_i', & \text{if } 1 \leq i \leq n; \\ 0, & \text{if } i > n. \end{cases}$$

By virtue of point B) of Proposition 6.4.12 the sequence $A'' = (a_1'', a_2'' \dots)$ is $\mathcal{U}$–admissible, where:

$$a_k'' = \begin{cases} a_i, & \text{if } k = 2i - 1 \quad \text{for } i \in \mathbb{N}; \\ a_i', & \text{if } k = 2i \quad\quad\ \text{for } i \in \mathbb{N}. \end{cases}$$

We can consider that $m \leq n$. Put $p = 2n$, and for $1 \leq k \leq p$ put

$$h_k = \begin{cases} g_i, & \text{if } k = 2i - 1 \quad \text{for } 1 \leq i \leq m; \\ e, & \text{if } k = 2i - 1 \quad \text{for } m < i \leq n; \\ g_i', & \text{if } k = 2i \quad\quad\ \text{for } 1 \leq i \leq n. \end{cases}$$

Then we get

$$f + f' = \sum_{i=1}^{m} b_i \cdot g_i + \sum_{i=1}^{n} b_i' \cdot g_i' = \sum_{k=1}^{p} a_k'' \cdot h_k.$$

Hence, $f + f \in W_{\mathcal{U}}$, i.e. $W_{\mathcal{U}'} + W_{\mathcal{U}'} \subseteq W_{\mathcal{U}}$. Thus, the fulfillment of the condition (BN3) is verified.

Let us verify the fulfillment of the condition (BN4). Since the sequence $A = (a_1, a_2, \ldots)$ is $\mathcal{U}$–admissible if and only if the sequence $A' = (-a_1, -a_2, \ldots)$ is $\mathcal{U}$–admissible, then $f = \sum_{i=1}^{n} a_i \cdot g_i \in W_{\mathcal{U}}$ if and only if $-f = \sum_{i=1}^{n} -a_i \cdot g_i \in W_{\mathcal{U}}$, i.e. $-W_{\mathcal{U}} = W_{\mathcal{U}}$. The fulfillment of the condition (BN4) is verified.

Let us verify the fulfillment of condition (BN5). We consider at first the case when R is a commutative ring.

If $\mathcal{U} = (V_1, V_2, \ldots) \in \Delta$ then for any $i \in \mathbb{N}$ there can be found such neighborhood $V_i' \in \mathcal{B}_0$ that $V_i' \subseteq \bigcap_{j=1}^{3i^2} V_j$.

Let $\mathcal{U}' = (V_1', V_2', \ldots)$ and $f, f' \in W_{\mathcal{U}'}$. Then there exist $\mathcal{U}'$–admissible sequences $A = (a_1, a_2, \ldots)$ and $A' = (a_1', a_2', \ldots)$ of elements of the ring R, natural numbers m, n, and sets $\{g_1, \ldots, g_m\}$ and $\{g_1', \ldots, g_n'\}$ of elements of G such that $f = \sum_{i=1}^{m} a_i \cdot g_i$ and $f' = \sum_{j=1}^{n} a_j' \cdot g_j'$. By virtue of statement C) of Proposition 6.4.12 the sequence $A'' = (a_1'', a_2'' \ldots)$ is $\mathcal{U}$–admissible, where:

$$a_k'' = \begin{cases} a_i \cdot a_j', & \text{if } \quad k = 3\rho(i,j) - 2, \quad \text{for } \quad 1 \le i \le m, \quad \text{and } 1 \le j \le n; \\ 0, & \text{otherwise}. \end{cases}$$

We can consider that $m < n$. Putting $p = 3n^2$ and

$$h_k = \begin{cases} g_i \cdot g_j', & \text{if } \quad k = 3\rho(i,j) - 2, \quad \text{for } \quad 1 \le i \le m, \quad \text{and } \quad 1 \le i \le n; \\ e, & \text{otherwise,} \qquad\qquad \text{for } \quad \le k \le p, \end{cases}$$

we obtain

$$f \cdot f' = \left(\sum_{i=1}^{m} a_i \cdot g_i\right) \cdot \left(\sum_{j=1}^{n} a_j' \cdot g_j'\right) = \sum_{i=1}^{m} \sum_{j=1}^{n} (a_i \cdot a_j') \cdot (g_i \cdot g_j') = \sum_{k=1}^{p} a_k'' \cdot h_k.$$

Hence, $f \cdot f \in W_{\mathcal{U}}$, i.e. $W_{\mathcal{U}'} \cdot W_{\mathcal{U}'} \subseteq W_{\mathcal{U}}$. Thus, the condition (BN5) is verified in the case when R is a commutative ring.

Let now (R, τ_0) be a ring locally bounded from left. Taking into account Theorem 1.6.46, we can consider, that $\mathcal{B}_0$ consists of right ideals of some neighborhood V_0 of zero in (R, τ_0), which is a sub-semigroup of the multiplicative semigroup of the ring R.

If $\mathcal{U} = (V_1, V_2, \ldots)$, then for any $i \in \mathbb{N}$ there can be found a neighborhood $V_i' \in \mathcal{B}_0$ such that $V_i' \subseteq \bigcap_{j=1}^{3i+1} V_j$. Let $\mathcal{U}' = (V_1', V_2', \ldots)$ and $f, f' \in W_{\mathcal{U}'}$. Then there exist $\mathcal{U}'$-admissible sequences $B = (b_1, b_2, \dots)$ and $B' = (b'_1, b'_2, \dots)$ of elements of the ring R, natural numbers m, n, and sets $\{g_1, \dots, g_m\}$, $\{g'_1, \dots, g'_n\}$ of the elements of G such that $f = \sum_{i=1}^{m} b_i \cdot g_i$ and $f' = \sum_{j=1}^{n} b'_i \cdot g'_i$.

From Definition 6.4.11 follows that the sequences $A = (a_1, a_2, \dots)$ and $A' = (a'_1, a'_2, \dots)$ are also $\mathcal{U}'$-admissible, where:

$$a_i = \begin{cases} b_i, & \text{if } 1 \le i \le m; \\ 0, & \text{if } i > m, \end{cases}$$

and

$$a'_i = \begin{cases} b'_i, & \text{if } 1 \le i \le n; \\ 0, & \text{if } i > n. \end{cases}$$

By virtue of statement D) of Proposition 6.4.12 the sequence $A'' = (a''_1, a''_1, \dots)$ is $\mathcal{U}''$-admissible, where $a''_k = a_i \cdot a'_j$ if $k = \rho(i, j)$, for $i, j \in \mathbb{N}$. We can consider that $m \le n$. Putting $p = n^2$ and

$$h_k = \begin{cases} g_i \cdot g'_j, & \text{if } k = \rho(i, j), \text{ for } 1 \le i \le m \text{ and } 1 \le j \le n; \\ e, & \text{otherwise,} \qquad \text{for } 1 \le k \le p, \end{cases}$$

we get

$$f \cdot f' = \left(\sum_{i=1}^{m} b_i \cdot g_i\right) \cdot \left(\sum_{j=1}^{n} b'_i \cdot g'_i\right) = \sum_{i=1}^{m} \sum_{j=1}^{n} (a_i \cdot a'_j) \cdot (g_i \cdot g'_j) = \sum_{k=1}^{p} a''_k \cdot h_k.$$

Hence, $f \cdot f' \in W_\mathcal{U}$, i.e. $W_{\mathcal{U}'} \cdot W_{\mathcal{U}'} \subseteq W_\mathcal{U}$.

Thus, the fulfillment of the condition (BN5) is verified in the case when (R, τ_0) is a ring locally bounded from left.

Let us verify the fulfillment of the condition (BN6). Let $f' = \sum_{i=1}^{n} d_i \cdot \tilde{g}_i \in RG$ and $\mathcal{U} = (V_1, V_2, \dots) \in \Delta$. For each $i \in \mathbb{N}$ there can be found a neighborhood $V'_i \in \mathcal{B}_0$ such that

$$V'_i \cup d_k \cdot V'_i \cup V'_i \cdot d_k \subseteq \bigcap_{j=1}^{(n+1) \cdot i} V_j$$

for $1 \le k \le n$. Let $\mathcal{U}' = (V'_1, V'_2, \dots)$ and $f \in W_{\mathcal{U}'}$. Then there exist $\mathcal{U}'$-admissible sequence $A = (a_1, a_2, \dots)$, a number $m \in \mathbb{N}$ and elements $g_1, g_2, \dots, g_m \in G$ such that $f = \sum_{j=1}^{m} a_j \cdot g_j$. By virtue of statement E) of Proposition 6.4.12 the sequence $A'' = (a''_1, a''_2, \dots)$ is $\mathcal{U}$-admissible, where:

$$a''_k = \begin{cases} d_i \cdot a_j, & \text{if } k = (n+1) \cdot (j-1) + i \quad \text{for } 1 \le i \le n \text{ and } j \in \mathbb{N}; \\ 0, & \text{otherwise.} \end{cases}$$

Putting $p = (n+1) \cdot m$ and

$$
h_k = \begin{cases} \tilde{g}_i \cdot g_j, & \text{if} \quad k = (n+1) \cdot (j-1) + i \quad \text{for} \quad 1 \leq i \leq n, \quad \text{and} \quad j \in \mathbb{N}; \\ e, & \text{otherwise}, \qquad\qquad\qquad\qquad\quad \text{for} \quad 1 \leq k \leq p, \end{cases}
$$

we get

$$
f \cdot f' = \left(\sum_{i=1}^{n} d_i \cdot \tilde{g}_i \right) \cdot \left(\sum_{j=1}^{m} a_j \cdot g_j \right) = \sum_{i=1}^{n} \sum_{j=1}^{m} (d_i \cdot a_j) \cdot (\tilde{g}_i \cdot g_j) = \sum_{k=1}^{p} a_k'' \cdot h_k .
$$

Hence, $f' \cdot f \in W_{\mathcal{U}}$, i.e. $f' \cdot W_{\mathcal{U}'} \subseteq W_{\mathcal{U}}$.

Analogously it is verified that $W_{\mathcal{U}'} \cdot f' \subseteq W_{\mathcal{U}}$.

Thus, the fulfillment of the condition (BN6) is verified.

Then, according to Theorem 1.2.5, for the cases a) and b) the family $\{W_{\mathcal{U}} | \mathcal{U} \in \Delta\}$ defines on RG a ring topology τ in which it is a basis of neighborhoods of zero.

Let us show now that $\tau|_R = \tau_0$.

From Definition 6.4.11 it easily follows that for any sequence $\mathcal{U} = (V_1, V_2, \dots) \in \Delta$ and any element $a \in V_1$ the sequence $A = (a_1, a_2, \dots)$ is $\mathcal{U}$–admissible, where $a_1 = a$ and $a_i = 0$ for $i \geq 2$. Then $a = a \cdot e \in R \cap W_{\mathcal{U}}$, where e is the identity element of the monoid G, i.e. $V_1 \subseteq R \cap W_{\mathcal{U}}$. Hence, $\tau|_R \leq \tau_0$.

Let now V_0 be a neighborhood of zero in (R, τ_0). There exists a sequence $\mathcal{U} = (V_1, V_2, \dots) \in \Delta$ such that

$$
V_i + V_i + V_i \cdot V_i + V_i \cdot V_i \subseteq V_{i-1}
$$

for all $i \in \mathbb{N}$. If now $r \in R \cap W_{\mathcal{U}}$, then there exist $\mathcal{U}$–admissible sequence $A = (a_1, a_2, \dots)$, a number $n \in \mathbb{N}$ and elements $g_1, g_2, \dots, g_n \in G$ (possibly, $g_i = g_j$ for some $i \neq j$) such that $r = \sum_{i=1}^{n} a_i \cdot g_i$. Then, according to statement F) of Proposition 6.4.12, $r = \sum_{g_i = e} a_i \in V_0$, i.e. $W_{\mathcal{U}} \cap R \subseteq V_0$. Hence, $\tau_0 \leq \tau|_R$, and it means that $\tau_0 = \tau|_R$.

Let us verify now that $\tau|_G = \tau_1$, i.e. that for any $h \in G$ there exists a sequence $\mathcal{U} \in \Delta$ such that $\{h\} = (h + W_{\mathcal{U}}) \cap G$.

Assume the contrary, i.e. that $(h + W_{\mathcal{U}}) \cap G \neq \{h\}$ for some $h \in G$ and any $\mathcal{U} \in \Delta$. There exists a neighborhood V_0 of zero in (R, τ_0) such that $-1 \notin V_0$. Let $\mathcal{U} = (V_1, V_2, \dots)$ be a sequence from Δ such that

$$
V_i + V_i + V_i \cdot V_i + V_i \cdot V_i \subseteq V_{i-1}
$$

for any $i \in \mathbb{N}$. If now $g \neq h$ and $g \in (h + W_{\mathcal{U}}) \bigcap G$, then there exist $\mathcal{U}$–admissible sequence $A = (a_1, a_2, \dots)$, a natural number n and elements $g_1, g_2, \dots, g_n \in G$ such that $g = h + \sum_{i=1}^{n} a_i \cdot g_i$. Then, according to statement F) of Proposition 6.4.12, $-1 = \sum_{g_i = h} a_i \in V_0$. We have the contradiction to the choice of neighborhood V_0. Hence, $\tau|_G = \tau_0$.

To complete the proof of the theorem for the cases a) and b) it remains to verify that the constructed topology on RG is Hausdorff.

Assume the contrary, and let $0 \neq f \in \bigcap_{\mathcal{U} \in \Delta} W_{\mathcal{U}}$. Then there can be found a natural number m, non-zero elements $r_1, \dots, r_m \in R$ and pairwise distinct elements $\tilde{g}_1, \dots, \tilde{g}_m \in G$ such that $f = \sum_{i=1}^{m} r_i \cdot \tilde{g}_i$.

Since τ_0 is a Hausdorff topology on the ring R, there can be found $V_0 \in \mathcal{B}_0$ and $\mathcal{U} = (V_1, V_2, \dots) \in \Delta$ such that $r_1 \notin V_0$ and $V_i + V_i + V_i \cdot V_i + V_i \cdot V_I \subseteq V_{i-1}$ for all $i \in \mathbb{N}$. Since $f \in W_{\mathcal{U}}$, there exist $\mathcal{U}$-admissible sequence $A = (a_1, a_2, \dots)$, a number $n \in \mathbb{N}$ and elements $g_1, \dots, g_n \in G$ (possibly, $g_i = g_j$ for some $i \neq j$) such that $f = \sum_{i=1}^{n} a_i \cdot g_i$. Then $r_1 = \sum_{g_i = \tilde{g}_1} a_i$. By virtue of statement F) of Proposition 6.4.12, we get that $r_1 \in V_0$, which contradicts the choice of neighborhood V_0.

This completes the proof for the cases a) and b).

Let us now verify the statement c). Let (R, τ_0) be a ring locally bounded from right. We define operation of multiplication " $*$ " on the additive group $R(+)$ of the ring R as: $a * b = b \cdot a$ for any $a, b \in R$.

It is easy to see that $R(+)$ with respect to the defined operation of multiplication is a ring. Let us denote this ring over R'. Then (R', τ_0) is a topological ring locally bounded from left and, according to the case b), the topology τ_0 we can extend to the ring topology τ on the ring $R'G$.

It is clear that the additive groups of the rings RG and $R'G$ coincide, i.e. $R'G = RG$. From the fact that $(R'G, \tau)$ is a topological ring it follows that $(R'G, \tau)$ is a topological ring.

Since $\tau|_R = \tau'|_R = \tau_0$, then the statement c) is valid.

This completes the proof of the theorem.

6.4.14. Corollary.* *Let (R, τ_0) be a Hausdorff topological ring with the unitary element, and (G, τ_1) be a Hausdorff topological Abelian group possessing a basis of neighborhoods of zero consisting of subgroups. Then the pair (τ_0, τ_1) is extendable in each case below:*

a) R is a commutative ring ;

b) (R, τ_0) is a ring, locally-bounded from left (right).

The proof of this corollary is carried out analogously to the proof of Corollary 6.4.10 with the only difference that Theorem 6.4.13 is used instead of Theorem 6.4.9.

6.4.15. Lemma. *Let G be an Abelian group, $\{S_1, \ldots, S_n\}$ be a finite family of subsets of the group G, and let ξ be the binary relation on G, defined in the following way:*

$x \xi y$ if and only if $x = y$ or there exist not necessarily pairwise distinct natural numbers $i_1, i_2, \ldots, i_k$ such that $1 \leq i_j \leq n$ for $1 \leq j \leq k$, and $x \in S_{i_1}$, $y \in S_{i_k}$, $S_{i_j} \cap S_{i_{j+1}} \neq \emptyset$ for $1 \leq j \leq k - 1$.

Then ξ is an equivalence relation on G (see A.15) and, moreover, if $x \xi y$, then there exist pairwise distinct numbers $i_1, \ldots, i_m$ such that $i_j \leq n$ for $1 \leq j \leq m$, and

$$x \cdot y_1^{-1} \in \left(S_{i_1} \cdot S_{i_2}^{-1}\right) \cdot \ldots \cdot \left(S_{i_m} \cdot S_{i_m}^{-1}\right).$$

PROOF. The fact that ξ is an equivalence relation readily follows from the definition of this relation. Let now $x \xi y$. If $x = y$, then $x \cdot y^{-1} \in S_{i_1} \cdot S_{i_1}^{-1}$. If however $x \neq y$, then there exist numbers $i_1, i_2, \ldots, i_m$ such that $x \in S_{i_1}$, $y \in S_{i_m}$ and $S_{i_j} \cap S_{i_{j+1}} \neq 0$ for $1 \leq j \leq m - 1$.

Without loss of generality we can consider that m is the least number from such ones. Then the numbers $i_1, \ldots, i_m$ are pairwise distinct, since if $i_s = i_t$ for $1 \leq s < t \leq m$, then the sequence $i_1, \ldots, i_s, i_{t+1}, \ldots i_m$ fulfills the required conditions too, but its length is less than m.

---

* The present result is also valid for non-Abelian Hausdorff topological groups possessing a basis of neighborhoods of the unitary element consisting of normal subgroups, due to the same reasons, which were mentioned within the footnote to Corollary 6.4.10.

For each $1 \leq j \leq m - 1$ we choose some element b_j from non-empty subset $S_{i_j} \cap S_{i_{j+1}}$. Then

$$x \cdot y^{-1} = (x \cdot b_1^{-1}) \cdot (b_1 \cdot b_2^{-1}) \cdot \ldots \cdot (b_{m-1} \cdot y^{-1}) \in (S_{i_1} \cdot S_{i_1}^{-1}) \cdot (S_{i_2} \cdot S_{i_1}^{-1}) \cdot \ldots \cdot (S_{i_m} \cdot S_{i_m}^{-1}).$$

Thus, the lemma is completely proved.

6.4.16. THEOREM.* *Let (G, τ_1) be a Hausdorff topological Abelian group, (R, τ_0) be a locally bounded topological ring with the unitary element 1. If (R, τ_0) is a union of countable number of bounded sets, then the pair (τ_0, τ_1) is extendable.*

PROOF. According to Theorem 1.6.46, in (R, τ_0) exist a neighborhood V_0 of zero and a basis $\mathcal{B}_0$ of neighborhoods of zero such that V_0 is a sub-semigroup of the multiplicative semigroup of the ring R, and any neighborhood $V \in \mathcal{B}_0$ is an ideal in the semigroup V_0.

Since R is a union of countable number of bounded sets, then from Proposition 1.6.19 and Corollary 1.6.22 follows that there exist bounded symmetrical subsets $D_1, D_2, \ldots$ in the ring (R, τ_0) such that $V_0 \subseteq D_1$, $D_i \cdot D_i \subseteq D_{i+1}$ and $D_i + D_i \subseteq D_{i+1}$ for any $i \in \mathbb{N}$. Let $\mathcal{B}_1$ be a basis of neighborhoods of the unitary element in (G, τ_1), then we consider:

$$\mathfrak{M} = \left\{ \mathcal{U} = (U_1, U_2, \ldots) \middle| U_i \in \mathcal{B}_0 \quad \text{for} \quad i \in \mathbb{N} \right\};$$
$$\mathfrak{N} = \left\{ \mathcal{V} = (V_1, V_2, \ldots) \middle| V_i \in \mathcal{B}_1 \quad \text{for} \quad i \in \mathbb{N} \right\}.$$

For any $\mathcal{U} = (U_1, U_2, \ldots) \in \mathfrak{M}$ and $\mathcal{V} = (V_1, V_2, \ldots) \in \mathfrak{N}$ we define the set $W(\mathcal{U}, \mathcal{V}) = \left\{ z \in RG \middle| z = \sum_{i=1}^m p_i \cdot f_i + \sum_{i=1}^n q_i \cdot (g_i - h_i), \text{ where: } f_i, g_i, h_i \in G; \ g_i \cdot h_i^{-1} \in V_i; \ p_i \in U_i; \ q_i \in D_i \text{ for } i \in \mathbb{N} \right\}$.

Let us verify that the family $\left\{ W(\mathcal{U}, \mathcal{V}) \middle| \mathcal{U} \in \mathfrak{M}, \ \mathcal{V} \in \mathfrak{N} \right\}$ defines on RG a certain ring topology τ in which it is a basis of neighborhoods of zero. For doing this, according to Theorem 1.2.5, it is sufficient to verify that this family fulfills the conditions (BN1) – (BN6).

The fulfillment of the condition (BN1) is obvious, since $0 = 0 \cdot e + 1 \cdot (e - e)$, where e is the identity element of the group G.

---

* The present result is also valid for the larger class of topological groups (see [49]).

Let now $\mathcal{U} = (U_1, U_2, \dots)$, $\mathcal{U}' = (U'_1, U'_2, \dots) \in \mathfrak{M}$ and $\mathcal{V} = (V_1, V_2, \dots)$, $\mathcal{V}' = (V'_1, V'_2, \dots) \in \mathfrak{N}$. Let $\mathcal{U}'' = (U''_1, U''_2, \dots) \in \mathfrak{M}$ and $\mathcal{V}'' = (V''_1, V''_2, \dots) \in \mathfrak{N}$ be such that $U''_i \subseteq U_i \cap U'_i$ and $V''_i \subseteq V_i \cap V'_i$ for $i \in \mathbb{N}$, then from the definition of the sets $W(\mathcal{U}, \mathcal{V})$ it easily follows that

$$W(\mathcal{U}'', \mathcal{V}'') \subseteq W(\mathcal{U}, \mathcal{V}) \cap W(\mathcal{U}', \mathcal{V}').$$

Hence, the fulfillment of the condition (BN2) is verified.

Let $\mathcal{U} = (U_1, U_2, \dots) \in \mathfrak{M}$ and $\mathcal{V} = (V_1, V_2, \dots) \in \mathfrak{N}$. There exist $\mathcal{U}' = (U'_1, U'_2, \dots) \in \mathfrak{M}$ and $\mathcal{V}' = (V'_1, V'_2, \dots) \in \mathfrak{N}$ such that $U'_i \subseteq U_{2i-1} \cap U_{2i}$ and $V'_i \subseteq V_{2i-1} \cap V_{2i}$ for $i \in \mathbb{N}$. If now $z, z' \in W(\mathcal{U}', \mathcal{V}')$, and

$$z = \sum_{i=1}^{m} p_i \cdot f_i + \sum_{i=1}^{n} q_i \cdot (g_i - h_i),$$

$$z' = \sum_{i=1}^{m'} p'_i \cdot f'_i + \sum_{i=1}^{n'} q'_i \cdot (g'_i - h'_i)$$

are the presentations, mentioned above in the definition of the sets $W(\mathcal{U}, \mathcal{V})$, then

$$z + z' = \sum_{i=1}^{m} p_i \cdot f_i + \sum_{i=1}^{m'} p'_i \cdot f'_i + \sum_{i=1}^{n} q_i \cdot (g_i - h_i) + \sum_{i=1}^{n'} q'_i \cdot (g'_i - h'_i)$$

$$= \sum_{i=1}^{m''} p''_i \cdot f''_i + \sum_{i=1}^{n''} q''_i \cdot (g''_i - h''_i),$$

where:

$$m'' = 2\max\{m, m'\} \quad \text{and} \quad n'' = 2\max\{n, n'\};$$

$$p''_k = \begin{cases} p_i, & \text{if } k = 2i - 1 \text{ for } 1 \leq i \leq m; \\ p'_i, & \text{if } k = 2i \quad\quad\text{ for } 1 \leq i \leq m'; \\ 0, & \text{otherwise}; \end{cases}$$

$$f''_k = \begin{cases} f_i, & \text{if } k = 2i - 1 \text{ for } 1 \leq i \leq m; \\ f'_i, & \text{if } k = 2i \quad\quad\text{ for } 1 \leq i \leq m'; \\ 0, & \text{otherwise}; \end{cases}$$

$$g''_k = \begin{cases} g_i, & \text{if } k = 2i - 1 \text{ for } 1 \leq i \leq n; \\ g'_i, & \text{if } k = 2i \quad\quad\text{ for } 1 \leq i \leq n'; \\ 0, & \text{otherwise}; \end{cases}$$

$$q_k'' = \begin{cases} q_i, & \text{if } k = 2i - 1 \quad \text{for } 1 \le i \le n; \\ q_i', & \text{if } k = 2i \qquad \text{for } 1 \le i \le n'; \\ e, & \text{otherwise}; \end{cases}$$

$$h_k'' = \begin{cases} h_i, & \text{if } k = 2i - 1 \quad \text{for } 1 \le i \le n; \\ h_i', & \text{if } k = 2i \qquad \text{for } 1 \le i \le n'; \\ e, & \text{otherwise}; \end{cases}$$

As: $U_t' \subseteq U_{2t-1} \cap U_{2t}$; $V_t' \subseteq V_{2y-1} \cap V_{2t}$; $D_t \subseteq D_{2t-1} \subseteq D_{2t}$ for $t \in \mathbb{N}$, then $p_k'' \in U_k$ for $1 \le k \le m''$, $(g_j'') \cdot (h_j'')^{-1} \in U_j$ and $q_j \in D_j$ for $1 \le j \le n''$, i.e. $z + z' \in W(\mathcal{U}, \mathcal{V})$. Thus, the fulfillment of the condition (BN3) is verified.

From the symmetry of the sets $U \in \mathcal{B}_0$ and D_i for $i \in \mathbb{N}$ it follows that $W(\mathcal{U}, \mathcal{V}) = -W(\mathcal{U}, \mathcal{V})$ for any $\mathcal{U} \in \mathfrak{M}, \mathcal{V} \in \mathfrak{N}$. Thus, the fulfillment of the condition (BN4) is verified.

Let $\mathcal{U} = (U_1, U_2, \dots) \in \mathfrak{M}$ and $\mathcal{V} = (V_1, V_2, \dots) \in \mathfrak{N}$. For each $i \in \mathbb{N}$ by virtue of boundedness of the sets D_i, there exist $U_i' \in \mathcal{B}_0$ such that

$$U_i' \cdot D_i \cup D_i \cdot U_i' \subseteq \bigcap_{j=1}^{5i^2+4} U_j,$$

and there exist $V_i' \in \mathcal{B}_1$ such that $V_i' \subseteq \bigcap_{j=1}^{4i^2+3} V_j$.

Then $\mathcal{U}' = (U_1', U_2', \dots) \in \mathfrak{M}$ and $\mathcal{V}' = (V_1', V_2', \dots) \in \mathfrak{N}$.

Let $z, z' \in W(\mathcal{U}'\mathcal{V}')$. If $z = \sum_{i=1}^{m} p_i \cdot f_i + \sum_{i=1}^{n} q_i \cdot (g_i - h_i)$ and $z' = \sum_{i=1}^{m'} p_i' \cdot f_i' + \sum_{i=1}^{m'} q_i' \cdot (g_i' - h_i')$ are the presentations, mentioned above in the definition of the sets $W(\mathcal{U}, \mathcal{V})$, then adding, if necessary, the summands of the form $0 \cdot e$ and $0 \cdot (e - e)$, we can consider that $m = m' = n' = n$. Then

$$z \cdot z' = \left(\sum_{i=1}^{m} p_i \cdot f_i + \sum_{i=1}^{m} q_i \cdot (g_i - h_i) \right) \left(\sum_{j=1}^{m} p_j' \cdot f_j' + \sum_{j=1}^{m} q_j' \cdot (g_j' - h_j') \right)$$

$$= \sum_{i=1}^{m} \sum_{j=1}^{m} (p_i \cdot p_j') \cdot (f_i \cdot f_j') + \sum_{i=1}^{m} \sum_{j=1}^{i} (p_i \cdot q_j') \cdot (f_i \cdot g_j')$$

$$+ \sum_{i=1}^{m} \sum_{j=1}^{i} (-p_i \cdot q_j') \cdot (f_i \cdot h_j) + \sum_{j=1}^{m} \sum_{i=1}^{j} (q_i \cdot p_j') \cdot (g_i \cdot f_j')$$

$$+ \sum_{j=1}^{m} \sum_{i=1}^{j} (-q_i \cdot p_j') \cdot (h_i \cdot f_j') + \sum_{i=1}^{m} \sum_{j=i+1}^{m} (p_i \cdot q_j') \cdot (f_i \cdot g_j' - f_i \cdot h_j')$$

$$+ \sum_{j=1}^{m} \sum_{i=j+1}^{m} (q_i \cdot p'_j) \cdot (g_i \cdot f'_j - h_i \cdot f'_j) + \sum_{i=1}^{m} \sum_{j=1}^{i} q_i \cdot q'_j \cdot (g_i \cdot g'_j - h_i \cdot g'_j)$$

$$+ \sum_{i=1}^{m} \sum_{j=1}^{i} (-q_i \cdot q'_j) \cdot (g_i \cdot h'_j - h_i \cdot h'_j) + \sum_{j=1}^{m} \sum_{i=1}^{j-1} (q_i \cdot q'_j) \cdot (g_i \cdot g'_j - g_i \cdot h'_j)$$

$$+ \sum_{j=2}^{m} \sum_{i=1}^{j-1} (-q_i \cdot q'_j) \cdot (h_i \cdot g'_j - h_i \cdot h'_j).$$

There exists a bijection $\rho : \mathbb{N} \times \mathbb{N} \to \mathbb{N}$ such that $\max\{i,j\} \le \rho(i,j) \le max\{i^2, j^2\}$ (see A.14). Put:

$$p''_k = \begin{cases} p_i \cdot p'_j, & \text{if} \quad k = 5\rho(i,j) & \text{for} \quad 1 \le i \le m \quad \text{and} \quad 1 \le j \le m; \\ p_i \cdot q'_j, & \text{if} \quad k = 5\rho(i,j) + 1 & \text{for} \quad 1 \le j \le i \le m; \\ -p_i \cdot q'_j, & \text{if} \quad k = 5\rho(i,j) + 2 & \text{for} \quad 1 \le j \le i \le m; \\ q_i \cdot p'_j, & \text{if} \quad k = 5\rho(i,j) + 3 & \text{for} \quad 1 \le i \le j \le m; \\ -q_i \cdot p'_j, & \text{if} \quad k = 5\rho(i,j) + 4 & \text{for} \quad 1 \le i \le j \le m; \\ 0, & \text{otherwise}; \end{cases}$$

$$q''_k = \begin{cases} p'_i \cdot q'_j, & \text{if} \quad k = 4\rho(i,j) & \text{for} \quad 1 \le i < j \le m; \\ q_i \cdot p'_j, & \text{if} \quad k = 4\rho(i,j) + 1 & \text{for} \quad 1 \le j \le i \le m; \\ q_i \cdot q'_j, & \text{if} \quad k = 4\rho(i,j) + 2 & \text{for} \quad 1 \le i \le m \quad \text{and} \quad 1 \le j \le m; \\ -q_i \cdot q'_j, & \text{if} \quad k = 4\rho(i,j) + 3 & \text{for} \quad 1 \le i \le m \quad \text{and} \quad 1 \le j \le m; \\ 0, & \text{otherwise}; \end{cases}$$

$$f''_k = \begin{cases} f'_i \cdot f'_j, & \text{if} \quad k = 5\rho(i,j) & \text{for} \quad 1 \le i \le m \quad \text{and} \quad 1 \le j \le m; \\ f_i \cdot g'_j, & \text{if} \quad k = 5\rho(i,j) + 1 & \text{for} \quad 1 \le j \le i \le m; \\ f_i \cdot h'_j, & \text{if} \quad k = 5\rho(i,j) + 2 & \text{for} \quad 1 \le j \le i \le m; \\ g_i \cdot f'_j, & \text{if} \quad k = 5\rho(i,j) + 3 & \text{for} \quad 1 \le i \le j \le m; \\ h_i \cdot f'_j, & \text{if} \quad k = 5\rho(i,j) + 4 & \text{for} \quad 1 \le i \le j \le m; \\ e, & \text{otherwise}; \end{cases}$$

$$g''_k = \begin{cases} f_i \cdot g'_j, & \text{if} \quad k = 4\rho(i,j) & \text{for} \quad 1 \le i < j \le m; \\ g_i \cdot f'_j, & \text{if} \quad k = 4\rho(i,j) + 1 & \text{for} \quad 1 \le j < i \le m; \\ g_i \cdot g'_j, & \text{if} \quad k = 4\rho(i,j) + 2 & \text{for} \quad 1 \le i \le m \quad \text{and} \quad 1 \le j \le m; \\ q_i \cdot h'_j, & \text{if} \quad k = 4\rho(i,j) + 3 & \text{for} \quad 1 \le j \le i \le m; \\ h_i \cdot g'_j, & \text{if} \quad k = 4\rho(i,j) + 3 & \text{for} \quad 1 \le i < j \le m; \\ e, & \text{otherwise}; \end{cases}$$

$$
h_k'' = \begin{cases}
f_i \cdot h_j', & \text{if } k = 4\rho(i,j) & \text{for } 1 \le i < j \le m; \\
h_i \cdot q_j', & \text{if } k = 4\rho(i,j)+1 & \text{for } 1 \le j < i \le m; \\
h_i \cdot q_j', & \text{if } k = 4\rho(i,j)+2 & \text{for } 1 \le j \le i \le m; \\
g_i \cdot h_j', & \text{if } k = 4\rho(i,j)+2 & \text{for } 1 \le i < j \le m; \\
h_i \cdot h_j', & \text{if } k = 4\rho(i,j)+3 & \text{for } 1 \le i \le m \quad \text{and} \quad 1 \le j \le m; \\
e, & \text{otherwise.}
\end{cases}
$$

Then, examining each case separately, and taking into account the choice of neighborhoods U_i' and V_i' for $i \in \mathbb{N}$, the properties of the sets D_j for $j \in \mathbb{N}$, and belonging of the elements: $p_i, p_i' \in U_i'$; $q_i, q_i' \in D_i$; $g_i \cdot h_i^{-1} \in V_i'$; $(g_i') \cdot (h_i')^{-1} \in V_i'$ for $i \in \mathbb{N}$, and the fact that $\max\{i,j\} \le \rho(i,j) \le \max\{i^2, j^2\}$ for $i, j \in \mathbb{N}$, we derive:

$$
p_k'' \in U_k; \quad q_k'' \in D_i \cdot D_j \subseteq D_{\max\{i,j\}+1} \subseteq D_k; \quad (g_k'') \cdot (h_k'')^{-1} \in V_k
$$

for $k \in \mathbb{N}$. Therefore, $z \cdot z' \in W(\mathcal{U}, \mathcal{V})$. From the arbitrariness of the choice of the elements z and z' it follows that $W(\mathcal{U}', \mathcal{V}') \cdot W(\mathcal{U}', \mathcal{V}') \subseteq W(\mathcal{U}, \mathcal{V})$, i.e. the condition (BN5) is satisfied.

Let $\mathcal{U} = (U_1, U_2, \dots) \in \mathfrak{M}$ and $\mathcal{V} = (V_1, V_2, \dots) \in \mathfrak{N}$. Let $a \in RG$ and $a = \sum_{i=1}^{k} r_i \cdot \tilde{g}_i$ be its presentation, where $r_i \in R$, $\tilde{g}_i \in G$ for $1 \le i \le k$, then there exists $s \in \mathbb{N}$ such that $r_i \in D_s$ for $1 \le i \le k$. Since D_i is a bounded set, then for each $j \in \mathbb{N}$ there exist neighborhoods $U_j' \in \mathcal{B}_0$ such that $(U_j' \cdot D_s) \bigcup (D_s \cdot U_j') \subseteq \bigcap_{i=1}^{k \cdot j} U_i$ and there exist $V_j \in \mathcal{B}_1$ such that $V_j' \subseteq \bigcap_{i=1}^{s+k \cdot j} V_i$. Let $\mathcal{U}' = (U_1', U_2', \dots)$, and $\mathcal{V}' = (V_1', V_2', \dots)$. If $z \in W(\mathcal{U}', \mathcal{V}')$ and $z = \sum_{i=1}^{n} p_i \cdot f_i + \sum_{j=1}^{m} q_i \cdot (g_i - h_i)$ is its presentation, mentioned above in the definition of the set $W(\mathcal{U}, \mathcal{V})$, then

$$
a \cdot z = \left(\sum_{i=1}^{k} r_i \cdot \tilde{g}_i \right) \cdot \left(\sum_{j=1}^{n} p_j \cdot f_j + \sum_{j=1}^{m} q_j \cdot (g_j - h_j) \right)
$$

$$
= \sum_{j=1}^{n} \sum_{i=1}^{k} (r_i \cdot p_j) \cdot (\tilde{g}_i \cdot f_j) + \sum_{j=1}^{m} \sum_{i=1}^{k} (r_i \cdot q_j) \cdot (\tilde{g}_i \cdot g_j - \tilde{g}_i \cdot h_j)
$$

$$
= \sum_{t=1}^{n \cdot k} p_t' \cdot f_t' + \sum_{t=1}^{s+m \cdot k} q_t' \cdot (g_t' - h_t'),
$$

where:

$p'_t = r_i \cdot p_j \in D_s \cdot U'_j \subseteq U_t$, if $t = k \cdot (j-1) + i$, for $1 \le j \le n$ and $1 \le i \le k$;

$q'_t = r_i \cdot q_j \in D_s \cdot D_j \subseteq D_t$, if $t = s + k \cdot (j-1) + i$, for $1 \le j \le n$ and $1 \le i \le k$;

$q'_t = 0 \in D_t$, if $1 \le t \le s$;

$g'_t = \tilde{g}_i \cdot g_j$, if $t = s + k \cdot (j-1) + i$, for $1 \le j \le n$ and $1 \le i \le k$;

$h'_t = \tilde{g}_i \cdot h_t$, if $t = s + k \cdot (j-1) + i$, for $1 \le j \le n$ and $1 \le i \le k$;

$g_t = h'_t = e$, if $1 \le t \le s$.

Since: $(g'_t) \cdot (h'_t)^{-1} = g_j \cdot h_j^{-1} \in V'_j \subseteq V_t$, if $t = s + k \cdot (j-1) + i$, for $1 \le j \le n$ and $1 \le i \le k$, and $(g'_t) \cdot (h'_t)^{-1} = e \in V_t$, if $1 \le t \le s$, then $a \cdot z' \in W(\mathcal{U}, \mathcal{V})$. From the arbitrariness of the choice of the element z it follows that $z \cdot W(\mathcal{U}', \mathcal{V}') \subseteq W(\mathcal{U}, \mathcal{V})$.

Analogously, it can be proved that $W(\mathcal{U}', \mathcal{V}') \cdot a \subseteq W(\mathcal{U}, \mathcal{V})$.

Thus, the fulfillment of the condition (BN6) is verified.

Therefore, the family $\{W(\mathcal{U}, \mathcal{V}) | \mathcal{U} \in \mathfrak{M}, \mathcal{V} \in \mathfrak{N}\}$ defines on RG a certain ring topology τ.

It remains to verify that $\tau|_R = \tau_0$ and $\tau|_G = \tau_1$ and the fact that τ is a Hausdorff topology.

Before completing the proof of the theorem, we prove the following lemma.

6.4.17. LEMMA. *Let:* $U_0 \in \mathcal{B}_0$; $V_0 \in \mathcal{B}_1$; $\tilde{r}_1, \tilde{r}_2, \dots, \tilde{r}_k \in R$ *and* $\tilde{r}_1 \notin U_0$; $\tilde{g}_1, \tilde{g}_2, \dots, \tilde{g}_k \in G$ *and* $\tilde{g}_1 \cdot \tilde{g}_j^{-1} \notin V_0$ *for* $2 \le j \le k$. *If* $\mathcal{U} = (U_1, U_2, \dots) \in \mathfrak{M}$ *and* $\mathcal{V} = (V_1, V_2, \dots) \in \mathfrak{N}$, *and, moreover,* $\sum_{i=1}^t U_i \subseteq U_0$ *and* $V_1 \cdot V_2 \cdot \dots \cdot V_t \subseteq V_0$ *for any* $t \in \mathbb{N}$, *then* $\sum_{i=1}^k \tilde{r}_i \cdot \tilde{g}_i \notin W(\mathcal{U}, \mathcal{V})$.

PROOF. Assume the contrary, i.e. that $\sum_{i=1}^k \tilde{r}_i \cdot \tilde{g}_i \in W(\mathcal{U}, \mathcal{V})$, and let

$$\sum_{i=1}^k \tilde{r}_i \cdot \tilde{g}_i = \sum_{i=1}^m p_i \cdot f_i + \sum_{i=1}^n q_i \cdot (g_i - h_i)$$

be the presentation mentioned above in the definition of the sets $W(\mathcal{U}, \mathcal{V})$. We consider the relation ξ generated by the family of pairs $\{g_1, h_1\}, \{g_2, h_2\}, \dots, \{g_k, h_k\}$ (see Lemma 6.4.15). Let us show that the element $\tilde{g}_1$ is not equivalent to any element of the set

$\{\tilde{g}_2, \dots, \tilde{g}_k\}$. Indeed, if $\tilde{g}_1 \xi \tilde{g}_{j_0}$ for some $2 \leq j_0 \leq k$, then, according to Lemma 6.4.15, there exist pairwise distinct numbers $i_1, i_2, \dots, i_t$ such that $1 \leq i_j \leq n$ for $1 \leq j \leq t$, and

$$\tilde{g}_1 \cdot (\tilde{g}_{j_0})^{-1} \in (\{g_{i_1}, h_{i_1}\} \cdot \{g_{i_1}, h_{i_1}\}^{-1}) \cdot (\{g_{i_2}, h_{i_2}\} \cdot \{g_{i_2}, h_{i_2}\}^{-1})$$
$$\cdots \cdot (\{g_{i_t}, h_{i_t}\} \cdot \{g_{i_t}, h_{i_t}\}^{-1}) \subseteq V_{i_1} \cdot V_{i_2} \cdot \dots \cdot V_{i_t}.$$

Since G is an Abelian group and the numbers $i_1, i_2, \dots, i_t$ are pairwise distinct, then $\tilde{g}_1 \cdot (\tilde{g}_{j_0})^{-1} \in V_{i_1} \cdot V_{i_2} \cdot \dots \cdot V_{i_t} \subseteq V_1 \cdot V_2 \cdot \dots \cdot V_s \subseteq V_0$, where $s = \max\{i_1, i_2, \dots, i_t\}$. We have the contradiction with the condition of the lemma. Hence, $\tilde{g}_1$ and $\tilde{g}_j$ are not equivalent with respect to ξ for any $2 \leq j \leq k$.

Let $\bar{G}$ be the factor-set of the set G by the equivalence relation ξ. We also denote by ξ the natural mapping from G onto $\bar{G}$ (see A.15). Let M be a free R–module, generated by the set $\bar{G}$, then the mapping $\hat{\xi} : RG \to M$ functioning by the rule

$$\hat{\xi}(\sum_{i=1}^{l} r_i \cdot \tilde{h}_i) = \sum_{i=1}^{l} r_i \cdot \xi(\tilde{h}_i)$$

for any $r_i \in R$, and $\tilde{h}_i \in G$, $1 \leq i \leq l$, $l \in \mathbb{N}$, is a homomorphism of R–module RG onto R–module M, moreover $\hat{\xi}(\tilde{g}_1) \neq \hat{\xi}(\tilde{g}_j)$ for $2 \leq j \leq k$, and $\hat{\xi}(g_i) = \hat{\xi}(h_i)$ for $1 \leq i \leq n$. Then

$$\sum_{i=1}^{k} \tilde{r}_i \cdot \hat{\xi}(\tilde{g}_i) = \hat{\xi}(\sum_{i=1}^{k} \tilde{r}_i \cdot \tilde{g}_i) = \hat{\xi}(\sum_{i=1}^{m} p_i \cdot f_i + \sum_{i=1}^{n} q_i \cdot (g_i - h_i))$$
$$= \sum_{i=1}^{m} p_i \cdot \hat{\xi}(f_i) + \sum_{i=1}^{n} q_i \cdot (\hat{\xi}(g_i) - \hat{\xi}(h_i)) = \sum_{i=1}^{m} p_i \cdot \hat{\xi}(f_i).$$

Since $\hat{\xi}(\tilde{g}_1) \neq \hat{\xi}(\tilde{g}_j)$ for $2 \leq j \leq k$, then

$$r_1 = \sum_{\xi(f_i) = \tilde{g}_1} p_i \in \sum_{i=1}^{s} U_i \subseteq U_0,$$

where $s = \max\{i | \hat{\xi}(f_i) = \hat{\xi}(\tilde{g}_1)\}$. We have the contradiction to the condition of the lemma. Thus, the lemma is completely proved.

Let us pass now to the completion of the proof of Theorem 6.4.16.

Let us show that $\tau\big|_R = \tau_0$ and $\tau\big|_G = \tau_1$. Let $\mathcal{U}(U_1, U_2, \dots) \in \mathfrak{M}$ and $\mathcal{V} = (V_1, V_2, \dots) \in \mathfrak{N}$. From the definition of the sets $W(\mathcal{U}, \mathcal{V})$ it follows that $r = r \cdot e \in W(\mathcal{U}, \mathcal{V})$ and $g - h \in W(\mathcal{U}, \mathcal{V})$ for any $r \in U_1$ and $g, h \in G$ such that $g \cdot h^{-1} \in V_1$. Then

$$U_1 \subseteq W(\mathcal{U}, \mathcal{V}) \cap R \quad \text{and} \quad V_1 \cdot h \subseteq \big(h + W(\mathcal{U}, \mathcal{V})\big) \cap G.$$

Hence, $\tau\big|_R \leq \tau_0$ and $\tau\big|_G \leq \tau_1$.

Let $U_0 \in \mathcal{B}_0$. By induction we can construct a sequence $\mathcal{U} = (U_1, U_2, \dots)$ of neighborhoods from $\mathcal{B}_0$ such that $U_i + U_i \subseteq U_{i-1}$ for any $i \in \mathbb{N}$. Then by induction by the number k it is easily proved that $\sum_{i=1}^{k} U_i + U_k \subseteq U_0$ and, hence, $\sum_{i=1}^{s} U_i \subseteq U_0$ for any $s \in \mathbb{N}$.

Let now $\mathcal{V} = (V_1, V_2, \dots)$, where $V_i = G$ for $i \in \mathbb{N}$ and $r \in W(\mathcal{U}, \mathcal{V}) \cap R$, then

$$r = \sum_{i=1}^{m} p_i \cdot f_i + \sum_{i=1}^{n} q_i \cdot (g_i - h_i),$$

where $p_i \in U_i$ for $i \in \mathbb{N}$. Let $\hat{\psi} : RG \to R$ be a ring homomorphism, which is the extension of the constant mapping $\psi : G \to \{e\}$. Then

$$r = \hat{\psi}(r) = \sum_{i=1}^{m} p_i \in \sum_{i=1}^{m} U_i \subseteq U_0.$$

From the arbitrariness of the choice of the element r it follows that $W(\mathcal{U}, \mathcal{V}) \cap R \subseteq U_0$. This proves the fact that $\tau\big|_R = \tau_0$.

Let now $h \in G$ and V be a neighborhood of the element h in (G, τ_1). There exists a neighborhood $V_0 \in \mathcal{B}_1$ such that $V_0 \cdot h \subseteq V$. By induction we can construct a sequence $\mathcal{V} = (V_1, V_2, \dots)$ of neighborhoods from $\mathcal{B}_1$ such that $V_i \cdot V_i \subseteq V_{i-1}$ for $i \in \mathbb{N}$. Then by induction by the number k it is easy to prove that $V_1 \cdot V_2 \cdot \ldots \cdot V_k \cdot V_k \subseteq V_0$ and, hence, $V_1 \cdot V_2 \cdot \ldots \cdot V_s \subseteq V_0$ for any $s \in \mathbb{N}$. Since τ_0 is a Hausdorff topology, there could be found a neighborhood $U_0 \in \mathcal{B}_0$ such that $1 \notin U_0$. We can construct a sequence $\mathcal{U} = (U_1, U_2, \dots)$ of neighborhoods from $\mathcal{B}_0$ such that $U_i + U_i \subseteq U_{i-1}$ for $i \in \mathbb{N}$. Then, as it was mentioned above, $\sum_{i=1}^{s} U_i \subseteq U_0$ for any $s \in \mathbb{N}$.

Let us show that $(W(\mathcal{U}, \mathcal{V}) + h) \cap G \subseteq V$. Let g be an element of $G \setminus V$. Since $1 \notin U_0$ and $g \cdot h^{-1} \notin V_0$, then applying Lemma 6.4.17 to $r_1 = 1$, $r_2 = -1$, $\tilde{g}_1 = g$, and $\tilde{g}_2 = h$, we obtain $g - h \notin W(\mathcal{U}, \mathcal{V})$. Hence, $g \notin W(\mathcal{U}, \mathcal{V}) + h$. From the arbitrariness of the choice of the element g it follows that $(W(\mathcal{U}, \mathcal{V}) + h) \bigcap G \subseteq V$. In summary, we have shown that $\tau\big|_G = \tau_1$.

To complete the proof of the theorem it remains to verify that the constructed topology τ is Hausdorff.

Let $\sum_{i=1}^{k} \tilde{r}_i \cdot \tilde{g}_i$ be a non-zero element of RG. Without loss of generality we can consider that $\tilde{r}_1 \neq 0$ and $\tilde{g}_1 \neq \tilde{g}_j$ for $2 \leq j \leq k$.

If $k = 1$, then there exists a neighborhood $U_0 \in \mathcal{B}_0$ such that $r_1 \notin U_0$. Since $\tau\big|_R = \tau_0$ and (RG, τ) is a topological ring, there exist neighborhoods $W(\mathcal{U}, \mathcal{V})$ and $W(\mathcal{U}', \mathcal{V}')$ of zero in (RG, τ) such that $W(\mathcal{U}, \mathcal{V}) \bigcap R \subseteq U_0$ and $W(\mathcal{U}', \mathcal{V}') \cdot \tilde{g}_1^{-1} \subseteq W(\mathcal{U}, \mathcal{V})$. Since $\tilde{r}_1 = \tilde{r}_1 \cdot \tilde{g}_1 \cdot \tilde{g}_1^{-1}$ and $\tilde{r}_1 \notin W(\mathcal{U}, \mathcal{V})$, then $\tilde{r}_1 \cdot \tilde{g}_1 \notin W(\mathcal{U}', \mathcal{V}')$.

Let now $k \geq 2$. As τ_0 and τ_1 are Hausdorff topologies, there exist $U_0 \in \mathcal{B}_0$ and $V_0 \in \mathcal{B}_1$ such that $r_1 \notin U_0$ and $\tilde{g}_1 \cdot \tilde{g}_i^{-1} \notin V_0$ for $2 \leq i \leq k$. As above, we can construct sequences $\mathcal{U} = (U_1, U_2, \dots) \in \mathfrak{M}$ and $\mathcal{V} = (V_1, V_2, \dots) \in \mathfrak{N}$ such that $U_i + U_i \subseteq U_{i-1}$ and $V_i \cdot V_i \subseteq V_{i-1}$ for any $i \in \mathbb{N}$. Then, as it was mentioned above, $\sum_{i=1}^{s} U_i \subseteq U_0$ and $V_1 \cdot V_2 \cdot \ldots \cdot V_s \subseteq V_0$ for any $s \in \mathbb{N}$, and, hence, according to Lemma 6.4.17, $\sum_{i=1}^{k} \tilde{r}_i \cdot \tilde{g}_i \notin W(\mathcal{U}, \mathcal{V})$.

Thus, the theorem is completely proved.

6.4.18. COROLLARY. *Let (G, τ_1) be a Hausdorff topological Abelian group, then there exists a Hausdorff topological ring (R, τ) such that (G, τ_1) is a sub-semigroup of the multiplicative semigroup of the topological ring (R, τ).*

PROOF. Indeed, let $(\mathbb{Z}, \tau_0)$ be the ring of integers with the discrete topology, then on the semigroup ring $R = \mathbb{Z}G$ exists a Hausdorff ring topology τ such that $\tau\big|_G = \tau_1$.

Open Problems

Several open problems of the theory of topological rings and modules that have a certain relation to the questions considered in this book were taken from [413] and are presented here, some of them in a renewed from and with certain comments.

1. Does any infinite associative ring admit a non-discrete Hausdorff ring topology (this is true for countable rings and for commutative rings, see Chapter 5)? V.I.Arnautov.

2. Does any countable skew field admit a non-discrete Hausdorff topology in which it is a topological skew field? (According to Paragraph 5.3, any countable ring, and, hence, any countable skew field, admits a non-discrete Hausdorff ring topology, but in this topology the operation of taking the inverse element can be not continuous). V.I.Arnautov.

3. Let the multiplicative group of non-zero elements of a skew field admit a non-discrete Hausdorff group topology. Is it true that then the skew field itself admits a Hausdorff non-discrete topology in which it is a topological skew field? A.D.Taimanov.

4. Could any ring topology on a skew field R be weakened to a topology, in which R is a topological skew field? (This is true for fields, see Theorem 1.7.8. It is clear that the positive solution of this problem gives also a positive answer to problem 2, too). V.I.Arnautov.

5. Let τ be a non-weakenable Hausdorff ring topology on a skew field R. Is it true that (R, τ) is a topological skew field? (The negative answer to this problem gives the negative solution of the problem 4). V.I.Arnautov.

6. Let R^* be the multiplicative group of non-zero elements of a skew field R and let τ be a ring topology on R such that $\tau|_{R^*}$ could be weakened to a topology in which R^* is a topological group. Is it true that in this case τ could be weakened to a ring topology such that R is a topological skew field? V.I.Arnautov.

7. Let (R, τ) be a topological ring in which the quasi-component of zero (i.e. the intersection of all open-closed subsets containing zero) is equal to zero. Is it true that the topology τ can be weakened to a zero-dimensional ring topology? V.I.Arnautov.

8. Let R be a field (a skew field) and τ a ring topology in which (R, τ) is a zero-dimensional ring. Is it true that the topology τ can be weakened to a zero-dimensional topology in which R is a topological field (skew field)? V.I.Arnautov.

9. Let (R, τ) be a totally disconnected topological field (skew field). Is it true that the topology τ can be weakened to a zero-dimensional topology in which R is a topological field (skew field)? (Since in a totally disconnected skew field the quasi-component of zero equals zero, then the positive solution of problems 7 and 8 gives the positive answer to this problem, too). V.I.Arnautov.

10. Is it true that a totally disconnected Hausdorff topological field (skew field) with unweakenable topology is zero-dimensional? (The negative answer to this problem gives the negative solution of the problems 7 and 9). V.I.Arnautov.

11. Does there exist a non-locally bounded topological field whose topology is unweakenable in the class of all Hausdorff ring topologies? A well known problem.

12. Does on the ring $\mathbb{Z}$ of integers exist a ring topology which is different from the p–adic one and is minimal in the class of all ring Hausdorff topologies? (Since in any ring topology any closed non-zero ideal in $\mathbb{Z}$ is open, then in such topology $\mathbb{Z}$ does not contain non-trivial closed ideals, and therefore, taking into account Theorems 1.6.70 and 3.5.3, the positive solution of this problem gives the positive solution of problem 11, too). V.I. Arnautov.

13. Let F be the algebraic closure of a finite field. Is it true that on F there exists a minimal Hausdorff ring topology? (Since there exist no locally bounded topologies on F, then the positive solution of this problem gives the positive answer to problem 11, too). V.I.Arnautov.

14. Do there exist rings, admiting a finite number, different from 2^n, $n \in \mathbb{N}$, of distinct linearly compact Hausdorff ring topologies? V.I.Arnautov, M.I.Ursul.

15. Does there exist an associative ring without divisors of zero admitting more than one Hausdorff compact ring topology? V.I.Arnautov, M.I.Ursul.

16. Does there exist a ring (a group) admitting only a finite number of compact Hausdorff ring (group) topologies, but more than one? M.I.Ursul.

17. Describe the structure of rings and semigroups whose semigroup ring admits:

a) compact ring topology;

b) non-discrete locally compact ring topology;

c) countably compact ring topology.

A.V.Kliushin, I.B.Kojuhov.

18. Does any infinite Abelian group admit a Hausdorff non-discrete group topology in which it is complete and any of its endomorphism is continuous? From the proof of Theorem 5.1.2 and from [271] it follows that for each of these conditions there exists such Hausdorff non-discrete topology). P.I.Kirku.

19. Is it true that any countable torsion-free Abelian group admits a group topology without non-trivial characters? I.V.Protasov.

20. Is it true that for any Hausdorff topological Abelian group (H, τ_0) there exists a Hausdorff topological Abelian group (G, τ) such that C-$\mathrm{Hom}((G, \tau), (H, \tau))$ is trivial, i.e. it is the one-element group? V.I.Arnautov.

21. Does any infinite commutative ring admit a non-discrete ring topology in which it is a complete ring? (This is true for Abelian groups, see, for instance, [271] and for countable rings [38]). V.I.Arnautov.

22. Can any non-discrete Hausdorff ring topology, satisfying the first axiom of countability, be strengthened to a complete non-discrete ring topology? (This is so if the ring is countable, see [38]). V.I.Arnautov.

23. Does any commutative ring admit a maximal non-discrete ring topology in which it is a complete ring? (The positive solution of this problem gives the positive answer to the problem 21, too). V.I.Arnautov.

24. Does any countable ring admit a maximal Hausdorff non-discrete ring topology in which it is a complete ring? V.I.Arnautov.

25. Does there exist a ring which is complete in any maximal non-discrete ring topology? (In any ring which is a direct sum of a countable number of the same finite field, under continuum hypothesis, among maximal ring topologies there exist both complete and non-complete topologies, see [39]). V.I.Arnautov.

26. Can any Hausdorff non-discrete ring topology on a countable ring, satisfying the first axiom of countability, be strengthened to a maximal non-discrete complete ring topology? (It is known (see [38]) that in this case the topology can be strengthened to a complete topology which is possibly not maximal. The positive solution of this problem gives the positive answer to problem 22, as well). V.I.Arnautov.

27. Can any Hausdorff non-discrete ring topology on a commutative ring, satisfying the first axiom of countability, be strengthened to a maximal complete ring topology? (The positive solution of this problem gives the positive answer to problem 23, as well). V.I.Arnautov.

28. Does there exist an Abelian group with a countable basis (in terms of topology) such that any Abelian group with a countable basis is a subgroup of this group? (This problem is solved completely for non-Abelian groups). A.V. Arhangelskii.

29. Does there exist a Hausdorff topological group which is not embeddable into the multiplicative groups of topological rings? (Such topological semigroups were constructed, see [47]). V.I.Arnautov, A.V.Mikhalev.

30. Let (G, τ) be a Hausdorff topological group. Are the following conditions equivalent:

a) (G, τ) is a sub-semigroup of a multiplicative semigroup of some Hausdorff topological ring;

b) for any discrete ring R the topology τ is extendable to a Hausdorff ring topology on the semigroup ring RG;

c) the topology τ is extendable to a Hausdorff ring topology on the semigroup ring $\mathbb{Z}G$;

d) the topology τ is extendable to a Hausdorff ring topology on the semigroup ring FG, where F is the two-element field.

V.I.Arnautov, A.V.Mikhalev.

31. Let (R, τ) be a Hausdorff topological ring. Is it true that the topology τ can be extended to a Hausdorff ring topology on the semigroup ring RG for an arbitrary discrete monoid G? (This is so if R is a commutative ring or (R, τ) is a ring, locally bounded from left, see Theorem 6.4.13. V.I.Arnautov, A.V.Mikhalev.

32. Let (R, τ) be a topological ring with the unitary element, G a group and τ_1, τ_1' group topologies on G such that $\tau_1 \leq \tau_1'$. Is it true that if the pair (τ_0, τ_1) is extendable on the semigroup ring RG, then the pair (τ_0, τ_1') is extendable, too. V.I.Arnautov, A.V.Mikhalev.

33. Let (R, τ_0) be a topological ring with the unitary element, $\{(G_\alpha, \tau_\alpha) | \alpha \in A\}$ some family of topological groups and (G, τ) the direct sum (product) of these groups with Tychonoff topology. Is it true that if for each α the pair (τ_0, τ_α) is extendable on the group ring RG, then the pair (τ_0, τ) is also extendable on the group ring RG? V.I.Arnautov,

A.V.Mikhalev.

34. Let (R, τ) be a locally bounded topological commutative ring, containing no topological divisors of zero. Is τ extendable to a ring topology on the quotient field of the ring R? (This is so if R is a countable ring, see Corollary 6.5.2). V.I.Arnautov.

35. Is any discrete skew field embeddable into a connected topological skew field? (This is true for fields, see [295]). V.I.Arnautov.

36. Is any topological field embeddable into a connected topological field? (This is true for the fields with the discrete topology, see [295]). V.I.Arnautov.

37. Does there exist a connected but not linearly connected topological field? E.M.Vechtomov.

38. Which topological spaces can be realized as subspaces of a topological skew field? (From Proposition 1.7.9 it follows that not all Tychonoff spaces, and even compact ones, have this property). V.I.Arnautov.

39. Let τ_1 and τ_2 be two Tychonoff topologies on a set X such that $\tau_1 \leq \tau_2$. Is it true that if (X, τ_1) is embeddable into a topological skew field, then the space (X, τ_2) is embeddable into some topological skew field, as well? D.B.Shahmatov.

40. Does there exist a topological space which is embeddable into a topological skew field, but is not embeddable into a topological field? D.B.Shahmatov.

41. Is it true that if a topological space X is embeddable into a topological field of a finite characteristic p, then it is embeddable into a topological field of the characteristic $q \neq p$? D.B.Shahmatov.

42. Is every normal space embeddable as a closed subspace into a topological Abelian group (a topological ring) whose topological space is normal. A.V.Arhangelskii.

43. Is every Lindeloef space embeddable as a closed subspace into a topological group (a topological ring) whose topological space is Lindeloef? A.V.Arhangelskii.

44. Could in every topological skew field (a topological field) it be found a linearly ordered (by inclusion) family of open sets with zero intersection? (From Proposition 1.7.9 it follows that the answer is positive for a skew field containing a non-discrete countable subset). A.V.Arhangelskii.

45. Let $\hat{R}$ be the completion of a ring R. Find conditions on the ring R under which

any basis of a filter of $\hat{R}$ consisting of cosets by closed left ideals has non-zero intersection. (If R possesses a basis of neighborhoods of zero consisting of left ideals, then the required condition is: for any open left ideal J of the ring R the factor-module R/J is a linearly compact R–module). V.I.Arnautov.

46. Let $\mathfrak{m}$ be a cardinal number and R be a topological ring (group) such that for any neighborhood V of zero in R there exists a subset $S \subseteq R$ such that $S+V = R$ and $|S| < M$. Is it true that the completion $\hat{R}$ of the topological ring (group) R is $\mathfrak{m}$–compact, i.e. from any open cover of the ring (group) R it can be chosen a subcover with the cardinality less than $\mathfrak{m}$? (For the case $m = \aleph_0$ this is true, see Proposition 3.4.4 and Theorem 3.4.16). V.I.Arnautov.

References

1. Abels H., Normen auf freien topoloschen Gruppen., Math. Z., 129, N1:25–42 (1972).

2. Abouabdillan D., Topologies de corps A lineaires., Pacif. J. Math., 107, N2:257–266 (1983).

3. Abouabdillan D., Topologies lineaires minimales sur un group Abelien., Lect. Not. Math., 1006:582–588 (1983).

4. Ajtai M., Havas J., Komlos. J., Every group admits a bad topology, Stud. Pure Math. Mem. Poul. Tután, Budapest, 21–34 (1983).

5. Alajbegović J., On Cauchy sequences of valued field, Algebraic Conf. Novi Sad 1981, Novi Sad, 39–46 (1982).

6. Albert A.A., Absolute valued real algebras, Ann. Math., 48: 495-501 (1947).

7. Alexandrov P.S., Introduction to the set theory and general topology, Moscow, Nauka, (1977).

8. Alexey S.F., On connectedness of free topological modules, Mat. Issled., 8, N3:136–139 (1973).

9. Alexey S.F., On boundedness of a free topological module, Mat. Issled., 9, N(33):3–14 (1974).

10. Alexey S.F., Free topological modules in certain classes, Mat. Issled., 10, N2:3–14 (1975).

11. Alexey S.F., Arnautov V.I., On free topological modules, Mat. Issled., 7, N3:3–18 (1972).

12. Alexey S.F., Arnautov V.I., Vodinchar M.I., On boundedness of free topological modules, Mat. Issled., 10, N2:16–27 (1975).

13. Alexey S.F., Kalistru R.K., On completeness of a free topological module, Mat. Issled., 44:164-173 (1977).

14. Alexey S.F., Kalistru R.K., Necessary conditions of completeness of free topological module, Mat. Issled., 65:3-8 (1982).

15. Amuraritei Gh., Siruri si serii in algebre topologice, Bul. Univ. Brasov., 21:3-8 (1979).

16. Andrunakievich V.A., Arnautov V.I., Invertability in topological rings, DAN SSSR, 170:755-758 (1966).

17. Anthony J.M., Topologies for quotient fields of commutative integral domains, Pacif. J. Math., 36:585–601 (1971).

18. Anzai H., On compact topological rings, Proc. Jap. Acad. Tokyo, 19:613-615 (1943).

19. Arens R.F., Linear topological division algebras, Bull. Amer. Math. Soc., 53:623–630 (1947).

20. Arens R.F., Inverse-producting extensions of normed algebras, Trans. Amer. Math. Soc., 88:536-548 (1958).

21. Arens R.F., Hoffman K., Algebraic extension of normed algebras,. Proc. Amer. Math. Soc. 7, N2:203–210 (1956).

22. Arhangelskii A.V., Algebraic objects generated by topological structure, Itogi Nauki i Techniki. Algebra, Topology, Geometry, 25:141–198 (1987).

23. Arhangelskii A.V., Ponomarev V.I., Bases of general topology in tasks and exercises, Nauka, (1974).

24. Arnautov V.I., Topological rings of certain local weight, Sb. Issled. po Obsch. Alg., 25–36 (1965).

25. Arnautov V.I., Criterion of pseudo-normalizability of topological rings, Algebra i Logika, V. N4:3–24 (1965).

26. Arnautov V.I., On topologizations of the ring of integers, Izv. AN MSSR, Ser. Phys.-Tech. Math., 1:3–15 (1968).

27. Arnautov V.I., On topologizations of countable rings, Sib. Mat. J., 9, N6:1251–1262 (1968).

28. Arnautov V.I., Semitopological isomorphism of topological rings, Mat. Issled., 4, 2(12):3–16 (1969).

29. Arnautov V.I., An example of infinite ring admitting only the discrete topologization, Mat. Issled., 5, N3:182–185 (1970).

30. Arnautov V.I., Non-discrete topologizability of countable rings, DAN SSSR, 191: 747–750 (1970).

31. Arnautov V.I., Non-discrete topologizability of infinite commutative rings, DAN SSSR, 194:991–994 (1970).

32. Arnautov V.I., Non-discrete topologizability of infinite commutative rings, Mat. Issled., 5, N4:3–15 (1970).

33. Arnautov V.I., On topologization of infinite rings, Mat. Issled., 7, N1:3–15 (1972).

34. Arnautov V.I., On topologization of infinite modules, Mat. Issled., 7, N4:241–243 (1972).

35. Arnautov V.I., Extension of topology on a commutative ring to its quotient rings, Mat. Issled., 48:3–13 (1978).

36. Arnautov V.I., Extension of locally bounded topology on a field to its simple transcendental extension, Algebra i Logika, 20, N5:511–521 (1981).

37. Arnautov V.I., On extension of ring topology of δ–bounded field to its simple transcendental extension, Mat. Sb., 133(175):275–292 (1987).

38. Arnautov V.I., On strengthening of a topology of a countable ring to complete, Thes. VI Symp. on Theor. of Rings, Alg. and Mod., 8 (1990).

39. Arnautov V.I., On completeness of rings in a maximal ring topologies, Thes. Int. Algebr. Conf. Barnaul 1991, 9 (1991).

40. Arnautov V.I., On extension of pseudonorm on a ring to its quotient ring, Izv. AN RM, Ser. Mat., 2(8):45–58 (1992).

41. Arnautov V.I., Isomorphisms of free topological groups, rings and modules that are generated by topological spaces, Izv. AN RM, Ser. Mat., N2(12):63–71 (1993).

42. Arnautov V.I., Beidar K.I., Glavatsky S.T., Mikhalev A.V., On the intersection property of radicals of topological algebras, Trudy sem. im. Petrovskogo, 15:178–188 (1991).

43. Arnautov V.I., Beidar K.I., Glavatsky S.T., Mikhalev A.V., Intersection property in radical theory of topological algebras, Contemporary Mathematics, AMS, 131 (Part 2):205–225 (1992).

44. Arnautov V.I., Kabanova E.I., On strengthening of group topology of a countable group to complete, Sib. Mat. J., 31, N1:3–13 (1990).

45. Arnautov V.I., Marin E.I., Mikhalev A.V., Necessary conditions of extension of

topologies of group and field to their group algebra, Vestnik MGU, 6:58–61 (1984).

46. Arnautov V.I., Mikhalev A.V., Topologies of polynomial rings and their topological analogs. Hilbert theorems on a basis, Mat. Sb., 116:467-482 (1981).

47. Arnautov V.I., Mikhalev A.V., Example of commutative fully regular topological semigroup which does not admit embeddings into multiplicative semigroups of topological rings, Vestnik MGU, 6:68–70 (1981).

48. Arnautov V.I., Mikhalev A.V., Extension of topologies to group and semigroup rings, Proc.Leningr. Int. Topol. Conf. 23–27 Aug. 1982, 27–31 (1983).

49. Arnautov V.I., Mikhalev A.V., Sufficient conditions of extension of topologies of group and ring to their group ring, Vestnik MGU, Ser.I, 5:25–33 (1983).

50. Arnautov V.I., Mikhalev A.V., To the question of extension of topologies of a group and of a ring to their group ring, Usp. Math. Nauk, 40, N4:135–136 (1985).

51. Arnautov V.I., Mikhalev A.V., On extension of topologies of a group and of a ring to their group ring, Mat. Issled., 85:8–20 (1985).

52. Arnautov V.I., Mikhalev A.V., Compact groups and their group rings, Mat. Zametki, 46, N6:3–9 (1989).

53. Arnautov V.I., Mikhalev A.V., Extension of topologies of a ring and of a discrete monoid to the semigroup ring, Mat. Issled., N 111:9–23 (1989).

54. Arnautov V.I., Mikhalev A.V., Questions on the possibility of extending the topologies of a ring and of a semigroup to their semigroup ring, Trudy Matem. Inst. of the Steclov RAN, 193:22–27 (1992).

55. Arnautov V.I., Ursul M.I., On the uniquness of a linearly compact topology on rings, Mat. Issled., 53:6–14 (1979).

56. Arnautov V.I., Ursul M.I., Embedding of topological rings into connected ones, Mat. Issled., 49:11–15 (1979).

57. Arnautov V.I., Ursul M.I., Quasi-components of topological rings and modules, Izv. Akad. Nauk Mold. SSR, Ser. Phys.-Tech. Mat. Mauk, N1:9–13 (1984).

58. Arnautov V.I., Vizitiu V.N., Extension of locally bounded topology of a field to its algebraic extensions, Izv. AN MSSR, Ser. Phys.-Tech. i Mat. Nauk, 9, N2:29–43 (1974).

59. Arnautov V.I., Vizitiu V.N., Extension of locally bounded topology of a field to its

algebraic extension, DAN SSSR, 216, N3:477–480 (1974).

60. Arnautov V.I., Vodinchar M.I., On local compactness of free topological module, Mat. Issled., 10, N3:3–14 (1975).

61. Arnautov V.I., Vodinchar M.I., Glavatsky S.T., Mikhalev A.V., Constructions of topological rings and modules, Kishinev, Stinitsa, (1988).

62. Arnautov V.I., Vodichar M.I., Mikhalev A.V., Introduction to the theory of topological rings and modules, Kishinev, Stiintsa (1981).

63. Arnautov V.I., Zelenjuk E.G., To the problem of compeleteness of maximal groups, Ukr. Mat. J., 43:21–27 (1991).

64. Artin E., Geometric algebra, Princeton Univ., (1957).

65. Aurora S., On normed rings with monothone multiplication, Pacif. J. Math., 33:15–20 (1970).

66. Aurora S., Normed fields which extend normed rings of integers, Pacif. J. Math. V.33, N1:21–28 (1970).

67. Bagley R.W., Wu T.S., Topological groups with equal left and right uniformities, Proc. Amer. Math. Soc., 18, N1:142–147 (1967).

68. Banaszczyk W., Countable products of LCA groups: their closed subgroups, quotients and duality properties, Colloq. Math., 59:53–57 (1990).

69. Beidar K.I., Glavatsky S.T., Mikhalev A.V., Semisimple classes and lower radicals of topological non–associative algebras, Trudy sem. im. Petrovskogo, 14:250–261 (1989).

70. Beidar K.I., Glavatsky S.T., Mikhalev A.V., Varieties of topological ω–groups, Izv. Vuzov, Ser. Mat., 6(325):40–42 (1989).

71. Bel'nov V.K., Some theorems about free Abelian metrizable groups, Sib. Mat. Zh., 13, N6:1213–1228 (1972).

72. Bel'nov V.K., On metrization of polynomial rings, Bull. Acad. Pol. Sci., Ser. Sci. Math., Astron. et Phys., 22:1227–1233 (1974).

73. Bel'nov V.K., On topologization of Abelian groups, Vestnik MGU, Ser. Mat. Mech., N1:43–49 (1979).

74. Berhanu S., Comfort W.W., Reid J.D., Counting subgroups and topological group topologies, Pacif. J. Math., 116, N2:217–241 (1985).

75. Berrondo F., Corps topologiquement henseliens, C.R. Acad. Sci., Paris, 181, N10, Ser A.:305–307 (1975).

76. Berrondo F., Corps topologiques lineaires non localement bornes it complets, C.R. Acad. Sci., 284, N15:A343–A346 (1977).

77. Bicknell K., Morris S.A., Norms on free topological groups. Bull. London Math. Soc. 10, N3:280–284 (1978).

78. Borevich Z.I., Shafarevich I.R., Number theory (2–nd edition), Nauka, (1972).

79. Borges C.R., Quotient topologies on free topological groups, Math. Jap., 30, N6:835–837 (1985).

80. Bourbaki N., Elements de mathematique: Les strutures fondamentales de l'analise livre I. Theorie des ensembles. Chapitre III. Ensembles ordonnes–carinaus nombres entiers. Hermann, (1956).

81. Bourbaki N., Elements de mathematique. Premiere partie. Les structures fondamentales de l'analyse. Livre II, Algebre, Hermann & c. Editeurs 6, Rue de la Sorbonne, 6.

82. Bourbaki N., Elements de mathematique. Premiere partie. Livre III. Topologie generale, Hermann.

83. Bourbaki N., Elements de mathematique. Premiere partie. Les structures fondamentales de analyse. Livre V. Espaces vectoriels topologiques, Paris, Hermann & C. editeurs 6, Rue de la Sorbonne 6.

84. Bourbaki N., Elements de mathematique. Algebre commutative, Chapitre 6. Valuations, Hermann, (1964).

85. Braconier J., Sur les groups topologiques localement compacts, J. Math. Pures Appl., 27:1–85 (1948).

86. Brown R., Hardy J.P.L., Subgroups of free topological groups and free topological products of topological groups, J. London. Math. Soc. 10, N4:431–440 (1975).

87. Bulgakov D. N., On the conditions of a complete regularity of the space of a continuous loup, Tkani i Quasi-gr., Kalinin, 8–12 (1981).

88. Burgin M.S., Topological algebras with continuous signatures, Dan SSSR, 213, N3:505–508 (1973).

89. Čarin V.S., Topological groups. Algebra, 123–160 (1964). Acad. Nauk SSSR, Inst. Naučn. Informacii, Moscow, 1966.

90. Cohen J., Locally bounded topologies on the ring of integers of a global field, Can. J. Math., 33, N3:571–584 (1981).

91. Cohen J., Topologies on the ring of integers of a global field, Pacif. J. Math., 93, N2:269–276 (1981).

92. Cohen J., On the number of locally bounded field topologies, Proc. Amer. Math. Soc., 90, N2:207–210 (1984).

93. Cohen J., Topologies on the quotient field of a Dedekind domain, Pacif. J. Math., 117, N1:51–67 (1985).

94. Cohen J., Extensions of valuation and absolute valued topologies, Pasif. J. Math., 125, N1:39–44 (1986).

95. Cohen G.E., Profinite modules, Can. Math. Bull., 16: 405-415 (1973).

96. Cohn P.M., An invariant characterization of pseudo-valuations of field, Proc. Cambridge Phil. Soc., 50:159-177 (1954).

97. Cohn P.M., On extending valuations in division algebras, Stud. Sci. Math. Hung., 16, N1–2:65–70 (1981).

98. Comfort W.W., Soundararajan T., Pseudocompact group topologies and totally dense subgroups, Pacif. J. Math., 100:61-84 (1982).

99. Comfort W.W., Remus D., Long Chains of Hausdorff topological group topologies, J. Pure and Appl. Algebra, 70, N1-2:53–72 (1991).

100. Correl E., Topologies on quotient fields, Duke Math. J., 35, N1:175–178 (1968).

101. Cutler D.O., Stringall R.W., A topology for primary Abelian groups, Etudes groupes Abeliens, Paris–Berlin–Heidelberg–New Jork, 93–100 (1968).

102. De Marco G., Orsatti A., Complete linear topologies on Abelian groups, Symp. math. Jst. naz. alta mat. Conv. nov.–die., 1972, Vol. 13, London–New York, 153–161 (1974).

103. D'Este G., Sui gruppi abeliani il cui anello degli endomorfismi e numerubile e non discreto nella topologia finita, Atti. Ist. veneto Sci. Let. Ed. Cb. Sci. Mat. e Natur., 134:239-243 (1975-1976).

104. D'Este G., The $\oplus_c$-topology on Abelian p-groups, Ann. Scn. norm. super. Pisa Cl., sci 7, N2:241–256 (1980).

105. Dikranjan D., Minimal precompact topologies in rings, God. Sof. Univ. Fac. Math. i Mech., 1972-1973, 67:391-397 (1976).

106. Dikranjan D., Minimal ring topologies and Krull Demension, Coll. Math. Soc. Janos Bolyai 23. Topology Budapest (Hungary), 357–366 (1978).

107. Dikranjan D., One class of topological rings, Dokl. VII Prolet. Conf. Soyuza Matem. Bulg., Slynchev Bryag 1978, Sofia, 305–313 (1978).

108. Dikranjan D., Minimal topological rings, Serd. Bul. Math., 8:149-165 (1982).

109. Dikranjan D., Extension of minimal ring topologies, Gen. Topol. and Relat. Mod. Anal. and Algebra, 5:132- 144 (1983).

110. Dikranjan D., Minimalizable topological groups, Rend.Circ.Mat. Palermo, 33:43-57 (1984).

111. Dikranjan D., Sur la minimalite des produits de groupes topologiques abeliens, C. R. Acad. Sci., Ser 1, 299:303-306 (1984).

112. Dikranjan D., Divisible Abelian groups admitting minimal topologies, Lect. Not. Math., 1060:217-226 (1984).

113. Dikranjan D., On a conjecture of Prodanov, Dokl. Bulg. AN, 38:1117-1120 (1985).

114. Dikranjan D., Prodanov Iv., Totally minimal topological groups, Annuaire de l'Univ de Sofia, Fac. Math. Mech.69: 5–11 (1974/75).

115. Dikranjan D., Prodanov I.R., Stojanov L.N., Topological groups. Characters dualities. Minimal group topologies, Marcel Dekker, (1990).

116. Dikranjan D., Ronido N., Quotients of minimal topological groups, Mat. i Mat. Obrazov., Dokl. 14 Prolet. Conf. Soyuza Matem. Bulg., Slynchev Bryag 6–9 Apr. 1985, Sofia, 234–237 (1985).

117. Dikranjan D., Shakhmatov D.B., Pseudocompact topologizations of groups, Zb. rad. fil. fak. Nišu., Ser. mat., 4:83–93 (1990).

118. Dikranjan D., Stoyanov L.N., Criterion for minimality of all subgroups of topological Abelian group, Dokl. Bulg. An. 34: 635–638 (1981).

119. Dikranjan D.N., Stoyanov L.N., A-classes of minimal Abelian group, Sof. Un.

Fac. Mat. i Mach. 1976–1977, 71, N2:53–62 (1986).

120. Dikranjan D., Wieslaw W., Rings with only ideal topologies, Comm. Math. Univ. Santi Pauli, 29, N2:157–167 (1980).

121. Dixon J.D., Formanek E.W., Poland J.C., Ribes L., Profinite completions and isomorphic finite quotients, J. Pure and Appl. Algebra, 23:227-231 (1982).

122. Doitchinov D., Produits de groups topologique minimaux, Bull. Soc. Math., 2-e Ser., 96:59–64 (1972).

123. Douwen E.K. van, The maximal totally bounded group topology on G and the biggest minimal G-space, for Abelian groups G, Topol. and Appl, 34:69-91 (1990).

124. Dumitrashku S.S., Choban M.M., On free topological algebras with continuous signature, Mat. Issled., N65:27–53 (1982).

125. Eberhardt V., Schwanengel U., $(\mathbb{Q}/\mathbb{Z})^N$ est un groupe topologique minimal, Rev. roum. math. pures et appl., 27, N9:957–964 (1982).

126. Edelman I., Larotomda A., Casi-normas en anillos, Revista de la Union Mathematica Argentina, 25:357–361 (1971).

127. Endo M., On the embedding of topological rings into quotient rings, J. Fac. Sci. Univ. Tokyo, 10:196-214 (1964).

128. Endo M., On the embedding of topological rings into quotient rings II, Scient. Papers Coll. Gen. Educ. Univ. Tokyo, 14:51-54 (1964).

129. Endo M., A note on locally compact division rings., Commentary Math. Univ. Saneti Payli, Tokyo, 14:57–64 (1966).

130. Endo M., On the topologies of the rational number field, Comm. Math. Univ. St. Pauli, 16:11-20 (1967).

131. Engelking R., General Topology, Warsaw, (1977, 1985).

132. Ezau P.F., An example of infinite commutative non-associative ring admitting only discrete topologization, Proc. VI Symp. Theor. Rings, Alg. Mod., 153 (1990).

133. Faith C., Algebra: Rings, modules and categories I, Springer-Verlag, 1973.

134. Fay T.H., Walls G.L., Maximal functorial topologies on Abelian groups, Arch. Math, 38:167-174 (1982).

135. Filippov K.M., On completeness of free topological module generated by discrete

space, Mat. Issled., 118:114-120 (1990).

136. Filippov K.M., On completeness of free topological module over discrete ring, Moscow, VINITI, 5281-B90 (08.10.90).

137. Filippov K.M., Properties of topological modules in certain maximal topologies, Moscow, VINITI, 3930-B91 (11.10.91).

138. Fleischer I., Sur le corps topologiques et les valuations, C. R. Acad. Sci., 236, N 13:1320-1322 (1953).

139. Fleischer I., Sur les corps localement bornés, C.R. Acad. sci. 237, N11:546–548 (1953).

140. Fleischer I., Characterisation topologique des corps valués, Semin. Krasner 1953–1954, Fac. Sci. Paris, Paris, Exp 7 (1954).

141. Fleischer I., Espaces vectoriels sur un corps valué de Krull, Semin. Krasner 1953–1954, Fac. Sci. Paris, Paris, Exp 15 (1954).

142. Fuchs L., Note on linearly compact Abelian groups., J. Austral. Math. Soc., 9, N3–4:433–440 (1969).

143. Fukawa Masami, On the theory of valuations, J. Fac. Sci. Univ. Tokyo, Sec. 1, 7, N1:57–79 (1965).

144. Gelbaum B.R., Kalisch G., Olmsted J., On the embedding of topological semi-groups and integral domains, Proc. Amer. Math. Soc., 2:805–821 (1951).

145. Gelfand I.M., Normierte Ringe, Math. Sb., 9:3–24 (1941).

146. Gelfand I.M., Raikov D.A., Shilov G.E., Commutative normed rings, Physmatiz-dat, (1960).

147. Gilmer R., On ideal–adic topologies for a commutative ring, Enseignement Math., 18:201–204 (1972).

148. Glavatsky S.T., Topological quasi-Frobenius modules, Moscow, VINITI, 2448:1–50 (20.07.78).

149. Glavatsky S.T., Topological quasi-Frobenius modules II, Moscow, VINITI, 3411:1–22 (31.10.78).

150. Glavatsky S.T., Topological quasi–invective and quasi–projective modules, Moscow, VINITI, 3412: 1–20 (31.10.1978).

151. Glavatsky S.T., Compact power of locally compact modules, Thes. IV Tir. Symp. Gen. Topol. and Appl., Kishinev, 27–28 (1979).

152. Glavatsky S.T., Topological quasi–invective and quasi–projective modules, Abel. Group. i Moduli, Tomsk, 31–44 (1980).

153. Glavatsky S.T., Locally compact quasi–Frobenius modules and duality, Thes. Int. Algebr. Conf. Barnaul 1991, 150 (1991).

154 Goldman O., A Wedderburn-Artin-Jacobson structure theorem, J. Algebra 34, N 1, 64 - 73 (1975).

155. Goldman O., Sah Chin–Han, On a special class of locally compact rings, J. Algebra, 4, N1:71–95 (1966).

156. Gould G.G., Locally unbounded topological fields and box topologies on products of vector spaces, J. London. Math. Soc., 36, N3:273–281 (1961).

157. Grayev M.I., To the theory of complete direct products of groups, Mat. Sb., 17:85–104 (1945).

158. Grayev M.I., Free topological groups, Izv. AN SSSR, Ser. Math., 12:279–324 (1948).

159. Gruber B., Topological groups and global properties, Rept. Inst. Math. Sci. (India), S.a., N63, 100 pp.

160. Guerindon J., Espaces vectoriels normes sur un corps valué complete, et unicite du prolongement de sa valuation dans ses extensions algebriques de degre fini, Semin. Kracner 1953–1954, Fac. Sci Paris., Paris, Exp 15 (1954).

161. Guerindon J., Sur la topologie spectrale d'un anneau, C. R. Acad. Sci, 254:3066–3068 (1962).

162. Hacque M., Completion in trinseques et co–intrinseques, Comm. Alg., 12, N15–16:1931–1988 (1984).

163. Haley D.K., On compact commutative Noetherian rings, Math. Ann., 189:272–274 (1970).

164. Haley D.K., Equationally compact Noetherian rings, Queen's Pap. Pure and Appl. Math., 25:258–267 (1970).

165. Haley D.K., Equational compactness and compact topologies in rings satisfying

A.C.C., Pacif. J. Math., 62:99–115 (1976).

166. Halter-Koch F., Idealtheorie der offen einbettbaren kommutativen topologischen Ringe, J. Reine und Angew. Math., 256:168–172 (1972).

167. Hanson J.R., An infinite groupoid which admits only trivial topologies, Amer. Math. Month., 74, N5:568–569 (1967).

168. Hardy J.P.L., The free topological group on a cell complex, Bull. Austral. Math. Soc., 11, N3:455–463 (1974).

169. Hardy J.P.L., Morris Sidney A., The free topological group on a simply connected space, Proc. Amer. Math. Soc., 55, N1:155–159 (1976).

170. Hartman S., Mycielsky J., On the embedding of topological groups into connected topological groups, Colloq. Math., 5:167–169 (1958).

171. Heine J., Existence of locally unbounded topological fields and field topologies which are not the intersection of bounded ring topologies, J. London Math. Soc., 5, N3:481–487 (1972).

172. Heine J., Ring topologien auf nichtalgebraischen Korpern, Mat. Ann., 199, N3:205–211 (1972).

173. Heine J., Ring topologies of type N on fields, Gen. Topol. and its Appl., 3:135–148 (1973).

174. Heine J., Warner S., Ring topologies on the quotient field of Dedekind domain, Duke Math. J., 40:473-486 (1973).

175. Hejcman J., Topological vector group topologies for the real line, Gen. Topol. and Relat. Mod. Anal. and Algebra 6; Proc. 6-th Praque topol. Symp. Aug 25–29 1986, Berlin, 241–248 (1988).

176. Helemskii A.Ya., Topologies in Banach and topological algebras, Moscow, Moscow Univ., (1986).

177. Helemskii A.Ya., Homology in Banach and topological algebras, Moscow Univ., (1986).

178. Helemskii A.Ya., Banach and normed algebras, Nauka, (1989).

179. Herstein I.N., Noncommutative rings, Published by the Math. Assoc. of Amer., Distribuded by John Wiley and Sons., (1968).

180. Hewitt E., Ross K.A., Abstract harmonic analysis, Vol. I. Structure of topological groups, integration theory, group representations, Springer-Verlag, (1963).

181. Higgins P.J., An introduction to topological groups., London Math. Soc., Lect. Note Ser., N15, 109 pp. (1974).

182. Hinrichs L.A., Integer topologies, Proc. Amer. Math. Soc., 15:991–995 (1964).

183. Hinrichs L.A., The existence of topologies of field extensions, Trans. Amer. Math. Soc., 113:397–406 (1964).

184. Hochster M., Ring with non–discrete ideal topologies, Proc. Amer. Math. Soc., 21:357–362 (1969).

185. Hochster M., Existence of topologies for commutative rings with indentity, Duke Math. J., 38:551–554 (1971).

186. Hochster M., Kiltinen J.O., Commutative rings with identity have ring topologies, Bull. Amer. Math. Soc., 76:419–420 (1970).

187. Hoffmann K.H., Mostert P., Splitting in topological groups, Met. Amer. Math. Soc. N43, 75 pp. (1963).

188. Hromulyak O.M., Topological filters on semigroups, Ukr. NIINTI, 1873 UK-90 (22.11.90).

189. Hromulyak O.M., T-filters on rings, Ukr. NIINTI, 1871, UK-90 (22.11.90).

190. Hulanicki A., Algebraic characterization of Abelian groups which admits compact topologies, Bull. Acad. Polon. Sci., 4, N7:405–406 (1956).

191. Hulanicki A., Algebraic characterization of Abelian divisible groups which admit compact topologies, Fund. Math., 44:192–197 (1957).

192. Hulanicki A., Algebraic structure of compact Abelian groups, Bull. Acad. Polonsci, Ser. Sci. Math. Astron. et Phys., 6, N2:71–73 (1958).

193. Iseki K., On 0-dimensional compact ring, Math. Jap., 3:37–40 (1953).

194. Jacobson N., Totally disconnected locally compact rings, Amer. J. Math., 56:433–449 (1936).

195. Jacobson N., Basic algebra I, Freeman, (1970).

196. Jacobson N., Basic algebra II, Freeman, (1980).

197. Janakiraman S., Topologies in locally compact groups II, III, J. Mat., 17:177–197 (1973).

198. Janakiraman S., Rajagopalan M., Topologies in locally compact groups, J. Math. Ann., 176, N3:169–180 (1968).

199. Jebli A., Sur certaines topologies lineaires, C. R. Acad. Sci., 274, N 6:A444–A446 (1972).

200. Johson R.L., Rings of quotiens of topological rings, Math. Ann., 179:203–211 (1969).

201. Kabenjuk M.J., Some topologies on Abelian group related to the ring of its endomorphisms, Topol. 4th Colloq. Budapest 1978, 2:705–712 (1980).

202. Kakol J., On inductive limits of topological algebras, Colloq. Math., 47:71–78 (1982).

203. Kalistru R.K., On completeness of a free topological module generated by a discrete space, Mat. Issled., 53:98–103 (1979).

204. Kalistru R.K., On completeness of a free topological module generated by a zero-dimensional space, Mat. Issled., 62:57–64 (1981).

205. Kalistru R.K., On completeness of topological ring of polynomials, Mat. Issled., 85:71–82 (1985).

206. Kaplansky I., Topological methods in valuation theory, Duke Math. J., 14, N3:527–541 (1947).

207. Kaplansky I., Topological rings, Amer. J. Math., 69, N1:153–183 (1947).

208. Kaplansky I., Locally compact rings I, Amer. J. Math., 70:447–452 (1948).

209. Kaplansky I., Locally compact rings II, Amer. J. Math., 73:20–24 (1951).

210. Kaplansky I., Topological algebra, Notes Mat. Inst. Mat. Pure Apl., N 16 (1959).

211. Katz E., Morris S.A., Nickolas P., Free Abelian topological groups on spheres, Quart. J. Math. 35, N138:173–181 (1984).

212. Katz E., Morris S.A., Nickolas P., Free subgroups of free Abelian topological groups, Math. Proc. Cambridge Phil. Soc., 100, N2:347–353 (1986).

213. Kelley J.L., General Topology, Van Nostrand Comp. Inc., (1957).

214. Kertesz A., Szele T., On existence of non-discrete topologies on infinite Abelian groups, Publ. Math., 3:187–189 (1953).

215. Kiltinen J., Inductive ring topologies, Trans.Amer. Mat. Soc., 134, N1:149–169 (1968).

216. Kiltinen J., Embedding a topological domain in a countably generated algebraic ring extension, Duke Math. J., 37, N4:647–654 (1970).

217. Kiltinen J., On the number of field topologies on an infinite field, Proc. Amer. Math. Soc., 40:30–36 (1973).

218. Kiltinen J. O., Infinite Abelian groups are highly topologizable, Duke Math. J., 41, N1:151–154 (1974).

219. Kirku P.I., On precompact topologizability of modules, Mat. Issled., 154, N65:88–90 (1982).

220. Kirku P.I., On compact and locally compact topologizability of modules over topological ring of integers, Mat. Issled., 74:69–79 (1983).

221. Kirku P.I., On locally compact topologizability of modules over finite rings, Izv. AN MSSR, 1:3–7 (1983).

222. Kirku P.I., On locally compact topologizability of modules over Dedekind rings, Mat. Issled., 76:47–72 (1984).

223. Kirku P.I., On existence and uniqueness of locally compact group topology on Abelian groups, Mat. Issled., 85:96–103 (1985).

224. Kirku P.I., On locally compact groups in which non-unitary closed subgroups are open, Mat. Zametki, 40, N2:238–242 (1986).

225. Kirku P.I., Abelian torsionless groups admitting only a finite number of locally compact group topologies, Mat. Issled., 90:76–89 (1986).

226. Kirku P.I., Characterization of Abelian torsionless groups in which the number of non-isomorphic locally compact topologizations is finite, Mat. Issled., 91:15–28 (1987).

227. Kirku P.I., A number of locally compact group topologies on divisible Abelian torsionless groups, Mat. Issled., 105:93–104 (1988).

228. Kirku P.I., Criterion of the connected topologizability of torsion Abelian groups of finite period, Mat. Issled., 118:66–73 (1990).

229. Kiyek K., Pseudobetragsfunktionen auf Quotientekorpern von Dedekindringen, J. Reine und Angew. Math., 274–275:244–257 (1975).

230. Koh K., On compact prime rings and their rings of quotients, Can. Math. Bull., 11, N4:563–568 (1968).

231. Kokorin A.I., Kopytov V.M., Linearly ordered groups, Nauka, (1972).

232. Kolmogoroff A.N., Zur Normierberkeit eines allgemeinen topologischen linearen Raumes, Studia Math., 5:29–33 (1934).

233. Kontolatou A., Stabakis J., Completion in a G–valued topological field, Rend. Circ. Mat. Palermo,ˆ37, suppl. 1, N18:319–326 (1988).

234. Kopytov V.M., Stucturely ordered groups, Nauka, (1984).

235. Kortesi P., About some separation properties in topological groups, Proc. Nat. Colloq. Geom. and Topology organiz. Occas. 180 th. Anniv. Bolyai Janos and 150 Geors since Publ. Work Append. Cluj – Napoca; Tr̂gu – Mures 7–10. Sept 1982. Babis – Bolyai s.a. 72–74.

236. Kostrikin A.I., Introduction to algebra, Nauka, 1970.

237. Kowalsky H., Zur topologischen Kennzeichnung von Korpern, Math.Nachr., 9:261–268 (1953).

238. Kowalsky H., Beiträge sur topologischen Algebra, Math. Nachr. 11, N3:143–185 (1954).

239. Kowalsky H., Dürbaum H., Arithmetische Kennzeichnung von Korpertopologien, J. Rein. Angew. Math., 191, N1/2:136–152 (1953).

240. Krasner M., Unicite de prolongement des valuations dans les extensions algebraiques des corps values mit a complets de Krull, Semin. Krasner 1953–1954, Fac. Sci. Paris, Paris, Exp. 15 (1954).

241. Kriǎniče F. Topološke grupe., Durštvo mat., fiz., in astron. S.R. Slovenije, Ljubljana, 132 pp., (1974).

242. Kurosh A.G., Kombinatorischer Aufbau der bikompakten topologischen Raumen, Composito Math., 4:471–476 (1935).

243. Kurosh A.G., Lectures on general algebra, Nauka, (1962).

244. Kurosh A.G., Theory of groups, Nauka, (1967).

245. Kurosh A.G., Course of general algebra, Nauka, (1968).

246. Lambek J., Lectures on rings and modules, Blaisdell Publ.Co., (1966).

247. Lang S., Algebra, Addison-Wesley, (1965).

248. Leptin H., Linear Kompakte Moduln und Ringe, Math. Z., 62:241–267 (1955).

249. Lipkina Z.S., On pseudo-normazability of topological rings, Sib. Math. J., 6:1046–1052 (1965).

250. Lipkina Z.S., Locally bi-compact rings without divisors of zero, DAN SSSR, 161:523–525 (1965).

251. Lipkina Z.S., Locally bi-compact rigs without divisors of zero, Izv. Acad. Nauk SSSR, Ser. Math., 31, N6:1239–1262 (1967).

252. Llorente P., Grupos topologicos I, Estructuras, Cursos semin y tesis PEAM maracaibo Ed. La Univ. Zulia, 1, 134 pp., (1975).

253. Llorente P., Grupos topologicas II. Teoria de dualidad, Maracaibo, Ed. Ja Univ. Zulia, 2, 111 pp., (1976).

254. Loonstra F., Topologisch subdirekte Produkte, Proc. Koninkl. Nederl. Akad. Wet., A62, N4:434–438 (1959).

255. Lucke J.B., Warner S., Structure theorems for certain topological rings, Trans. Amer. Math. Soc., 186:65–90 (1973).

256. Luedeman J.K., On the embedding of topological domains into quotient fields, Mans. Math., 3:213–226 (1970).

257. Luedeman J.K., On the embedding of compact domains, Math. Z., 115:113–116 (1970).

258. Mader A., Duality and completions of linearly topologized modules, Math. Z., 179, N3, 325–335 (1982).

259. Mader A., The $\oplus_c$–topology is not completable. Ann. Sc. norm. super. Pisa. Cl. sci., 10, N4:579–586 (1983).

260. Mader A., Mines R., Completions of linearly topologized vectorspaces, J. Algebra, 74, N2:317–327 (1982).

261. Mader A., Vinsonhaler Ch., Minimal Hausdorff topologies on Abelian groups, Abelian Group Theory: Proc. 3-rd Conf. Oberwolfach Aug. 11-17 1985, 399–416 (1987).

262. Magnin L., Deux exemples de pathologies pour les groups topologiques non localement compacts, Expos. Math., 3:169–174 (1985).

263. Mal'cev A.I., Free topological algebras, Izv. AN SSSR, Ser. Mat., 21, N2:171–198
(1957).

264. Manchanda P., Singh S., Locally bounded topologies on $F(x)$. Indian J. Pure and
Appl. Math., 22, N4:313–321 (1991).

265. Marin E.I., Thrifty complete topologies on countable Abelian torsionless groups,
Moscow, VINITI, 281-B (13.01.86).

266. Marin E.I., Thrifty complete topologies on Abelian torsionless groups, Moscow,
VINITI, 1331-B (26.02.86).

267. Marin E.I., Thrifty topologies on Abelian groups, Izv. AN MSSR, 1:9–17 (1986).

268. Marin E.I., Complete thrifty topologies on Abelian groups, Izv. AN MSSR, 1:53–54
(1986).

269. Marin E.I., Thrifty topologies on the group of integers, Usp. Mat. Nauk, 41,
N3:193–194 (1986).

270. Marin E.I., Thrifty topologies on Abelian groups, Mat. Issled., 91:29–43 (1987).

271. Marin E.I., Strengthening of a group topology on Abelian group to complete, Mat.
Issled., 105:105–119 (1988).

272. Markov A.A., On free topological groups, Izv. AN SSSR, 9:3–64 (1945).

273. Markov A.A., On absolutely closed sets, Mat. Sb., 18:3–28 (1946).

274. Marxen D., Neighborhoods of the identity of the free Abelian topological groups,
Math. Slovaca (CSSR), 26, N3:247–256 (1976).

275. Mathiak K., Zur Bewertungstheorie nicht kommutativer Korper, J. Algebra, 73,
N2:586–600 (1981).

276. Maurer I., Topologization of the infinite substituon group, Bull. Stiint. Acad.
R.P.Romana, Sec. Mat i Fiz., 8, N2:265–272 (1956).

277. Maurer I., Sur la topologisation des anneaux, Studii si cercetări Mat. Acad. RPR
Fil., Cluj 8, N1–2:177–178 (1957).

278. Maurer I., Esine Topologisierung der Permutationsgruppen einer beliebigen un-
emdlichen Menge, Bull. Sci. Math. et Phys., RPR 2, N1:55–59 (1958).

279. Maurer I., Topologische Unensuchung gewisser Substitutionsgruppen, Publs. Math., 8, N3–4:262–273 (1961).

280. Mazur S., Sur les anneaux lineaires, C.R. Acad. Sci. Paris. Ser A–B, 207:1025–1027 (1938).

281. Menini Cl. Linearly compact rings and selfcogenerators, Rend. Sem. mat.Univ. Padova 72, 99 - 116 (1984).

282. Menini Cl. A characterization of linearly compact modules, Math.Ann. 271, N 1, 1 - 11 (1985).

283. Mil'man D.P., Normalizability of topological rings, DAN SSSR, 47:162–164 (1945).

284. Miller C.B., Rajagopalan M., Topologies in locally compact groups III, Proc. Lond. Math. Soc., 31, N1:55–78 (1975).

285. Morikuni G., An introduction to topological groups, Lect. Notes. Ser. Mat., Inst. Aurhus. Univ., N40, 62 pp., (1974).

286. Morris S.A., Pontryagin duality and the structure of locally compact Abelian groups, London. Math. Soc., Lecture Note Series, 29 (Cambrige University Press, Cambridge), 128 pp., (1977).

287. Morris S.A., Free Abelian Topological Groups, Proc. Int. Conf. Toledo Ohio, Aug. 1–5, 1983, 375–391 (1984).

288. Morris S.A., Locally compact topologies on Abelian groups, Math. Proc. Cambridge Phil. Soc., 101, N2:233–235 (1987).

289. Morris S.A., Thompson H.B., Free topological groups with no small subgroups, Proc. Amer. Math. Soc., 46, N3:431–437 (1974).

290. Moskalenko Z.I., On the question of locally compact topologization of a group, Ukr. Mat. Zh., 30, N2:257–260 (1970).

291. Moskalenko Z.I., On existence of non-isomorphic locally compact topologization of Abelian group with coinsiding sets of closed subgroups, Ukr. Mat. Zh., 33, N6:820–823 (1981).

292. Mukhin Ju. N., Topological groups, Itogi nauki i tekhniki, Alg. Topol. Geom., Moscow, VINITI, 20, (1982).

293. Mukhin V.V., On topologization of semigroups with invariant mesure, Ukr. Mat.

Zh., 35, N1:103–106 (1983).

294. Mutylin A.F., The example of a non-trivial topologization of the field of rational numbers. Complete locally bounded fields, Izv. AN SSSR, 30:873–890 (1966).

295. Mutylin A.F., On embedding of discrete fields into connected, DAN SSSR, 168:1005–1008 (1966).

296. Mutylin A.F., Non-normalized extension of p-adic field, Mat. Zametki, 2, N1:11–14 (1967).

297. Mutylin A.F., Connected complete locally bounded fields. Complete not locally bounded fields, Mat. Sb., 76:454–472 (1968).

298. Mutylin A.F., Corrections to paper "The example of a non-trivial topologization of the field of rational numbers. Complete locally bounded fields", Izv. AN SSSR, 32:245–246 (1968).

299. Mutylin A.F., Completely simple commutative topological rings, Mat. Zametki, 5:161–171 (1969).

300. Nachbin L., On strictly minimal topological division rings, Bul. Amer. Math. Soc., 55, N12:1128–1136 (1949).

301. Najmark M.A., Normed Rings, Nauka, (1968).

302. Nakano Takeo, On the locally bounded fields, Comment. Math. Univ. St. Pauli, 9, N2:77–85 (1961).

303. Nakano Takeo, Some examples of the locally bounded fields, Comment. Math. Univ. St. Pauli, 10, N2:63–66 (1961/1962).

304. Nazarov A.O., Open embeddings of topological rings, Vestnik MGU. Ser. Mat. i Mech., 1:51–54 (1983).

305. Nichols E., Cohen J., Locally bounded topologies on global fields, Duke Math. J., 44, N4:853–862 (1977).

306. Nicholson W.K., Sarath B., Ring with a largest linear topology, Comm. Alg., 13:769–780 (1985).

307. Nickolas P., Subgroups of the topological group on [0,1], J. London Math. Soc., 12, N2:199–205 (1976).

308. Nienhuys J.W., Not locally compact monothetic groups I, Proc. Kon. ned. Akad.

wetensch, A73, N4:295–310 (1970); Indag math. 32, N4:295–310 (1970).

309. Nienhuys J.W., Not locally compact monothetic groups II, Proc. Kon. ned. Akad. wetensch, A73, N4:311–326 (1970); Indag math. 32, N4:311–326 (1970).

310. Nienhuys J.W., Construction of group topologies on Abelian groups, Fund. Math., 75:101–116 (1972).

311. Nienhuys J.W., Some examples of monothetic groups, Fund. Math., 88, N2:163–171 (1975).

312. Nikodym O.M., Sur l'extension des corps algebraiques abstraits par le procédé généralisé de Cantor, basé sur les suites generales de Moore–Smith qui contiennent un châin confinale., C. R. Acad. sci., 241, N19:1249–1250 (1955).

313. Nyikos P.J., Reichel H.C., Topologically orderable groups, Gen. Topol. and Appl., 5, N3:195–204 (1975).

314. Ol'shansky A.Yu., Remark on countable non-topologizable group, Vestnik MGU. Ser. Mat. i Mech., 3:103 (1980).

315. Ordman E.T., Free k-groups and free topological groups., Gen. Topol. and Appl., 5, N4:205–219 (1975).

316. Ordman E.T., Smith-Thomas Barbara V., Sequential conditions and free topological groups, Proc. Amer. Math. Soc., 79:319–326 (1980).

317. Paljutin E.A., Seese D.Q., Taimanov A.D., A remark on the topologization of algebraic structures, Rev. Roum. Math. Pures et Appl., 26:617–618 (1981).

318. Pestov V.G., The coincidence of the dimensions Dim of l-equivalent topological spaces, DAN SSSR, 266, N3:553–556 (1982).

319. Pestov V.G., To the theory of topological groups: free groups, extensions and compact coverability, Moscow, VINITI, N2207 (01.04.1985).

320. Pestov V.G., Free Banach spaces and representations of topological groups, Funct. Anal. i Prilozh., 20, 1:81–82 (1986).

321. Pestov V.G., Free topological Abelian groups and Pontrjagin duality, Vestik MGU, Ser. Math. i Mech., 1:3–5 (1986).

322. Pestov V.G., On absolutely closed sets and one hypothesis of A.A. Markov, Sib. Mat. J., 29, N2:124–132 (1988).

323. Podewski K.P., The number of field topologies on countable field, Proc. Amer. Math. Soc., 39:33–38 (1973).

324. Podewski K.P., Transcendental extensions of field topologies on countable fields, Proc. Amer. Math. Soc., 39:39–43 (1973).

325. Podewski K.P., Topologisierung algebraischer Strukturen, Rev. Roum. Math. Pures et Appl., 22, N9:1283–1290 (1977).

326. Pontrjagin L.S., Continuous Groups, Nauka, (1973).

327. Porst H.E., On free topological algebras, Can. topol. et geom. differ. categor., 28, N3:235–253 (1987).

328. Porst H.E., Struktur freier topologischer Gruppen, Rostock Math. Kolloq., N44, 5–20 (1991).

329. Prodanov I., Precompact minimal group topologies and p-adic numbers, Godish. Sofiisc. Univ. fac. Math., 1971–1972, 66:249–266 (1974).

330. Prodanov I., Some minimal group topologies are precompact, Math. Ann., 227, N2:117–125 (1977).

331. Prodanov I., Minimal topologies in Abelian groups, Matem. i Matem. Obrazov., Dokl. VII Prolet. Conf. Soyuza Matem. Bulg., Slynchev Bryag, 1978, Sofia, 107–113, (1978).

332. Prodanov I., Maximal and minimal topologies on Abelian groups, Topology 4th Colloq. Budapest (1978), 2:985-997 (1980).

333. Prodanov I., Elementary proof of Peter-Weil theorem, Dokl. Bolg. AN, 34, N3:315–318 (1981).

334. Prodanov I., Minimal topologies on countable Abelian groups, Sof. Un. Fac. Mat. i Mech. 1975-1976, 70:107–118 (1981).

335. Prodanov I., Stojanov L.N., Every minimal Abelian group is precompact, Dokl. Bolg. AN, 37:23–26 (1984).

336. Pyartli S.A., Topological analog of polycyclic bi-finite groups and problem of extension of topologies, Vestn. Mosk. Univ. Ser. 1, N3:57–59 (1992).

337. Pyartli S.A., Pseudonormability of topological semigroup rings, Usp. Math. Nauk, 47, N3 (285):171–172 (1992).

338. Pyartli S.A., To the question of simultanuos extensions of a topology and a psedonorm on a monoid to its semigroup ring, Vestnik NGU, Ser.1, N1:38–42 (1993).

339. Pyartli S.A., Concerning of simultanuos extensions of a topology and a psedonorm on a monoid to its semigroup ring, Mat. Zametki 53, N 2 110 - 113 (1993).

340. Remus D., The number of T_2-precompact group topologies on free groups, Proc. Amer. Math. Soc., 95, N2:315–319 (1985).

341. Remus D., Die Anzahl von T-prakompakten Gruppentopologien auf unendichen Abelschen Gruppen, Rev. Roum. Math. Pures et Appl., 31, N9:803–806 (1986).

342. Remus D., On the number of Hausdorff group topologies on infinite Abelian and free groups, Bolg. Math. J., 13, N2, 194–196 (1987).

343. Remus D., Topological groups without non-trivial characters., General Topology and Relations to Modern Analysis and Algebra VI, Proc. Sixth Prague Topological Symposium (1986), Z. Frolin (ed), Copyright Heldermann Verlag Berlin, 477–484 (1988).

344. Remus D., A short solution of Markov's problem on connected group topologies, Proc. Amer. Math. Soc., 110, N4:1109–1110 (1990).

345. Remus D., Minimal and precompact group topologies on free groups, J. Pure and Appl. Algebra, 70, N1–2:147–157 (1991).

346. Rickart C.E., The singular elements of Banach algebra, Duke Math. J., 14:1063–1077 (1947).

347. Rickart Ch. E., General theory of Banach algebras, New York, (1960).

348. Rickert N.W., Locally compact topologies for groups, Trans. Amer. Math. Soc., 126, N2:225–235 (1967).

349. Rigo Th., Warner S., Topologies extending valuations, Can. J. Math., 30, N1:164–169 (1978).

350. Roelke W., Dierols S., Uniform Structures on Topological Groups and their Quotients, McGraw-Hill, (1981).

351. Rossignol A., Prolongement d'un anneau topologique dans un anneau topologique unitaire, Application aux modules topologiques, C. R. Acad. Sci., 274, N7:A543 (1972).

352. Schiffels G., Stenzel M., Einbettung lokal-beschrankter topologischer Ringe in Quotietenringe, Result. Math., 7:234–248 (1984).

353. Schinkel F., Existence and number of maximal precompact topologies on Abelian groups, Result. Math., 17:282–286 (1990).

354. Shafarevich I.R., On normability of topological fields, DAN SSSR, 40:149–151 (1943).

355. Shakhmatov D.B., Embeddings into topological fields and construction of a field with non-normal space, Comment. Math. Univ. Carolinal, 24:525–540 (1983).

356. Shakhmatov D.B., Cardinal invariants of topological fields, Dokl. AN SSSR, 271, N6:1332–1336 (1983).

357. Shakhmatov D.B., Minimal topologizations of free groups, Alg. Log. and Numb. Theory, Conf. Mech.-Math. Fac. MSU, Feb.-March 1985, MSU, 97–100 (1986).

358. Shanks M.E., Warner S., Topologies on the rational field, Bull. Amer. Math. Soc., 79:1281–1282 (1973).

359. Shanks M.E., Warner S., Locally bounded topologies on the rational field, Fund. Math., 92:57–61 (1976).

360. Sharma P.L., Hausdorff topologies on groups I, Mat. Jap., 26, N5:555–556 (1981).

361. Sheelter W., Topological rings of quotients., Can. J. Math., 26, N5:1228–1233 (1974).

362. Shell N., Maximal and minimal ring topologies., Proc. Amer. Math. Soc. 68, N1:23–26 (1978).

363. Shell N., Bounded and relatively bounded sets, Bull. Austral. Math. Soc., 36, N3:493–498 (1987).

364. Shell N., Point at infinity, Bull. Austral. Math. Soc., 38:321–323 (1988).

365. Shell N., Algebraically closed completions, Math. Stud., 59, N1–4:73–76 (1992).

366. Shestopal V.E., Completions of topological algebras, Preprint ITEF-13, (1977).

367. Singh S., Seminormability of certain ring topologies on Dedekind domains, Indian. J. Pure Appl. Math., 16(12):1465–1471 (1985).

368. Singh S., Saluja K.S., Locally bounded ring topologies on global fields, Indian. J. Pure and Appl. Math., 14, N7:919–929 (1983).

369. Skornjakov L.A., Locally bi-compact bi-regular rings, Mat. Sb., 62(104):3–13 (1963).

370. Skornjakov L.A. Einfache lokal bicompakte ringe, Math. Zeitschr 87, 241 - 251 (1965).

371. Skornjakov L.A., Locally bi-compact bi-regular rings (addition), Mat. Sb., 69(111): 663 (1966).

372. Skornjakov L.A., To structure of locally bi-compact bi-regular rings, Mat. Sb., 104(146):652-656 (1977).

373. Skornjakov L.A., Elements of general algebra, Nauka, (1983).

374. Slin'ko A.M., On topological rings with involution, Usp. Mat. Nauk, 41, N5:197–198 (1986).

375. Soundararajan T., The topological group of the p–adic integers, Publs. Math., 16, N1–4:75–78 (1969).

376. Stănăsilă O., Certain observations on topologizing modules, Stud. Cere Mat., 18:681–686 (1966).

377. Staum R., The completion of an algebraically closed valued field, Boll. Unione Mat. Ital., 10, N3:553–555 (1974).

378. Stavskii M.Sh., On algebraic closeness of a completion of algebraically closed field, Novosib. Gosud. Pedag. Inst., Nauchn. Trudy, 28, Mat. i Phys., 50–52 (1968).

379. Stephenson R. M. Jr., Minimal topological groups, Math. Ann., 192, N3:193–195 (1971).

380. Stewart T.F., Uniqueness of the topology in certain compact groups, Trans. Amer. Math. Soc., 97, N3:487–494 (1960).

381. Stojanov L.N., Weak periodicity and minimality of topological groups, Annuaire de l'Univ. de Sofia Kliment ohridski, Faculté de Math. et Mec., 73:155–167 (1979).

382. Stojanov L.N., A property of precompact minimal Abelian groups, Sof. Un. Fac. Mat. i Mech., 70:253–260 (1981).

383. Stojanov L.N., Cardinalities of minimal Abelian groups, Dokl. Conf. Matem. Bulg., 203–208 (1981).

384. Stojanov L.N., Some minimal Abelian groups are precompact, Dokl. Bolg. AN, 34, N4:473–475 (1981).

385. Stojanov L.N., On products of minimal and totally minimal groups, Dokl. Bolg.

Conf. 6-9 April 1982, Sofia, 287–294 (1982).

386. Stojanov L.N., Precompactness of minimal metrizable periodic Abelian groups, Sof. Un. Fac. Mat. i Mech. 1976-1977, 71, N2:45–52 (1986).

387. Subramanian H., Ideal neighbourhoods in a ring, Pacif. J. Math., 24:173–176 (1968).

388. Suvorov N.M., Commutative IP-loup admitting only discrete topology, Sib. Mat. J., 32, N5:193 (1991).

389. Suvorov N.M., Kryuchkov N. I., Examples of certain quasi-groups and loups, admitting only discrete topologization, Sib. Mat. J., 17, N2:471–473 (1976).

390. Suvorov N.M., Ufnarovskaya T.P., Example of finite semigroup admitting only discrete topologization, Issled. po teorii quasi-group i loup, Kishinev, Stiintsa, 163–165 (1973).

391. Swierczkowski S., Topologies in free algebras, Proc. London Math. Soc., 14, N3:566–576 (1964).

392. Szász F., On topological algebras and rings, I, Mat. Zapok., 13:256–278 (1962).

393. Szász F., On topological algebras and rings, II, Mat. Zapok., 14:74–87 (1963).

394. Szele T., On a topology in endomorphism rings of Abelian groups, Publs. Math. S., N1–2 (1957).

395. Tagdir H., Introduction to topological groups, Philadelphia, W.B. Saunders Co., 218 pp. (1966).

396. Taimanov A.D., Example of semigroup admitting only discrete topology, Algebra i logika, 12:114–116 (1973).

397. Taimanov A.D., On topologizable groups, Sib. Matem. J., 18, N4:947–948 (1977).

398. Taimanov A.D., On topologizable groups, II, Sib. Mat. J., 19, N5:1201–1203 (1978).

399. Taimanov A.D., On topologizability of countable algebras, Math. Anal. and Adj. Quest. Math., 254–275 (1978).

400. Thomas B., Smith W., Free topological groups., Gen. Topol. and Appl., 4, N1:51–72 (1974).

401. Thompson H.B., A remark on free topological groups with no small subgroups, J.

Austral. Math. Soc., 18, N4:482–484 (1974).

402. Tkachenko M.G., On zero-dimensional topological groups, Tr. Leningr. mezhdunar. topol. konf. 23-27 avg. 1982, 113–118 (1983).

403. Tkachenko M.G., On Suslin's property in free topological groups over bi-compacts, Mat. zametki, 34:601–607 (1983).

404. Tkachenko M.G., On completeness of free Abelian topological groups, DAN SSSR, 269, N2:299–303 (1983).

405. Tkachenko M.G., On topology of free groups over bi-compacts, Maps and Functors, M., MGU, 122–137 (1984).

406. Tkachenko M.G., On topologies of free groups, Czechosl. Math. J., 34:541–551 (1984).

407. Tkachenko M.G., On some properties of free topological groups, Mat. Zametki, 37:110–118 (1985).

408. Tkachenko M.G., On countably compact and pseudocompact topologies on free Abelian groups, Izv. Vuzov, Ser. Mat., 5:68–75 (1990).

409. Tkachuk V.V., Homomorphisms of free topological groups do not keep compactness, Matem. Zametki, 42, N43:455–462 (1987).

410. Tsarelunga B.I., On certain class of rings with single compact topology, Moscow, VINITI, 2335-B89 (11.04.89).

411. Tsarelunga B.I., On uniqueness of complete Hausdorff metrizable topologies on rings, Mat. Issled., 118:121–125 (1990).

412. Tsarelunga B.I., On strenthening of topologies on compact rings, Izv. AN MSSR, Ser. Math., 2:49–52 (1990).

413. Unsolved problems of topological algebra, Kishinev, Stiintsa, (1985).

414. Ursul M.I., Products of hereditary linearly compact rings, Mat. Issled., 9, N4(34):137–149 (1974).

415. Ursul M.I., Locally hereditary linearly compact bi–regular rings, Mat. Issled., 48:146–160 (1978).

416. Ursul M.I., On the product of hereditary linearly compact rings, Usp. Mat. Nauk., 35, N3:230–233 (1980).

417. Ursul M.I., Locally finite and locally projectively nilpotent ideals of topological rings, Mat. Sb., 125(167), N3(11):291–305 (1984).

418. Ursul M.I., Embedding of a topological group into another group as a quasi-component, Mat. Issled., 91:92–102 (1987).

419. Ursul M.I., On uniqueness of compact topologies on rings, Mat. Issled., 105:142–152 (1988).

420. Ursul M.I., Some remarks on countably compact Boolean rings, Izv. AN MSSR, 2:70–75 (1990).

421. Ursul M.I., On strengthening of topologies of countably compact rings, Mat. Issled., 118:126–136 (1990).

422. Ursul M.I., Embedding of precompact groups into pseudo-compact ones, Izv. AN SSSR, Ser. Mat., N1:88–89 (1991).

423. Ursul M.I., Compact rings and their generalizations, Kishinev, Stiintsa (1991).

424. Ursul M.I., Boundedness in locally compact rings, Topol. and its Appl. 55,17–21 (1994).

425. Ursul M.I., Tsarelunga B.I., Uniqueness of compact topologies on semidirect products of rings, Moscow, VINITI, 4023-B90 (17.07.90).

426. Uspenskii V.V., Universal topological group with a countable base, Func. Anal. i yego Pril., 20, N2:86–87 (1986).

427. Uspenskii V.V., Free topological groups of metrizable spaces, Izv. AN SSSR, Ser. Mat., 54, N6:1295–1319 (1990).

428. Vasilach S., Ensemles bornes dans les modules topologiques, C. R. Acad. Sci., 267:A681–A683 (1968).

429. Vasilach S., Sur les ensembles bornes dans les modules topologiques, Atti. Acad. naz Lincei. Rend. Cl. Sci Fis., Mat. e Natur, 49:258–260 (1971).

430. Venkataraman M., Rajagopalan M., Soudararajan T., Orderable topological spaces, Gen. Topol. and Appl., 2, N1:1–10 (1972).

431. Venkataraman R., Interval of group topologies satisfying Pontryagin duality, Math. Z., 155, N2:143–149 (1977).

432. Vilenkin N.Ya., Theory of topological groups II, direct products, direct sums of

goups of rank 1, locally bicompact Abelian groups. Fibrous and weakly separable groups, Usp. Mat. Nauk, 5, 4:19–74 (1950).

433. Vilenkin N.Ya., Vector cpaces over topological fields, Mat. Sb., 32(74), N1:195–208 (1953).

434. Vizitiu V.N., Extension of m–topologies of a field, Mat. Issled., 10, N2:64–78 (1975).

435. Vizitiu V.N., Topological rings with m–unweakable topology, Mat. Issled., 38:97–115 (1976).

436. Vizitiu V.N., On extensions of m–topologies of commutative rings, Mat. Issled., 48:24–39 (1978).

437. Vizitiu V.N., Uniqueness of extension of m–topologies of a field to its algebraic extensions, Mat. Issled., 53:27–42 (1979).

438. Vizitiu V.N., Extension of pseudo–norms of fields, Mat. Issled., 62:3–8 (1981).

439. Vizitiu V.N., Extension of certain topologies of rings to their extensions, Mat. Issled., 65:17–26 (1982).

440. Vodinchar M.I., Criterion of pseudo-normalizability of topological algebras, Mat. Issled., 2, N1:133–140 (1967).

441. Waerden van der B.L., Algebra II. Funfte Auflage, Springer-Verlag, (1967).

442. Waerden van der B.L., Algebra I. Achte Auflage der Modernen Algebra, Springer-Verlag, (1971).

443. Warner S., Polinomial completeness in locally multiplicatively-convex algebras, Duke Math. J., 23:1–11 (1956).

444. Warner S., Weakly topologized algebras, Proc. of the Amer. Math. Soc., 8:314–316 (1957).

445. Warner S., Characters of cartezian products of algebras, Canad. J. Math., 11:70–79 (1959).

446. Warner S., Compact Noetherian rings, Math. Ann., 141, N2:161–170 (1960).

447. Warner S., Compact rings and Stone-Cech compactifications, Arch. Math., 11, fase.5:327–332 (1960).

448. Warner S., Compact rings, Math. Ann., 145, N1:52–63 (1962).

449. Warner S., Locally compact simple rings having minimal left ideals, Trans. Amer. Math. Soc., 125, N3:395–405 (1966).

450. Warner S., Compactly generated algebras over discrete fields, Bull. Amer. Math. Soc., 73, N2:227–230 (1967).

451. Warner S., Locally compact semilocal rings, Bull. Amer. Math. Soc., 73:906–908 (1967).

452. Warner S., Locally compact vector spaces and algebras over discrete fields, Trans. Amer. Math. Soc., 130:463–493 (1968).

453. Warner S., Locally compact equicharacteristic semilocal rings, Duke Math. J., 35, N1:179–190 (1968).

454. Warner S., Linearly compact Noetherian rings, Math. Ann., 178, N1:53–61 (1968).

455. Warner S., Two types of locally compact rings, Bull. Amer. Math. Soc., 74, N5:926–930 (1968).

456. Warner S., Compact and finite-dimensional locally compact vector spaces, Ill. J. Math., 13:383–393 (1969).

457. Warner S., Locally compact rings having a topologically nilpotent unit, Trans. Amer. Math. Soc., 139:145–154 (1969).

458. Warner S., Locally compact principal ideal domains, Math. Ann. 188:317–334 (1970).

459. Warner S., Sheltered modules and rings, Proc. Amer. Math. Soc., 30:8–14 (1971).

460. Warner S., Metrizability of locally compact vector spaces, Proc. Amer. Math. Soc., 27:511–513 (1971).

461. Warner S., Locally compact commutative Artinian rings, Ill. J. Math., 16:102–115 (1972).

462. Warner S., Normability of certain topological rings, Proc. Amer. Math. Soc., 33, N2:423–427 (1972).

463. Warner S., Openly embedding local Noetherian domains, J. Reine und Angew. Math., 253:146–151 (1972).

464. Warner S., Linearly compact rings and modules, Math. Ann., 197:29–43 (1972).

465. Warner S., A topological characterization of complete discretely valued fields,

Pacif. J. Math., 48, N1:293–298 (1973).

466. Warner S., Topological rings that generalize Artinian rings, J. Reine und Angew. Math., 293/294:99–108 (1977).

467. Warner S., Generalizations of a theorem of Mutylin, Proc. Amer. Math. Soc., 78:327–330 (1980).

468. Warner S., Topological fields, North-Holland Math. Stud., 157, Notas de Math. [Math. Notes], 126, North-Holland Publishing Co. Amsterdam-New York, 563 pp., (1989).

469. Warner S., Topological rings, North-Holland Math. Stud., 178, 508 pp., (1993).

470. Waterman A.G., Berman G.M., Connected fields of arbitrary characteristic, J. Math. Kyoto Univ., 5, N2:177–184 (1966).

471. Weber H., Ringtopologien auf Z and Q, Math., Z., 155:287–298 (1977).

472. Weber H., Zur einem problem von H.J. Kowalsky, Abn. Braunschweig Wiss. Gesellsch, 29:127–134 (1978).

473. Weber H., Charakterisierung der lokalbeschränkten Ringtopologien auf globalen Korpern., Mat Ann., 239:193–205 (1979).

474. Weber H., Topologische Charakterisierung globaler körper und algebraischer Funktionenkörper in einer Variablen, Math. Z., 169, N2:167–177 (1979).

475. Weber H., Unabhängige Topologien, Zerlegung vol Ringtopologien, Math. Z., 180, N3:379–393 (1982).

476. Weber H., Bestimmung aller Integritätsbereiche, deren einzige lokal-beschräkte Ringtopologien Idealtopologien sind, Arch. Math., 41:328–336 (1983).

477. Weil A., L'integration dans les groupes topologiques et ses applications, Paris, (1940).

478. Weiss E., Boundedness in topological rings, Pacif. J. Math., 6:149–158 (1956).

479. Weiss E., Zieler N., Locally compact division rings. Pacific J. Math., 8:369–371 (1958).

480. Wieslaw W., On some characterizations of the complex number field, Bull. de l'Acad. Polonaise de Sciences. Ser. Sci. Math., Astron. et Phis., 19:353–354 (1971).

481. Wieslaw W., A remark on complete and connected rings, Bull. de l'Acad. Polonase des Sciences, Ser. Sci. Math., Astr. et Phis., 19, N11:981–982 (1971).

482. Wieslaw W., On some characterizations of the complex number field, Coll. Math., 24:139–145 (1970/1971).

483. Wieslaw W., A characterization of locally compact fields of zero characteristic, Fund. Math., 76:149–155 (1972).

484. Wieslaw W., Errata to the paper "A characterization of locally compact fields of zero characteristic", Fund. Math., 83, N1:97–98 (1973).

485. Wieslaw W., Extensions of topological fields, Colloq. Math., 28:203–204 (1973).

486. Wieslaw W., On topological fields, Colloq. Math., 29, N1:119–146 (1974).

487. Wieslaw W., A remark on complete and connected rings II, Bull. de l'Acad Polonase des Seiences, Ser Sci. Math., Astr. et Phis., 22, N1:15–17 (1974).

488. Wieslaw W., An example of a locally unbounded complete extension of the p-adic number field, Colloq. Math., 30:105–108 (1974).

489. Wieslaw W., A characterization of locally compact fields II, Fundam. Math., 88, N2:121–125, (1975).

490. Wieslaw W., Corrections to the papers "An example of locally unbounded complete extension of the p–adic number field" and "On topological fields", Colloq. math., 34, N2:293–294 (1976).

491. Wieslaw W., Independent topologies on fields, Comment. Math., Univ. St. Pauli, 26, N2:201–208 (1978).

492. Wieslaw W., A characterization of locally compact fields III, Colloq. Math., 39:313–317 (1978).

493. Wieslaw W., Locally bounded topologies on fields and rings, Mitt. Math. Genellschaft Hamburg Band X, 6:481–508 (1978).

494. Wieslaw W., Locally bounded topologies on Euclidean rings, Arch. Math., 31:33–37 (1979).

495. Wieslaw W., Locally bounded topologies on some Dedekind rings, Arch. Math., 33:41–44 (1979/1980).

496. Wieslaw W., Embeddings of fields in complete ones, Math. Nachr., 93:133–137 (1979).

497. Wieslaw W., Topological fields, Acta Univ. Wratisl. Mat., Fiz., Astron., 43, 219

pp., (1982).

498. Wieslaw W., Topological Fields, Marcel Dekker, (1988).

499. Williamson J.H., On topologizying the field $C(t)$, Proc. Amer. Math. Soc., 5:729–734 (1954).

500. Yunusova D.I., On extension of a norm of a group and a ring to their group ring, Mat. Issled., Algebras and rings, 90:137- 146 (1986).

501. Yunusova D.I., On extension of pseudo-norm of a group and a ring to their group ring, Moscow, VINITI, 7485–B (30.10.86).

502. Yunusova D.I., On extension of pseudo-norm of a ring, seminorm and metric of a semigroup to their semigroup ring, Mat. Issled., Rings and topologies, 105:172–194 (1988).

503. Yunusova D.I., Extension of non-Archimedean real–valued pseudonorm and metric to semigroup ring of free matroid, Moscow, VINITI, 1655–B88 (01.03.88).

504. Yunusova D.I., On extension of pseudonorm and metric to semigroup ring, Moscow, VINITI, 1670–B88 (01.03.88).

505. Zelazko W., Example of an algebra which is non–topologizable as a locally convex topological algebra, Proc. Amer. Math. Soc., 110, N4:947–949 (1990).

506. Zelenjuk E.G., Example of non-monothetic Abelian group all proper quotients of which are finite and cyclic, Ukr. INTEI, N933 (1992).

507. Zelenjuk E.G., Protasov I.V., Topologies on Abelian groups, Izv. AN SSSR, Ser. Math., 54, N5:1090–1107 (1990).

508. Zelenjuk E.G., Protasov I.V., Potentially compact groups, Mat. Sb., 181:278–286 (1990).

509. Zelenjuk E.G., Protasov I.V., Hromulyak O.M., Topologies on countable groups and rings, Dokl. AN UkrSSR, 8:8–11 (1991).

510. Zelinsky D., Complete fields from local rings, Proc. Nation. Acad. Sci., 37:379–381 (1951).

511. Zelinsky D., Rings with ideal nucleus, Duke Math. J., 18:431-442 (1951).

512. Zelinsky D., Linearly compact modules and rings, Amer. J. Math., 75:79-90 (1953).

513. Zhalilova H.Zh., Class of all non-trivially topologizable groups and its basic sub-classes, Moscow, VINITI, N43000 (18.12.1979).

514. Zhalilova H.Zh., Topologizability of groups, Moscow, VINITI, 2023-81 (7.05.81).

515. Zhalilova H.Zh., Correct topologizability of groups from certain classes, Moscow, VINITI, 178-81 (14.01.81).

516. Zhalilova N.Zh., Topologization of groups from different classes, Moscow, VINITI, 1511–83 Dep., (24.03.83).

517. Zobel R., Gruppentopologien auf nichtabelschen abzahlbaren gruppen, Man. Math., 14, N3: 207–216 (1974).

Notation

$\{a\}$, A.1

$\{x|\mathcal{P}\}$, A.1

$\emptyset$, A.1

$A \subseteq B$, A.2

$A \subset B$, A.2

$x \in A$, A.2

$A \bigcap B, \bigcap_{i \in I} A_i$, A.3

$A \bigcup B, \bigcup_{i \in I} A_i$, A.3

$A \backslash B$, A.3

$A \times B$, A.3

$\varphi(A)$, A.4

$\varphi|_A$, A.4

$\varphi^{-1}(B)$, A.4

$\psi \circ \varphi$, A.5

$\aleph_0$, A.7

c, A.7

$2^{|X|}$, A.7

$\omega(\mathfrak{m})$, A.7

ω_0, A.7

$|S|$, A.7

$\mathbb{C}$, A.8

$\mathbb{H}$, A.8, B.19

$\mathbb{N}$, A.8

$\mathbb{Q}$, A.8

$\mathbb{R}$, A.8

$\mathbb{Z}$, A.8

$[a;b]$, A.9

$(a; b)$, A.9

$(E, \leq)$, A.9

nb, B.2

$A + B$, B.3

$A - B$, B.3

$-A$, B.3

nA, B.3

$n \cdot A$, B.3

$\langle D \rangle$, B.3

$a \circ b$, B.4, 5.2.14

$Hom(A, B)$, B.8

$\ker \varphi$, B.8

A/B, B.11

AC, B.12

$A \cdot C$, B.12

A^n, B.12

$A^{(n)}$, B.12

$(A : B)_R$, B.12

$(A : \{r\})_R$, B.12

$(D : \{m\})_M$, B.12

$(A : r)_R$, B.12

$(D : A)_M$, B.12

$(D : m)_M$, B.12

RG, B.17

R_S, B.18

(X, τ), C.1

(X, ρ), C.4

$[A]_{(X, \tau)}$, C.5

$[A]_X$, C.5

$[A]$, C.5

$\inf\{\tau_\omega \mid \omega \in \Omega\}$, C.9

$\sup\{\tau_\gamma \mid \gamma \in \Gamma\}$, C.9

$\tau|_Y$, C.10

C_x, C.13

$\lim_{\gamma \in \Gamma} x_\gamma$, 3.1.11

$\lim \mathcal{F}$, 3.1.14

$\prod_{i \in I} X_i$, 4.1.1, 4.1.8, 4.1.14

$\sum_{i \in I} X_i$, 4.1.20

$\sum_{i=1}^n X_i$, 4.1.20

$\sum_{j=1}^\infty X_j$, 4.1.20

$X_1 \oplus X_2 \oplus \ldots \oplus X_n$, 4.1.20

τ_B, 4.1.24

τ_c, 4.1.68

$\varprojlim S$, 4.2.1

$\prod_{i \in I}(M_i : N_i)$, 4.3.9, 4.3.13, 4.3.15

$\prod_{i \in I}(f_i : M_i \to M_0)$, 4.3.17

$M_{\mathcal{L}}^I$, 4.3.30

$S \rtimes R$, 4.4.2

$R[x]$, 5.2.1

$\overline{R[x]}$, 5.2.1

$F_k(S_1, S_2, \ldots, S_k; V_1, V_2, \ldots, V_k)$, 5.3.1

$F_R(S; V)$, 5.4.1

F_X, 6.1.1

$R[X]$, 6.1.1

$\deg z$, 6.1.2

Index of Terms

Abelian group, B.2

absolutely convex subset of vector space, 2.3.12

accumulation point of a sequence in a topological space, C.6

Addition Continuity Condition (AC), 1.1.1

additive group of a ring, B.5

Additive Inversion Continuity Condition (AIC), 1.1.1

admissible sequence, 6.4.11

algebraic subset over ring, 5.2.11

algebraic subset over ring with respect to other subset, 5.2.11

anti-discrete topological space, C.1

appropriate sequence, 6.4.11

Archimedean group, B.15

axioms of separation T_0, T_1, T_2, T_3, T_4, T_5, C.11

$\mathcal{U}$–admissible sequence, 6.4.11

basis of a filter, A.12

basis of neighbourhoods of an element of a topological space, C.3

basis of open subsets of a topological space, C.2

basis of a topological space, C.2

bijection, A.5

bounded from left subset of a topological ring, 1.6.2

bounded from left topological ring, 1.6.2

bounded from right subset of a topological ring, 1.6.2

bounded from right topological ring, 1.6.2

bounded subset of a topological module, 1.6.1

bounded subset of a topological ring, 1.6.2

bounded topological module, 1.6.1

bounded topological ring, 1.6.2

brick topology, 4.1.24

image of a subset, A.4

induced topology, C.10

inductive partially ordered set, A.11

injection, A.5

interval in an ordered set, A.9

interval topology, C.1

interval topology, defined by pseudonorm on a module, 2.3.5

inverse bounded from left subset in a topological ring, 1.6.60

inverse bounded from right subset in a topological ring, 1.6.60

inverse bounded subset in a topological ring, 1.6.60

inverse limit, 4.2.1

inverse spectrum, 4.2.1

invertible element, B.1, B.4

isolated element in a topological space, C.3

isomorphism of Abelian groups, B.7

isomorphism of modules, B.7

isomorphism of rings, B.7

kernel of homomorphism, B.8

$\mathcal{L}$-boundary, 4.3.34

$\mathcal{L}$-bounded power of a module, 4.3.30

$\mathcal{L}$-bounded power of a ring, 4.3.36

least upper bound of family of a topologies, C.9

left ideal of a ring, B.9

left ideal of a semigroup, B.9

left module over a ring, B.6

left quasi-invertible element of a ring, B.4

left quasi-regular element, B.4

left topological divisor of zero, 1.8.1

left zero divisor, B.4